Advanced Analysis of Steel Frames

Advanced Analysis of Steel Frames

Theory, Software, and Applications

Edited by
W. F. Chen
Purdue University
West Lafayette, Indiana

S. Toma
Hokkai-Gakuen University
Sapporo, Japan

CRC Press
Boca Raton   Ann Arbor   London   Tokyo

Library of Congress Cataloging-in-Publication Data

Advanced analysis of steel frames : theory, software, and applications / edited by W. F. Chen and S. Toma.
    p. cm.
    Includes bibliographical references and index.
    ISBN 0-8493-8281-5
    1. Steel, Structural.   2. Structural frames.   I. Chen, Wai-Fah, 1936-   II. Toma, S.
TA684.A25   1994
624.1′773—dc20   93-29822
                                                                                                                                                                                                CIP

    This book contains information obtained from authentic and highly regarded sources. Reprinted material is quoted with permission, and sources are indicated. A wide variety of references are listed. Reasonable efforts have been made to publish reliable data and information, but the author and the publisher cannot assume responsibility for the validity of all materials or for the consequences of their use.
    Neither this book nor any part may be reproduced or transmitted in any form or by any means, electronic or mechanical, including photocopying, microfilming, and recording, or by any information storage or retrieval system, without prior permission in writing from the publisher.
    All rights reserved. Authorization to photocopy items for internal or personal use, or the personal or internal use of specific clients, may be granted by CRC Press, Inc., provided that $.50 per page photocopied is paid directly to Copyright Clearance Center, 27 Congress Street, Salem, MA 01970 USA. The fee code for users of the Transactional Reporting Service is ISBN 0-8493-8281-5/94/$0.00+$.50. The fee is subject to change without notice. For organizations that have been granted a photocopy license by the CCC, a separate system of payment has been arranged.
    CRC Press, Inc.'s consent does not extend to copying for general distribution, for promotion, for creating new works, or for resale. Specific permission must be obtained in writing from CRC Press for such copying.
    Direct all inquiries to CRC Press, Inc., 2000 Corporate Blvd., N.W., Boca Raton, Florida 33431.

© 1994 by CRC Press, Inc.

No claim to original U.S. Government works
International Standard Book Number 0-8493-8281-5
Library of Congress Card Number 93-29822
Printed in the United States of America 1 2 3 4 5 6 7 8 9 0
Printed on acid-free paper

New Directions in Civil Engineering
Series Editor: W. F. Chen, *Purdue University*

Response Spectrum Method in Seismic Analysis and Design of Structures
by Ajaya Kumar Gupta, *North Carolina State University*

Stability Design of Steel Frames
by W. F. Chen, *Purdue University*
and E. M. Lui, *Syracuse University*

Concrete Buildings: Analysis for Safe Construction
by W. F. Chen, *Purdue University*
and K. H. Mosallam, *Minister of Interior, Saudi Arabia*

Unified Theory of Reinforced Concrete
by Thomas T. C. Hsu, *University of Houston*

Stability and Ductility of Steel Structures under Cyclic Loading
by Yuhshi Fukumoto, *Osaka University*
and George C. Lee, *State University of New York at Buffalo*

Advanced Analysis of Steel Frames: Theory, Software, and Applications
by W. F. Chen, *Purdue University*
and S. Toma, *Hokkai-Gakuen University*

Analysis and Software of Tubular Members
by W. F. Chen, *Purdue University*
and S. Toma, *Hokkai-Gakuen University*

Flexural-Torsional Buckling of Structures
by N. S. Trahair, *University of Sydney*

Water Treatment Processes: Simple Options
by S. Vigneswaran, *University of Technology, Sydney*
and C. Visvanathan, *Asian Institute of Technology, Bangkok*

Buckling of Thin-Walled Structures
by J. Rhodes, *University of Strathclyde*

Fracture Processes in Concrete
by Jan G. M. van Meir, *Delft University of Technology*

Fracture Mechanics of Concrete
by Zdenek P. Bazant, *Northwestern University*
and Jaime Planas, *Technical University, Madrid*

Preface

This book presents a concise encapsulation of recent developments in the second-order analysis for steel frame design in the US. The first part of the book deals with the direct second-order elastic analysis of frames with rigid and semi-rigid beam-to-column building connections. An advanced analysis method, based on second-order inelastic analysis for frame design, is then presented. Issues involved with the direct use of advanced analysis in engineering practice are also discussed. The main advantage of advanced analysis is that separate checks of member design equations are unnecessary, since the analysis accounts for member and system stability in a direct manner.

The advanced analysis methods are used to predict the behavior and strength of some calibration frames where their behavior has been accurately established through extensive analytical studies and full-scale tests on members and frames. Computer programs for each method of analysis are also provided in the form of a floppy disk for easy implementation. Sample problems are described and solved, and user's manuals for each program are carefully documented and illustrated. For calibration and verification of computer programs for advanced analysis, benchmark problems from around the world are provided at the end of the book.

West Lafayette, IN   W. F. Chen
December 1992   S. Toma

Contents

1 TRENDS TOWARD ADVANCED ANALYSIS 1
J. Y. Richard Liew and W. F. Chen
1.1 Introduction 1
1.2 Design Formats 2
    1.2.1 Allowable Stress Design (ASD) 3
    1.2.2 Plastic Design (PD) 3
    1.2.3 Load and Resistance Factor Design (LRFD) 4
    1.2.4 Advanced Analysis Format 4
1.3 Organization 6
1.4 Elastic Methods of Analysis 6
    1.4.1 AISC-LRFD Beam-Columns Interaction Equations 6
    1.4.2 The Effective Length Factor 8
    1.4.3 Second-Order Elastic Analysis for Frame Design 12
            Recommended Design Procedure 13
1.5 Semi-Rigid Frame Design 16
    1.5.1 Design Provisions and Connection Classifications 16
    1.5.2 Connection Data Base 19
            Goverdhan Data Base 19
            Nethercot Data Base 19
            Kishi and Chen Data Base 20
    1.5.3 Simplified Analysis/Design Method 22
    1.5.4 Direct Second-Order Analysis Methods 25
1.6 Second-Order Inelastic Analyses 26
    1.6.1 Plastic-Zone Method 26
    1.6.2 Elastic-Plastic Hinge Method 27
    1.6.3 Notional Load Plastic Hinge Method 28
    1.6.4 Refined-Plastic Hinge Analysis 31
            Analysis of Simple Portal Frames 33
            Analysis of a Six-Story Frame 35
1.7 Benchmarking Verification 40
1.8 Conclusions 42
References 43

2 SECOND-ORDER ELASTIC ANALYSIS OF FRAMES 47
Yoshiaki Goto
2.1 Introduction 47
2.2 Second-Order Theory for In-Plane Frames 48

2.3 Stiffness Equations for Beam-Column Member 54
2.4 Modeling of Semi-Rigid Connections 60
2.5 Modified Secant Stiffness Equation with
    Connection Flexibility 63
2.6 Solution Procedure for Nonlinear Stiffness Equations 67
2.7 Numerical Examples 69
2.8 Computer Program 75
2.9 User's Manual 77
Acknowledgment 83
References 89

3 SEMI-RIGID CONNECTIONS 91
Norimitsu Kishi
3.1 Introduction 91
3.2 Types of Semi-Rigid Connections 92
    3.2.1 Single Web-Angle Connections/Single
           Plate Connections 92
    3.2.2 Double Web-Angle Connections 93
    3.2.3 Top- and Seat-Angle Connections with Double
           Web Angle 93
    3.2.4 Top- and Seat-Angle Connections 93
    3.2.5 Extended End-Plate Connections/Flush End-Plate
           Connections 93
    3.2.6 Header-Plate Connections 98
3.3 Modeling of Connections 98
    3.3.1 General Remarks 98
    3.3.2 Frye-Morris Polynomial Model 99
    3.3.3 Modified Exponential Model 100
    3.3.4 Three-Parameter Power Model 101
3.4 Connection Data Base 104
3.5 Parameter Definition for Connection Type 105
    3.5.1 Single Web-Angle Connections/Single
           Plate Connections 105
    3.5.2 Double Web-Angle Connections 107
    3.5.3 Top- and Seat-Angle Connections with
           Double Web Angle 107
    3.5.4 Top- and Seat-Angle Connections 114
    3.5.5 Extended End-Plate Connections 119
    3.5.6 Flush End-Plate Connections 123
    3.5.7 Header-Plate Connections 125
3.6 Steel Connection Data Bank Program 127
    3.6.1 Outline of SCDB 127
    3.6.2 User's Manual for Program SCDB 129
    3.6.3 Examples 130
References 135

4 SECOND-ORDER PLASTIC HINGE ANALYSIS OF FRAMES 139
J. Y. Liew and W. F. Chen
Notations 139
4.1 Introduction 140
4.2 Assumptions and Scope 142
4.3 Modeling of Elastic Frame Elements 144
4.4 Modeling of Elastic Truss Elements 148
4.5 Second-Order Elastic-Plastic Hinge Analysis 149
    4.5.1 Cross-Section Plastic Strength 150
    4.5.2 Modification of Element Stiffness for the Presence of
          Plastic Hinges 151
    4.5.3 Illustrative Example 153
4.6 Second-Order Refined Plastic Hinge Analysis 154
    4.6.1 Tangent Modulus Approach 155
    4.6.2 Two-Surface Stiffness Degradation Model 158
          Effect of Plastification at End A Only 159
          Effect of Plastification at End B Only 161
          Effect of Plastification at Both Ends 161
    4.6.3 Illustrative Examples 162
4.7 Analysis of Semi-Rigid Frames 163
    4.7.1 Modeling of Connections 163
    4.7.2 Modification of Element Stiffness to Account for
          End Connections 164
4.8 Numerical Implementation 167
4.9 PHINGE — A Second-Order Plastic Hinge Based
    Analysis Program 168
    4.9.1 Program Overall View 168
    4.9.2 Input Instructions 169
    4.9.3 Examples 174
References 193

5 PLASTIC-ZONE ANALYSIS OF BEAM-COLUMNS AND
PORTAL FRAMES 195
S. P. Zhou and W. F. Chen
5.1 Introduction 195
5.2 Analysis of In-Plane Beam-Columns 196
    5.2.1 Principle of Analysis 196
          Analytical Conditions 196
          Analytical Method 198
    5.2.2 Analytical Steps 200
    5.2.3 Flow Chart of BCIN 207
    5.2.4 User's Manual of BCIN 208
          Input Data 208
          Output Data 210
          How to Determine the Input Data 212

	5.2.5	Sample Problems of BCIN 213
5.3	Analysis of Portal Braced Frames 215	
	5.3.1	Principle of Analysis 215
	5.3.2	Analytical Steps 218
	5.3.3	Flow Chart of FRAMP 220
	5.3.4	User's Manual of FRAMP 220
		Input Data 221
		Output Data 222
		Frame Failures 225
		The Case with BAT = 0 227
	5.3.5	Sample Problems of FRAMP 227
5.4	Analysis of Portal Unbraced Frames 232	
	5.4.1	Principle of Analysis 232
		Structure and Load of the Unbraced Frames 232
		Second-Order Elastic Analysis of the Unbraced Frames 232
		Second-Order Inelastic Analysis 235
	5.4.2	Analytical Steps 236
	5.4.3	Flow Chart of FRAMH 237
	5.4.4	User's Manual of FRAMH 238
		Input Data 238
		Output Data 240
		Frame Failures 241
	5.4.5	Sample Problems for FRAMH 241
References 243		
Appendix A: Sample Problems of BCIN (File "OBC") 244		
Appendix B: Sample Problems of FRAMP (File "OFP") 246		
Appendix C: Sample Problems of FRAMH (File "OFH") 252		

6 PLASTIC-ZONE ANALYSIS OF FRAMES 259
Murray J. Clark

6.1	Introduction 259	
6.2	Historical Sketch 260	
6.3	Finite Element Formulation 262	
	6.3.1	General 262
	6.3.2	Strain-Displacement Relations 264
	6.3.3	Stress-Strain Relations 266
		Stress Resultants 267
	6.3.4	The Principle of Virtual Displacements 269
	6.3.5	Discretization of the Virtual Work Equation 270
		Element Geometric Description 270
		Strain-Displacement Matrices 274
		Equilibrium Equations 276
	6.3.6	Solving the Nonlinear Equilibrium Equations 276

		6.3.7	Transformation, Condensation, and Recovery of Nodal Variables 280

                      Transformation of Nodal Variables 280
                      Condensation of Nodal Variables 281
                      Recovery of Nodal Variables 282
                      Internal Rotational Releases 282
  6.4 Cross-Sectional Analysis 283
        6.4.1 General 283
        6.4.2 Numerical Integration Procedure 284
        6.4.3 Numerical Integration of Stress Resultants 285
        6.4.4 Numerical Integration of the Tangent Modulus Matrix ... 286
  6.5 Some Aspects of the Computer Program Implementation 286
        6.5.1 General 286
        6.5.2 Node Numbering and Numerical Integration 287
        6.5.3 Nonlinear Formulations and Incremental-Iterative Strategies 287
        6.5.4 Material Inelasticity 288
        6.5.5 Computer Requirements 289
  6.6 Investigation of Analysis Parameters 290
        6.6.1 Introductory Comments 290
        6.6.2 I-Section Column 290
        6.6.3 Beam with Fully Constrained Ends 292
  6.7 Analysis of Beam-Columns 294
  6.8 European Calibration Frames 296
        6.8.1 General 296
        6.8.2 Rectangular Portal Frame with Fixed Bases 296
                Comments on Sensitivity of Analysis Results to Modeling Assumptions 298
        6.8.3 Pitched-Roof Portal Frame 299
        6.8.4 Six-Story Two-Bay Frame 300
  6.9 A North American Calibration Frame 303
  6.10 Australian Calibration Frames 304
        6.10.1 Rigid-Jointed Truss 304
        6.10.2 Stressed-Arch Frame 308
  6.11 Application to Engineering Practice 312
        6.11.1 Current Practice 312
        6.11.2 Future Research 315
  6.12 Summary and Concluding Remarks 316
References 317

7 BENCHMARK PROBLEMS AND SOLUTIONS 321
Shouji Toma and W. F. Chen
  7.1 Introduction 321

7.2	Requirements for Benchmark Problems	322
	7.2.1 Physical Attributes and Behavioral Phenomena of Frames	322
	7.2.2 Criteria for Selecting Calibration Frames	323
	7.2.3 Required Information	324
	7.2.4 Some Comments on the Analytical Assumptions	325
	Constitutive Relations of Materials	325
	Residual Stresses (Material Imperfection)	325
	Geometrical Imperfections	327
	Joint Conditions	328
7.3	North American Calibration Frames	328
	7.3.1 Introduction	328
	7.3.2 Beam-Columns	329
	7.3.3 Portal Frames	333
	7.3.4 Interaction Curves for Portal Frames	336
7.4	European Calibration Frames	338
	7.4.1 Introduction	338
	7.4.2 Analytical Assumptions	343
	Stress-Strain Relation	345
	Residual Stresses	345
	Geometrical Imperfections	345
	7.4.3 A Portal Frame	347
	7.4.4 A Gable Frame	348
	7.4.5 A Six-Story, Two-Bay Frame	349
7.5	Japanese Calibration Frames	351
	7.5.1 Introduction	351
	7.5.2 Full-Size Test of Portal Frame	352
	Test Specimens	352
	Loading Procedures	353
	Test Results	355
	Monotonic Loading	355
	Cyclic Loading	355
	Theoretical Analysis	355
	7.5.3 One-Quarter Scaled Test of Portal Frames	359
	Test Specimens	359
	Loading Procedures	365
	Test Results of One-Story Frames	365
	Test Results of Two-Story Frames	365
	Theoretical Analysis	370
References		372
Index		375

1: Trends Toward Advanced Analysis

J. Y. Richard Liew, *Department of Civil Engineering, National University of Singapore, Singapore*

W. F. Chen, *School of Civil Engineering, Purdue University, West Lafayette, Indiana*

1.1 Introduction

Since the publication of the two-volume book on *Theory of Beam-Columns* (Chen and Atsuta, 1976 and 1977), and the subsequent books and monograph related to *Stability Design of Frames* (Chen and Lui, 1986 and 1991; SSRC, 1992), our understanding of certain aspects of the behavior and design of steel members and frames has increased considerably and many extensions and advancements have been made during the past 10 years.

The advent of limit-states specifications has resulted in more explicit and more rational consideration of the combined effects of inelasticity and stability at maximum strength levels. Since limit-states design is based directly on factored loads and limits of resistance, it is expected that structural systems and their members will behave nonlinearly before their capacity is reached. Of course, the most direct approach for structural design is to model all the significant nonlinear effects in the analysis. However, until recently, rigorous consideration of system as well as member strength and stability in the analysis of large-scale structural system were not feasible and practical. As a result, contemporary specification provisions have been based primarily on simpler methods of analysis and member interaction equations which account approximately for the interaction of strength and stability between the member and structural system.

Recently, the advancement in computer hardware, particularly in the computing and graphics performance of personal computers and workstations, is making advanced methods of analysis more and more feasible for design use. This advancement has made it possible for the engineer to adopt the limit-states design philosophy in a wider perspective. Advanced analysis techniques hold the promise of more realistic prediction of load effects and overall structural performance, and therefore in certain cases, yield greater economic and more uniform safety. The two task groups in the U.S. — the American Institute of Steel Construction (AISC) Technical Committee 117 on *Inelastic Analysis and Design* and the Structural Stability Research Council (SSRC) Task Group 29 on *Second-Order Inelastic Analysis for Frame Design* — are working toward the implementation of advanced analysis for practical design use. Similar developments also have been observed worldwide including contributions from several specification committees such as Australia (AS4100, 1990) and Europe (EC 3, 1990), among others.

Advanced analysis is referred to any method of analysis that sufficiently represents the strength and stability behavior such that separate specification member capacity checks are not required. In recent years, there has been an intense interest in the development of advanced analysis methods suitable for design use. However, the current state-of-the-art of advanced analysis is still largely fragmented and disjointed. This book, which consists of seven chapters, is a compendium of research papers and reports, and computer programs specially catered for engineers and researchers who want to improve their knowledge and to explore the use of advanced analysis techniques for frame design. The main purposes of the book are to report the recent development in advance analysis and to provide educational type of computer software for engineers to perform planar frame analysis for a more realistic prediction of system's strength and stability.

This chapter presents a succinct encapsulation of various analysis approaches for frame design ranging from simplified member-by-member design approach to more sophisticated computer-based analysis/design approach. Its aim is to provide a smooth transition from the current state-of-the-art knowledge on the analysis and design of steel frames to upcoming new technology and developments in engineering practice. Finally, trends and directions of advanced analysis in steel building design are identified and addressed.

1.2 Design Formats

The steel building design specifications in the U.S. has undergone a metamorphosis from the *Allowable Stress Design* (ASD) approach through the *Plastic Design* (PD) approach to the present stage of the *Load and Resistance Factor Design* (LRFD) approach. One fundamental feature that demarcates the LRFD approach to steel design from ASD and PD approaches is that it is a probability based design methodology that uses statistical evidenec and reliability theory to prove the logistic and rationale of the design. Design considerations are guided by the identification of a set of limit states. A design is considered acceptable if none of the limit states is violated. Since a probabilistic mathematical model was used in the development of the load and resistance factors, it is possible to give proper weight to the degree of importance for each load effect and resistance. The design based on this approach often yields a greater economic and more uniform margin of safety.

Recently, an emerging technology called advanced analysis has been proposed. This analysis approach has the potential to unlock the power of the computer for direct assessment of system and member strength and stability without the need of specification member capacity checks. The advanced analysis approach and the various design approaches upon which current practice is based are discussed separately in the following sections.

1.2.1 Allowable Stress Design (ASD)

Under the ASD philosophy, a design is said to be satisfactory if the stresses computed under service loads (or working loads) do not exceed some predesignated allowable values. It has the format of

$$\frac{R_n}{F.S} \geq \sum Q_{ni} \tag{1.1}$$

where

R_n = nominal resistance of the structural member expressed in units of stress;
Q_n = nominal working stresses computed under working load conditions;
$F.S$ = factor of safety;
i = type of load (i.e., dead load, live load, and wind load, etc.);
n = number of load types.

These so-called allowable stresses are usually calculated by dividing either the yield or ultimate stress of the material by a safety factor. The left-hand side of Eq. (1.1) represents the allowable stress of the structural member or component under various loading conditions (e.g., tension, compression, bending, and shear). The right-hand side of the equation represents the computed combined stresses produced by various load combinations (e.g., dead load, live load, and wind load). The stress computation is based on a first-order elastic analysis, and any geometrical nonlinear effects are implicitly accounted for in the member interaction equations. However, one should realize that, in ASD, the factor of safety is applied to the resistance term, and safety is evaluated in the service load range.

1.2.2 Plastic Design (PD)

Under the PD philosophy, a design is said to be satisfactory if the load effects computed based on factored load combinations do not exceed the plastic resistance of the structural member or component. It has the format of:

$$R_n \geq \gamma \sum_{i=1}^{n} Q_{ni} \tag{1.2}$$

where R_n is the nominal plastic strength of the member, and γ is a load factor.

The load factor γ is taken as 1.7 in AISC (1978, 1989) if the nominal load effect Q_{ni} consists of only dead and live gravity loads, or as 1.3 if Q_{ni} consists of dead and live gravity loads plus wind or earthquake loads. In carrying out a plastic design, a first-order plastic hinge analysis or a rigid-plastic analysis is required. The main difference between these two methods of analysis is that rigid-plastic analysis is based on direct determination of the final collapse mechanism for a structure, whereas elastic-plastic analysis gives, in additional to the collapse load of the frame, additional information about the redistribution process prior to reaching the collapse load.

The main advantage of plastic design is that it allows the engineer to design a more economical structure by allowing inelastic force redistribution to occur in the structure. However, the analysis relies on the ability of steel to accept considerable deformation in excess of yield. Therefore, the regions of the structure in which significant plasticity (plastic hinges) develops must have adequate deformation capacity to carry load. To achieve this, members containing plastic hinges must satisfy the limitations on the flange and web proportions for plastic design sections. In addition, sufficient lateral and torsional restraints must be provided to prevent the member from buckling laterally. Although material nonlinearity has been considered in first-order plastic hinge analysis, the effects of spread of yielding and member stability are not accounted for. Thus, like ASD, any significant nonlinear effects have to be accounted for approximately in the member strength equations. In other words, member capacity checks are required when plastic analysis is used so as to prevent instability type of failure.

1.2.3 Load and Resistance Factor Design (LRFD)

Under the LRFD philosophy, a design is said to be satisfactory if the factored resistance exceeds the factored design loads. It has the format of

$$\phi R_n \geq \sum_{i=1}^{n} \gamma_i Q_{ni} \tag{1.3}$$

In LRFD, the load factors γ_i are derived based on statistical analyses, and the resistance factor ϕ is developed based on probability theory and calibration. These factors and the load combinations used in LRFD are summarized in Tables 1.1 and 1.2, respectively. An important feature of LRFD is that it is based on the limit-state concept in which various limit states that pertain to the strength, integrity, and serviceability of the structure are incorporated into the design in a systematic fashion.

In the LRFD specifications, a second-order elastic analysis that directly considers geometrical nonlinearity is recommended for the design of structural frames. The design interaction equations implicitly account for the ultimate strength behavior of a beam-column member. This treatment, unfortunately, does not satisfy compatibility between the actual inelastic member, as loaded by the design forces and accounted for in the interaction equation, and the elastic structural system as assumed in the frame analysis (Liew et al., 1991). Nevertheless, the introduction of the LRFD concept has laid the groundwork for a more rational approach for structural steel design.

1.2.4 Advanced Analysis Format

An upcoming technology for limit-states design is termed *advanced analysis* which is defined as any method of analysis that sufficiently represents the strength and

Table 1.1 Load Factors and Load Combinations

1.4 D
1.2 D + 1 6 L + 0.5 (Lr or S or R)
1.2 D + 1.6 (L$_r$ or S or R) + (0.5 L or 0.8 W)
1.2 D + 1.3 W + 0.5 L + 0.5 (L$_r$ or S or R)
1.2 D + 1.5 E + (0.5 L or 0.2 S)
0.9 D − 1.3 W or 1.5 E

Note: D = dead load, L = live load, L$_r$ = roof load, W = wind load, S = snow load, E = earthquake load, and R = nominal load due to initial rainwater on ice exclusive of the ponding contribution.

Table 1.2 Resistance Factors

Member type and limit state	φ
Tension member, limit state: yielding	0.90
Tension member, limit state: fracture	0.75
Pin-connected member, limit state: tension	0.75
Pin-connected member, limit state: shear	0.75
Pin-connected member, limit state: bearing	1.00
Column, all limit states	0.85
Beams, all limit states	0.90
High strength bolts, limit state: tension	0.75
High strength bolts, limit state: shear	
A307 bolts	0.60
others	0.65

stability behavior such that specification member capacity checks are not required. The advanced analysis design concept, when posted in the LRFD format, may be viewed as moving a number of resistance effects from the left to right-hand side of Eq. (1.3) and thus, eliminating a number of equation checks from the design. In particular, member capacity checks can be waived when an advanced analysis is performed. The reason for this is that, since advanced analysis can directly assess the strength and stability of the overall structural system as well as interdependence of member and system strength and stability, separate member capacity checks are therefore not required.

It is expected that use of advanced analysis techniques will simplify the design process considerably. However, this task cannot be accomplished fully in the present PD and LRFD format, even if all the significant nonlinear effects are considered directly in the analysis. The current LRFD specifications require the engineer to perform beam-column capacity checks even when an inelastic analysis is used. In contrast, the provisions for advanced analysis in the Eurocode (EC 3, 1991) and Australian limit-states specifications (AS4100, 1990) postulate that if all the significant planar behavioral effects are modeled properly in the analysis, the checking of conventional beam-column equations is not required. Thus, for advanced analysis to provide most benefit in frame design, specification provisions of this type are required. The main distinction between advanced analysis and other simplified analysis methods is that advanced analysis combines, for the first time, the theories of plasticity and stability in the limit states design of structural steel frameworks. Other analysis and design methods treat stability and plasticity separately — usually through the use of beam-column interaction equations and member effective length factors (Liew et al., 1991).

1.3 Organization

In an attempt to set the stage for the remaining sections in this chapter, Section 1.4 reviews the background behind the development of the current AISC-LRFD beam-column interaction equations. This is followed by a discussion of the use of effective length factors for beam-column design. Design procedures using second-order elastic analysis are then outlined.

Section 1.5 reviews the recent work on the design of semi-rigid frames based on the extension of the current AISC-LRFD B_1/B_2 approach. The design implications using second-order analysis are also discussed.

A general trend for steel frame design is toward a direct application of second-order inelastic analysis for proportioning members and connections. Section 1.6 provides a state-of-the-art summary on contemporary methods of advanced analysis techniques that are suitable for design use. Methods that are essential for representing initial imperfections and distributed plasticity effects on the side-sway instability of frames and local instability of members are discussed. The implications of these advanced analysis approaches with respect to limit-states specification and code provisions are highlighted.

Section 1.7 discusses the general issues pertinent to verification and benchmarking of second-order inelastic analyses for use as advanced analysis. Section 1.8 provides a conclusion and directions for further research are also indicated.

1.4 Elastic Methods of Analysis

This section first presents the AISC-LRFD method for the design of beam-columns in steel frameworks using a first-order elastic analysis. This is followed by a discussion of the design of braced frames, unbraced frames, and possible leaned-column frames using a direct second-order elastic analysis.

1.4.1 AISC-LRFD Beam-Columns Interaction Equations

The AISC-LRFD Specification (1986) provides the following interaction equations for designing steel beam-columns:

$$\frac{P_u}{\phi_c P_n} + \frac{8}{9}\left[\frac{M_{ux}}{\phi_b M_{nx}} + \frac{M_{uy}}{\phi_b M_{ny}}\right] = 1.0 \quad for \quad \frac{P_u}{\phi_c P_n} \geq 0.2 \qquad (1.4)$$

$$\frac{P_u}{2\phi_c P_n} + \left[\frac{M_{ux}}{\phi_b M_{nx}} + \frac{M_{uy}}{\phi_b M_{ny}}\right] \leq 1.0 \quad for \quad \frac{P_u}{\phi_c P_n} < 0.2 \tag{1.5}$$

where P_u is the required axial strength, P_n is the nominal axial compressive strength, M_{ux} and M_{uy} are the required flexural strength within the unbraced length of the member, M_{nx} and M_{ny} are the nominal flexural strengths, and ϕ_c and ϕ_b are the resistance factors for compression and flexure.

For the evaluation of P_n, the LRFD Specification has adopted the following column strength formulas:

$$P_n = \begin{cases} P_y\left(0.658^{\lambda_c^2}\right) & for \ \lambda_c \leq 1.5 \\ P_y\left(\dfrac{0.877}{\lambda_c^2}\right) & for \ \lambda_c > 1.5 \end{cases} \tag{1.6}$$

The LRFD specification suggests the following procedure to estimate the second-order elastic moment M_u in lieu of a direct second-order elastic analysis:

$$M_u = B_1 M_{nt} + B_2 M_{lt} \tag{1.7}$$

In this equation, M_{nt} and M_{lt} are the maximum first-order elastic moments in the member based on "no translation" (NT) and "lateral translation" (LT) analyses of the frame, respectively. The term B_1 is a P-δ moment amplification factor which accounts for the amplification of the first-order NT moments associated with member curvature effects. B_1 may be expressed as

$$B_1 = \frac{C_m}{1 - \dfrac{P_u}{P_{ek}}} \tag{1.8}$$

The term P_{ek} in Eq. (1.8) is the member buckling load defined as $\pi^2 EI/(K_n L)^2$, where K_n is an effective length factor in the plane of bending based on the assumption that side-sway is prevented. For braced frame members having transverse loading between supports, C_m is an integral part of the moment magnifier B_1. This value also may be determined by rational analysis (Chen and Lui, 1986). For braced-frame members subjected only to end moments and axial force, C_m, which converts the linear varying moment to an equivalent uniform moment, is given as:

$$C_m = 0.6 - 0.4 M_1/M_2 \tag{1.9}$$

where M_1 and M_2 are the smaller and larger end moments obtained from the first-order NT analysis, respectively. M_1/M_2 is positive when the member is bent in reverse curvature and negative when it is bent in single curvature.

8 ADVANCED ANALYSIS OF STEEL FRAMES

B_2 is a P-Δ moment amplification factor which amplifies the first-order member end moments associated with lateral translation of the story. This term is expressed by the LRFD Specification as

$$B_2 = \frac{1}{1 - \frac{\sum P \Delta_{oh}}{\sum HL}} \tag{1.10}$$

or alternately as

$$B_2 = \frac{1}{1 - \frac{\sum P_u}{\sum P_{ek}}} \tag{1.11}$$

where $\sum H$ is the sum of the story horizontal forces that induce the first-order story translation Δ_{oh}, L is the story height, and P_{ek} is the member buckling load defined as $\pi^2 EI/(K_n L)^2$, where K_n is an effective length factor in the plane of bending if the frame is free to side sway. The AISC-LRFD (1986) provides two different alignment charts for the evaluation of effective length factors for braced and unbraced frames.

1.4.2 The Effective Length Factor

In the development of the interaction equations, much discussion has been focused on the need and validity of using the effective length factor K in the equations. Although attempts were made to formulate general interaction equations without using K factor, it was found that this was almost impossible if the interaction equations were to be versatile enough for a wide range of slenderness ratios and load combinations (Liew et al., 1991). The final decision was to retain effective length factor in the interaction equations. Figure 1.1 compares the results obtained using the AISC-LRFD beam-column design approach with the "exact" solutions (Kanchanalai, 1977) for a simple portal frame. The example frame has a rigid girder and pinned bases. Here, the LRFD requires the calculation of column effective length which is twice the actual height of the column (i.e., K = 2). An attempt is also made to compute the axial resistance term P_n by using an effective length factor of K = 1. However, this approach does not correlate well with "exact" inelastic strength curves from Kanchanalai (1977). On the other hand, the LRFD equations give a good fit to the more exact solution when an effective length factor of K = 2 is used.

The specification describes a procedure by which the effective length factor is determined from the alignment charts (Julian and Lawrence, 1959) that correspond to the side-sway-inhibited and side-sway-permitted cases. However, it is the authors' belief that a more accurate method for determining K should be used. One significant drawback of the alignment chart methods is that they do not account for the fact that

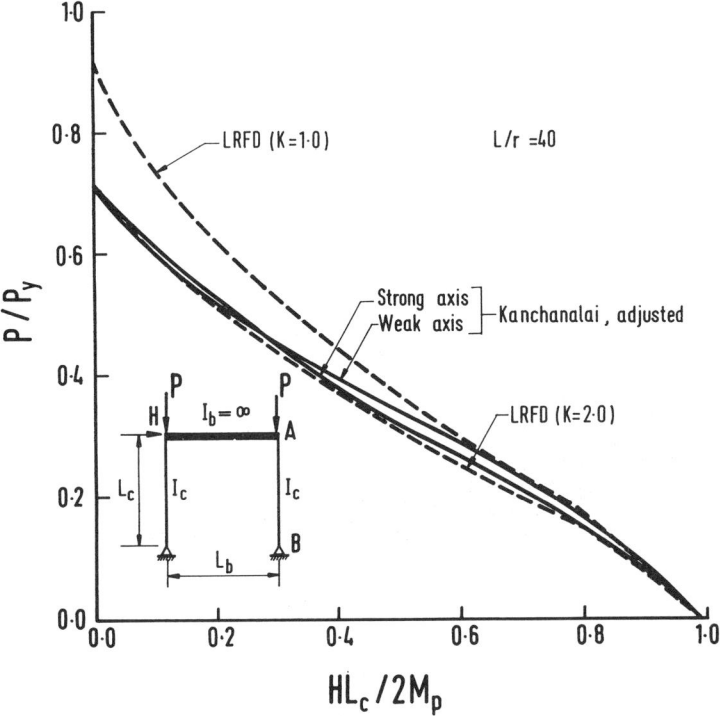

FIGURE 1.1. Comparison of the AISC-LRFD beam-column equations with and without the use of K factor with plastic-zone curves (Kanchanalai, 1977) for frames with imperfections.

a stronger column can provide effective bracing to a weaker column during buckling of a story assemblage. A more accurate approach by which K can be determined is given by LeMessurier (1977). In this approach, the effective length is expressed as

$$K_i = \sqrt{\frac{\pi^2 I_i}{P_{ui}}\left[\frac{\sum P_u + \sum(C_L P_u)}{\sum(\beta I)}\right]} \qquad (1.12)$$

where the subscript i refers to the i-th column of the story and

$\sum P_u$ = sum of vertical forces acting on the story at factored load level
P_{ui} = axial force on column i calculated based on first-order analysis
I_i = moment of inertia of column i
C_L = stiffness correction factor for a column which accounts for P-δ effects and can be written as $\left(\beta\dfrac{K_n^2}{\pi^2}-1\right)$
β = column end restraint coefficient defined as $\dfrac{6(G_A+G_B)+36}{2(G_A+G_B)+G_A G_B+3}$

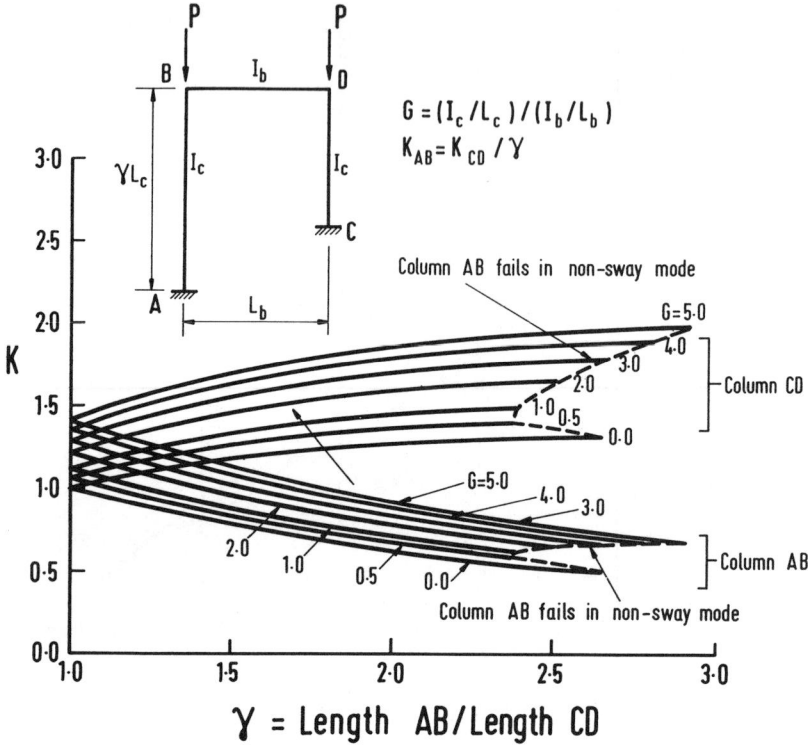

FIGURE 1.2. Characteristic of K-factor results for portal frames with unequal column lengths.

In the LeMessurier equation, β is a factor to account for the end-restraint effects on the column, and C_L is a factor to account for the P-δ effects in the column.

Alternatively, a simplified form of the LeMessurier equation may be written assuming that (1) the sum of the gravity load that causes sway buckling of a story is equal to the sum of the individual buckling loads of columns that provide story sidesway resistance, and (2) the individual column buckling loads are equal to the buckling load determined using the alignment chart K factor. This results in a story-buckling K factor equation which has the form:

$$K_i = \sqrt{\frac{\pi^2 EI_i \sum P}{L_i^2 P_i \sum P_{ek}}} \tag{1.13}$$

in which P_{ek} is the elastic buckling load, $\pi^2 EI/(K_n L)^2$, based on a value of K_n obtained from the alignment chart.

Figure 1.2 shows an unbraced frame with columns of unequal lengths. A buckling analysis of this frame has been performed for column length ratios, $\gamma = L_{AB}/L_{CD}$, ranging from 1.0 to 3.0. For this frame, Eqs. (1.12) and (1.13) predict essentially the same elastic K factors for all the cases considered. The use of the alignment chart is

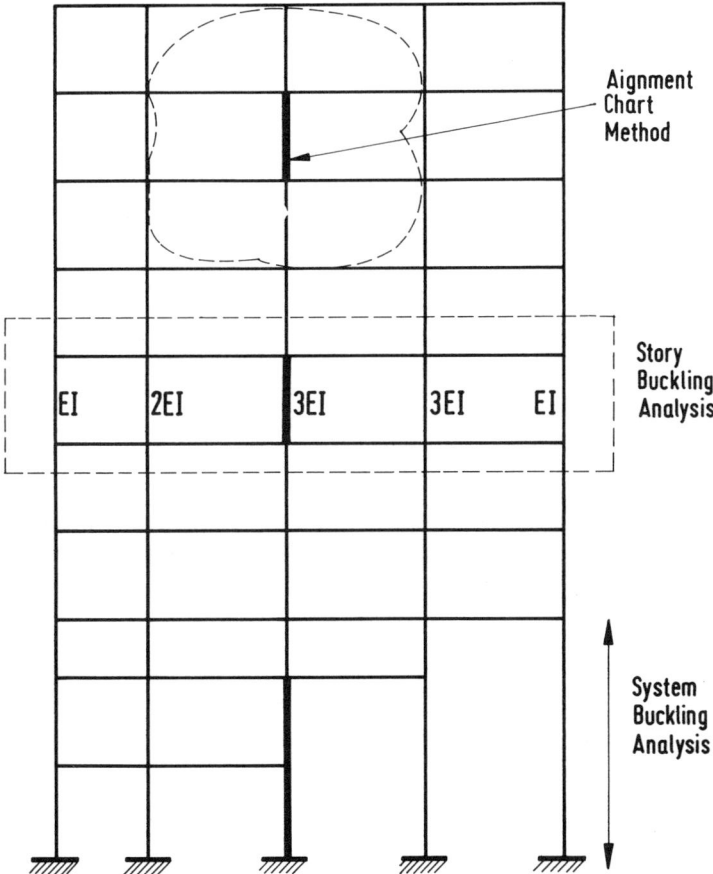

FIGURE 1.3. Different methods of calculating K factors.

strictly applicable only when the stiffness parameters of both columns are the same (i.e., $\gamma = 1.0$). It can be observed from Figure 1.2 that the direct use of the alignment chart will underestimate the K factor for column CD and overestimate the K factor for column AB. Therefore, the ultimate strength of column CD will be over-predicted whereas the strength for column AB will be under-predicted. The use of Eq. (1.12) or (1.13) is recommended by the authors over the alignment chart method since they are applicable for cases such as leaned column frames and for cases in which the flexural stiffnesses of the columns in a story differ significantly, as shown in Figure 1.3.

Recently, a practical approach for determining the effective length factor for unbraced frames was proposed by Lui (1992). The K factor equation is written as:

$$K_i = \sqrt{\left(\frac{\pi^2 EI_i}{P_i L_i^2}\right)\left(\sum \frac{P}{L}\right)\left(\frac{1}{5\sum \eta} + \frac{\Delta_{oh}}{\sum H}\right)} \qquad (1.14)$$

where

$$\eta = \frac{(3 + 4.8m + 4.2m^2)EI}{L^3} \tag{1.15}$$

is the member stiffness index. P_i is the compressive axial force in the member. $\Sigma(P/L)$ is the sum of the axial force to length ratio of all members in a story, ΣH is the sum of the story lateral forces producing the first-order inter-story deflection, Δ_{oh}, and EI_i and L_i are the flexural rigidity and length of the member, respectively. In Eq. (1.15), m is the ratio of the smaller to larger end moments of the member; and m is positive if the member bends in reverse curvature and is negative if the member bends in single curvature. It should be noted that Eq. (1.14) considers explicitly both the story instability (P-Δ effect) and member instability (P-δ effect) effects in its derivation. The P-δ effect is reflected by the term $\Sigma(P/L)/(5\Sigma\eta)$ in Eq. (1.14). This equation requires the use of first-order analysis and no special charts or iterations are required for K factor solutions. Its use will provide sufficiently accurate K-factor results for columns in unbraced frames with unequal distribution of lateral stiffnesses, unequal distribution of gravity loads, and for frames with leaner columns. The validity of the K factor equation when applied to these cases has been demonstrated by Lui (1992).

For nonrectangular and irregular frames and structures with a mixture of elements providing restraint against side-sway, a system buckling analysis is necessary for accurate determination of the effective length factors to be used in the P_n term of the beam-column interaction formulas. The K factor can be calculated using the equation

$$K_i = \sqrt{\frac{P_e}{P_{cr}}} = \sqrt{\frac{\pi^2 EI/L^2}{P_{cr}}} \tag{1.16}$$

where P_e is the Euler buckling load of column, and P_{cr} is the axial force in column at incipient buckling of the frame. An example for application of system buckling analysis is the lower portion of the framework shown in Figure 1.3, in which the concept of "story" buckling is virtually absent.

1.4.3 Second-Order Elastic Analysis For Frame Design

The need for a second-order elastic analysis of steel frame is increasing in view of the AISC-LRFD Specifications which give explicit permission for the engineer to compute load effects from a direct second-order elastic analysis. Second-order elastic analysis is also the "preferred" method in the Canadian Limit States design specification (CSA, 1989), and in many other limit-states code provisions. Although there are several other approximated methods (LeMessurier, 1976 and 1977; Yura, 1971; Stevens, 1967; Cheong, 1977; Nixon et al., 1975) which are based on same philosophy as the B_1/B_2 analysis in the LRFD Specification, these methods do not always give satisfactory results because they are derived based on simplified assumptions and are applicable to rectangular frames with rigid connections and smaller displacements. Furthermore, it is rather cumbersome and tedious to calculate the amplification factors applied to a first-order analysis. In view of this, it is more convenient and straightforward to use computer-based method to calculate member forces for beam-column design as long as the method can be easily implemented.

A comprehensive second-order elastic analysis theory with a computer program that accounts for both P-δ and P-Δ effects in general types of plane frame structures is presented in Chapter 2 of this book. It is not surprising to discover that the comprehensive second-order elastic analysis can be as efficient, and in certain respects, more simple than other more approximate moment-amplification methods. The computer program presented in Chapter 2, coupled with the design procedure described below, may be used in the design of both simple structural systems and for systems with irregular geometry and leaned columns.

Recommended Design Procedure

The analysis and design of tall building frameworks involve the following sequence of calculations:

(1) preliminary selection of a structural system;
(2) preliminary selection of member sizes;
(3) approximate analysis of the entire framework or substructures at various levels for an initial assessment of maximum strength and drift; and
(4) reanalysis and redesign the structure in which the initial sizes are refined to meet the code limit-states requirements.

Below are the recommended analysis/design approaches that utilize second-order elastic analysis approach to design lateral-framing systems according to the AISC-LRFD specification provisions.

(1) ***Perform preliminary analysis/design steps.*** In preliminary analysis and design of the structural system, the assumed member sizes may not be appropriate for use in the second-order analysis, since the second-order effects estimated for these members may be unrealistically large or small (White and Hajjar, 1991). Therefore, first-order analysis with approximate calculation of second-order effects using moment amplifiers given by Eqs. (1.8) and (1.10 or 1.11) is considered to be more appropriate than use of direct second-order analysis procedure at this preliminary stage. The preliminary member sizing is usually dependent on the engineer's experience to arrive at an initial design that is close to the final design solution.

(2) ***Make a distinction between braced and unbraced structures.*** This step is necessary for estimating proper effective length factor for member design. Simple rules for classification of frames into sway and nonsway, braced and unbraced frames can be found in the ECCS design manual (ECCS, 1991). In general, a frame that is laterally restrained by a bracing system may be classified as a brace frame when the sway stiffness of the bracing system is approximately five times the sway stiffness

14 ADVANCED ANALYSIS OF STEEL FRAMES

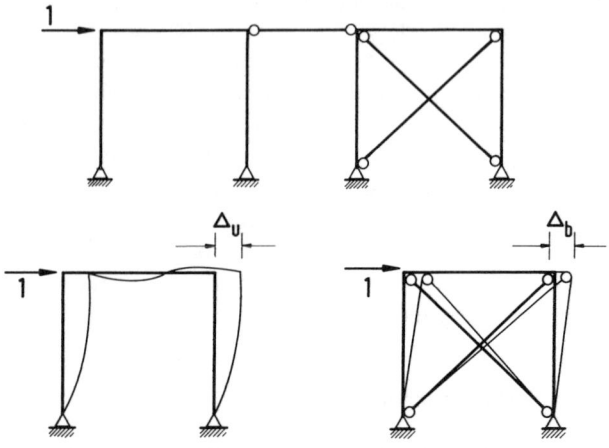

FIGURE 1.4. Classification of braced and unbraced frames with and without lateral braces.

of the frame under consideration (see Figure 1.4). For frames that are not restrained by any bracing system, P-Δ amplification factor $B_2 \leq 1.1$ may be used to classify a frame as nonsway. The B_2 factor used in this calculation may be evaluate from Eq. (1.10) which does not involve a K factor. If B_2 is less than 1.1, the beam-column strength is not likely to be sensitive to the effective length factor; therefore, the use of $K = 1$ for member design would be sufficient. The classification of sway and nonsway frames that are laterally supported and unsupported by bracing systems is summarized as shown in Figure 1.4.

(3) *Determine the appropriate effective length factors for member design.* For nonsway and braced frames, the column effective length factors may be conservatively taken as 1.0. For unbraced or sway frames, the story buckling K factor equations (Eq. (1.12), (1.13), or (1.14)) should be used to calculate the effective length factors for column members that are participated in the lateral-force resisting frame. For leaned columns, an effective length factor of unity should be used for the calculation of P_n. Detailed treatment of K factors for members in semi-rigid frames is discussed in Section 1.5.3.

(4) ***Analyze the structure for each load combination as shown in Table 1.1.*** The structure may be analyzed using member center-line to center-line dimensions (Chen and Liew, 1992). The analysis step should be carried out for all different distribution and magnitudes of loading considered. If the structure is asymmetric, wind load from different directions must be considered in the analysis. Live load reduction may be performed for each load combination. A more comprehensive scheme for incorporating live load reduction into a second-order analysis has been developed by Ziemian (1990), and it may be applied in the present step of analyses. The analyses may be carried out using a simple step increment procedure, since this solution method is generally well behave and it often exhibits good computational efficiency. The more "exact" iterative solution methods would be useful at the final state of design at which the more "exact" force distribution and response are required for the evaluation of member and system strength and stability.

(5) ***Determine maximum moment in the member.*** Although a comprehensive second-order elastic analysis program may represent the P-δ and P-Δ effects adequately, the analysis provides the member forces only at the element ends. Therefore, additional calculations are necessary to determine the location and magnitude of maximum second-order elastic moment, M_u, within the member. For beam-columns with in-span loading, the maximum moment may be determined from the "exact" analytical expressions based on the end moments and axial force from the analysis and the in-span loading on the isolated member. Analytical maximum moment expressions, M_u, for beam-columns subjected to different in-span loading and axial force can be derived from the member governing differential equations, and the solutions are given in Chen and Lui (1986, 1991). For a beam-column element loaded only with end forces, the analytical expression given below may be used for general axial load cases.

$$M_u = |M_2|\sqrt{\frac{(M_1/M_2)^2 + 2(M_1/M_2)\cos kL + 1}{\sin^2 kL}} \tag{1.17}$$

for $0 \leq X_c \leq L$, and

$$M_u = |M_2| \tag{1.18}$$

for $X_c < 0$ or $X_c > L$. The term X_c is the location of the maximum moment along the length of the member and can be calculated as

$$\tan(kX_c) = -\frac{(M_1/M_2)\cos kL + 1}{(M_1/M_2)\sin kL} \tag{1.19}$$

where $k = \sqrt{P/EI}$, P is the axial force, and M_1 and M_2 are the smaller and larger of the two end moments, respectively. M_1/M_2 is positive when the member is bent in double curvature, and negative when bent in single curvature, as shown in Figure 1.5. The above calculations may be easily programmed for use in second-order structural

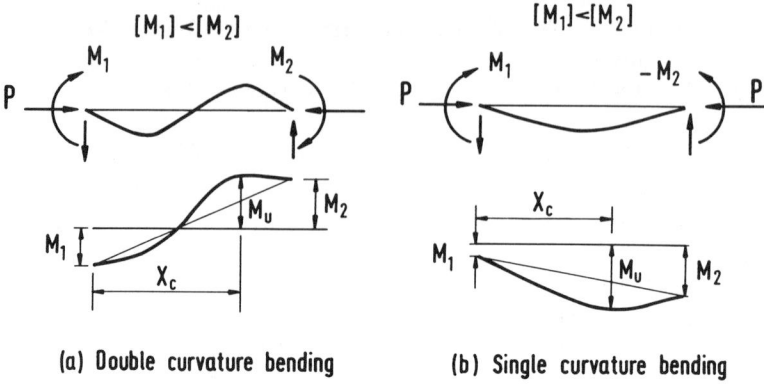

FIGURE 1.5. Maximum moment location in a beam-column member subjected to end loading.

analysis and do not need to be performed by hand. However, it should be noted that for axial load P less than the Euler's buckling load P_e, B_1 from Eq. (1.8), in which P_{ek} is evaluated based on $K = 1$, may be used to amplify the second-order end moment to obtain the maximum second-order elastic moment for design.

(6) *Check the members against the specification provisions.*
(7) *Resize the members if necessary.*

The authors believe that the current technology for direct analysis of second-order effects is well developed and is within the reach of current profession. With the availability of more and more commercial software programs that can provide a comprehensive second-order elastic analysis capability (i.e., one that can account for both member curvature, P-δ, and chord rotation, P-Δ, effects), this design approach eventually will gain wide acceptance among the design community.

1.5 Semi-Rigid Frame Design

1.5.1 Design Provisions and Connection Classifications

Much research and development work has focused primarily on the analysis and design of frames based on idealization of the joints as either fully rigid or pinned. In reality, the actual behavior of the structure is as much dependent on the connection and joint characteristics as on the individual component elements making up the structure. Ample research evidence exists which establishes that the observed joint behavior is substantially different from the assumed idealized models. Depending on the stiffness, strength, and deformation capacity, the connections in a structural framework can influence the behavior of the structure in several ways. Under static loads, the connection deformations contribute to the vertical deflections of the beams and lateral drift of the frame. The moment resistance of the connections will influence the internal forces distribution and the local and global stability of the frame.

Realizing the potential influence of connections on frame performance, the American Institute of Steel Construction (AISC, 1986, 1989) has introduced provisions to allow designers to consider explicitly the behavior of connections in the design of structural steel frames.

The ASD Specifications (AISC, 1989) list three types of constructions for designing a multi-story frame:

1. *Type 1 or "rigid framing"*. This construction assumes that the beam-to-column connections have sufficient rigidity to maintain the original geometric angle between intersecting members. Rigid connections are assumed for elastic structural analysis. Type 1 connections are sometimes referred to as moment connections.

2. *Type 2 or "simple framing"*. This construction assumes that, when the structure is loaded with gravity loads, the beam and girder connections transfer only vertical shear reactions without bending moment. The connections are allowed to rotate freely without any restraint. This type of connection is also called a shear connection.

3. *Type 3 or "semi-rigid framing"*. This construction assumes that the connections can transfer vertical shear and also have adequate stiffness and capacity to transfer some moment.

The AISC-LRFD Specifications (1986) designate two types of constructions in their provisions: *Type FR (fully restrained)* and *Type PR (partially restrained)*. Type FR corresponds to ASD Type 1. Type PR includes ASD Types 2 and 3. If Type PR construction is used, the effect of connection flexibility must be considered in the analysis and design of a structure.

To be pertinent, the rigidity of the connection should be defined with respect to the rigidity of the connecting member (Colson, 1991). For general application to a wide range of beam-to-column connections, Bjorhovde et al. (1990) introduce a nondimensional system of classification that compares the connection stiffness to the beam stiffness. In defining the beam stiffness, a reference beam length of 5d is used, where d is the beam depth to which the connection is attached.

The nondimensional parameters used in the classification of connections are:

$$\overline{m} = \frac{M}{M_p} \quad \overline{\theta} = \frac{\theta_r}{\theta_p} \tag{1.20}$$

in which θ_r is the relative deformation angle of the connection, $\theta_p = M_p/(EI_b/5d)$, I_b and L_b are the moment of inertia and length of the beam, and M_p is the full plastic moment capacity of the beam. The classification is based on the strength and stiffness of the connections with the boundary regions shown in Figure 1.6. The three different regions in Figure 1.6 are defined as:

(1) Rigid connection
   In terms of strength: $\quad \overline{m} \geq 0.7$
   In terms of stiffness: $\quad \overline{m} \geq 2.5\overline{\theta}$ \hfill (1.21)

(2) Semi-rigid connection
   In terms of strength: $\quad 0.7 > \overline{m} > 0.2$
   In terms of stiffness: $\quad 2.5\overline{\theta} > \overline{m} > 0.5\overline{\theta}$ \hfill (1.22)

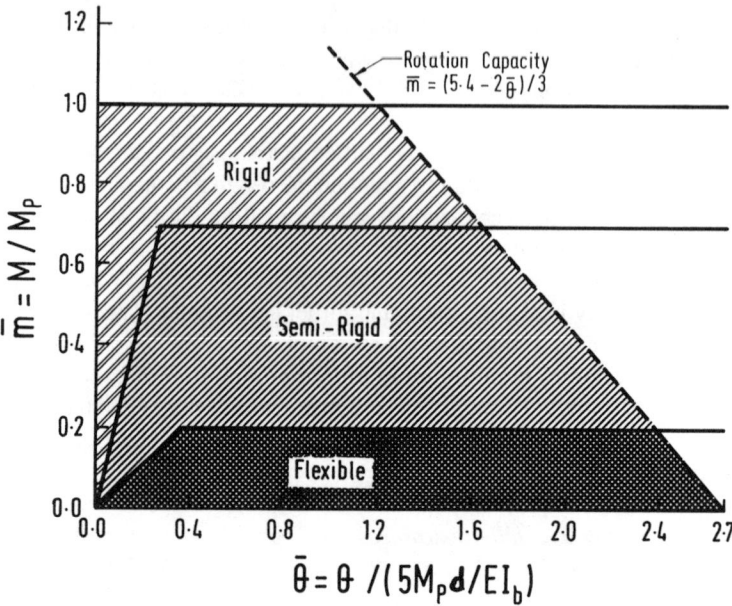

FIGURE 1.6. Classification of connections according to Bjorhovde et al. (1990).

(3) Flexible connection
In terms of strength: $\bar{m} \leq 0.2$
In terms of stiffness: $\bar{m} \leq 0.5\bar{\theta}$ (1.23)

Bjorhovde et al. (1990) also have proposed an expression for calculating the required rotation capacity of the connection based on a reference beam length and by curve fitting with test data. This simplified expression is written as:

$$\bar{m} = \frac{(5.4 - 2\bar{\theta})}{3} \qquad (1.24)$$

According to this formula, the required rotation capacity of a beam-to-column connection depends on the ratio of the ultimate moment capacity of the connection to the fully plastic moment of the beam, and it is inversely proportional to the initial connection stiffness, R_{ki}. In other words, the smaller the initial connection stiffness, the larger the necessary rotation capacity. Equation (1.24) is plotted and shown in Figure 1.6. This connection classification system may be used for the selection of connections for use in the analysis and design of semi-rigid frames (Kishi et al., 1992; Liew et al. 1993a and b). EC 3 (1990) also provides a different scheme for classification of beam-to-column connections. Comparison and discussion of these two schemes are given in Liew (1992), Bijlaard and Steenhuis (1991), and Colson (1991).

To properly apply the LRFD specifications for the design of semi-rigid frames, it is necessary to develop practical means for modeling the moment-rotation behavior of semi-rigid connections. Also, it is necessary to provide the means for designers to execute the analysis and design quickly and accurately. The first step to achieve these is to collect connection data relevant to their general behavior and characteristics. As a result of this effort, a comparative assessment of the performance of the different types of connections, in terms of initial stiffness and moment capacity, could be undertaken.

1.5.2 Connection Data Base

The connection data base is a collection of experimental tests for several types of beam-column connections. These data are compiled with the corresponding details and dimensions of the beam, column, and connection. Specific details such as the type of steel used, the names of the researchers, and the date of the tests conducted are also included. Several moment-rotation prediction models are also incorporated in the data base to simulate the test results, and to generate moment-rotation $(M - \theta_r)$ curves suitable for design use.

Goverdhan Data Base

Goverdhan (1983) collected extensive connection data from experimental tests conducted after 1950. The moment-rotation data were compiled in the computer in the form of a data base. Several prediction equations were presented for each type of connection. The experimental moment-rotation curves available were compared with the available moment-rotation prediction equations for each connection type. The validity and the drawbacks of the equations were discussed and recommendations regarding their use in design were given. This collection covered the following types of connection.

(1) Double web angle.
(2) Single web angle and single plate.
(3) Header plate.
(4) Endplate.
(5) Top and seat angles with or without web angles.

Nethercot Data Base

Nethercot (1985) reviewed data of over 70 separate experimental studies on steel beam-column connections. Out of more than 700 individual tests examined, Nethercot selected the useful data for his analysis. The curve-fitting of the experimental data was conducted in the study and some preliminary comparative studies of the role of different joint parameters on moment-rotation curves were incorporated. The study covered the following ten types of connections:

(1) Single web angle.
(2) Single web plate.
(3) Double web angle.

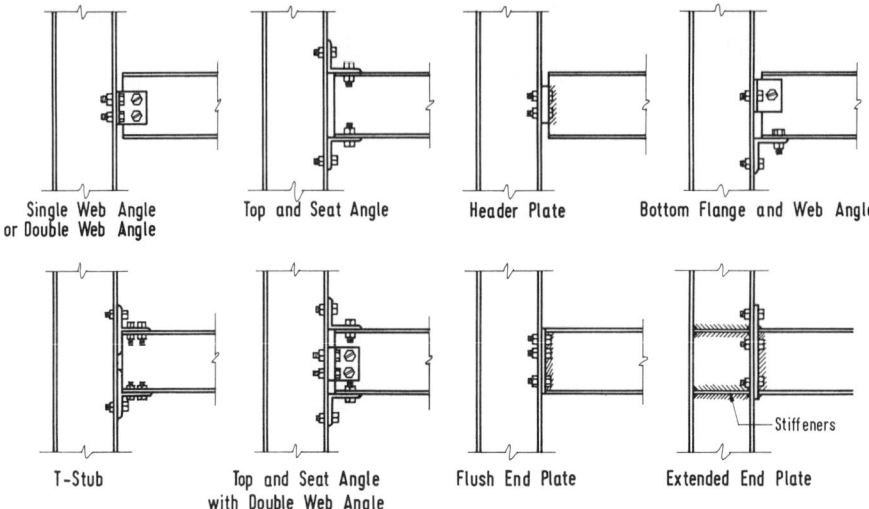

FIGURE 1.7. Typical beam-to-column connections.

(4) Flange angle.
(5) Header plate.
(6) Flush/extended end plate.
(7) Combined web and flange angles.
(8) T-stubs.
(9) Top and seat angle.
(10) T-stubs and web angles.

Some of the most commonly used connection types reported in Goverdhan (1984) and Nethercot (1985) are shown in Figure 1.7. Typical moment-rotation curves for these connections are illustrated in Figure 1.8. It should be noted that for any connection type, the stiffness and strength are dependent upon the geometric parameters such as plate or angle thickness, bolt size, method of tightening, and connection depth. The more flexible ones, such as single and double web angle connections, may be approximated as pin connections represented by the horizontal axis in the figure. These types of connections might be classified as "flexible." The stiffest ones, such as the extended end plate connection, might be classified as "rigid," represented by the vertical axis. Those with intermediate stiffness might be classified as "semi-rigid." All welded connections, which are not shown in Figure 1.7, tend to be extremely stiff and they are usually designed to transmit full moment capacity of the adjoining members. For the purpose of design, they may be assumed as perfectly rigid.

Kishi and Chen Data Base

In 1986, Kishi and Chen conducted a comprehensive search on beam-column connection data. Characteristic results on moment-rotation response and the correspond-

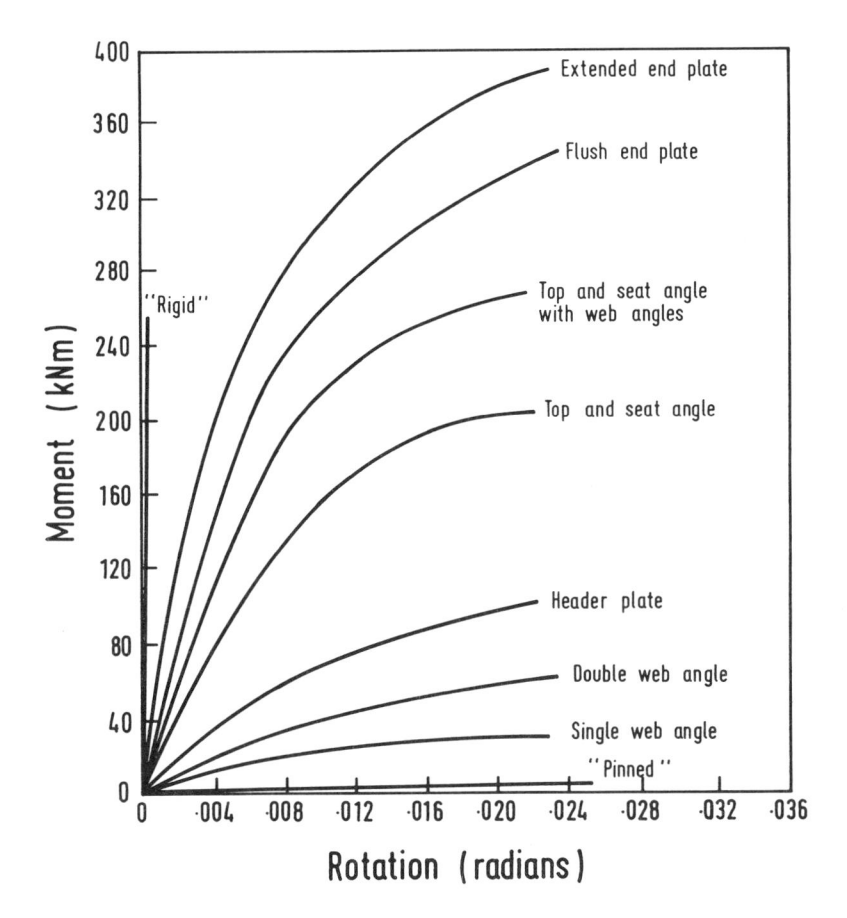

FIGURE 1.8. Typical moment-rotation curves.

ing parameters of beam-to-column connections used frequently in steel construction were collected and stored in a data base (Kishi and Chen, 1986, Kishi et al., 1992). The data base consisted of experimental data on rivet, bolted, and welded connections that were published from 1936 to 1986. The experimental data collected were compared with some prediction equations in order to develop a rational method of analysis for semi-rigid frame design.

In particular, three prediction equations were discussed in detail. The first was the analytical polynomial equation proposed by Frye and Morris (1976). The second was a curve-fitting equation using a modified exponential model proposed by Kishi and Chen (1986). The third model utilizes a three-parameter power model proposed by Kishi and Chen (1990), which is a simplified analytical model suitable for use in the analysis and design of steel frames.

The Kishi and Chen connection data base is made available to the reader in Chapter 3. This data base includes connection types with semi-rigid moment-rotation characteristics similar to those reported in Goverdhan's data bank. Available test data on steel beam-to-column connections and comparison between experiment and theoretical results calculated using the three prediction models are available by running the data base program. This program is an essential tool of performing structural analysis of frames comprising semi-rigid joints. The following two sections provide recommended procedures for designing semi-rigid frames based on the use of simplified and second-order analysis techniques.

1.5.3 Simplified Analysis/Design Method

A simplified design procedure for unbraced semi-rigid frames was introduced recently by Barakat and Chen (1990). The procedure is largely based on the philosophy of the B_1/B_2 amplification factor method recommended by the AISC-LRFD Specification (1986). While the amplification factor method is developed for elastic rigid frames, the simplified design procedure suggests a number of modifications to accommodate the presence of semi-rigid connections. These include (1) two linearized moment-rotation relationships (expressed by R_{ko} and R_{kb}) for modeling the connection behavior, and (2) a modified relative stiffness factor for determining the effective length of elastically restrained columns. The modified initial stiffness (R_{ko}) is used for implementation in the first-order analysis of the non sway frame configuration which produces M_{nt} moments. The secant stiffness R_{kb} is used for implementation in the first-order analysis of the sway frame which yields M_{lt} moments. The maximum design moment M_u in a member may be then determined from Eq. (1.7) in lieu of a direct second-order elastic analysis.

The analysis/design procedure is described below:

(1) Determine the connection ultimate moment, M_u, and initial stiffness, R_{ki}. If the connection moment-rotation curve is the type of standard connection available in the data base, R_{ki} and M_u can be obtained directly by generating the data base program described in Chapter 3. If the connection is not the standard type, R_{ki} and M_u for some selected types of connections may be evaluated analytically using simple connection models which are also available in the data base program.

(2) Determine R_{ko} which is the secant stiffness corresponding to a rotation of θ_o. θ_o is obtained as the intersection between the lines of the initial stiffness R_{ki} and the ultimate moment M_u (Figure 1.9). R_{ko} is used instead of R_{ki} as a representative connection stiffness for calculating M, because it was felt that R_{ki} was too high a value to be used for analysis recognizing the fact that the connection stiffness degrades as the moment in the connection increases. The secant stiffness R_{ko} determined by the above procedure is used in a first-order frame analysis to obtain M_{nt}.

(3) Determine R_{kb}. As loading progresses, the connection sustains increasing rotations and may sequentially exhibit declining stiffness values. For sway frames, the connection is presumed to undergo noticeable deformation when the effect of lateral loads is added to that of gravity loads. When referred to the AISC-LRFD B_1/B_2 method, this situation may be viewed as the phase in

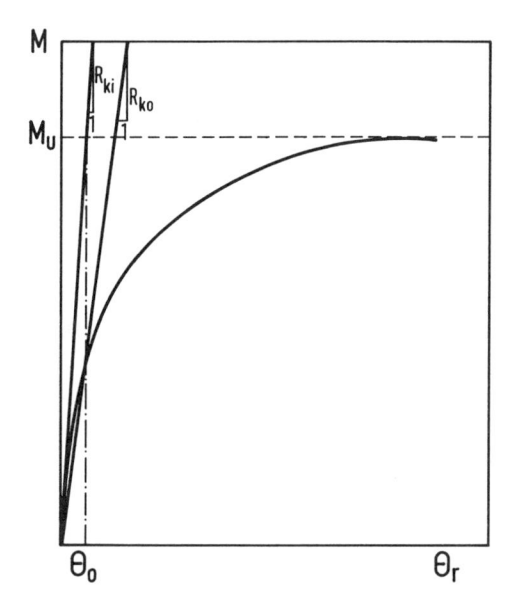

FIGURE 1.9. Determination of R_{ko}.

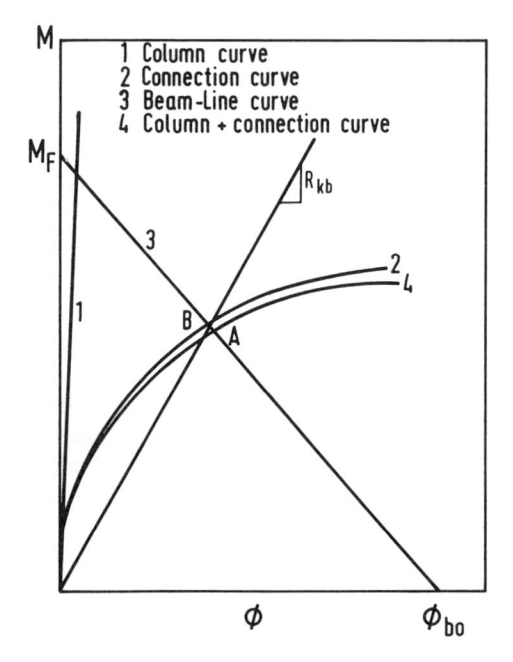

FIGURE 1.10. Determination of R_{kb}.

which M_{lt} is determined. The design connection stiffness to be used in this phase should, therefore, be less than that used for determining M_{nt}. Barakat and Chen (1990) proposed a stiffness value of R_{kb} for the sway analysis. The determination of R_{kb} is shown schematically in Figure 1.10. In the figure, curve 1 represents the deformation due to column rotation, curve 2 represents the deformation due to connection flexibility, curve 3 is the so-called beam line, and curve 4 represents the combined effect of column rotation and connection deformation. Compatibility of rotational deformation at a joint is satisfied at the intersection of the beam line, represented by curve 3 and curve 4 at point A in Figure 1.10. However, for design purposes, it is reasonable to assume that the effect of column rotation is negligible compared to that of connection deformation. Consequently, curve 2 rather than curve 4 is recommended for determining R_{kb}. From Figure 1.10, it can be observed that R_{kb} is obtained as the secant stiffness corresponding to a rotation defined by the intersection of curve 2 and curve 3 (point B in the figure).

(4) Determine the K factor for column design. The column effective length factor K may be obtained from the story-buckling analysis based on Eq. (1.12), (1.13), or (1.14).

The design procedure is summarized as follows:

(1) Determine R_{ko} and R_{kb}.
(2) Perform a first-order analysis incorporating the connection stiffness R_{ko} for M_{nt}.
(3) Perform another first-order analysis incorporating the connection stiffness R_{kb} for M_{lt}.
(4) Determine G factors that account for the presence of connection at the girder's ends, where $G = \Sigma(EI_c/L_c)/\Sigma(EI_b/L_b)$.
(5) Determine the column effective length factor from a story-buckling analysis using Eq. (1.12) or (1.13). These equations require the calculation of alignment chart K_n factor, which is obtained in the usual manner using the modifications for elastically restrained beam ends as described in Barakat (1989). However, Eq. (1.14) proposed by Lui (1992) may be used instead. This effective length factor equation requires a first-order deflection analysis of the semi-rigid frame, which has been carried out in steps (2) or (3).
(6) Evaluate B_1 and B_2 according to AISC-LRFD equations given in Eq. (1.8) and Eq. (1.10) or (1.11).
(7) Obtain the design moment M_u using Eq. (1.7).
(8) Use the beam-column interaction equations (Eqs. 1.4 and 1.5) to design the member.

The implementation of the above procedure for semi-rigid frame design on personal computers can be made in a spreadsheet format. The spreadsheet program has been implemented in Barakat and Chen (1991) and it was found to be very efficient for many types of frame design problems.

It is important to note that the LRFD method, as all other methods, is strength based, which means that the design is based on strength, not serviceability. Since excessive frame drift is often a problem for semi-rigid frames, due consideration must be given to check the resulting design for serviceability requirements, and proper measures must be taken to limit the drift of semi-rigid frames (Gerstle and Ackroyd, 1990).

1.5.4 Direct Second-Order Analysis Methods

Instead of using the modified B_1/B_2 method, a direct second-order elastic analysis may be used to estimate the second-order moment for member design. Chapter 2 of this book presents a computer-based second-order elastic analysis method that can be applied to design practice for the type of PR construction using a small computer. In the development of this computer program, special attention is paid to efficient computer implementation, and to simplicity of the analytical procedures without losing the numerical stability and accuracy of a rigorous solution. At a research level, direct analysis of second-order elastic effects in semi-rigid steel frames is well developed and is within the reach of many engineering professions. Moreover, significant developments also have been made to tackle important practical aspects of applying second-order analysis with comprehensive procedures for actual design use (Liew et al., 1993a and b; Kishi et al., 1991 and 1992; White and Hajjar, 1991). The contemporary limit-state specifications have given explicit permission for the engineer to compute load effects from a direct second-order elastic analysis. This type of analysis is beginning to gain wide acceptance among the design community. Several computer programs, which will be described in Chapters 2 and 3, provide a "comprehensive" second-order analysis capability in which both member curvature, P-δ, and chord rotation, P-Δ, effects are accounted for in a general fashion.

A direct second-order elastic analysis can be used as efficiently, and in certain respects is more simple to perform than approximate methods which involve the amplification of the loads or member forces from linear elastic analysis, especially for the analysis and design of a complex semi-rigid structural system. Under the limit-state design concepts, and coupled with reasonable computing facility, direct second-order elastic analysis is certainly the emerging technology that will become the common practice in the not too distant future.

However, it is important to emphasis that although second-order elastic analysis eliminates the need to perform B_1/B_2 analysis as discussed in the earlier section, an effective length factor is still required to determine the axial resistance for design by the current ASD and LRFD provisions. The advantages of the second-order elastic analysis procedure are its simple concept and the ease with which complex frameworks can be analyzed using standard programs. The disadvantages are that limits are thought to be required to guard against onset of significant inelastic response through the use of member interaction equations which involve estimation of effective length factors. Because of these approximations, the maximum strength of the system is not determined. The only way to assess the real performance of a structural system is through direct second-order inelastic analysis.

With the availability of more powerful computer workstations, it is now becoming more realistic and straightforward to develop advanced analysis method which directly considers the effects of semi-rigid connections, and material and geometrical nonlinearities in a general fashion. With this analysis technique in place, the need of estimating various factors for specification member capacity checks can be totally avoided. Advanced analysis method, when coupled with integrated graphic analysis and design system, allows the engineer to exercise greater freedom in structural design. Its use can assist in providing both efficient and cost-effective design solutions.

1.6 Second-Order Inelastic Analyses

Inelastic analysis refers to any method of analysis in which the effects of material yielding are accounted for. The different types of inelastic models may be generalized into three main groups:

(1) the plastic-zone or distributed plasticity,
(2) the elastic-plastic hinge, and
(3) the refined- or modified-plastic hinge.

This generalization is based on the degree of refinement in representation of yielding effects. The elastic-plastic hinge model is the simplest approach while the plastic-zone model exhibits the greatest refinement. The refined-plastic hinge model is an improvement over the elastic-plastic hinge model, and it requires less computational effort and is less costly than the plastic-zone model.

1.6.1 Plastic-Zone Method

There are generally two types of plastic-zone analyses. The first involves the use of three-dimensional finite shell elements in which the elastic constitutive matrix, in the usual incremental stress-strain relations, is replaced by an elastic-plastic constitutive matrix once yielding is detected. Based on a deformation theory of plasticity, the effects of the combined effects of normal and shear stresses may be accounted for. This analysis approach typically requires modeling of structures using a large number of finite three-dimensional shell elements, and numerical integration for the evaluation of the elastic-plastic stiffness matrix. The three-dimensional spread-of-plasticity analysis when combined with second-order theory which deals with frame stability is computational intensive and, therefore, best suited for analyzing small-scale structures, or if the detailed solutions for member local instability and yielding behavior are required. Since detailed analysis of local effects in realistic building frames is not the common practice in engineering design, this approach is considered over expansive for practical use.

The second approach for second-order plastic-zone analysis is based on the use of beam-column theory, in which the member is discretized into many line segments, and the cross-section of each segment is further subdivided into a number of finite elements. Inelasticity is typically modeled by the consideration of normal stress only. When the computed stresses at the centroid of any fibers reach the uniaxial normal strength of the material, the fiber is considered as yielded. Also, compatibility is treated by assuming that full continuity is retained throughout the volume of the structure in the same manner as for elastic range calculations. Although there may exist quite sharp curvature in the vicinity of inelastic portions of the structure, the so-called "plastic hinges" can never develop. In plastic-zone analysis, the calculation of forces and deformations in the structure after yielding requires iterative trial-and-error processes because of the nonlinearity of the load-deformation response, and the change in cross-section effective stiffness at inelastic regions associated with the increase in the applied loads and the change in structural geometry. Although most of the plastic-zone analysis methods have been developed for planar analysis (Clarke et al., 1992; White, 1985; Vogel, 1985; El-Zanaty et al., 1980; Alvarez and Birnstiel, 1967), three-dimensional plastic-zone techniques are also available involving various degrees of refinements (White, 1988; Wang, 1988; Chen and Atsuta, 1977).

A plastic-zone analysis that includes the spread of plasticity, residual stresses, initial geometric imperfections, and any other significant second-order behavioral effects, would certainly eliminate the need for checking individual member capacities in the frame. Therefore, this type of method is generally classified as advanced inelastic analysis in which the checking of beam-column interaction equations is not required. In fact, the member interaction equations in the modern limit-states specifications were developed, in part, by curve-fitting to the results from this type of analysis. However, plastic-zone analysis is generally too computationally intensive for routine design use. Its applications are limited only for research studies, generation of design charts, and very special design problems (ECCS, 1984; White and Chen, 1990). In reality, some significant behavioral effects such as joints' and connection's performances tend to defy precise numerical and analytical modeling. In such cases, a simpler method of analysis that adequately represents the significant behavior would be sufficient for engineering application.

The detailed formulation of plastic-zone analysis techniques and the computer programs capable of analyzing beam-column subassemblies and multi-story frames are described in Chapters 5 and 6, respectively. These computer programs may be utilized to generate benchmark problems for verification of second-order elastic and inelastic programs intended for routine design use.

1.6.2 Elastic-Plastic Hinge Method

A more simple and efficient approach for representing inelasticity effects in frames is to employ the elastic-plastic hinge method. This method assumes that the element remains elastic except at its ends where zero-length plastic hinges are allowed to form. This analysis approach typically involves the use of one beam-column element for each frame member, and thus it is efficient for analyzing large building frameworks. This is particularly true for structures in which the axial forces in the component members are small and the predominate behavior is associated with bending actions. In such cases, second-order elastic-plastic hinge analysis may be applied to describe the inelastic behavior with sufficient accuracy, assuming that lateral-torsional and local buckling modes of failure are prevented (Liew, 1992).

In general, second-order elastic-plastic hinge analysis is only an approximate method. When used to analyze a single beam-column element subjected to combined axial load and moment, this method often overestimates the strength and stiffness of the element when it is loaded into the inelastic range. Although elastic-plastic hinge approaches can provide essentially the same load-displacement predictions as plastic-zone methods for many frame problems, they are not adequate to be classified as advanced analysis methods in general (Liew et al., 1992a; Liew and Chen, 1991; White, 1991).

However, research by Ziemian* (Ziemian et al., 1990; Ziemian, 1990) has shown that the elastic-plastic hinge analysis can be classified as an advanced inelastic analysis since this method is rather accurate for matching the strength and load-displacement response of several building frames from plastic-zone analysis. But many cases considered in Ziemian's work, especially when the axial load is less than $0.5P_y$, are not sensitive benchmarks for determining the accuracy and the possible limitations of the elastic-plastic hinge method. Therefore, suitable benchmark problems should be used to provide more in-depth study on the qualities and limitations of the second-order elastic-plastic hinge method before it can be readily accepted as a legitimate tool in the design of steel structures.

Chapter 7 of this book presents a wide range of theoretical and experimental benchmark problems which may be used for verification of second-order inelastic computer programs intended for use as advanced analysis tools. Additional test problems also can be found in Liew (1992) and SSRC (1993). A computer program capable of performing second-order elastic-plastic hinge analysis of planar frames is given in Chapter 4. The following two sections propose the two approaches of refining plastic hinge analysis for designing steel framing structures without the need of specification member capacity checks. Their design implications with respect to limit state specification formats are discussed.

1.6.3 Notional Load Plastic Hinge Method

One possible approach to advance the use of second-order elastic-plastic hinge analysis for frame design is to specify artificially large values of frame imperfections (i.e., initial out-of-plumbness) in the analysis. This is the general approach adopted by EC 3 (1990) for frame design using second-order analysis. In additional to accounting for standard erection tolerance for out-of-plumbness, these artificial large imperfections are intended to account for the effects of residual stresses, frame imperfections, and distributed plasticity that are not considered in the frame analysis.

In Liew et al. (1992a), an approach, which does not involve any modifications to the elastic-plastic hinge theory and which promises to satisfy the planar strength

---

* Ziemian uses a tangent-modulus-plastic hinge model which is identical to the elastic-plastic hinge model when the column axial load P is less than $0.5P_y$.

Frame Imperfections : Eurocode 3 and Notional Load Plastic Hinge

$$k_c = \sqrt{[0.5 + 1/n_c]} \leq 1.0$$

$$k_s = \sqrt{[0.2 + 1/n_s]} \leq 1.0$$

$$\Psi_0 = \frac{1}{200} k_c k_s$$

n_c = number of columns per plane
n_s = number of stories per plane

Member Imperfections δ_0 modeled only if $P > 0.26\pi^2 EI/L^2$

Eurocode 3	
δ_0	Column curve
L/400	a
L/250	b
L/200	c
L/150	d

Notional Load Plastic Hinge
$\delta_0 = L/400$

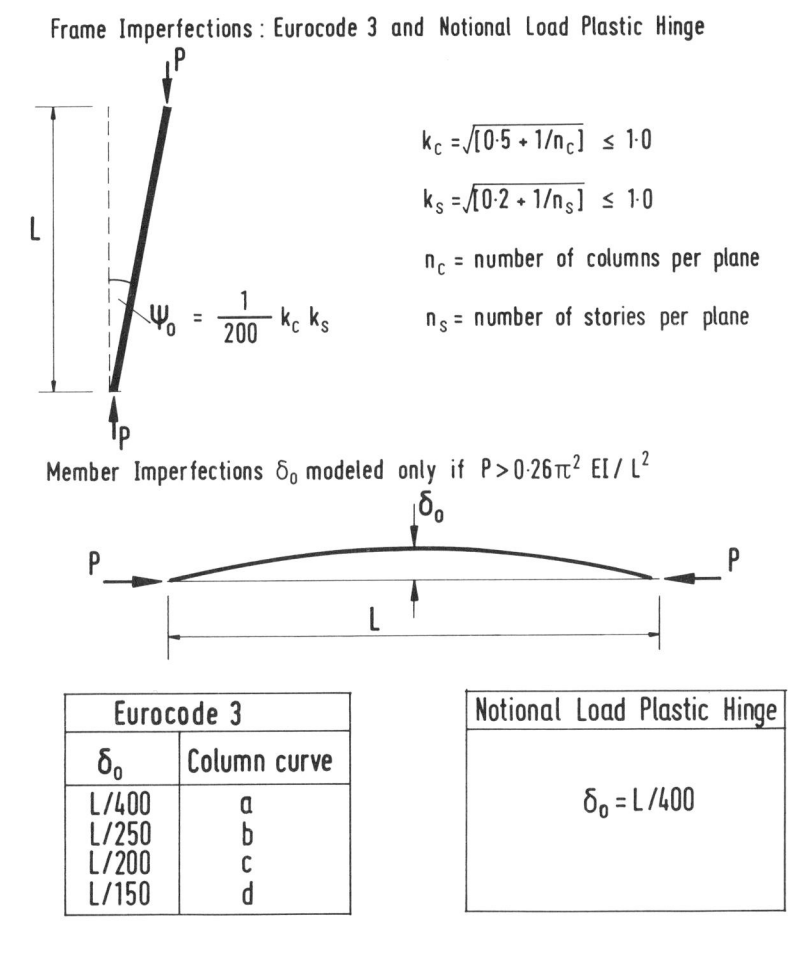

FIGURE 1.11. Comparison between the notional load plastic hinge method with Eurocode 3 procedures (EC 3, 1990).

requirements for advanced analysis in AISC-LRFD, is presented. This method of analysis utilizes a set of equivalent notional lateral loads to account approximately for the influence of member imperfections and distributed plasticity effects on the overall system strength. These notional lateral loads, which are expressed as a fraction of the story gravity loads or member axial force, are chosen to allow adequately for frame and member instability when they are combined with the applied lateral loads used for the global frame analysis.

The notional load plastic hinge approach is similar in concept as the "enlarged" geometric imperfection approach adopted by EC 3. The similarities and differences between these two methods are summarized as shown in Figure 1.11 (White et al., 1992). Basically, the specification of frame imperfections is the same in both approaches, with the maximum out-of-plumbness being defined as $\Psi_o = L/200$. Furthermore, the modeling of member initial out-of-straightness imperfections is not required when the column axial load P is less than $0.26P_e$ in both approaches. However,

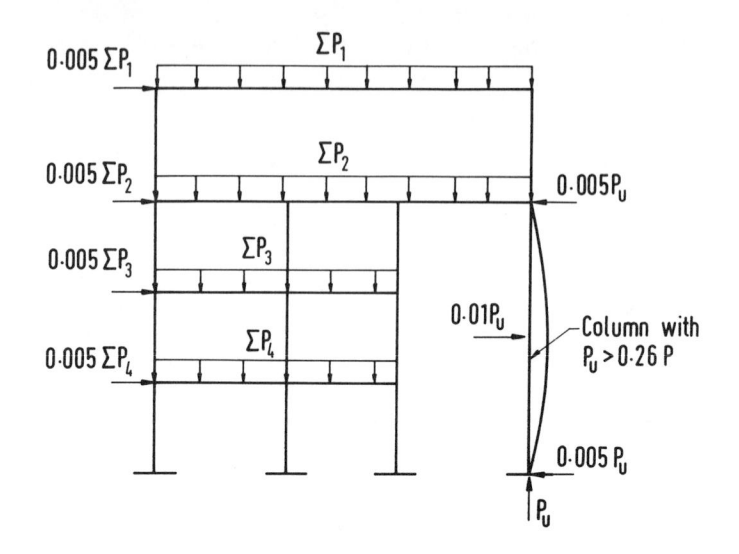

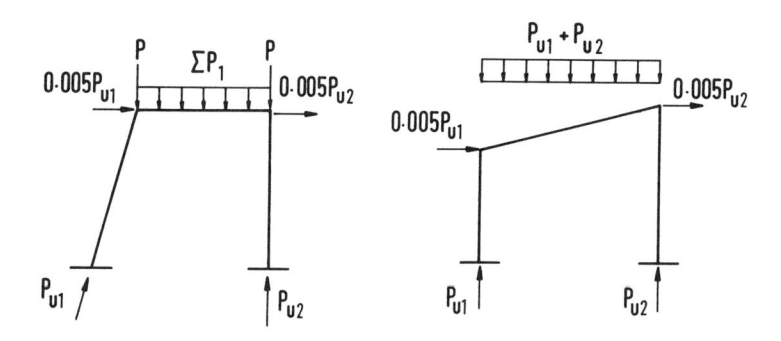

FIGURE 1.12. Examples on application of notional loads for second-order elastic-plastic hinge analysis.

the modeling of member imperfection effects for cases in which this limit is exceeded is different in the two procedures.

EC 3 specifies different equivalent initial out-of-straightness imperfections based on the relevant column curve for the member under consideration. The maximum out-of-straightness magnitudes range from L/400, for rolled I-sections under strong-axis bending, to L/150, for heavy I-sections under strong- or weak-axis bending. The equivalent maximum out-of-straightness magnitude assumed in the notional load plastic hinge approach is L/400. This value is consistent with the use of a single-column curve in AISC-LRFD and also, it is equivalent to application of the maximum equivalent out-of-plumbness rule to an individual member with a maximum mid-span deflection of (L/2)/200. Of course, the equivalent notional load corresponding to this member imperfection is 0.01P, and the equivalent notional load corresponding to the maximum out-of-plumbness of 1/200 is 0.005P. The application of these notional loads to several example frames is illustrated in Figure 1.12.

The general application of the notional load analysis to beam-column and frame cases has been studied by Liew et al. (1992a). Figure 1.13 illustrates the interaction

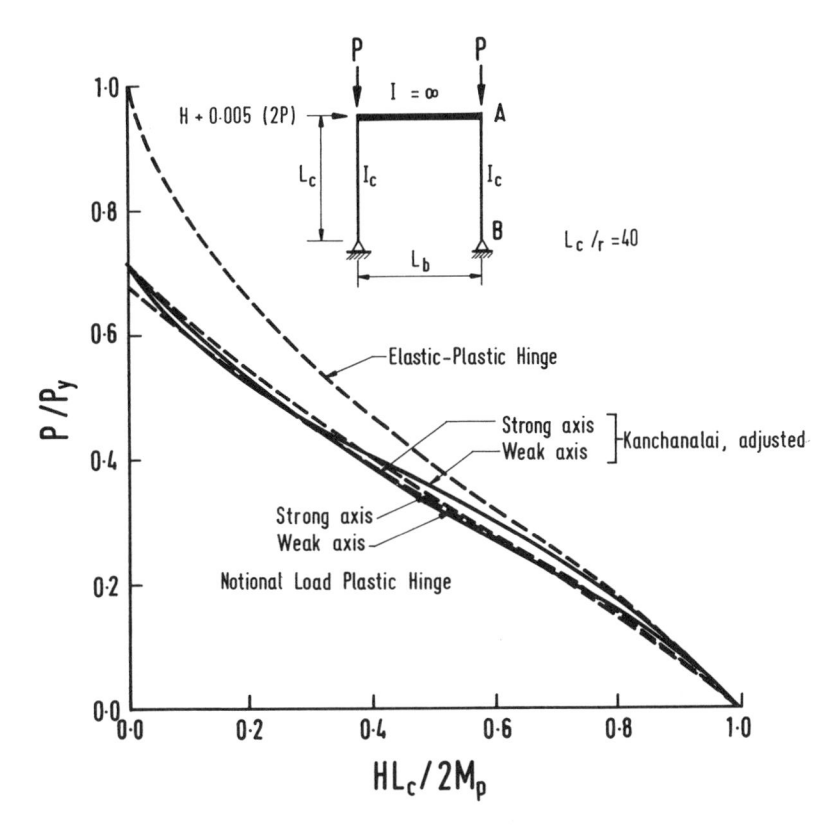

FIGURE 1.13. Comparison of strength curves from notional load plastic hinge analysis with adjusted strength curves (Kanchanalai, 1977).

strength curve obtained by application of this approach. The results are compared with the adjusted plastic-zone solutions generated by Kanchanalai (1977). The "exact" solutions are obtained by subtracting a "used-up" P-Δ moment from the results for perfect frame geometry, as described by Liew et al. (1991). This procedure results in the adjusted "exact" strength that is consistent with the AISC-LRFD column strength for the case of pure column action. It is observed that the notional load plastic hinge approach compares well with the plastic-zone solutions for all range of axial force versus moment. If the notional load is not employed, the maximum unconservative error can be as high as 40%. The frame example shown in Figure 1.13 is only one of the worst cases from the more comprehensive set of problems considered in Liew (1992).

1.6.4 Refined-Plastic Hinge Analysis

In recent work by Abdel-Ghaffar et al. (1991), Al-Mashary and Chen, (1991), King, et al. (1991), Liew and Chen (1991), Liew et al. (1992b-c), and White et al. (1992) among others, an inelastic analysis approach based on simple refinements of the elastic-plastic hinge model has been proposed for planar frame analysis. This type of analysis seeks to represent the effects of distributed plasticity through the cross-section, assuming that the plastic hinge stiffness degradation is smooth. The inelastic behavior in the member is modeled in terms of member forces instead of the detailed level of stresses and strains as used in the plastic-zone analysis model. The principal merits of the refined-plastic hinge model are that the method is as simple and as efficient as the elastic-plastic hinge analysis approach, and it is sufficiently accurate for the assessment of members and system strength and stability. The investigations by Liew et al. (1992b and c) further suggest that the refined-plastic hinge method has the most promise for use analyzing a large-scale frame structure without the risk of overestimating the strength and stiffness of individual members in the frame.

In Chapter 4, an analysis approach has been developed by which connection flexibility also can be accounted for in a second-order refined-plastic hinge analysis of steel frames. The connections are modeled as rotational springs which are physically tied to the ends of the element by enforcing equilibrium and compatibility conditions between the connection element and the beam-column element. The refined-plastic hinge model is based on simple modification to the elastic-plastic hinge method by considering the actual distribution of plasticity in a beam-column element loaded by arbitrary end forces, and then attempting to model the effective stiffness of the element by approximating the effects of the distributed yielding. Distributed yielding in a member is accounted for by using an effective tangent modulus and a two-surface stiffness degradation. For members subjected to compression, the column tangent modulus, E_t, is evaluated based on the inelastic stiffness reduction procedure given in the AISC-LRFD manual for the calculation of inelastic column strength (AISC, 1986). Since this E_t model is derived from the LRFD column strength formula, it implicitly includes the effects of residual stresses and initial out-of-straightness in modeling the member effective stiffness (Liew et al. 1992b and c).

The recent research by the authors (Liew et al. 1992b and c) concludes that for members subjected to significant in-plane moment, the tangent-modulus approach is not sufficient to represent the gradual stiffness degradation as yielding progress through the volume of the member. Additional distributed plasticity effects in the member are related to the bending action. These effects can be represented by modifying the basic elastic-plastic hinge model such that the member stiffness degrades gradually from the stiffness associated with the onset of yielding, to that associated with the cross-section plastic strength. A gradual stiffness reduction scheme with initial yielding assumed to occur at a yield surface that is the same shape of and equal to one-half the size of the plastic strength surface is chosen for representing the inelastic behavior of beam-columns subjected to a combined action of bending and axial forces (see Figure 1.14). This approach is superior to the conventional elastic-plastic hinge approach in representing the inelastic behavior of many types of frame structures.

In load and resistance factor design, the cross-section plastic strength used in the analysis may need to be reduced by the appropriate resistance factors. Figure 1.14 shows an approach in which the nominal plastic strength surface is reduced to a design yield surface based on a resistance factor of ϕ for both flexural bending and axial compression. However, different resistance factors for flexural bending and

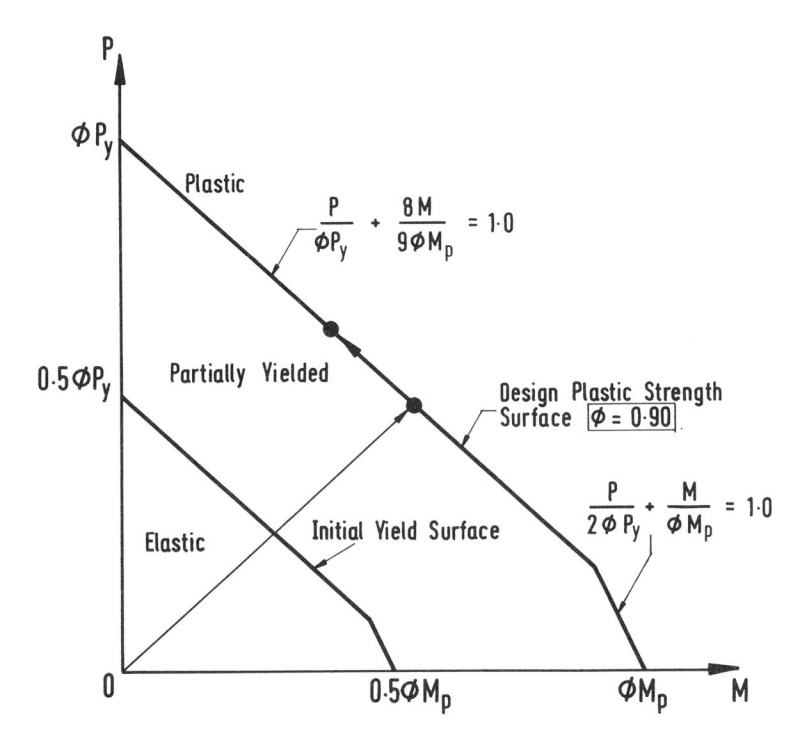

FIGURE 1.14. Two-surface stiffness degradation model for refined-plastic hinge analysis.

axial compression, such as those currently used in AISC-LRFD for beam-column design, may be used, if required, in the refined-plastic hinge analysis.

Analysis of Simple Portal Frames

Figure 1.15 compares the in-plane strength curves obtained by the refined-plastic hinge analysis with the results from second-order elastic-plastic hinge analysis and Kanchanalai's "exact" plastic-zone analysis (Kanchanalai, 1977). The frame in Figure 1.15 is one of the benchmark frames studied by Liew (1992) for verification of the refined-plastic hinge approach for advanced analysis. In Kanchanalai's plastic-zone analyses, both the in-plane strength curves for strong- and weak-axis bending are presented, whereas in the hinge based analyses, only the strength curves for the strong-axis bending are shown. The weak-axis strength curves from these analyses are identical to the strong-axis strength curves, since the results are presented in a nondimensional form, and only one plastic strength curve (see Figure 1.14) is used for both the strong- and weak-axis section strengths. The comparisons of strength curves in Figure 1.15 show that the onset of yielding in the frame members occurs long before the ultimate strength of the members is reached. This observation is true for almost all range of loading except when the frame is subjected to gravity load only. The refined plastic hinge method predicts a smaller load capacity for this frame

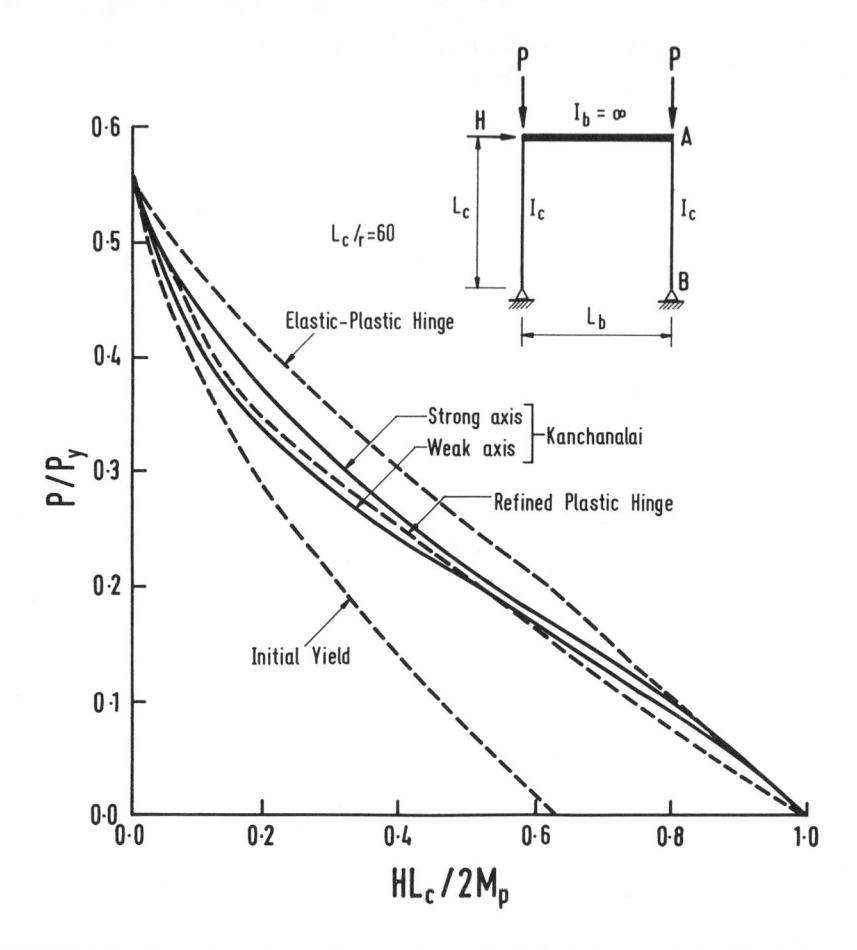

FIGURE 1.15. Comparison of strength curves from refined-plastic hinge analysis with plastic-zone strength curves (Kanchanalai, 1977).

compared to the elastic-plastic hinge method. The results are closer to the strength curves obtained by the plastic-zone method.

Figures 1.16 and 1.17 show the load-deflection curves of two portal frames bent about their strong- and weak-axis, respectively. Again, the load-displacement curves obtained from the refined-plastic hinge method are compared with the elastic-plastic hinge response curves and the "exact" plastic zone results by Vogel (1985) and Kanchanalai (1977) in Figures 1.16 and 1.17., respectively. The comparisons of the response curves show that the refined-plastic hinge approach is significantly more accurate in predicting the inelastic load-displacement behavior of these frames than the elastic-plastic hinge method. It also gives conservative and accurate predictions of the maximum strengths of these frames compared to the "exact" plastic zone results. However, it should be pointed out that the portal frame examples in Figures 1.15 through 1.17 are rather critical for testing the accuracy of a second-order inelastic analysis. A multi-story and multi-bay frame is cited below to give a better ideal of the accuracy of various analysis methods in representing the yield effects.

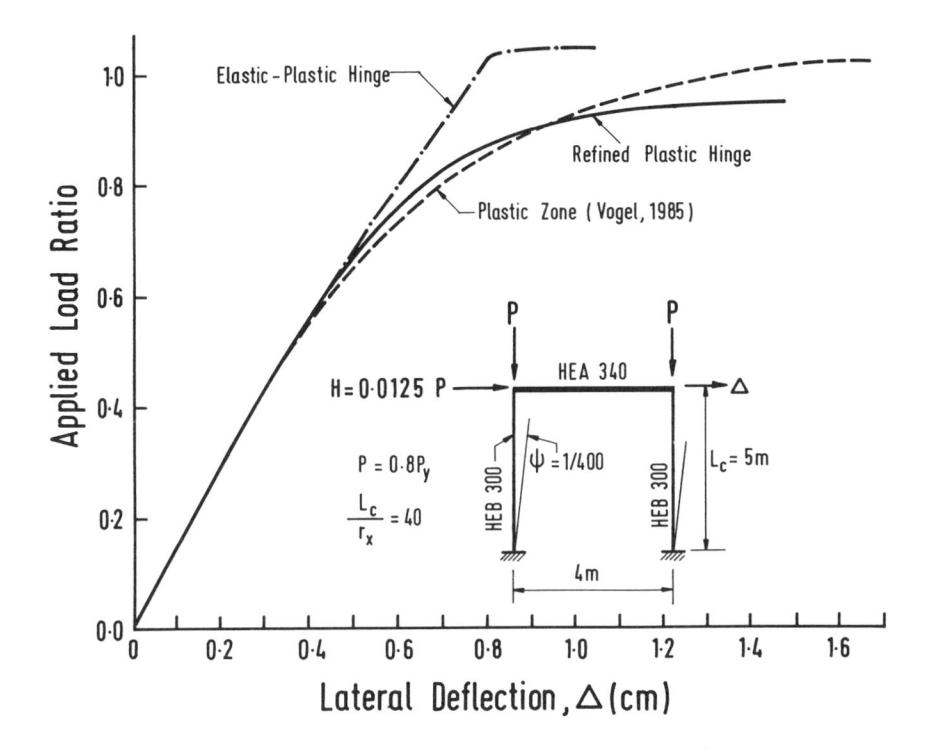

FIGURE 1.16. Comparison of load-displacement characteristics for a portal frame bending about the strong axis.

Analysis of a Six-Story Frame

The frame in Figure 1.18 is proposed by Vogel (1985) as a calibration frame for nonlinear inelastic analysis. The frame has initial out-of-plumbness associated with the value recommended by EC 3 (1990). Both gravity and lateral loads are applied proportionally until failure occurs. The applied load versus top- and fourth-story lateral displacement curves are shown in Figure 1.18.

The following methods, which are discussed in the earlier sections, are used to assess the strength limit states of the calibration frames.

(1) *LRFD-elastic analysis* — Member forces are obtained using second-order elastic analysis and with all members satisfying the AISC-LRFD strength limit-state equations. The ultimate strength of a system is obtained when the maximum force in any member of the frame reaches the limiting strength specified by the member equations. Side-sway imperfections in the calibration frames are not modeled in this analysis approach.

(2) *Notional load plastic hinge analysis* — A notional lateral load equal to 0.5% of the total gravity loads acting on the story is applied at the top of the column at each floor level. The analysis details are the same as the elastic-plastic hinge method except that explicit modeling of initial imperfections is not required in this approach.

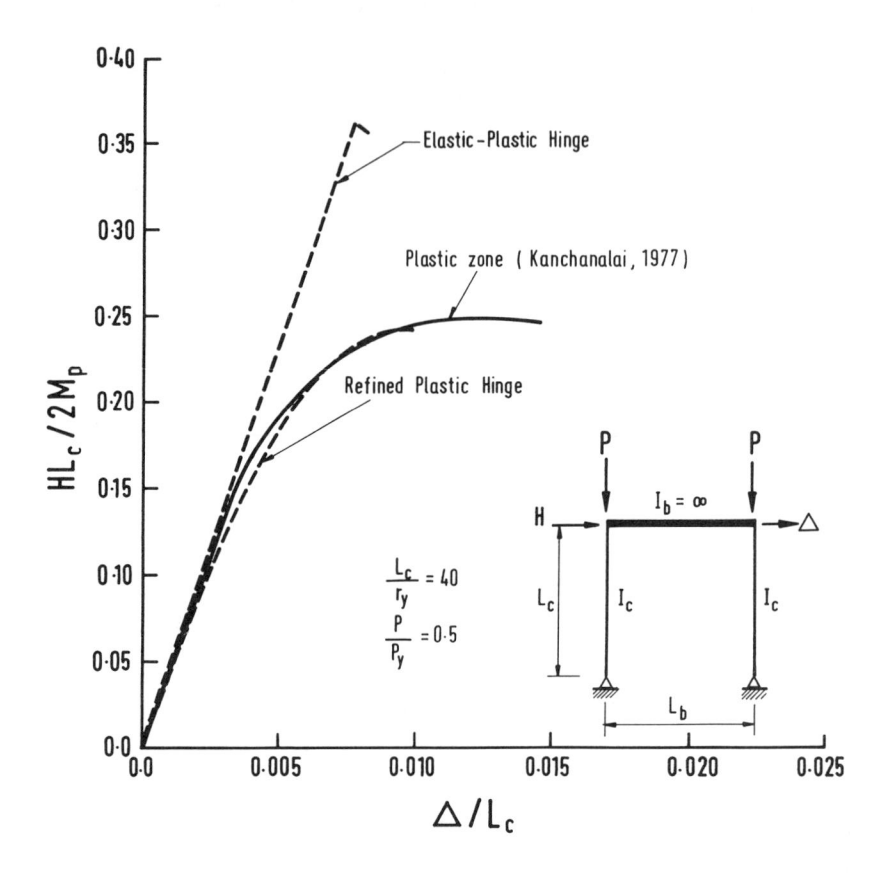

FIGURE 1.17. Comparison of load-displacement characteristics for a portal frame bending about the weak axis.

(3) *Refined-plastic hinge analysis* — The tangent modulus and parabolic stiffness degradation function are used to model the effective stiffness of the member (Liew, 1992). The analysis employs the same inelastic stiffness reduction scheme for all members in the structure. This is contrasted with the more "exact" plastic-zone analysis in which different compressive residual stress magnitudes have been assumed for members of different cross-section dimensions (Vogel, 1985). Initial side-sway imperfections similar to the plastic-zone analysis are modeled explicitly in the refined-plastic hinge analysis and the CRC tangent modulus is used for stiffness degradation (see Chapter 4).

For this frame, all the inelastic analysis methods predict essentially the same limit load. The maximum frame resistance is reached at a load parameter of 1.111 in Vogel's plastic-zone study, at 1.118 for the refined-plastic hinge analysis, at 1.097 for the notional load plastic hinge method, and 1.124 for the elastic-plastic hinge analysis (not shown in the figure). The maximum difference between these limit

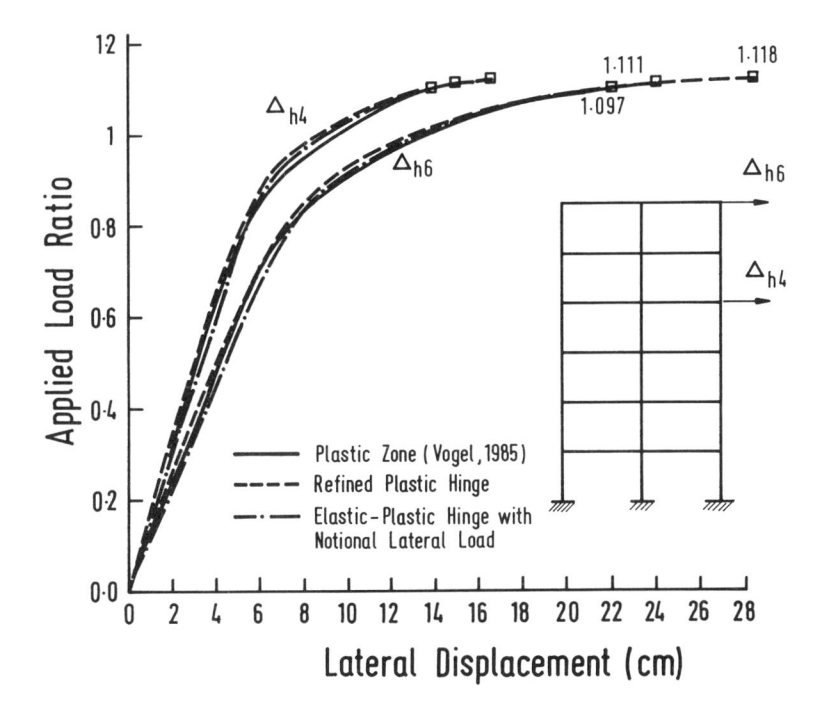

FIGURE 1.18. Comparison of load-displacement curves for six-story frame.

loads is less than 2%. These results support earlier studies by Al-Mashary and Chen (1991), King et al. (1991), and Ziemian (1989) which concluded that when the overall nonlinear behavior of the frame is dominated by inelastic action in the beams, the plastic-hinge based models generally give sufficient representation of the overall frame behavior.

Bending moments at selected locations in the frame, computed based on the three inelastic analyses at the frame's limit of resistance, are compared in Figure 1.19. The "exact" solutions shown in this figure are based on Ziemian's plastic-zone analysis in which the frame's limit of resistance is reached at a load factor of 1.180, which is slightly higher than the limit load predicted by Vogel's plastic-zone analysis. The distribution of forces predicted by all the inelastic analyses are remarkably close. Included in Figure 1.19 are the locations of plastic hinges detected by the refined-plastic hinge and notional load plastic hinge analyses. The first plastic hinge predicted by the notional load plastic hinge analysis occurs at an applied load ratio of 0.752, whereas in the refined-plastic hinge analysis, no plastic hinge is observed until the applied load ratio reaches a value of 0.873. The plastic-hinge analysis detects a total of 21 plastic hinges in comparison with the refined-plastic hinge approach, which detects only 15 plastic hinges in the frame.

For high-rise building frames, both plastic-hinge analyses should compare well with the plastic-zone method as shown by the closeness of the load-displacement curves in Figure 1.18 and the force distributions in Figure 1.19. The inability of the conventional elastic-plastic hinge method to represent member strength (such as for

38   ADVANCED ANALYSIS OF STEEL FRAMES

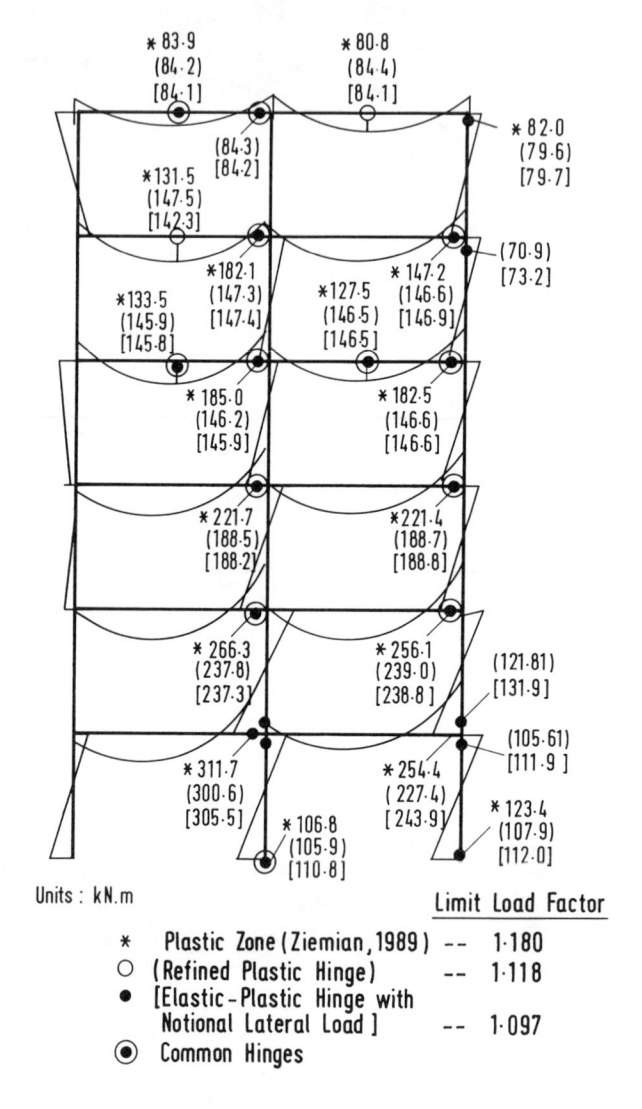

FIGURE 1.19. Comparison of bending diagrams and plastic hinge locations in the six-story frame.

the simple portal frames shown in Figures 1.15 through 1.17) can be attributed as "local effects." That is, the conventional-plastic hinge method is accurate in predicting the system response, but may not be accurate in predicting the strength and stability of isolated beam-column elements or subassemblies. The inadequacy of the elastic-plastic hinge approach can be improved by using the proposed notional load approach or the refined-plastic hinge method.

An attempt is also made to evaluate the limit of resistance of this frame by the LRFD-elastic analysis approach. For this six-story frame, the effective lengths of the columns are evaluated based on the modified K factor formula in Eq. (1.13). The

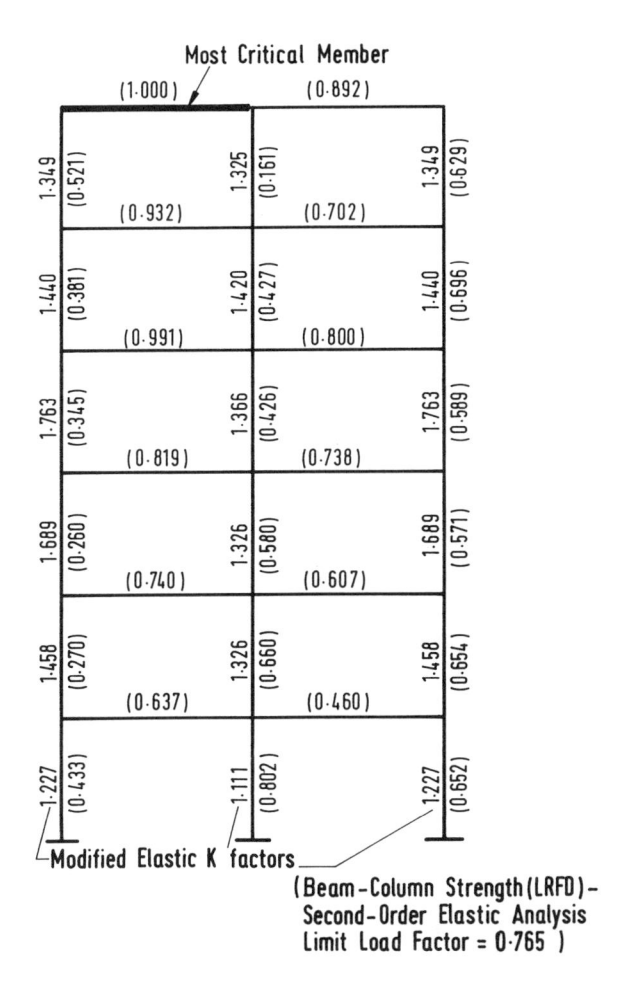

FIGURE 1.20. K factors and beam-column strengths from LRFD second-order elastic/design approach.

results of K factors are shown in Figure 1.20. Also included in this figure are the beam-column strengths evaluated based on the LRFD beam-column expressions. The most critically loaded member in the frame is the roof beam shown in Figure 1.20. The limit load of this frame is reached at a load parameter of 0.765 at which the most critically loaded member reaches its limiting strength of 1.0. The results shown in Figure 1.20 suggest that the LRFD-elastic analysis procedure is always conservative. It can be observed that several lower-floor beams and almost all the columns in the frame still have significant capacities that have not been fully utilized for the evaluation of the overall system strength. The reserve capacity of the frame, which is dependent primarily on inelastic force redistribution after the strengths of the most critically loaded members are reached, can be quite substantial for a highly redundant system. Of course, if no reliable inelastic force redistribution exists, the beam-column strength implied by the LRFD-elastic analysis procedure will generally be about the same as the true system strength. This would be the case for the one-story portal frame in which the LRFD-elastic analysis procedure gives results that are comparable with the plastic-zone solutions (Liew, 1992).

One may argue that if first-order plastic hinge theory is used to evaluate member forces, than the procedures based on member capacity checks will generally account for inelastic redistribution of forces in the system, and the results obtained from this approach would not be unduly conservative. However, procedures for determining the moment amplification and the effective length factors for members that are connected to a hinged subassemblage are complicated and the accuracy of these factors for representing the true member and system behavior is questionable. Clearly, the more rational way of estimating the member and system strength, which also avoids the complex issues of evaluating various factors for member capacity checks, is through advanced analysis.

1.7 Benchmarking Verification

Any numerical methods intended to be classified as advanced analysis require validation to establish their correctness and limits of applicability. An extensive list of possible attributes of computer programs and implementation-dependent aspects has been reported by Clarke et al. (1992) and White et al. (1992). Included in these papers are several benchmark problems intended to test many aspects of the formulation, implementation, and capability of computer programs for the analysis of two-dimensional frames. A wide collection of benchmark problems has also been reported by Toma et al. (1991, 1992) and Toma and Chen (1992) in which calibration frames involving both numerical and experimental results from Europe, North America, and Japan are reported.

One basic requirement of an advanced analysis method is that it must be applicable to a wide range of structural problems. The possible types of structural components that may be included in the analysis of a structural system consist of (1) beams; (2) columns; (3) beam-columns; (4) connections; (5) joints; (6) stiffeners; (7) struts or bracing members; (8) members employed primarily for shear transfer, such as shear links in eccentrically braced frames; (9) transfer girders or trusses employed for transmission of gravity loads to alternate column lines; (10) structural walls; (11) floor slabs; and (12) secondary systems such as cladding and partitions, among others. However, in practice, many of these types of components would not be included in the analysis of the overall structural system.

To establish benchmarks for verification of advanced analysis models and to represent adequately the behavioral phenomena of structures under static loading, the following attributes may be considered in the nonlinear analysis of plane frame structures:

Frames
  (1) regular and irregular in shapes
  (2) fully braced, unbraced, or partially braced

(3) with leaned columns or inclined members
(4) rigid, semi-rigid, or simply connected

Members
 (1) strong and weak axes bending
 (2) wide range of slenderness ratio, say L/r from 20 to 150
 (3) various cross-section geometry such as wide-flange sections, tubular sections, or built-up sections
 (4) prismatic and nonprismatic
 (5) different cross-section classifications, say compact, semi-compact, or slender.

Loading
 (1) loads producing significant P-δ or P-Δ effects
 (2) members subjected to wide range of combined axial force versus moment
 (3) conservative or nonconservative loading
 (4) load-sequence effects
 (5) proportional or nonproportional loading
 (6) modeling and orientation of loads

At first glance, these requirements seem rather restrictive and a wide range of benchmark solutions are necessary for general verification of an analysis model. However, it should be noted that, although the characteristics of advanced analysis solutions should be investigated for large, redundant structural systems with regular and irregular geometry, the basic requirements of advanced analysis can be studied thoroughly by consideration of simple isolated members and frame subassemblies. That is, if the performance of an isolated member is predicted properly for general loading cases, the behavior of the system will be captured accurately by following the general principles of equilibrium and compatibility.

Also, within the current scope of engineering practice, the development of an advanced analysis program should focus first on two-dimensional frames composed of prismatic, double-symmetric members. Consideration of spatial beam-column behavior including local buckling effects is ultimately desirable. It should be emphasized that two-dimensional analysis is still of great practical value for many design situations. Any members whose response characteristics are likely to be governed by spatial failure can always be proportioned or prevented in such a manner to preclude these effects. This can be accomplished by using specification interaction equation checks for out-of-plane buckling or by providing adequate bracing to prevent out-of-plane flexural buckling.

Chapter 7 of this book presents a wide range of numerical and experimental benchmark problems which may be used for verification of second-order inelastic computer programs intended to be classified as advanced analysis. These benchmark problems are limited in scope to in-plane behavior of isolated members and two-dimensional frames. In fact, the technology presently exists for analysis of three-dimensional small structures and assemblies using shell three-dimensional solid finite elements. Active research is also being conducted on three-dimensional advanced analysis models based on beam-column element discretizations and plastic-hinge type representation of yielding effects. It is expected that, in the near future, this type of analysis will be available for engineering practice.

1.8 Conclusions

Today technology has reached such a stage that many of the behavioral phenomena and structural attributes can be considered directly in the analysis without much difficulty. This chapter conveys an important message, that is, if the engineer employs an analysis method that adequately represents a limit state, then the checks of corresponding specification rules would not be required. In fact, the rationality of the analysis/design approach may be improved upon in certain instances by accurate analysis of behavioral phenomena which, in current practice, can only be approximated by specification formulas. For complicated frameworks, second-order inelastic analysis can provide this type of improvement over elastic analysis which requires specification beam-column capacity checks. However, for second-order inelastic analysis to be used in this way in actual design practice, the capability of different analysis approaches for representing beam-column and connection performance must be clearly identified. Also, any limit states that are not properly modeled in the analysis must be well-understood such that appropriate specification checks can be performed. The authors believe that the full consideration of member and system in-plane strength and stability by second-order analysis is entirely achievable. This chapter attempts to address how this can be accommodated and discusses several analysis approaches that are feasible for design use.

Although much work has been done on the analysis and design of rigid framework, connections' nonlinear behavior and its effect on the overall frame response requires special attention. In order to implement semi-rigid frame design in practice, engineers need to be assured that they understand the effects of connections on the structure's performance as a whole. To achieve this, it is essential to develop a readily understood analysis/design approach that can provide reliable, economical, and safe design. Also, this design approach should provide an economic trade-off when compared with other more conventional design methods. The main obstacles encountered in the design implementation of semi-rigid frames are (1) classifications of connection uncertainties, (2) need for a reliable and general connection moment-rotation model, (3) development of efficient analysis methods, and (4) concerns on strength, stability, and serviceability limit state conditions. This chapter has attempted to address many of these issues, and the information is now adequate for engineers to implement semi-rigid frame design with confidence.

The subsequent chapters of this book are arranged in such a way as to provide a smooth transition from the current state-of-the-art knowledge on frame design using second-order elastic analysis to upcoming technology and developments based upon advanced analysis/design techniques. Chapters 2 through 6 contain essential background information, basic theories, numerical formulations, and also computer source codes for self learning and for implementation within the framework of two-dimensional frame analysis and design. Clearly, methods such as these will become feasible for the design of many types of frame structures as computing technology progresses in parallel with our improving knowledge on member and connection nonlinear behavior.

With the computer-based analysis techniques in place, future refinements in frame analysis and design will likely focus more on the overall system response and less on individual member response. In particular, Chapters 4 through 7 focus on the development of advanced analysis methods and the verification of second-order inelastic analysis for use in design practice. Advanced analysis holds many answers to real behavior of steel structures and, as such, the authors commend these methods to all engineers seeking to improve their knowledge in the analysis and design of steel frame structures.

References

Abdel-Ghaffar, M., White, D. W., and Chen, W. F. (1991) Simplified second-order inelastic analysis for steel frame design, Special Volume of Session on Approximate Methods and Verification Procedures of Structural Analysis and Design, Proceedings at Structures Congress 91, ASCE, New York, 47–62.

Al-Mashary, F. and Chen, W. F. (1991) Simplified second-order inelastic analysis for steel frames, *J. Inst. Struct. Eng.*, 69(23), 395–399.

AISC (1978) *Specification for the Design, Fabrication and Erection of Structural Steel for Buildings*, American Institute of Steel Construction, Chicago.

AISC (1986) *Load and Resistance Factor Design Specification for Structural Steel Buildings*, American Institute of Steel Construction, Chicago.

AISC (1989) *Allowable Stress Design Specification for Structural Steel Buildings*, American Institute of Steel Construction, Chicago.

Alvarez, R. J. and Birnstiel, C. (1967) Elasto-Plastic Analysis of Plane Rigid Frames, School of Engineering and Science, Department of Civil Engineering, New York University, New York.

Barakat, M. A. (1989) Simplified Design Analysis of Frames with Semi-Rigid Connections, Ph.D. Dissertation, School of Civil Engineering, Purdue University, West Lafayette, IN, 211 pp.

Barakat, M. A. and Chen, W. F. (1990) Practical analysis of semi-rigid frames," *Eng. J.*, 27(2), 54–68.

Barakat, M. A. and Chen, W. F. (1991) Design analysis of semi-rigid frames: evaluation and implementation, *Eng. J.*, 28(2), 55–64.

Bijlaard, F. S. K. and Steenhuis, C. M. (1991) Prediction of the influence of the connection behavior on the strength, deformations and stability of frames by classification of connections, Proceedings of the Second International Workshop, Pittsburgh, American Institute of Steel Construction, Chicago.

Bjorhovde, R., Brozzetti, J., and Colson, A. (1990) A classification system for beam to column connections, *J. Struct. Eng.*, 116(11), 3059–3076.

Chen, W. F. and Atsuta, T. (1976) *Theory of Beam-Column, Vol. 1, In-Plane Behavior and Design*, McGraw-Hill, New York, 513 pp.

Chen, W. F. and Atsuta, T. (1977) *Theory of Beam-Column, Vol. 2, Space Behavior and Design*, McGraw-Hill, New York, 732 pp.

Chen, W. F. and Kishi, N. (1989) Semi-rigid steel beam-to column connections: data base and modeling, *J. Struct. Eng.*, 115(1), 105–119.

Chen, W. F. and Liew, J. Y. R. (1992), Seismic resistance design of steel moment resisting frames considering panel-zone deformation, in *Design Implementation*, Fukumoto, Y. and Lee, C. G., Eds., CRC Press, Boca Raton, FL.

Chen, W. F. and Lui, E. M. (1986) *Structural Stability — Theory and Implementation*, Elsevier, New York, 490 pp.

Chen W. F. and Lui, E. M. (1992) *Stability Design of Steel Frames*, CRC Press, Boca Raton, FL, 380 pp.

Cheong-Siat-Moy, F. (1977) Consideration of secondary effects in frame design, *J. Struct. Div.*, 103(ST10), 2005–2019.

Clarke, M. J., Bridge, R. Q., Hancock, G. J., and Trahair, N. S. (1992) Benchmarking and verification of second-order elastic and inelastic frame analysis programs, in *SSRC TG 29 Workshop and Nomograph on Plastic Hinge Based Methods for Advanced Analysis and Design of Steel Frames*, White, D. W. and Chen, W. F., Eds., SSRC, Lehigh University, Bethlehem, PA.

Colson, A. (1991) Classification of Connection for Beam to Column Connections, Tentative Document for Circulation to Members of SSRC TG 25, SSRC, 19 Sept., 7 pp.

CSA (1989) *Limit States Design of Steel Structures*, CAN/CSA-S16.1-M89, Canadian Standards Association.

EC 3 (1990) *Design of Steel Structures: Part I — General Rules and Rules for Buildings*, Vol. 1, Eurocode edited draft, Issue 3.

ECCS (1984) Ultimate Limit State Calculation of Sway Frames with Rigid Joints, Technical Committee 8 — Structural Stability Technical Working Group 8.2 — System, Publication No. 33, 20 pp.

ECCS (1991) Essentials of Eurocode 3 Design Manual for Steel Structures in Building, ECCS–Advisory Committee 5, No. 65, 60 pp.

El-Zanaty, M., Murray, D., and Bjorhovde, R. (1980) Inelastic Behavior of Multistory Steel Frames, Structural Engineering Report No. 83, University of Alberta, Alberta, Canada.

Frye, M. J. and Morris, G. A. (1976) Analysis of flexibly connected steel frames, *Can. J. Civil Eng.*, 2(3), 280–291.

Gerstle, K. H. and Ackroyd, M. H. (1990) Behavior and design of flexibly-connected building frames, *Eng. J.*, 27(1), 22–29.

Goverdhan, A. V. (1983) A Collection of Experimental Moment-Rotation Curves and Evaluation of Prediction Equations for Semi-Rigid Connections, Master's thesis, Vanderbilt University, Nashville, TN, 490 pp.

Julian, O. G. and Lawrence, L. S. (1943) Notes on Julian and Lawrence Nomographs for Determination of Effective Length, unpublished report.

Kanchanalai, T. (1977) The Design and Behavior of Beam-Columns in Unbraced Steel Frames, AISI Project No. 189, Report No. 2, Civil Engineering/Structures Research Lab., University of Texas at Austin, 300 pp.

King, W. S., White, D. W., and Chen, W. F. (1991) On second-order inelastic methods for steel frame design, *J. Struct. Eng.*, 118(2), 408–428.

Kishi, N. and Chen, W. F. (1986) Data Base of Steel Beam-To-Column Connections, Structural Engineering Report No. CE-STR-86–26, School of Civil Engineering, Purdue University, West Lafayette, IN, 2 vol., 653 pp.

Kishi, N. and Chen, W. F. (1990) Moment-rotation relations of semi-rigid connections with angles, *J. Struct. Eng.*, 116(7), 1813-1834.

Kishi, N., Chen, W. F., Goto, Y., and Matsuoka, K. (1991) Analysis Program for Design of Flexibly Jointed Frames, Structural Engineering Report, CE-STR-91–26, Purdue University, West Lafayette, IN, 25 pp.

Kishi, N., Chen, W. F., Goto, Y., and Matsuoka, K. G. (1992) Design Aid of Semi-Rigid Connections for Frame Analysis, Structural Engineering Report, CE-STR-92–24, Purdue University, West Lafayette, IN, 34 pp.

LeMessurier, W. J. (1976) A practical method of second order analysis. I, *Eng. J.*, 13(4), 89–96.

LeMessurier, W. J. (1977) A practical method of second order analysis. II. Rigid frames, *Eng. J.*, 14(2), 49–67.

Liew, J. Y. R. (1992) Advanced Analysis for Frame Design, Ph.D dissertation, School of Civil Engineering, Purdue University, West Lafayette, IN, May, 393 pp.

Liew, J. Y. R. and Chen, W. F. (1991) Refining the plastic hinge concept for advanced analysis/design of steel frames, *J. Singapore Struct. Steel Soc., Steel Struct.*, 2(1), 13–30.

Liew, J. Y. R., White, D. W., and Chen, W. F. (1991) Beam-column design in steel frameworks — insight on current methods and trends, *J. Constructional Steel Res.*, 18, 269–308.

Liew, J. Y. R., White, D. W., and Chen, W. F. (1992a) Notional Load Plastic Hinge Method for Frame Design, Structural Engineering Report, CE-STR-92–4, Purdue University, West Lafayette, IN, 42 pp.

Liew, J. Y. R., White, D. W., and Chen, W. F. (1992b) Second-Order Refined Plastic Hinge Analysis of Frames, Structural Engineering Report, CE-STR-92-12, Purdue University, West Lafayette, IN, 35 pp.

Liew, J. Y. R., White, D. W., and Chen, W. F. (1992c) Second-Order Refined Plastic Hinge Analysis for Frame Design, Structural Engineering Report, CE-STR-92-14, Purdue University, West Lafayette, IN, 36 pp.

Liew, J. Y. R., White, D. W., and Chen, W. F. (1993a) Limit states design of semi-rigid frames using advanced analysis. I. Connection modelling and classification, *J. Construct. Steel Res.*, 26(1), 1–27.

Liew, J. Y. R., White, D. W., and Chen, W. F. (1993b) Limit states design of semi-rigid frames using advanced analysis. II. Analysis and design, *J. Construct. Steel Res.*, 26(1), 29–57.

Lui, E. M. (1992) A practical approach for K factor determination, Proc. Tenth Structures Congr., 92, ASCE, April 13–15, San Antonio, TX.

Nethercot, D. A. (1985) Steel Beam-to-Column Connections — A Review of Test Data and Its Applicability to the Evaluation of Joint Behavior in the Performance of Steel Frames, CIRIA Project Record, PR338.

Nixon, D., Beaulieu, D., and Adams, P. F. (1975) Simplified second-order frame analysis, *Can. J. Civil Eng.*, 2(4).

SSRC (1993) *Plastic Hinge Based Methods for Advanced Analysis and Design of Steel Frames, An Assessment of the State-of-the-Art,* White, D. W. and Chen, W. F., Eds., SSRC, Lehigh University, Bethlehem, PA, 299 pp.

Standards Australia (1990) AS4100–1990, Steel Structures, Sydney, Australia.

Stevens, L. K. (1967) Elastic stability of practical multistory frames, *Proc. Inst. Civil Eng.*, Vol. 36.

Toma, S., Chen, W. F., and White, D. W. (1991) Calibration Frames for Second-Order Inelastic Analysis in North America, Structural Engineering Report, CE-STR-91-23, Purdue University, West Lafayette, IN, 70 pp.

Toma, S., Chen, W. F., and White, D. W. (1992) European calibration frames for second-order inelastic analysis, in *Engineering Structures*, Butterworth-Heinemann, 114(1), 35–48.

Toma, S. and Chen, W. F. (1992) Calibration Frames for Second-Order Inelastic Analysis in Japan, Structural Engineering Report, CE-STR-92-16, Purdue University, West Lafayette, IN.

Vogel, U. (1985) Calibrating frames, *Stahlbau*, 10, 1–7.

Wang, Y. C. (1988) Ultimate Strength Analysis of 3-D Beam Columns and Column Subassemblages with Flexible Connections, Ph.D thesis, University of Sheffield, England.

White, D. W. (1985), Material and Geometric Nonlinear Analysis of Local Planar Behavior in Steel Frames Using Iterative Computer Graphics, M. S. thesis., Cornell University, Ithaca, NY, 281 pp.

White, D. W. (1988) Analysis of Monotonic and Cyclic Stability of Steel Frame Subassemblages, Ph.D. dissertation, Cornell University, Ithaca, NY.

White, D. W. and Chen, W. F. (1990) Second-order inelastic analysis for frame design, in Proc. National Symposium on Advances in Steel Structures, IIT, Madras, India, February 7–9.

White, D. W. and Hajjar, J. F. (1991) Application of second-order elastic analysis in LRFD: research to practice, *Eng. J.*, 28(4), 133–148.

White, D. W., Liew, J. Y. R., and Chen, W. F. (1991) Second-Order Inelastic Analysis for Fame Design: A Report to SSRC Task Group 29 on Recent Research and the Perceived Sate-of-the-Art, Structural Engineering Report, CE-STR-91-12, Purdue University, West Lafayette, IN, 116 pp.

White, D. W., Liew, J. Y. R., and Chen, W. F. (1993) Toward advanced analysis in LRFD, in *Plastic Hinge Based Methods for Advanced Analysis and Design of Steel Frames, An Assessment of the State-of-the-Art*, White, D. W. and Chen, W. F., Eds., SSRC, Lehigh University, Bethlehem, March, 95–173.

Yura, J. A. (1971) The effective length of columns in unbraced frames, *Eng. J.*, 8(2), 37–42.

Yura, J. A. (1988) Elements for Teaching Load and Resistance Factor Design: Combined Bending and Axial Load, University of Texas at Austin, 55–71.

Ziemian, R. D. (1990), Advanced Methods of Inelastic Analysis in the Limit States Design of Steel Structures, Ph.D dissertation, School of Civil and Environmental Engineering, Cornell University, Ithaca, NY, 265 pp.

Ziemian, R. D, White, D. W., Deierlein, G. G., and McGuire, W. (1990) One approach to inelastic analysis and design, Proceedings of the 1990 National Steel Conference, AISC, Chicago, 19.1–19.

2: Second-Order Elastic Analysis of Frames

Yoshiaki Goto, *Professor, Department of Civil Engineering, Nagoya Institute of Technology Gokiso-cho, Showa-ku, Nagoya 466, Japan*

2.1 Introduction

The steel framework is one of the most commonly used structural systems in modern construction. The analysis of such a structural system depends largely on the assumptions adopted in the modeling of its elements, especially those concerning the behavior of beam-to-column connections. Conventional methods of steel frame analysis use two highly idealized connection models: the rigid-joint model and the pinned-joint model. Since the actual behavior of joints in a frame always falls in between these two extremes, much attention has been focused in recent years toward a more accurate modeling of such connections. Recent research on beam-to-column connections has resulted in considerable progress and an understanding of the subject and hence changes in design provisions.

The Load and Resistance Factor Design (LRFD) Specification (AISC, 1986) designate two types of constructions in their provisions: Type FR (fully restrained) and Type PR (partially restrained). Type FR construction corresponds to the so-called "rigid-frame". Type PR construction, which assumes that the connections of beams and girders possess an insufficient rigidity to hold the original angles between intersecting members virtually unchanged, is usually referred to as "semi-rigid frames". In the design of Type FR construction based on the 1986 AISC-LRFD specification, the required flexural strength is evaluated by considering only the geometrical nonlinearity of structures known as the second-order effect. In Type PR construction, however, experiments have shown that connection moment-rotation curves are generally always nonlinear over the entire range. It is therefore necessary to calculate the required flexural strength in a PR construction by considering both the material nonlinearity of connections and the geometrical nonlinearity of structures.

This chapter presents a versatile computer-based method of second-order elastic analysis for both Type FR and Type PR constructions, that is numerically stable and also can be simply implemented in a computer. In this development, special attention has been paid to the necessary simplification of the analytical procedures without losing numerical stability and accuracy of a rigorous solution.

For the ease of computer implementation, the direct stiffness matrix method is adopted here as a numerical procedure in the discrete analysis. The stiffness equations for beam and column members were first derived in closed form from the governing differential equations of a finite displacement theory with the assumption that members undergo only moderately large rotations (Goto et al., 1987a, 1991b). This approximation is valid for the present applications, because extremely large displacements seldom occur in the design of framed structures. This derivation procedure of the stiffness equation will be presented in Section 2.3, followed by a brief description of the implementation of connection models and an iterative method to solve the nonlinear stiffness equations (Goto and Chen, 1987b).

Modeling of connections is an important subject that strongly influences the accuracy of analysis with respect to Type PR construction. Although several practical connection models have been developed in recent years, no single model has yet been widely accepted that can adequately describe the overall behavior of various connections used in engineering practice. Herein, we shall select the most appropriate connection model from the comprehensive connection data bank developed by Kishi and Chen (1986). This data bank includes a wide range of experimental data as well as several practical moment-rotation prediction equations. Chapter 3 will explain the details of this data bank.

The analysis program (Kishi et al., 1991) which is basically the same as FLFRM (Goto and Chen, 1986) will be introduced and developed in Sections 2.8 and 2.9 with some refinement made to the original program.

2.2 Second-Order Theory for In-Plane Frames

The governing differential equations for a second-order analysis are formulated through the principle of virtual work, introducing the usual beam assumptions. In this way, accurate equilibrium equations consistent with the compatibility equations can be easily derived by a purely mathematical manipulation. This method originally proposed by Washizu (1968) was refined extensively by Nishino et al. (1973, 1975, 1979) and Goto et al. (1985) in a very rigorous manner to formulate the geometrically nonlinear theories of thin-walled members.

Consider a plane frame member 1–2, as shown in Figure 2.1, subjected to a uniformly distributed lateral force p_y, acting perpendicular to the member axis before deformation. A Cartesian coordinate system (x, y) as well as displacement components (u, v) are defined at the initial configuration of the member.

First, we shall derive the nonlinear axial strain that will be used in the equation of virtual work (Goto and Chen, 1987a, 1989). For the present design applications the conditions of small strains as well as moderate rotations are introduced herein to simplify the nonlinear strain.

SECOND-ORDER ELASTIC ANALYSIS OF FRAMES

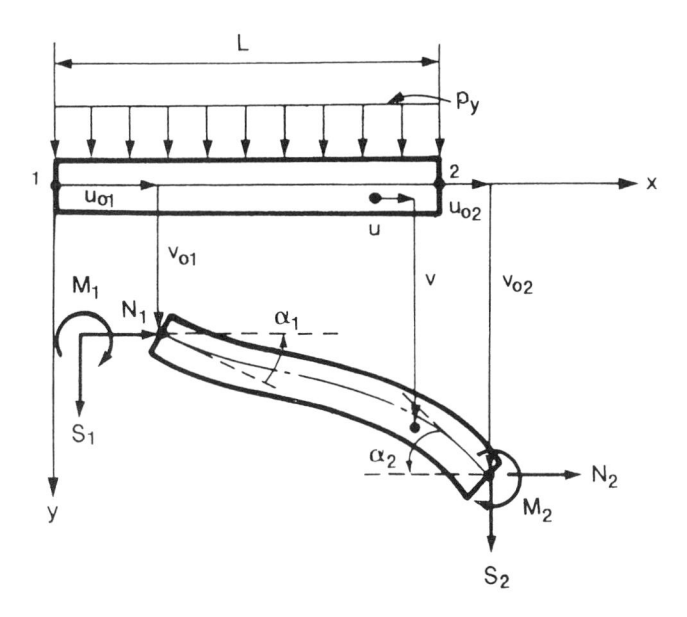

FIGURE 2.1. Coordinate system for a member.

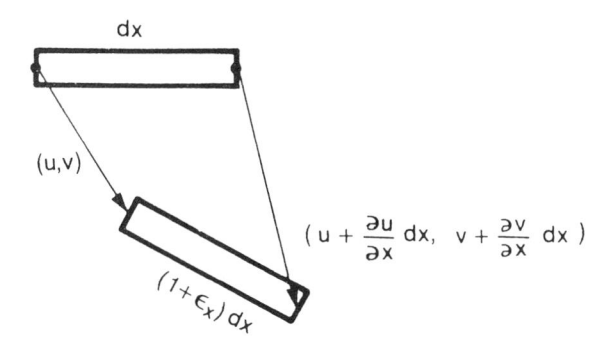

FIGURE 2.2. Deformation of an infinitesimal longitudinal fiber.

Consider an infinitesimal fiber dx of the member parallel to the centroidal axis as shown in Figure 2.2. The axial component of the Green strain tensor is given by

$$e_{xx} = \frac{\partial u}{\partial x} + \frac{1}{2}\left(\frac{\partial u}{\partial x}\right)^2 + \frac{1}{2}\left(\frac{\partial v}{dx}\right)^2 \tag{2.1}$$

e_{xx} can be related to the corresponding component ε_x of the so-called engineering strain as

$$e_{xx} = \varepsilon_x\left(1 + \frac{1}{2}\varepsilon_x\right) \tag{2.2}$$

where

$$\varepsilon_x = \sqrt{\left(1+\frac{\partial u}{\partial x}\right)^2 + \left(\frac{\partial v}{\partial x}\right)^2} - 1 \qquad (2.3)$$

Under the conditions of small strains mathematically expressed as

$$e_{xx} \ll 1 \quad or \quad \varepsilon_x \ll 1 \qquad (2.4)$$

the Green strain e_{xx} coincides with the engineering strain ε_x. Thus, in the customary design analysis of frames, it is unnecessary to take into account the difference of strains according to their definitions.

The nonlinear strain e_{xx} is simplified by introducing the condition of moderate rotations. This condition expressed by

$$\left(\frac{\partial v}{\partial x}\right)^2 \ll 1 \qquad (2.5)$$

is less restrictive when compared with the condition of small rotation, i.e., $(\partial v/\partial x) \ll 1$.

Using Eq. (2.5), the order of magnitude of $\partial u/\partial x$ can be known. For this purpose, Eq. (2.1) is solved for $(\partial u/\partial x)$ as

$$\frac{\partial u}{\partial x} = -1 + \sqrt{1+\lambda} \qquad (2.6)$$

$$\lambda = 2e_{xx} - \left(\frac{\partial v}{\partial x}\right)^2 \qquad (2.7)$$

From Eqs. (2.4), (2.5), and (2.6), $\partial u/\partial x$ is known to have the same order of magnitude as λ. This implies that

$$\frac{\partial u}{\partial x} = O(\lambda) \ll 1 \qquad (2.8)$$

The use of Eq. (2.8) reduces Eq. (2.1) to

$$e_{xx} = \frac{\partial u}{\partial x} + \frac{1}{2}\left(\frac{\partial v}{\partial x}\right)^2 \qquad (2.9)$$

Herein, we shall introduce the usual beam assumptions, that is, the assumptions of no change of cross-sectional shapes and the Bernoulli-Euler hypothesis where the transverse plane is assumed to be plane and normal to the beam axis throughout deformation. Based on these two assumptions, the displacement field can be expressed as follows using the displacement components (u_0, v_0) on the centroidal axis.

$$u = u_0 - y\sin\alpha, \quad v = v_0 - y(1 - \cos\alpha) \tag{2.10a,b}$$

where

$$\sin\alpha = \frac{\dfrac{dv_0}{dx}}{\sqrt{\left(1+\dfrac{du_0}{dx}\right)^2 + \left(\dfrac{dv_0}{dx}\right)^2}}, \quad \cos\alpha = \frac{1+\dfrac{du_0}{dx}}{\sqrt{\left(1+\dfrac{du_0}{dx}\right)^2 + \left(\dfrac{dv_0}{dx}\right)^2}} \tag{2.11a,b}$$

Equations (2.10a,b) can be simplified by the conditions (2.5) and (2.8) as

$$u = u_0 - y\frac{dv_0}{dx}, \quad v = v_0 \tag{2.12a,b}$$

Note that u_0 and v_0 are both functions of x.

Substituting Eqs. (2.12a,b) into Eq. (2.9), the axial component of the strain tensor is expressed by the displacement on the centroidal axis:

$$e_{xx} = \frac{du_0}{dx} - y\frac{d^2 v_0}{dx^2} + \frac{1}{2}\left(\frac{dv_0}{dx}\right)^2 \tag{2.13}$$

This nonlinear strain-displacement relationship is what is customarily used in a second-order finite element analysis for plane frames (Connor et al., 1968; Galambos, 1987; Mallet and Marcal, 1968). To obtain the closed-form solution corresponding to the customary finite element solution, the governing differential equations are derived through the principle of virtual work.

Using the strain tensor component e_{xx} of Eq. (2.13) and the corresponding stress component σ_{xx} in addition to the components of external forces and displacements, as shown in Figure 2.1, the equation of virtual work for the plane member is given by

$$\delta\Pi = \int_0^L \int_A \sigma_{xx} \delta e_{xx} dA\, dx - \int_0^L p_y \delta v_0$$
$$- \left[n_x\left(N_i \delta u_{0i} + S_i \delta v_{0i} + M_i \delta v'_{0i}\right)\right]_1^2 = 0 \tag{2.14}$$

in which $\int_A dA$ = integration over the cross-sectional area; $\int_0^L dx$ = integration over the length of the member; and n_x has the values of -1 and 1, respectively, at Nodes 1 and 2.

The virtual strain δe_{xx} can be calculated by taking the variation of Eq. 2.13:

$$\delta e_{xx} = \delta u_0' - y\delta v_0'' + v_0'\delta v_0' \tag{2.15}$$

in which the notation $(.)'$ is introduced for simplicity to express the differentiation with respect to x. Substituting Eq. (2.15) into Eq. (2.14) and integrating by parts leads to

$$\delta \Pi = \left[(N - n_x N_i)\delta u_{oi} + (Nv_0' + M' - n_x S_i)\delta v_{oi} \right.$$
$$\left. - (M + n_x M_i)\delta v_{oi}' \right]_1^2 - \int_0^L \left\{ N'\delta u_o + \left[(Nv_0' + M')' + p_y \right]\delta v_0 \right\} dx \tag{2.16}$$

in which

$$N = \int_A \sigma_{xx} dA \tag{2.17a}$$

$$M = \int_A \sigma_{xx} y dA \tag{2.17b}$$

These stress resultants correspond to axial force and bending moment, respectively.

Equilibrium equations and the associated boundary conditions are obtained from the necessary and sufficient conditions for Eq. (2.16) to hold for any virtual displacements. The terms in the largest brackets under the integral sign in Eq. (2.16) yield the equilibrium equations. Thus

$$N' = 0 \tag{2.18a}$$

$$(Nv_0' + M')' + p_y = 0 \tag{2.18b}$$

and the integrated terms give the associated boundary conditions at Nodes 1 and 2. Also

$$u_0 = u_{0i} \quad \text{or} \quad N = n_x N_i \tag{2.19a}$$

$$v_0 = v_{0i} \quad \text{or} \quad Nv_0' + M' = n_x S_i \tag{2.19b}$$

$$v_0' = \alpha_i \quad \text{or} \quad M = -n_x M_i \quad (i = 1, 2) \tag{2.19c}$$

in which S_i and α_i = vertical forces and end rotations of the member.

The stress resultant-displacement relations are obtained by substituting Eq. (2.13) into Eqs. (2.17a,b). If the x axis is selected such that it coincides with the centroidal axis of the member, these relations are simplified as

$$N = EA\left[u'_0 + \frac{1}{2}(v'_0)^2\right] \quad (2.20a)$$

$$M = -EIv''_0 \quad (2.20b)$$

Eqs. (2.18) to (2.20) are the consistent governing equations of beam-columns derived from the theorem of virtual work. It should be noted that in these equations the effect known as bowing or curvature shortening is included in Eq. (2.20a). We shall refer here to the theory with these governing equations as *the nonlinear beam-column theory*. This theory yields a symmetric tangent stiffness matrix consistent with the energy theory of elastic solids (Goto et al., 1991b). If the bowing effect is ignored in Eq. (2.20a), this equation is reduced to

$$N = EAu'_0 \quad (2.21)$$

The beam-column theory using Eq. (2.21) instead of Eq. (2.20a) is commonly used. Since Eq. (2.21) does not include nonlinear terms resulting from the bowing effect, this theory is here referred to as *the linearized beam-column theory*. Due to the exclusion of the nonlinear term, *the linearized beam-column theory* is mathematically easy to handle but the tangent stiffness matrix derived from this theory becomes asymmetric, unless the prebuckling deformation or the increment of axial force is ignored. This is against the elastic theory in a strict sense. As demonstrated by Goto et al. (1991b), the difference of the two theories in some cases can be significant for the critical behaviors such as bifurcation and limit-load instability. However, this difference is negligible for most design analyses of rectangular frames either rigidly or semi-rigidly connected (Goto and Chen, 1987a,b), and therefore, satisfactory results for design purposes can generally be obtained by using *the linearized beam-column theory*. Although both theories are implemented in the analysis program to be introduced in Sections 2.8 and 2.9, *the linearized beam-column theory* requires less time to achieve convergent solutions and is therefore recommended here for general use in the customary design analysis.

Last but not least, as can be seen from the assumptions of Eqs. (2.4) and (2.5) adopted in the derivation, the above theories should be used primarily for the case when frames undergo moderate rotation as in the design analysis. To deal with the case of extremely large rotations, the beam-column theories can be approximately used within the co-rotational framework (Saafan, 1963; Oran, 1973).

The exact geometrically nonlinear theories for two-dimensional beam-columns are reported by Nishino et al. (1975, 1979) and their general closed-form solutions using elliptic integrals are also given by Goto et al. (1987c, 1990). These solutions can be used as the benchmark checks for approximate numerical analyses. The

2.3 Stiffness Equations for Beam-Column Member

Closed-form stiffness equations for a beam-column member are derived from the governing equations in Section 2.2. Although it is possible and easier to obtain stiffness equations based on finite element techniques, the stiffness equations so derived are approximate and it is necessary to divide a member into finite elements to achieve more accurate solutions. This approach normally increases the number of input data, computational time, and required computer capacity. Furthermore, the finite element method using *the nonlinear beam-column theory* (Mallet and Marcal, 1968) sometimes cannot yield accurate solutions when the rotations become relatively large (Byskov, 1989). To avoid these inconveniences, here we shall use the closed-form stiffness equations.

From Eq. (2.18a), the axial force N is constant because there is no distributed forces in the axial direction. Therefore Eq. (2.18b) can be solved with respect to v_0 independent of Eq. (2.18a). This implies that the solution v_0 is the same regardless of which of the two beam-column theories is used. By using the mechanical and the geometrical boundary conditions at Node 1, the solutions can be expressed by

For $N_1 > 0$:

$$\frac{v_0}{L} = \frac{v_{01}}{L} + \frac{\alpha_1}{\gamma L}\sin\gamma x + \frac{\tilde{S}_1}{(\gamma L)^3}(\gamma x - \sin\gamma x) - \frac{\tilde{M}_1}{(\gamma L)^2}(1-\cos\gamma x)$$
$$+ \tilde{p}_y\left[\frac{(\gamma x)^2}{2(\gamma L)^4} + \frac{1}{(\gamma L)^4}(\cos\gamma x - 1)\right] \quad (2.22a)$$

For $N_1 < 0$:

$$\frac{v_0}{L} = \frac{v_{01}}{L} + \frac{\alpha_1}{\gamma L}\sinh\gamma x + \frac{\tilde{S}_1}{(\gamma L)^3}(\sinh\gamma x - \gamma x) - \frac{\tilde{M}_1}{(\gamma L)^2}(\cosh\gamma x - 1)$$
$$+ \tilde{p}_y\left[\frac{(\gamma x)^2}{2(\gamma L)^4} + \frac{1}{(\gamma L)^4}(1 - \cosh\gamma x)\right] \quad (2.22b)$$

in which

$$\gamma = \sqrt{\frac{|N_1|}{EI}}, \quad \tilde{S}_i = \frac{S_i L^2}{EI}; \quad \tilde{M}_i = \frac{M_i L}{EI}; \quad \tilde{p}_y = \frac{p_y L^3}{EI} \quad (2.23)$$

As for u_0, the solution is different according to the theories. For *the nonlinear beam-column theory*, u_0 can be obtained as follows by integrating Eq. (2.20a) after substitution of Eqs. (2.22a,b) and making use of the boundary conditions at Node 1.

$$u_0 = u_{01} + \frac{N_1}{EA}x - \frac{1}{2}\int_0^x v_0'^2 dx \qquad (2.24)$$

For $N_1 > 0$:

$$\int_0^x v_0'^2 dx = L\left\{\frac{1}{4\gamma L}\alpha_1^2(2\gamma x + \sin 2\gamma x) + \frac{1}{(\gamma L)^5}\tilde{S}_1^2[2(\gamma x - \sin\gamma x)\right.$$

$$\left. - \frac{(2\gamma x - \sin 2\gamma x)}{4}\right] + \frac{1}{4(\gamma L)^3}\tilde{M}_1^2(2\gamma x - \sin 2\gamma x)$$

$$+ \frac{2}{(\gamma L)^3}\alpha_1\tilde{S}_1\left[\sin\gamma x - \gamma x - \frac{(\sin 2\gamma x - 2\gamma x)}{4}\right]$$

$$+ \frac{1}{(\gamma L)^2}\alpha_1\tilde{M}_1\frac{(\cos 2\gamma x - 1)}{2} + \frac{1}{(\gamma L)^4}\tilde{S}_1\tilde{M}_1\left[2(\cos\gamma x - 1) - \frac{(\cos 2\gamma x - 1)}{2}\right]$$

$$+ \frac{1}{12(\gamma L)^7}\tilde{p}_y^2\left[4(\gamma x)^3 - 24(\sin\gamma x - \gamma x\cos\gamma x) + 3(2\gamma x - \sin 2\gamma x)\right]$$

$$+ \frac{1}{(\gamma L)^4}\tilde{p}_y\alpha_1\left[2(\cos\gamma x - 1) - (\sin\gamma x - \gamma x)^2 + (\gamma x)^2\right]$$

$$+ \frac{1}{(\gamma L)^6}\tilde{p}_y\tilde{S}_1(\gamma x - \sin\gamma x)^2$$

$$\left. + \frac{1}{(\gamma L)^5}\tilde{p}_y\tilde{M}_1\left[2(\gamma x - \sin\gamma x) + 2\gamma x(\cos\gamma x - 1) + \frac{(2\gamma x - \sin 2\gamma x)}{2}\right]\right\} \qquad (2.25a)$$

For $N_1 < 0$:

$$\int_0^x v_0'^2 dx = L\left\{\frac{1}{4\gamma L}\alpha_1^2(2\gamma x + \sinh 2\gamma x)\right.$$

$$+ \frac{1}{(\gamma L)^5}\tilde{S}_1^2\left[\frac{(\sinh 2\gamma x - 2\gamma x)}{4} - 2\sinh\gamma x + 2\gamma x\right]$$

$$+ \frac{1}{4(\gamma L)^3}\tilde{M}_1^2(\sinh 2\gamma x - 2\gamma x)$$

$$+\frac{2}{(\gamma L)^3}\alpha_1\tilde{S}_1\left[\frac{(\sinh 2\gamma x - 2\gamma x)}{4} - \sinh\gamma x + \gamma x\right]$$

$$+\frac{1}{(\gamma L)^2}\alpha_1\tilde{M}_1\frac{(1-\cosh 2\gamma x)}{2}$$

$$+\frac{1}{(\gamma L)^4}\tilde{S}_1\tilde{M}_1\left[2(\cosh\gamma x - 1) + \frac{(1-\cosh 2\gamma x)}{2}\right]$$

$$+\frac{1}{12(\gamma L)^7}\tilde{p}_y^2\left[4(\gamma x)^3 - 24(\gamma x\cosh\gamma x - \sinh\gamma x) + 3(\sinh 2\gamma x - 2\gamma x)\right]$$

$$+\frac{1}{(\gamma L)^4}\tilde{p}_y\alpha_1\left[2(\cosh\gamma x - 1) + (\sinh\gamma x - \gamma x)^2 + (\gamma x)^2\right]$$

$$+\frac{1}{(\gamma L)^6}\tilde{p}_y\tilde{S}_1(\gamma x - \sinh\gamma x)^2 + \frac{1}{(\gamma L)^5}\tilde{p}_y\tilde{M}_1\left[2(\gamma x - \sinh\gamma x)\right.$$

$$\left.+2\gamma x(\cosh\gamma x - 1) + \frac{(2\gamma x - \sinh 2\gamma x)}{2}\right]\Bigg\}\quad (2.25b)$$

Solution u_0 for *the linearized beam-column theory*, which is exactly the same as that of the small displacement theory can be derived as follows using Eq. (2.21).

$$u_0 = u_{0i} + \frac{N_1}{EA}x \quad (2.26)$$

From the analytical solutions obtained above, stiffness equations for a beam-column member can be derived. The coefficients of these equations are expressed either by hyperbolic functions or trigonometric functions according to whether the axial force N_1 is positive or negative. However, if we use the above functions, the coefficients of the stiffness equations become indefinite when the axial force is zero. This singularity at $N_1 = 0$ causes numerical instability in cases when the axial force approaches zero. To circumvent this situation as well as to avoid the use of different functions for compressive and tensile forces, a power series expansion is used (Goto and Chen, 1987a). As a result, we can remove the singularity and also unify the expressions.

Consider the following power series expansions for the trigonometric and the hyperbolic functions:

$$\left.\begin{array}{r}\sin\gamma L\\ \sinh\gamma L\end{array}\right\} = \gamma L + \gamma L\sum_{n=1}^{\infty}\frac{1}{(2n+1)!}\left(-\tilde{N}_1\right)^n \quad (2.27a)$$

$$\left.\begin{array}{l}\cos\gamma L\\ \cosh\gamma L\end{array}\right\} = 1 + \sum_{n=1}^{\infty} \frac{1}{(2n)!} \left(-\tilde{N}_1\right)^n \qquad (2.27b)$$

$$\tilde{N}_1 = \frac{N_1 L^2}{EI} = \frac{\pi^2 N_1}{P_e} \qquad (2.27c)$$

$$P_e = \frac{\pi^2 EI}{L^2} \qquad (2.27d)$$

where EI = sectional rigidity and L = length of the member. In the above equations N_1 is taken as positive for a tensile and negative for a compressive force.

The closed-form stiffness equations obtained from *the nonlinear beam-column theory* and *the linearized beam-column theory* are given respectively as (Goto et al., 1991b):

The nonlinear theory:

$$\tilde{F}_i = \sum_{j=1}^{6} \bar{K}_{ij}\left(\tilde{N}_1\right)\tilde{d}_j + \sum_{j=1}^{6}\sum_{k=1}^{6} \bar{K}_{ijk}\left(\tilde{N}_1\right)\tilde{d}_j\tilde{d}_k$$

$$+ \tilde{p}_y \sum_{j=1}^{6} \bar{L}_{ij}\left(\tilde{N}_1\right)\tilde{d}_j + \tilde{p}_y \bar{L}_i\left(\tilde{N}_1\right) + \tilde{p}_y^2 \bar{C}_i\left(\tilde{N}_1\right) \qquad (2.28)$$

The linearized theory:

$$\tilde{F}_i = \sum_{j=1}^{6} \bar{K}_{ij}\left(\tilde{N}_1\right)\tilde{d}_j + \tilde{p}_y \bar{L}_i\left(\tilde{N}_1\right) \qquad (2.29)$$

where

$${}^t\{\tilde{F}_i\} = \left(\tilde{N}_1, \tilde{S}_1, \tilde{M}_1, \tilde{N}_2, \tilde{S}_2, \tilde{M}_2\right), \quad {}^t\{\tilde{d}_i\} = \left(\tilde{u}_1, \tilde{v}_1, \alpha_1, \tilde{u}_2, \tilde{v}_2, \alpha_2\right) \qquad (2.30a,b)$$

$$\tilde{N}_\beta = N_\beta L^2/EI, \quad \tilde{S}_\beta = S_\beta L^2/EI, \quad \tilde{M}_\beta = M_\beta L/EI, \quad \tilde{p}_y = p_y L^3/EI \qquad (2.31a\text{-}d)$$

$$\tilde{u}_\beta = u_{0\beta}/L, \quad \tilde{v}_\beta = v_{0\beta}/L \quad (\beta = 1,2) \qquad (2.31e\text{-}f)$$

The coefficients of the stiffness equation $\bar{K}_{ij}(\tilde{N}_1)$, $\bar{K}_{ijk}(\tilde{N}_1)$, $\bar{L}_{ij}(\tilde{N}_1)$, $\bar{L}_i(\tilde{N}_1)$, $\bar{C}_i(\tilde{N}_1)$, are functions of the axial force $\tilde{N}_1$. Specifically, $\bar{K}_{ij}$ and $\bar{K}_{ijk}$ are symmetric with respect to subscripts (i,j) and (j,k) respectively. All the foregoing coefficients are shown in the following.

$$[\overline{K}_{ij}] = \begin{bmatrix} AI_z^2/I & 0 & 0 & -AI_z^2/I & 0 & 0 \\ & 12\phi_1 & 6\phi_2 & 0 & -12\phi_1 & 6\phi_2 \\ & & 4\phi_3 & 0 & -6\phi_2 & 6\phi_4 \\ & & & AL^2/I & 0 & 0 \\ & \text{sym.} & & & 12\phi_1 & -6\phi_2 \\ & & & & & 4\phi_3 \end{bmatrix} \quad (2.32)$$

ϕ_α ($\alpha = 1$ to 4) are what is called stability functions and their series expansions with respect the axial force $\tilde{N}_1$ are expressed in the following.

$$\phi_1 = \frac{\left[1 + \sum_{n=1}^{\infty} \frac{1}{(2n+1)!}(-\tilde{N}_1)^n\right]}{12\phi} \quad (2.33a)$$

$$\phi_2 = \frac{\left[\frac{1}{2} + \sum_{n=1}^{\infty} \frac{1}{(2n+2)!}(-\tilde{N}_1)^n\right]}{6\phi} \quad (2.33b)$$

$$\phi_3 = \frac{\left[\frac{1}{3} + \sum_{n=1}^{\infty} \frac{2(n+1)}{(2n+3)!}(-\tilde{N}_1)^n\right]}{4\phi} \quad (2.33c)$$

$$\phi_4 = \frac{\left[\frac{1}{6} + \sum_{n=1}^{\infty} \frac{1}{(2n+3)!}(-\tilde{N}_1)^n\right]}{2\phi} \quad (2.33d)$$

$$\phi = \frac{1}{12} + \sum_{n=1}^{\infty} \frac{2(n+1)}{(2n+4)!}(-\tilde{N}_1)^n \quad (2.33e)$$

$$\overline{K}_{1jk} = -\overline{K}_{4jk} = f_1 A_j A_k + f_2 \overline{K}_{2j}\overline{K}_{2k} + f_3 \overline{K}_{3j}\overline{K}_{3k} + f_4\left(A_j \overline{K}_{2k} + A_k \overline{K}_{2j}\right)/2$$
$$+ f_5\left(A_j \overline{K}_{3k} + A_k \overline{K}_{3j}\right)/2 + f_6\left(\overline{K}_{2j}\overline{K}_{3k} + \overline{K}_{3j}\overline{K}_{2k}\right)/2 \quad (2.34a)$$

$$\overline{K}_{2jk} = \overline{K}_{3jk} = \overline{K}_{5jk} = \overline{K}_{6jk} = 0 \quad (2.34b)$$

where ${}^t\{A_i\} = (0\ 0\ 1\ 0\ 0\ 0)$. f_b ($b = 1$ to 10) are expressed by the power series with respect to the axial force.

$$f_1 = 1 + \frac{1}{8}\sum_{n=1}^{\infty} \frac{4^{n+1}(-\tilde{N}_1)^n}{(2n+1)!} \quad (2.35a)$$

$$f_2 = \frac{1}{20} + 2\sum_{n=1}^{\infty} \frac{4^{n+1}\left(-\tilde{N}_1\right)^n}{(2n+5)!} - 2\sum_{n=1}^{\infty} \frac{\left(-\tilde{N}_1\right)^n}{(2n+5)!} \qquad (2.35b)$$

$$f_3 = \frac{1}{3} + \frac{1}{2}\sum_{n=1}^{\infty} \frac{4^{n+1}\left(-\tilde{N}_1\right)^n}{(2n+3)!} \qquad (2.35c)$$

$$f_4 = \frac{1}{3} - 2\sum_{n=1}^{\infty} \frac{\left(-\tilde{N}_1\right)^n}{(2n+3)!} + \sum_{n=1}^{\infty} \frac{4^{n+1}\left(-\tilde{N}_1\right)^n}{(2n+3)!} \qquad (2.35d)$$

$$f_5 = -1 - \frac{1}{2}\sum_{n=1}^{\infty} \frac{4^{n+1}\left(-\tilde{N}_1\right)^n}{(2n+2)!} \qquad (2.35e)$$

$$f_6 = -\frac{1}{4} + \sum_{n=1}^{\infty} \frac{2}{(2n+4)!}\left(-\tilde{N}_1\right)^n - 2\sum_{n=1}^{\infty} \frac{4^{n+1}}{(2n+4)!}\left(-\tilde{N}_1\right)^n \qquad (2.35f)$$

$$f_7 = \frac{1}{252} - 4\sum_{n=1}^{\infty} \frac{(n+3)\left(-\tilde{N}_1\right)^n}{(2n+7)!} + 2\sum_{n=1}^{\infty} \frac{4^{n+2}\left(-\tilde{N}_1\right)^n}{(2n+7)!} \qquad (2.35g)$$

$$f_8 = \frac{1}{12} + 2\sum_{n=1}^{\infty} \frac{1}{(2n+4)!}\left(-\tilde{N}_1\right)^n - \tilde{N}_1\left[\frac{1}{6} + \sum_{n=1}^{\infty} \frac{1}{(2n+3)!}\left(-\tilde{N}_1\right)^n\right]^2 \qquad (2.35h)$$

$$f_9 = \left[\frac{1}{6} + \sum_{n=1}^{\infty} \frac{1}{(2n+3)!}\left(-\tilde{N}_1\right)^n\right]^2 \qquad (2.35i)$$

$$f_{10} = -\frac{1}{15} + 4\sum_{n=1}^{\infty} \frac{n+2}{(2n+5)!}\left(-\tilde{N}_1\right)^n - \sum_{n=1}^{\infty} \frac{4^{n+2}}{(2n+5)!}\left(-\tilde{N}_1\right)^n \qquad (2.35j)$$

$$\begin{aligned}\bar{L}_{1j} = -\bar{L}_{4j} = &-f_2\bar{K}_{2j} - f_3\bar{K}_{3j}/\{2(2\phi_3 + \phi_4)\} - f_4 A_j/2 - f_5 A_j/\{4(2\phi_3 + \phi_4)\} \\ &- f_6\bar{K}_{3k}/2 - f_6\bar{K}_{2j}/\{4(2\phi_3 + \phi_4)\} + f_8 A_j + f_9\bar{K}_{2j} + f_{10}\bar{K}_{3j}\end{aligned} \qquad (2.36a)$$

$$\bar{L}_{2j} = \bar{L}_{3j} = \bar{L}_{5j} = \bar{L}_{6j} = 0 \qquad (2.36b)$$

$$\bar{L}_1 = \bar{L}_4 = 0, \quad \bar{L}_2 = \bar{L}_5 = -1/2, \quad \bar{L}_3 = -\bar{L}_6 = -1/\{4(2\phi_3 + \phi_4)\} \qquad (2.37\text{a-c})$$

$$\overline{C}_1 = -\overline{C}_4 = f_2/4 + f_3 / \{48(2\phi_3 + \phi_4)\} + f_6 / \{8(2\phi_3 + \phi_4)\}$$
$$+ f_7 - f_{9/2} - f_{10} / \{4(2\phi_3 + \phi_4)\} \qquad (2.38a)$$

$$\overline{C}_2 = \overline{C}_3 = \overline{C}_5 = \overline{C}_6 = 0 \qquad (2.38b)$$

The functions ϕ_i and f_i as given in Eqs. (2.33) and (2.35) are expressed mathematically by infinite series expansions in order to unify the different expressions according to the sign of the axial force as well as to avoid the numerical instability when the axial force approaches zero. However, for the practical numerical calculations, the necessary numbers of terms to be adopted in the series expansions are the first ten terms for ϕ_i and the first 14 terms for f_i (Goto and Chen, 1987a). These numbers are sufficient to achieve a numerical convergence.

In the present analysis, the secant stiffness method to be explained later in Section 2.6 is employed as a numerical iterative procedure and the stiffness equations given by Eqs. (2.28) and (2.29) are adequate for the present purpose. However, as displacements become large, the secant stiffness method loses its accuracy and it becomes difficult to obtain convergent solutions. In such a case, iterative procedures using the tangent stiffness such as the Newton-Raphson method combined with the arc-length controls (Riks, 1979) have to be used. Furthermore, in order to analyze the critical behaviors of semi-rigid frames known as *limit-load instability* and *bifurcation*, the tangent stiffness is essential as explained by Goto et al. (1991a and b). The closed-form tangent stiffness corresponding to both stiffness equations, Eqs. (2.28) and (2.29), are explicitly shown by Goto et al. (1991b). In the case of *the linearized beam-column theory*, the tangent stiffness equations coincide with the stiffness equations [Eq. (2.29)], if the prebuckling deformations are ignored. This is the reason why Eq. (2.29) is commonly used to calculate the bifurcation load in the practical applications where the prebuckling deformations are usually negligibly small.

2.4 Modeling of Semi-Rigid Connections

The flexible connection is represented herein by a discrete nonlinear rotational spring, whose moment-rotation ($M - \theta_r$) behavior under a monotonically increasing load is symbolically expressed by

$$M = k_c(\theta_r, M)\theta_r \qquad (2.39)$$

where $k_c(\theta_r, M)$, a function of θ_r and M, corresponds to the secant stiffness of the connection. It is assumed here that the moment-rotation relationship given by Eq. (2.39) is known for each connection in the frame either numerically or in some form of function. The connection models considered in the present analysis are summarized in Table 2.1. The details of these models are explained in Chapter 3.

In the case of unloading and reverse loading, the independent hardening model (Chen and Saleeb, 1982) is adopted for simplicity to express the connection behavior.

SECOND-ORDER ELASTIC ANALYSIS OF FRAMES 61

Table 2.1 Connection Models

Model	Form of function		Parameters
Modified exponential (Kishi-Chen, 1986)	$M = M_o + \sum_{i=1}^{m} A_i \left[1 - exp\left\{ -\frac{\theta_r}{2i\alpha} \right\} \right] + \sum_{j=1}^{n} R_j H(\theta_r - T_j)(\theta_r - T_j)$	M_o A_i, R_j T_j α $H(x)$	initial connection moment curve-fitting constants starting rotation of the linear components scaling factor Heaviside's step function
Polynominal (Frye-Morris, 1976)	$\theta_r = C_1(KM)^{p_1} \times 10^{n_1} + C_2(KM)^{p_2} \times 10^{n_2} + C_3(KM)^{p_3} \times 10^{n_3}$	C_i K	curve-fitting constants standardization parameter
Three-parameter (Kishi-Chen, 1990)	$M = \dfrac{R_{ki}\theta_r}{\left\{ 1 + \left[\theta_r / (M_u/R_{ki}) \right]^n \right\}^{1/n}}$	R_{ki} M_u n	initial stiffness ultimate moment capacity shape parameter

62 ADVANCED ANALYSIS OF STEEL FRAMES

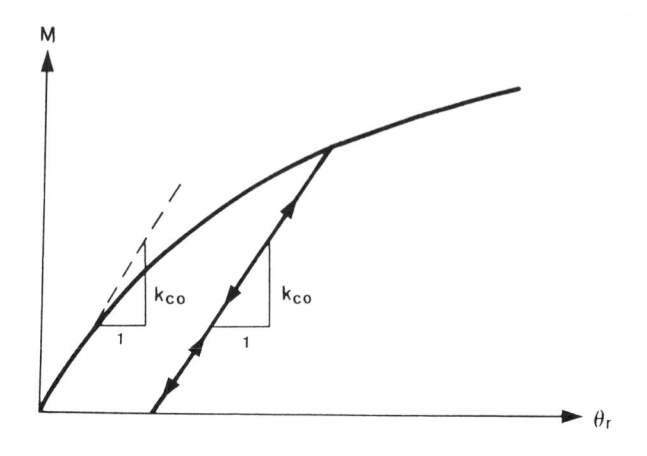

(a). Reloading from the Positive Range of M

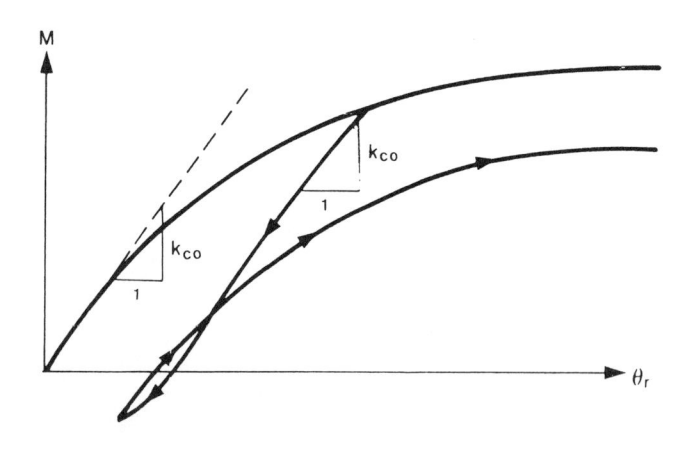

(b). Reloading from the Negative Range of M

FIGURE 2.3. Problems associated with the independent hardening model.

In the unloading process, the connection is assumed to unload linearly up to $M = 0$, following the initial stiffness k_{c0} of the connection as shown in Figure 2.3a. Hence the moment-rotation relationship on the unloading path initiating from the point (M_0, θ_{ro}) is given by

$$M = k_{co}(\theta_r - \theta_{ro}) + M_0 \tag{2.40}$$

When the sign of connection moment M changes in the unloading process, the connection enters into the reverse loading process. In this case, the connection behavior excluding the residual plastic rotation θ_r^p resulted from the previous loading process is assumed as follows to coincide with that of virgin connections under monotonic loading.

$$M = k_c\left(\left|\theta_r - \theta_r^p\right|, M\right)\left(\theta_r - \theta_r^p\right) \tag{2.41}$$

Equations (2.39), (2.40), and (2.41) can be generally expressed as

$$M = k_c \theta_r + M_c \tag{2.42}$$

On the loading path:

$$k_c = k_c(\theta_r, M), \quad M_c = 0 \tag{2.43a,b}$$

On the unloading path:

$$k_c = k_{co}, \quad M_c = M_0 - k_{co}\theta_{ro} \tag{2.44a,b}$$

On the reverse loading path:

$$k_c = k_c\left(\left|\theta_r - \theta_r^p\right|, M\right), \quad M_c = -k_c\theta_r^p \tag{2.45a,b}$$

As for the connection behavior under cyclic loading, few experimental data are known. Thus, it is difficult to use the curve-fitting technique to represent the moment-rotation relation. In the present formulation, the independent hardening model is used for simplicity to represent the inelastic behavior. Although this model can take care of the moment-rotation relation under one cycle of loading, unloading, and reverse loading, the connection behavior under the repetition of the above loading cycle may not be expressed with acceptable accuracy. As shown in Figure 2.3b, one problem associated with the independent hardening model is that the moment-rotation relation is completely different according to whether the connection is reloaded from the positive range of M or from the negative range. To overcome the deficiency of the independent hardening model, Goto et al. (1991a) introduced the bounding surface model (Dafalias and Popov, 1976) to describe the cyclic behavior after the first unloading occurs.

2.5 Modified Secant Stiffness Equation with Connection Flexibility

In order to reduce the number of degrees of freedom in the numerical analysis based on the secant stiffness method, a modified secant stiffness equation for a beam-column member with flexible connections (Figure 24) at member ends is derived. The derivation procedure is made herein specifically for a member with connections

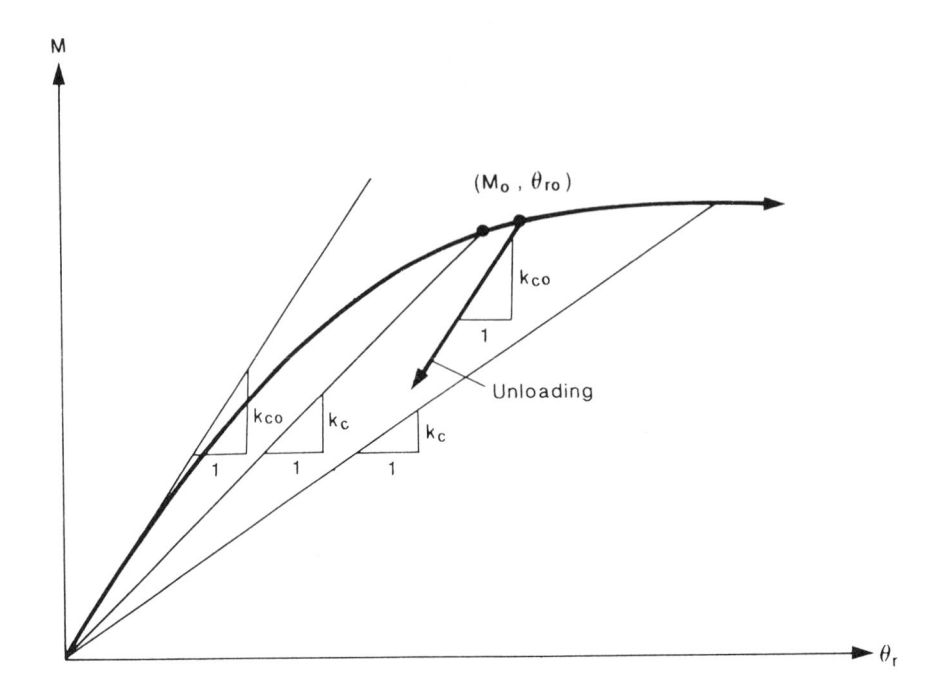

FIGURE 2.4. Connection behavior.

FIGURE 2.5. Beam-column member with flexible connections at both ends.

at both ends as illustrated in Figure 2.5. The modified stiffness equation with the flexible connection at either end of a member can be obtained in a similar manner.

To derive the modified stiffness equation, the rotational degrees of freedom of α_1 and α_2 for the beam-column member are eliminated. Relative rotation θ_{rm} at Node m is expressed as follows by the rotational degree of freedom of the beam-column member (α_m) as well as that of the connection (α_{cm}).

$$\theta_{rm} = \alpha_{cm} - \alpha_m \quad (m = 1, 2) \tag{2.46}$$

where subscript m indicates a quantity at Node m.

With Eqs. (2.42) and (2.46), the rotational degree α_m at the end of the beam-column member is given by

$$\alpha_m = \alpha_{cm} - \tilde{M}_m/\tilde{k}_{cm} + \tilde{M}_{cm}/\tilde{k}_{cm} \tag{2.47}$$

$$\tilde{M}_{cm} = M_{cm}L/EI, \quad \tilde{k}_{cm} = k_{cm}L/EI \tag{2.48}$$

For simplicity, the stiffness equations of Eqs. (2.28) and (2.29) are assumed to be expressed in the secant stiffness form as

$$\{\tilde{F}_i\} = [\overline{K}_{ij}^s]\{\tilde{d}_j\} + \{\tilde{p}_i^e\} \tag{2.49}$$

where the secant stiffness matrix $[\overline{K}_{ij}^s]$ as well as the equivalent member fixed end force vector $\{\tilde{p}_i^e\}$ can be expressed by

The nonlinear theory:

$$\overline{K}_{ij}^s = \overline{K}_{ij}(\tilde{N}_1) + \sum_{k=1}^{6}\overline{K}_{ijk}(\tilde{N}_1)\tilde{d}_k + \tilde{p}_y \overline{L}_{ij} \tag{2.50}$$

$$\tilde{p}_i^e = \tilde{p}_y \overline{L}_i(\tilde{N}_1) + \tilde{p}_y^2 \overline{C}_i(\tilde{N}_1) \tag{2.51}$$

The linearized theory:

$$\overline{K}_{ij}^s = \overline{K}_{ij}(\tilde{N}_1) \tag{2.52}$$

$$\tilde{p}_i^e = \tilde{p}_y \overline{L}_i(\tilde{N}_1) \tag{2.53}$$

Substituting Eq. (2.47) into Eq. (2.49) and solving the equation with respect to $\tilde{M}_m$ ($m = 1,2$), the modified stiffness equation for a beam-column member with flexible connections at both ends is obtained as

$$\{\tilde{F}_i^c\} = [\overline{K}_{ij}^c]\{\tilde{d}_j^c\} + \{\tilde{p}_i^{ec}\} \tag{2.54}$$

where the components of $[\overline{K}_{ij}^c]$ and $\{\tilde{p}_i^{ec}\}$ are shown in the following.

$$\overline{K}_{1j}^c = \overline{K}_{1j}^s \quad (j = 1 \text{ to } 6) \tag{2.55a}$$

$$\overline{K}_{22}^c = \overline{K}_{22}^s - \left\{\overline{K}_{23}^s\left(\overline{K}_{32}^s \tilde{k}_{c2} + \overline{K}_{32}^s \overline{K}_{66}^s - \overline{K}_{36}^s \overline{K}_{62}^s\right)\right.$$
$$\left. + \overline{K}_{26}^s\left(\overline{K}_{62}^s \tilde{k}_{c1} + \overline{K}_{62}^s \overline{K}_{33}^s - \overline{K}_{32}^s \overline{K}_{63}^s\right)\right\}/D \tag{2.55b}$$

$$\overline{K}^c_{23} = \tilde{k}_{c1}\left(\overline{K}^s_{23}\tilde{k}_{c2} + \overline{K}^s_{23}\overline{K}^s_{66} - \overline{K}^s_{63}\overline{K}^2_{26}\right)/D \tag{2.55c}$$

$$\overline{K}^c_{24} = \overline{K}^s_{24} \tag{2.55d}$$

$$\overline{K}^c_{25} = \overline{K}^s_{25} - \left\{\overline{K}^s_{23}\left(\overline{K}^s_{35}\tilde{k}_{c2} + \overline{K}^s_{35}\overline{K}^s_{66} - \overline{K}^s_{36}\overline{K}^s_{65}\right)\right.$$
$$\left. + \overline{K}^s_{26}\left(\overline{K}^s_{65}\tilde{k}_{c1} + \overline{K}^s_{65}\overline{K}^s_{33} - \overline{K}^s_{35}\overline{K}^s_{63}\right)\right\}/D \tag{2.55e}$$

$$\overline{K}^c_{26} = \tilde{k}_{c2}\left(\overline{K}^s_{26}\tilde{k}_{c1} + \overline{K}^s_{26}\overline{K}^s_{33} - \overline{K}^s_{23}\overline{K}^s_{36}\right)/D \tag{2.55f}$$

$$\overline{K}^c_{33} = \tilde{k}_{c1}\left(\overline{K}^s_{33}\tilde{k}_{c2} + \overline{K}^s_{33}\overline{K}^s_{66} - \overline{K}^s_{36}\overline{K}^s_{63}\right)/D \tag{2.55g}$$

$$\overline{K}^c_{34} = \overline{K}^s_{34} \tag{2.55h}$$

$$\overline{K}^c_{35} = \tilde{k}_{c1}\left(\overline{K}^s_{35}\tilde{k}_{c2} + \overline{K}^s_{33}\overline{K}^s_{66} - \overline{K}^s_{36}\overline{K}^s_{65}\right)/D \tag{2.55i}$$

$$\overline{K}^c_{36} = \tilde{k}_{c1}\tilde{k}_{c2}\overline{K}^s_{36}/D \tag{2.55j}$$

$$\overline{K}^c_{4j} = \overline{K}^s_{4j} \quad (j = 4 \, to \, 6) \tag{2.55k}$$

$$\overline{K}^c_{55} = \overline{K}^s_{55} - \left\{\overline{K}^s_{53}\left(\overline{K}^s_{35}\tilde{k}_{c2} + \overline{K}^s_{35}\overline{K}^s_{66} - \overline{K}^s_{36}\overline{K}^s_{65}\right)\right.$$
$$\left. + \overline{K}^s_{56}\left(\overline{K}^s_{65}\tilde{k}_{c1} + \overline{K}^s_{65}\overline{K}^s_{33} - \overline{K}^s_{35}\overline{K}^s_{63}\right)\right\}/D \tag{2.55l}$$

$$\overline{K}^c_{56} = \tilde{k}_{c2}\left(\overline{K}^s_{56}\tilde{k}_{c1} + \overline{K}^s_{56}\overline{K}^s_{33} - \overline{K}^s_{53}\overline{K}^s_{36}\right)/D \tag{2.55m}$$

$$\overline{K}^c_{66} = \tilde{k}_{c2}\left(\overline{K}^s_{66}\tilde{k}_{c1} + \overline{K}^s_{66}\overline{K}^s_{33} - \overline{K}^s_{36}\overline{K}^s_{63}\right)/D \tag{2.55n}$$

$$\tilde{f}^c_1 = \tilde{f}_1 \tag{2.56a}$$

$$\tilde{f}^c_2 = \tilde{f}_2 + \left(\tilde{M}_{c1}\overline{K}^s_{23}/\tilde{k}_{c1} + \tilde{M}_{c2}\overline{K}^s_{26}/\tilde{k}_{c2}\right)/D$$
$$- \left\{\overline{K}^s_{23}(\tilde{k}_{c2} + \tilde{k}_{66}) - \overline{K}^s_{63}\overline{K}^s_{26}\right\}\left(\tilde{f}_3 + \tilde{M}_{c1}\overline{K}^s_{33}/\tilde{k}_{c1} + \tilde{M}_{c2}\overline{K}^s_{36}/\tilde{k}_{c2}\right)/D$$
$$- \left\{\overline{K}^s_{26}(\tilde{k}_{c1} + \overline{K}^s_{33}) - \overline{K}^s_{23}\overline{K}^s_{36}\right\}\left(\tilde{f}_6 + \tilde{M}_{c1}\overline{K}^s_{63}/\tilde{k}_{c1} + \tilde{M}_{c2}\overline{K}^s_{66}/\tilde{k}_{c2}\right)/D \tag{2.56b}$$

$$\tilde{f}_3^c = \tilde{k}_{c1}\left(\tilde{k}_{c2} + \overline{K}_{66}^s\right)\left(\tilde{f}_3 + \tilde{M}_{c1}\overline{K}_{33}^s/\tilde{k}_{c1} + \tilde{M}_{c2}\overline{K}_{36}^s/\tilde{k}_{c2}\right)/D$$

$$- \tilde{k}_{c1}\overline{K}_{36}^s\left(\tilde{f}_6 + \tilde{M}_{c1}\overline{K}_{63}^s/\tilde{k}_{c1} + \tilde{M}_{c2}\overline{K}_{66}^s/\tilde{k}_{c2}\right)/D \qquad (2.56c)$$

$$\tilde{f}_4^c = \tilde{f}_4 \qquad (2.56d)$$

$$\tilde{f}_5^c = \tilde{f}_5 + \left(\tilde{M}_{c1}\overline{K}_{53}^s/\tilde{k}_{c1} + \tilde{M}_{c2}\overline{K}_{56}^s/\tilde{k}_{c2}\right)/D$$

$$- \left\{\overline{K}_{53}^s\left(\tilde{k}_{c2} + \overline{K}_{66}^s\right) - \overline{K}_{56}^s\overline{K}_{63}^s\right\}\left(\tilde{f}_3 + \tilde{M}_{c1}\overline{K}_{33}^s/\tilde{k}_{c1} + \tilde{M}_{c2}\overline{K}_{36}^s/\tilde{k}_{c2}\right)/D$$

$$- \left\{\overline{K}_{56}^s\left(\tilde{k}_{c1} + \overline{K}_{33}^s\right) - \overline{K}_{53}^s\overline{K}_{36}^s\right\}\left(\tilde{f}_6 + \tilde{M}_{c1}\overline{K}_{63}^s/\tilde{k}_{c1} + \tilde{M}_{c2}\overline{K}_{66}^s/\tilde{k}_{c2}\right)/D \qquad (2.56e)$$

$$\tilde{f}_6^c = \tilde{k}_{c2}\left(\tilde{k}_{c1} + \overline{K}_{33}^s\right)\left(\tilde{f}_6 + \tilde{M}_{c1}\overline{K}_{63}^s/\tilde{k}_{c1} + \tilde{M}_{c2}\overline{K}_{66}^s/\tilde{k}_{c2}\right)/D$$

$$- \tilde{k}_{c2}\overline{K}_{63}^s\left(\tilde{f}_3 + \tilde{M}_{c1}\overline{K}_{33}^s/\tilde{k}_{c1} + \tilde{M}_{c2}\overline{K}_{36}^s/\tilde{k}_{c2}\right)/D \qquad (2.56f)$$

$$D = \left(\tilde{k}_{c1} + \overline{K}_{33}^s\right)\left(\tilde{k}_{c2} + \overline{K}_{66}^s\right) - \overline{K}_{36}^s\overline{K}_{63}^s \qquad (2.57)$$

For simplicity, the explicit expressions are shown here only to the upper triangular matrix components of $[\overline{K}_{ij}^c]$ of Eq. (2.54). The lower triangular components can be given by just exchanging the order of the subscripts. That is, assume that an upper triangular matrix component $\overline{K}_{ij}^c$ be symbolically expressed as

$$\overline{K}_{ij}^c = F_{ij}\left(\overline{K}_{kl}^s, \overline{K}_{mn}^s, \ldots \overline{K}_{pq}^s\right) \qquad (2.58)$$

where $F_{ij}(a,b,\ldots c)$ denotes a function of $a, b, \ldots c$. The lower triangular matrix component can be given by

$$\overline{K}_{ji}^c = F_{ij}\left(\overline{K}_{lk}^s, \overline{K}_{nm}^s, \ldots \overline{K}_{qp}^s\right) \qquad (2.59)$$

2.6 Solution Procedure for Nonlinear Stiffness Equations

The load-control secant stiffness method is used here to solve the structural stiffness equation which is nonlinear in terms of the displacements as well as the axial force. This is because the secant stiffness method is not only simple in computer implementation but also stable regardless of the shape of the connection's moment-rotation curves, as long as the displacements are not very large. Indeed, the iterative procedure using the tangent stiffness such as the Newton-Raphson method, guarantees the second-order convergence, but this method requires the calculations of tangent stiffness as well as unbalanced force. Hence, the programming for the tangent stiffness method becomes more complicated than that of the secant stiffness method, that is almost like a simple repetition of the customary first-order analysis. Furthermore, the tangent stiffness is much influenced by a local variation of connection moment-rotation curves, and the iteration using tangent stiffness may encounter numerical difficulty if the slope of connection curves changes abruptly or irregularly. Such a problem can occur when connection moment-rotation curves are expressed by bi-linear, piecewise-linear, or polynominal models, or when these curves fitted with tests are not smooth.

When the secant stiffness method using the present closed-form stiffness equations is applied to the analysis of flexibly jointed frames, two kinds of iteration are generally required. One is the calculation of nodal displacements from the structural stiffness equation and the other is for the calculation of axial force $\tilde{N}_1$. The latter iteration has to be carried out at every cycle of a structural iteration using the member stiffness equation if the stiffness equation is nonlinear in terms of member forces. The member stiffness equation becomes nonlinear when the stiffness equation [Eq. (2.28)] derived from *the nonlinear beam-column theory* is used or a nonlinear flexible connection model is attached to the member under consideration. It should be noted, however, that member forces can be calculated without iteration for the stiffness equation [Eq. (2.29)] ignoring the bowing effect (i.e., equation is based on *the linearized beam-column theory*) and without nonlinear flexible connections.

The loading, unloading, and reloading behavior of connections is influenced by loading history and hence, loads have to be divided into smaller increments to account for this behavior. In each load increment, the secant stiffness method can be employed to obtain convergent solutions.

Since the present method is primarily used for design purpose, no special techniques are required in the iterative procedure. The secant stiffness procedure is no more than a repetition of the first-order analysis. The only difference is updating the nonlinear terms in the stiffness equations at every cycle of iteration, making use of the latest information obtained from the previous cycle of iteration.

To obtain convergent solutions, the iteration is continued until unbalanced force becomes within the prescribed tolerance. In updating the nonlinear terms, connection moments are compared with those in the previous load step to determine whether the connection is in a state of loading or unloading. To begin an iteration for each load increment, the solutions in the previous load's steps are used as an initial estimation of the nonlinear terms. The present iterative procedures have yielded satisfactory results regardless of the shape of connection moment-rotation curves as commonly used in practical examples. When the bowing effect is ignored, the convergence of iteration becomes fast and solutions can be obtained even under a relatively large displacement.

On the equilibrium path in the vicinity of limit point, it becomes difficult for the present load-control secant stiffness method to obtain convergent solutions. Furthermore, the present method cannot deal with critical behaviors of structural systems

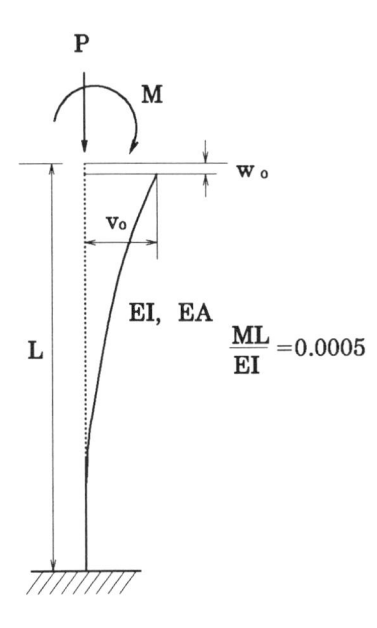

FIGURE 2.6. Cantilever column with small end moment.

such as limit-load instability and bifurcation. For second-order elastic analysis in frame design, however, the limit-load instability is seldom encountered. This is in contrast to the design using the second-order inelastic analysis. As a result, the present method should be adequate for design purposes.

To investigate critical behaviors more accurately, special formulations are required. Goto et al. (1991a,b) proposed an accurate, yet practical method for the analysis of critical and postcritical behaviors of semi-rigid frames with inelastic connections, where the incremental arc-length method was combined with the Newton-Raphson iterative procedure to deal specifically with the limit-load instability problem, while the inelastic bifurcation problems were analyzed by the use of Hill's bifurcation theory for elastic-plastic solids (Hill, 1958).

2.7 Numerical Examples

EXAMPLE 1: CANTILEVER COLUMN WITH SMALL END MOMENT

A cantilever column with small end moment shown in Figure 2.6 was analyzed numerically to demonstrate the accuracy and limitations of the second-order theories: *the nonlinear beam-column theory* and *the linearized beam-column theory*. Computed results of vertical load versus end displacement relations are summarized in Figure 2.7. Closed-form elliptic integral solutions for the exact theory (Goto, 1987c, 1990) are also shown in Figure 2.7 for comparison. This exact theory referred here to as *the theory of finite displacements with finite strains* was derived by Nishino et al. (1975), based on variational calculus. In its derivation, no approximations are

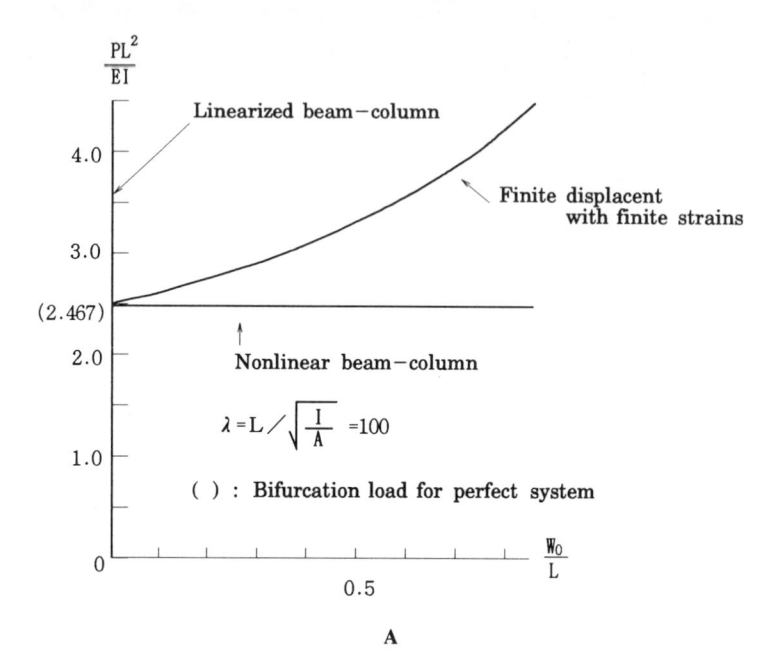

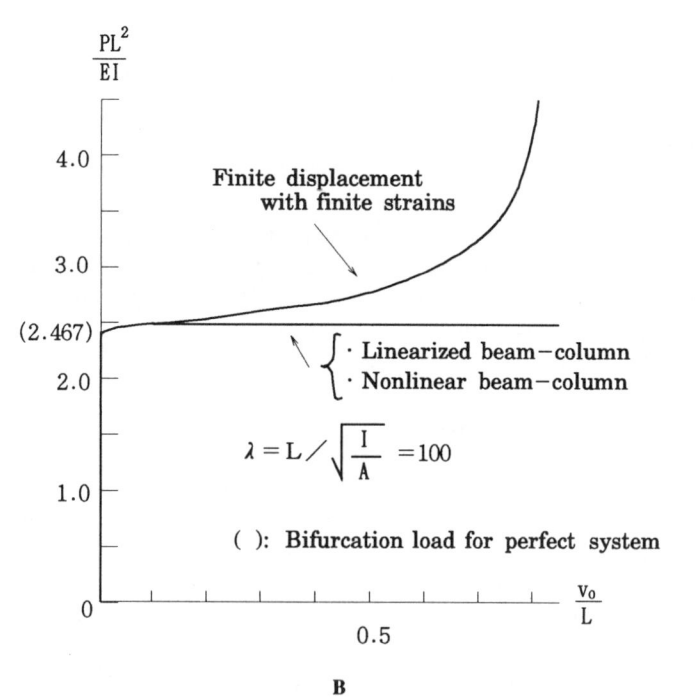

FIGURE 2.7. Load-displacement curves of cantilever column. (a) Vertical-displacement at column end. (b) Horizontal-displacement at column end.

introduced except for the usual beam assumptions: no change of cross sectional shapes and the Bernoulli-Euler hypothesis where the transverse plane is assumed to remain plane and normal to the beam axis. This theory coincides with the one shown by Reissner (1972). The above elliptic integral solutions are considered to be more accurate than the well known solutions of inextensional elastica (Timoshenko and Gere, 1961) which ignores the elongation of the centroidal axis. To solve the nonlinear stiffness equations in the post-buckling range, the incremental arc-length method combined with the Newton-Raphson iterative procedures was used.

From Figure 2.7, it can be seen that both *the nonlinear and linearized beam-column theories* yield accurate solutions at least up to the buckling load. So far as the vicinity of the buckling point is concerned, these theories can also be used to analyze the post-buckling behavior in terms of the horizontal displacement. However, as the displacements increase in the post-buckling range, the discrepancy between the exact theory and the beam-column theories become large. As can be seen from the linear constitutive equation given by Eq. (2.21), *the linearized beam-column theory* cannot simulate the buckling behavior with regard to the vertical displacement.

EXAMPLE 2: SEMI-RIGID RECTANGULAR FRAMES (GOTO AND CHEN, 1987b)

As seen from the results of Example 1, it is more accurate to use *the nonlinear beam-column theory* which considers the bowing effect in the second-order analysis. However, the stiffness equation including the bowing effect [Eq. (2.28)] becomes much more complicated than that ignoring this effect [Eq. (2.29)]. Furthermore, the inclusion of the bowing effect results in a slower convergence in the numerical iterative procedures. Therefore, it is desirable to ignore the bowing effect in design analyses where the structures seldom exhibits post-buckling behavior. For the rigid frames, it has been shown by numerical examples that the bowing effect is negligibly small for design purposes (Goto and Chen, 1987a; Korn, 1981). However, for PR construction, this bowing effect resulting from the rotational displacements of structures may have a significant influence on the results of analysis because the frames are more deformable and the nonlinear behavior of the connections is also more sensitive to the magnitude of applied moments. The bowing effect for the design analysis of PR construction is therefore investigated numerically.

Two typical rectangular frames designed as PR construction (Moncarz and Gerstle, 1981; Lindsey et al., 1985) are used here. These rectangular frames are illustrated in Figure 2.8 along with the loads considered in the analysis.

Model I is the two-story one-bay frame used by Moncarz and Gerstle (1981). They chose the connections from among the top and seat angles tested by Hechtman and Johnson (1947); specimens number 23 and 25 were selected, respectively, for the upper and the lower beams. Here, due to the lack of information about specimen number 25, specimen number 23 is used for all the connections. Further, to examine the effect of connection flexibility on bowing, two other kinds of connections, that is, specimens number 16 and 24 in the same experiment series are added for comparison as the two extreme cases in terms of connection flexibility. The frame was designed based on AISC-ASD so the load is multiplied by 1.4 to obtain the factored load shown in Figure 2.8.

72 ADVANCED ANALYSIS OF STEEL FRAMES

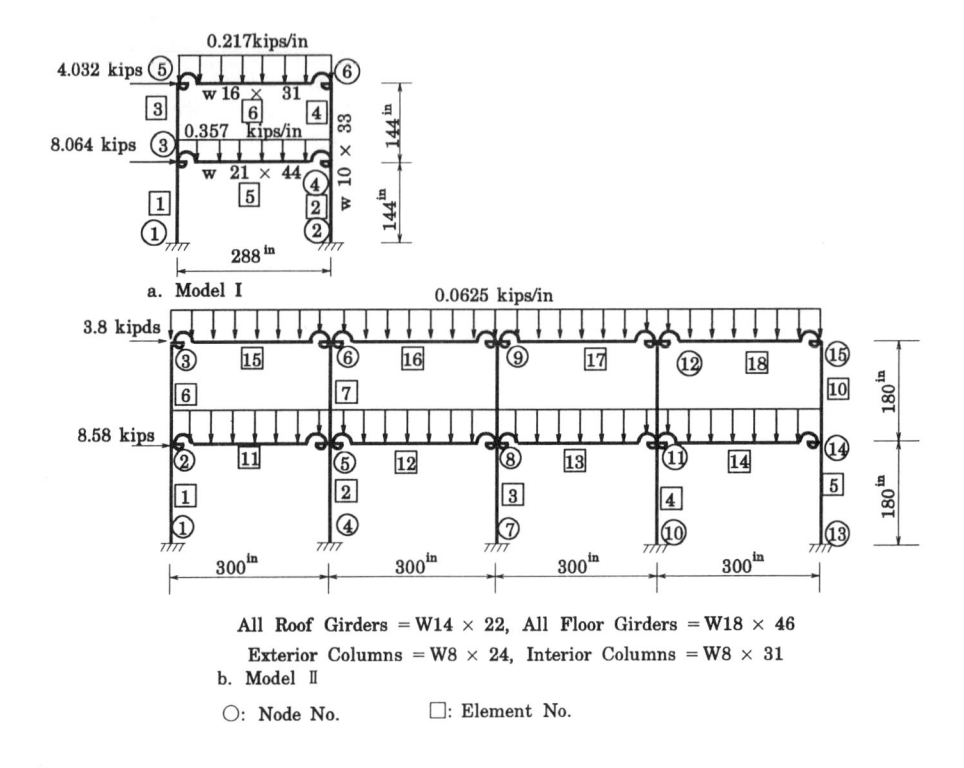

a. Model I

b. Model II

All Roof Girders = W14 × 22, All Floor Girders = W18 × 46
Exterior Columns = W8 × 24, Interior Columns = W8 × 31

◯: Node No.   ☐: Element No.

FIGURE 2.8. Type PR construction.

Model II is the two-story four-bay frame used by Lindsey et al. (1985), where the double web angles (4" × 3.5" × 3/8") were used for all connections. As the angles' heights are not indicated, the connections are selected from those shown by Rathbun (1936), considering the depth of beams; specimens A-4 and A-6 are used, respectively, for the lower and the upper beams. In addition, following the same philosophy as explained for Model I, specimens A1–1 and E1–3 are selected from the newer experiment (Thompson et al., 1970) as two extreme cases in terms of connection flexibility. In these two cases the same connection is used throughout the structure. The moment-rotation relationship of the connections used for Models I and II are shown, respectively, in Figures 2.9 and 2.10.

In the numerical analysis, the factored loads are divided equally into five steps to account for the unloading of connections and the loads are increased monotonically up to the factored loads. At each load step, the structural iteration is repeated until the ratio of an unbalanced force to the corresponding nodal force becomes less than 0.01%, while the member iteration required at every cycle of the structural iteration is also continued until the increment of a member force becomes within 0.01% of the corresponding member force.

The ratios of the physical quantities obtained by the second-order analysis to those by the first-order analysis are summarized for two types of frames in Tables 2.2 and 2.3, which are classified according to the method of analysis, as well as the connec-

FIGURE 2.10. Experimental moment-rotation relationship of connections for Model II.

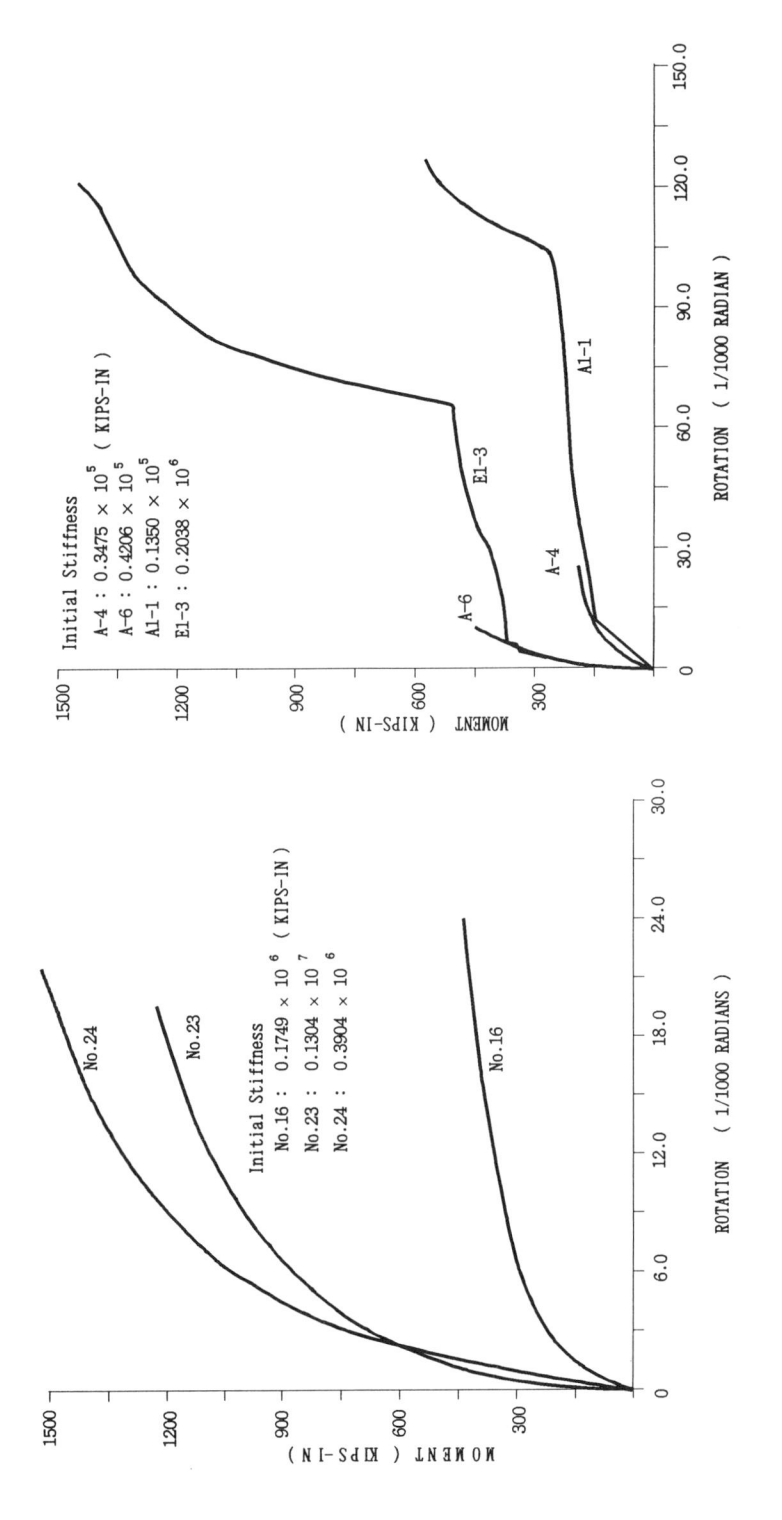

FIGURE 2.9. Experimental moment-rotation relationship of connections for Model I.

Table 2.2 Bowing Effect in Two-Story One-Bay Frame

	a. Drift				b. Maximum column moment in 1st story				c. Maximum column moment in 2nd story			
	Flexibly jointed frames			Rigid frame	Flexibly jointed frames			Rigid frame	Flexibly jointed frames			Rigid frame
Connection	No.16	No.23	No.24		No.16	No.23	No.24		No.16	No.23	No.24	
Location of maximum value	Node 5	Node 5	Node 5	Node 5	Node 2	Node 2	Node 2	Node 4	Node 6	Node 6	Node 6	Node 6
Considering bowing effect	1.113	1.089	1.081	1.042	1.048	1.041	1.072	1.008	1.014	1.015	1.014	1.007
Ignoring bowing effect	1.112	1.087	1.079	1.039	1.058	1.047	1.082	1.009	1.014	1.016	1.015	1.007

Table 2.3 Bowing Effect in Two-Story Four-Bay Frame

	a. Drift				b. Maximum column moment in 1st story				c. Maximum column moment in 2nd story			
	Flexibly jointed frames			Rigid frame	Flexibly jointed frames			Rigid frame	Flexibly jointed frames			Rigid frame
Connection	A-4/A-6	A1-1	E1-3		A-4/A-6	A1-1	E1-3		A-4/A-6	A1-1	E1-3	
Location of maximum value	Node 3	Node 3	Node 3	Node 3	Node 4	Node 4	Node 4	Node 14	Node 14	Node 15	Node 15	Node 14
Considering bowing effect	1.149	1.268	1.125	1.074	1.133	1.211	1.134	1.065	0.991	1.033	1.020	1.001
Ignoring bowing effect	1.140	1.262	1.121	1.072	1.125	1.202	1.124	1.064	0.981	1.033	1.018	1.000

SECOND-ORDER ELASTIC ANALYSIS OF FRAMES 75

Table 2.4 Connection Behavior of Model I

Connection	Method of analysis	Location of unloading connections		Load step				
		Element no.	Node no.	1	2	3	4	5
No. 16	Considering /ignoring bowing effect	5	3	←————→				
		6	5			←————→		
	1st order analysis	5	3			←————→		
		6	5					

←————→: Unloading

tion type. It should be noted that the first-order analysis used here only ignores the geometrical nonlinearity. For comparison, the results for a rigid frame are also added to Tables 2.2 and 2.3. Further, in Tables 2.4 and 2.5, it is shown how the difference in the methods has an effect on the loading/unloading characteristics of the connection in a steel frame.

It is recognized from Tables 2.2 and 2.3 that the bowing effect, along with the geometrical nonlinearity, generally has more influence on flexibly jointed frames than on rigid frames. The bowing also affects the loading/unloading behavior of connections, as shown in Table 2.5, although this effect is not as significant as that of the geometrical nonlinearity. Throughout Tables 2.2 to 2.5, the effect of the geometrical nonlinearity is significant in the drift as well as in the column moment of flexibly jointed frames. Hence, it can be said that the precise consideration on the geometrical nonlinearity may be more important in the design analysis of type PR construction. Indeed, the bowing effect on the column moment is more evident in flexibly jointed frames than in rigid frames, but it seems that the bowing effect can still be ignored in the design analysis for flexibly jointed frames.

2.8 Computer Program

This program, a refined version of FLFRM (Goto and Chen, 1986), is based on the analysis method explained in Sections 2.2 to 2.6 where the second-order frame analysis including the effect of connection flexibility is considered. It should be noted, however, that the member plastification is not considered in this program.

Two types of theory shown in Section 2.2 are used here for the second-order frame analysis. These are (1) *nonlinear beam-column theory* and (2) *linearized beam-column theory*. The *nonlinear beam-column theory* considers the bowing effect, while the *linearized beam-column theory* ignores it. In addition to the second-order analysis, the first-order analysis is also included in this program.

Three connection models described in Table 2.1 are considered in this program. These are the (1) *modified exponential model* (2) *polynominal model* and (3) *three-parameter model*. For the models mentioned above, either elastic or inelastic behav-

Table 2.5 Connection Behavior of Model II

Connection	Method of analysis	Location of unloading connections		Load step				
		Element no.	Node no.	1	2	3	4	5
Upper story: A—4	Considering bowing effect	11	2				←——→	
		12	5				←——→	
		14	11				←——→	
	Ignoring bowing effect	11	2				←—→	
		12	5					
		14	11					
Lower story: A—6								
	1st order analysis	11	2				←—→	
		12	5					
		14	11					
A1—1	Considering /ignoring bowing effect	11	2			←——→		
		12	5			←——→		
		13	8			←——→		
		14	11			←——→		
		15	3				←——→	
		16	6				←——→	
		17	9				←——→	
		18	12				←——→	
	1st order analysis	11	2				←—→	
		12	5				←—→	
		13	8				←—→	
		14	11				←—→	
		15	3				←—→	
		16	6					
		17	9					
		18	12					←—→
E1—3	Considering /ignoring bowing effect	11	2				←—→	
		12	5			←——→		
		13	8			←——→		
		14	11			←——→		
	1st order analysis	11	2					
		12	5					
		13	8					
		14	11					

←——→: Unloading

ior can be taken into account. To express the inelastic behavior, the independent hardening model as explained in Section 2.4 is used.

As an iterative procedure to solve the nonlinear stiffness equations, the load-control secant stiffness method explained in Section 2.6 is adopted. This procedure is found not only simple in programming, but also less influenced by the shape of the connection's moment-rotation curves. This iteration is usually continued until unbalanced force becomes within some prescribed tolerance. In the design analysis, it seldom happens that we have to analyze the post-buckling behavior of frames. Therefore, in most cases, the load-control secant stiffness method will yield satisfactory results.

2.9 User's Manual

The program FLFRM written in FORTRAN language is provided by the file named **flfrm.f** in an attached diskette. This program is tested by a Sun SPARC station IPX using Sun FORTRAN 77 compiler and a Sony RISC NEWS using RISC NEWS FORTRAN. **makefile** is also included in the diskette to show how an executable program can be made by the utility of UNIX system. Users can compile the source program and make the executable file by themselves.

In order to execute this program, two sets of data have to be given according to users' purpose. One is data called **input.d**. The other is **toframe**. **input.d** includes all the data except for the connection characteristics which are supplied by **toframe**. Both of these files should be in the same directory as the executable program. The computed results are written in **output.d**. If the connections in the data bank are used in the analysis, the values of the parameters which are to be put in **toframe** can be calculated by this data bank program. The details of the semi-rigid connection data bank are explained in Chapter 3. For users' convenience, **input.d**, **toframe**, and **output.d** corresponding to the semi-rigid frame shown later in **Example 3** are included in the attached diskette.

(1) input.d
Line 1 (4I5)
IDSI: Flag to select units used in analysis.
    0: United States system of units (Pounds, Inches)
    1: MKS system of unit (Kg, m)
NANL: Flag to specify the theory used for beam-column element.
    1: *small displacement theory* (1st order theory)
    2: *nonlinear beam–column theory* (bowing effect is considered)
    3: *linearized beam–column theory* (bowing effect is ignored)
MDLC: Flag to specify the connection model used in the analysis
    1: Elastic model including nonlinear elastic model
    2: Elastic-plastic model based on independent hardening rule

ITLM: Flag to control iterative procedure
   0: Iteration is continued until the relative error for displacements becomes within a prescribed tolerance given by Line 12.
   1: Iteration is stopped after one cycle of iteration. (The accuracy of this iterative procedure is guaranteed only when linearized beam-column theory is applied to rigid frames.)

Line 2 (4I5)
NE: Number of elements ≤ 210
ND: Number of nodes ≤ 121
NMAT: Number of element materials (element materials are specified in Line 6)
NMATC: Number of different flexible connections ≤ 100

Line 3 (3I5, F10.0)
NLSQ: Number of load sequences ≤ 20
NL: Largest number of all the "NCL(I)" and "NDLI(I)" for $1 \leq I \leq $ NLSQ ("NCL" and "NDLI" are given in Line 7) ≤ 100
KS: Number of terms adopted for power series to calculate stiffness equation. "KS" = 20 will be enough to obtain convergent solution. If "KS" = –1, number of the terms is determined so that the relative error becomes within the tolerance specified by "DE".
DE: Relative tolerance in the calculation of the power series. Leave the space blank, unless "KS" = –1. Generally "DE" = 10^{-5} will be enough.

Line 4 (9I5)
N: Element number.
IEB(1): Node number of the 1st end of an element.
IEB(2): Node number of the 2nd end of an element.
IEB(3): Element material identifier which corresponds to "KK" given in Line 6.
IEB(4): Element type identifier
   1: Frame element.
   2: Element with flexible connections at both ends.
   3: Frame element with a hinge attached to the 1st end.
   4: Frame element with a hinge attached to the 2nd end.
   5: Truss element with hinges at both ends.
   6: Frame element with a flexible connection attached to the 1st end.
   7: Frame element with a flexible connection attached to the 2nd end.
   8: Element with a flexible connection attached to the 1st end and a hinge to the 2nd end.
   9: Element with a hinge attached to the 1st end and a flexible connection to the 2nd end.
IEB(5): Flexible connection identifier for the 1st end of an element. This identifier corresponds to the DATA NO. defined in **toframe**. Leave this data space blank, unless IEB(4) is 2, 6, or 8.
IEB(6): Flexible connection identifier for the 2nd end of an element. This identifier corresponds to the DATA NO. defined in **toframe**. Leave this data space blank, unless IEB(4) is 2, 7, or 9.
IEB(7): Flag to use only the initial stiffness of the moment-rotation relation given by IEB(5) for the 1st connection. Leave this data space blank, unless IEB(4) is 2, 6, or 8.
   0: use only the initial stiffness.
   1: use full moment-rotation relation.
IEB(8): Flag to use only the initial stiffness of the moment-rotation relation given by IEB(6) for the 2nd connection Leave this data space blank, unless IEB(4) is 2, 7, or 9.

0: use only the initial stiffness.
1: use full moment-rotation relation.
Line 4 has to be supplied "NE" times.

Line 5 (I5,2(F8.0,I2))
KK: Element number.
DDX(N), IP: Horizontal projection of element "KK" before deformation. DDX(N) and IP denote the mantissa and the exponent, respectively.
DDY(N), JP: Vertical projection of element "KK" before deformation. DDY(N) and JP denote the mantissa and the exponent, respectively.
Coordinate systems are shown in Figure 2.11.
Line 5 must be supplied "NE" times.

Line 6 (I5,3(F8.0,I2))
KK: Element material identifier.
E(N),KP: Young's modulus. E(N) and KP denote the mantissa and the exponent, respectively.
BA(N),IP: Cross-sectional area. BA(N) and IP denote the mantissa and the exponent, respectively.
BI(N),JP: Second moment of inertia. BI(N) and JP denote the mantissa and the exponent, respectively.
Line 6 has to be supplied "NMAT" times.

Line 7 (4I5)
KK: Load sequence number.
NCL(I): Number of concentrated loads applied in the KKth load sequence.
NDLI(I): Number of the types of distributed loads applied in the KKth load sequence. As explained in Line 9, only the uniformly distributed load acting transverse to the beam-element axis can be considered in this program (Figure 2.13). Therefore, the types of distributed loads are classified just by the magnitude of their local y-component.
NDIN(I): Number of load increments in the KKth load sequence.
NDIN(I) has to be determined such that the convergent solution can be obtained in the nonlinear analysis.
When small displacement analysis is conducted on linear elastic frames, "NDIN(I)" = 1 is enough.

Line 8 (I5,3(F8.0,I2))
NNC(I): Node number where concentrated load is applied at the KKth load sequence.
FX(I,II), IP: X-component of concentrated load at the end of the KKth load sequence. FX(I,II) and IP denote the mantissa and the exponent, respectively.
FY(I,II), JP: Y-component of concentrated load at the end of the KKth load sequence. FY(I,II) and JP denote the mantissa and the exponent, respectively.
FM(I,II), KP: Concentrated moment at the end of the KKth load sequence. FM(I,II) and KP denote the mantissa and the exponent, respectively.
These force components follow the sign convention defined in Figure 2.12. This data must be supplied "NCL(I)" times. Without any concentrated load, this data should be omitted.

Line 9 (F8.0,I2,I5/10I5)
FDI(I,II), IP: Local y-component of the uniformly distributed load at the end of KKth load

sequence (Figure 2.13). FDI(I,II) and IP denote the mantissa and the exponent, respectively.

NEDN: The number of elements where the above uniformly distributed load is applied.

NEDI(J): Element number where the uniformly distributed load is applied. This data have to be supplied "NEDN" times, following the format specified as 10I5.

It should be noted that only the uniformly distributed load in the direction of local y-axis can be considered in this program. In case of variably distributed load or the load with local x-component, a member has to be divided into smaller elements in order to approximate the load distribution by piece-wise uniformly distributed loads or by concentrated loads applied at small intervals.

Line 9 has to be supplied "NDLI(I)" times. Without any distributed load, this data should be omitted.

Lines 7, 8, and 9 have to be supplied "NLSQ" times.

Line 10 (I5)
NBCSRT: Number of the types of restrained nodes. These types are explained in Line 11.

Line 11 (2I5/10I5)
NBTYPE: Identifier for restraining condition.
    0: Fixed.
    1: Hinged.
    2: Hinged and free to translate in the direction of X axis.
    3: Hinged and free to translate in the direction of Y axis.
NBCI: The number of nodes which have the restraining condition defined by "NBTYPE".
NNB(I): Node number where the above restraining condition is applied. This data has to be supplied "NBCI" times, following the format specified as 10I5.

Line 11 has to be supplied "NBCSRT" times.

Line 12 (2F10.0)
ERITE: Tolerance for structural iteration. "ERITE" = 10^{-3} will generally give a satisfactory result.

ERFL: Tolerance for element iteration. "ERFL" = 10^{-4} will generally give a satisfactory result. These data can be left blank in the case of the first-order elastic analysis.

The following convergence criterion is used in this program.

Structural iteration

$$\sum_{k=1}^{ndf} \left| \Delta d_k^i \right| \bigg/ \sum_{k=1}^{ndf} \left| d_k^i \right| \geq ERITE, \quad \Delta d_k^i = d_k^i - d_k^{i-1}$$

Element iteration

$$\sum_{k=1}^{6} \left| \Delta f_k^i \right| \bigg/ \sum_{k=1}^{6} \left| f_k^i \right| \geq ERFL, \quad \Delta f_k^i = f_k^i - f_k^{i-1}$$

where d_k^i and f_k^i are the kth component of displacement vector and sectional force vector, respectively, obtained at the ith cycle of iteration.

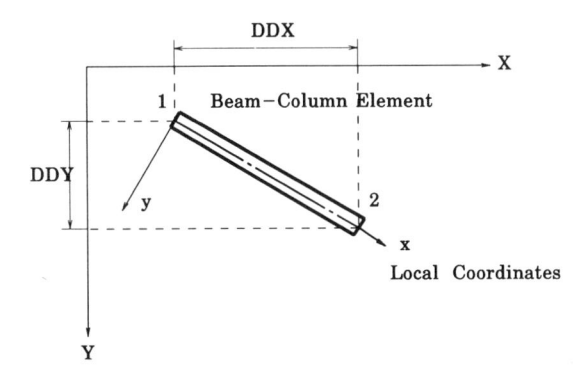

FIGURE 2.11. Coordinate systems.

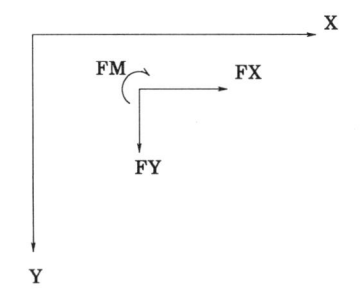

FIGURE 2.12. Sign conventions for concentrated loads.

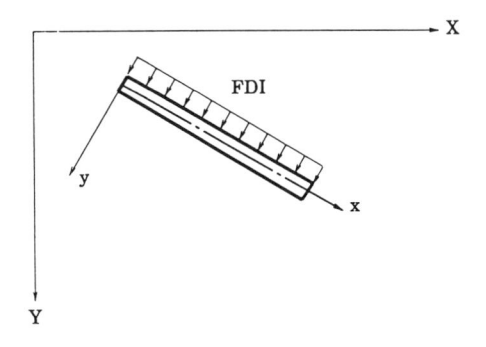

FIGURE 2.13. Sign convention for uniformly distributed loads.

(2) toframe
This data set is necessary to define the connection characteristics.

Line 1 (A9, I3)
Input the title "DATA NO." in format A9 and then input the data number in format I3. The data number must start from 1.

Line 2 (A8, I3)
Input the title "MODEL =" in format A8 and then input the connection model number in format I3. The numbers are allocated as follows to the connection models shown in Table 2.1.
   1: Modified Exponential Model
   2: Polynominal Model
   3: Three-Parameter Model

Input format after Line 2 is different according to the connection models. Notations used herein are exactly the same as those shown in Table 2.1. In case the connection data are made by the data-bank program explained in Chapter 3, these data are given in the U.S. system of units.

(Modified Exponential Model)
Line 4 (A8, I3)
This line specifies the number of linear terms adopted in this model. Input title "NLINEAR= " in format A8 and n in format I3 ($n \leq 3$).

Line 5 (6x, A5, E17.8, 2x, A4, E17.8, 2x, A5, I3)
This line specifies the values of constants α, M_0, m. First, input "AL=" in format A5 and the value of α in format E17.8. Second, input "BMO=" in format A4 and the value of M_0 in format E17.8. Third, input "NEXP=" in format A5 and the value of m in format E17.8 ($m \leq 6$).

Line 6 (6X, A5, 3E17.8)
This line specifies the values of constants A_i ($1 \leq i \leq 3$). First input "AI=" in format A5, followed by the values of A_1, A_2, A_3 in format 3E17.8.

Line 7 (11X, 3E17.8)
This line specifies the values of constant A_i ($4 \leq i \leq 6$). Input the values of A_4, A_5, A_6 in format 3E17.8.

Line 8 (6X, A5, 3E17.8)
This line specifies the values of T_j ($1 \leq i \leq n$). Input "TJ=" in format A5 and the values of T_1, T_2, T_3 in format 3E17.8.

Line 9 (6x, A5, 3E17.8)
This line specifies the values of R_j ($1 \leq i \leq n$). Input "RJ=" in format A5 and the values of R_1, R_2, R_3 in format 3E17.8.

Lines 8 and 9 must be omitted when $n = 0$.

(Polynominal Model)
Line 4 (3(3X, A4, F10.6,) 3X, A4, F10.6)
Input "C1=" in format A4 and the value of C_1 in format F10.6. Similarly, input "C2=", the value of C_2, "C3=" and the value of C_3. Then, input "K=" in format A4 and the value of K in format F10.6.

Line 5 (3(3X, A4, I10))
Input "P1=" in format A4 and the value of p_1 in format I10. Similarly, input "P2=", the value of p_2, "P3=" and the value of p_3.

Line 6 (3(3X, A4, I10))
Input "N1=" in format A4 and the value of n_1 in format I10. Similarly, input "N2=", the value of n_2, "N3=" and the value of n_3.

(Three-Parameter Model)
Line 4 (7X, A4, F6.3, 2(2X, A5, E17.8))
First, input "RN=" in format A4 and the value of n in format F6.3. Second, input "RKI=" in format A5 and the value of R_{ki} in format E17.8. Third, input "RMU=" in format A5 and the value of M_u in format E17.8.

EXAMPLE 3

The two-story one-bay frame illustrated in Figure 2.14 is selected as an example to demonstrate how to use this program. This frame is a semi-rigid frame with fixed bases. All the beams are connected to columns with same type of semi-rigid connections. For these connections, the modified exponential model is used to represent the moment-rotation relationship shown in Figure 2.15. All the loads are applied at the same time with one load increment.

Two sets of input data (**input.d** and **toframe**) for this example are given in Figures 2.16 and 2.17, respectively. Output data, **output.d**, is given in Figure 2.18. As computed results, Figure 2.19 shows a displacement diagram, a moment diagram, and a shear-force diagram. The aforementioned two sets of input data and the output data are provided in the attached diskette named as **inpex.d**, **tofrex**, and **outex.d**, respectively.

Acknowledgment

The author is grateful for Dr. N. Kishi's cooperation in preparing the revised version of FLFRM and this user's manual.

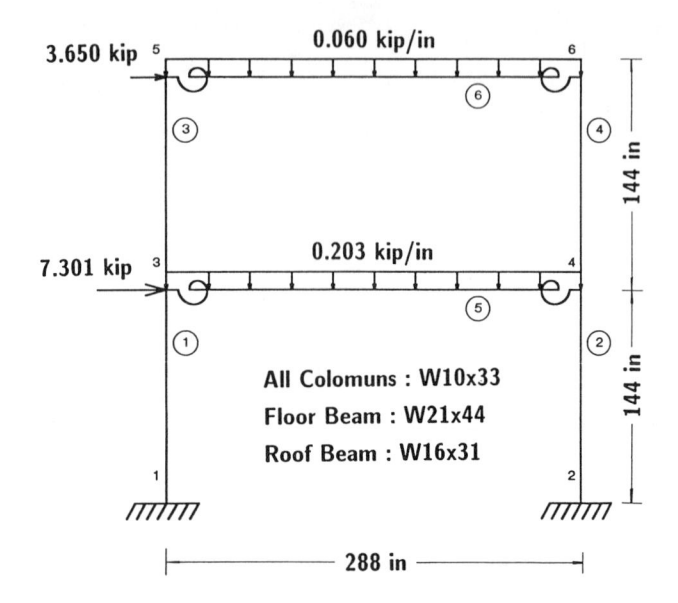

FIGURE 2.14. Two-story one-bay frame.

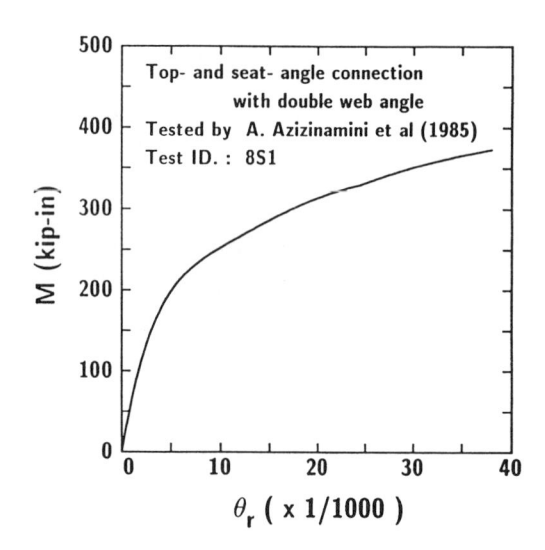

FIGURE 2.15. Moment-rotation relationship of connections.

```
         0    3    2    0
         6    6    3    1
         1    2   23          0. 0
         1    1    3    1    1    0    0    0    0
         2    2    4    1    1    0    0    0    0
         3    3    5    1    1    0    0    0    0
         4    4    6    1    1    0    0    0    0
         5    3    4    2    2    1    1    1    1
         6    5    6    3    2    1    1    1    1
         1    0. 0000 0  -1. 4400 2
         2    0. 0000 0  -1. 4400 2
         3    0. 0000 0  -1. 4400 2
         4    0. 0000 0  -1. 4400 2
         5    2. 8800 2    0. 0000 0
         6    2. 8800 2    0. 0000 0
         1    2. 900  4    9. 7100 0    1. 700  2
         2    2. 900  4    1. 3000 1    8. 430  2
         3    2. 900  4    9. 1200 0    3. 750  2
         1    2    2    1
         3  7. 3008   0    . 00000 0    . 00000 0
         5  3. 6504   0    . 00000 0    . 00000 0
      0. 2032  0    1
         5
      0. 0600  0    1
         6
         1
         0    2
         1    2
           0. 001      0. 0001
```

FIGURE 2.16. input.d.


```
DATA NO.   1
MODEL  =  1
NLINER =  1
       C =   0. 39499417E+00   BMO=    0. 00000000E+00   NEXP=   6
      AI = -0. 30968749E+02    0. 62361139E+03  -0. 50361980E+04
             0. 16127627E+05  -0. 20637821E+05   0. 93013973E+04
     RJ0 =                24. 7000
     RKJ =    0. 20517930E+01
```

FIGURE 2.17. toframe.

86 ADVANCED ANALYSIS OF STEEL FRAMES

```
*********************************************
****    SECOND ORDER FRAME ANALYSIS      ****
****      WITH/WITHOUT BOWING EFFECT     ****
****                                     ****
*********************************************

US UNIT IS USED IN THIS CALCULATION
LENGTH.......INCH
WEIGHT.......KIPS

FLAG FOR METHOD OF ANALYSIS..........  3
FLAG FOR CONNECTION MODEL............  2
FLAG FOR ITERATIVE PROCEDURE (ITLM)..  0

NO. OF ELEMENT (NE).................  6
NO. OF NODE (ND)....................  6
NO. OF ELEMENT MATERIAL (NMAT)......  3
NO. OF PREPARED CONNEC. DATA (NMATC)  1
NO. OF LOAD SEQUENCE (NLSQ).........  1
NO. OF LOADING POINTS (NL)..........  2

*******************CONNECTION PROPERTY********************
DATA NO. 1 FOR NONLINEAR FRAME ANALYSIS
MODIFIED EXPONENTIAL MODEL PARAMETERS
    C =  0.39499417E+00               BMO=   0.00000000E+00
  NEXP=  6    NLINER=  1
    AI = -0.30968749E+02   0.62361139E+03  -0.50361980E+04   0.16127627E+05  -0.20637821E+05   0.93013973E+04
   RJ0 =   24.7000
   RKJ =  0.20517930E+01

********************ELEMENT PROPERTY*********************

EL. NO.  ( KA, KB)       DX  ,    DY           L         E              A             BI        EL. TP  CON. TYPE (KA, KB)  INDXLN
   1     ( 1,  3)       0.00,  -144.00)     144.00  ( 0.29000E+05,  0.97100E+01,  0.17000E+03)    1       ( 0,  0)            0 0
   2     ( 2,  4)       0.00,  -144.00)     144.00  ( 0.29000E+05,  0.97100E+01,  0.17000E+03)    1       ( 0,  0)            0 0
   3     ( 3,  5)       0.00,  -144.00)     144.00  ( 0.29000E+05,  0.97100E+01,  0.17000E+03)    1       ( 0,  0)            0 0
   4     ( 4,  6)       0.00,  -144.00)     144.00  ( 0.29000E+05,  0.97100E+01,  0.17000E+03)    1       ( 0,  0)            0 0
   5     ( 3,  4)     288.00,     0.00)     288.00  ( 0.29000E+05,  0.13000E+02,  0.84300E+03)    2       ( 1,  1)            1 1
   6     ( 5,  6)     288.00,     0.00)     288.00  ( 0.29000E+05,  0.91200E+01,  0.37500E+03)    2       ( 1,  1)            1 1
```

FIGURE 2.18. output.d.

```
*****************LOADING CONDITION*********************
**** LOAD SEQUENCE =  1 ,  NO. OF LOAD STEPS =   1 *****
*** CONCENTRATED AT THE END OF LOAD SEQUENCE 1 ***
NODE NO.      FX          FY          BM
    3       7.3008      0.0000      0.0000
    5       3.6504      0.0000      0.0000
*** DISTRIBUTED LOAD AT THE END OF LOAD SEQUENCE 1 ***
DP : VERTICAL LOAD DEFINED IN TERMS OF LO. COORD
        DP        NEDI
    0.20320        1
ELE. NO.
    5
        DP        NEDI
    0.06000        1
ELE. NO.
    6

****************BOUNDARY CONDITION*********************
NBTY : NODE NUMBERS
  0  :    1     2
*******************************************************
***** ALLOWABLE ERROR IN CONVERGENCE *****
ERR. IN STRUCTURAL ITERATION (ERITE) :   0.10000E-02
ERR. IN ELEMENT ITERATION (ERFL)     :   0.10000E-03
*******************************************************

*******************************************************
NO. OF ITERATION (ITE)......     8
ERROR IN ITERATION (ERR) ....    0.364829E-03
               (SMRD) .......    0.237905E-02
               (SMD) ........    0.647391E+01
*******************************************************
LOAD SEQUENCE (NSL)..........     1
PRESENT TOTAL LOAD RATIO (CNINC)..   1.00000
NO. OF STRCTURAL ITERATION (ITE)..      8
NO. OF TOTAL STEPS (NTST).........      1
*******************************************************
```

FIGURE 2.18. *(continued)*

88 ADVANCED ANALYSIS OF STEEL FRAMES

```
*** NODAL DISPLACEMENTS ***    ( ) : VALUE FOR ONE CYCLE BEFORE COVERGENCE )
NOD NO       DX                      DY                    ROTATION OF CONNECTION
  1    0.000000E+00 ( 0.000000E+00)  0.000000E+00 ( 0.000000E+00)  0.000000E+00 ( 0.000000E+00)
  2    0.000000E+00 ( 0.000000E+00)  0.000000E+00 ( 0.000000E+00)  0.000000E+00 ( 0.000000E+00)
  3    0.944687E+00 ( 0.944438E+00)  0.217084E-01 ( 0.217028E-01)  0.936764E-02 ( 0.936446E-02)
  4    0.938088E+00 ( 0.937841E+00)  0.243112E-01 ( 0.243071E-01)  0.873637E-02 ( 0.873295E-02)
  5    0.222641E+01 ( 0.222616E+01)  0.311595E-01 ( 0.311376E-01)  0.774969E-02 ( 0.774479E-02)
  6    0.222181E+01 ( 0.222090E+01)  0.351601E-01 ( 0.351387E-01)  0.710918E-02 ( 0.710316E-02)

*** MEMBER FORCE ***

ELM NO  KA KB    AXIALA        SHEARA        MOMENTA       AXIALB         SHEARB        MOMENTB   ITER   ERROR
  1     1  3   0.35257E+02  -0.51091E+01  -0.70125E+03  -0.35257E+02   0.51091E+01  -0.67774E+02    2   0.19880E-05
  2     2  4   0.40537E+02  -0.58420E+01  -0.73447E+03  -0.40537E+02   0.58420E+01  -0.14480E+03    2   0.15542E-05
  3     3  5   0.72930E+01  -0.91080E+00  -0.15000E+03  -0.72930E+01   0.91080E+00  -0.12550E+03    2   0.74112E-05
  4     4  6   0.99642E+01  -0.27392E+01  -0.14811E+03  -0.99642E+01   0.27392E+01  -0.25913E+03    2   0.26450E-05
  5     3  4   0.31022E+01  -0.27957E+02   0.82747E+02  -0.31022E+01  -0.30566E+02   0.29286E+03    3   0.36682E-05
  6     5  6   0.27391E+01  -0.73049E+01   0.12545E+03  -0.27391E+01  -0.99751E+01   0.25904E+03    3   0.23439E-04

*** CONNECTION INFORMATION ***    (LOC=1 : LOADING, LOC=0 : UNLOADING PASS )
ELE NO KA  LOC RELATIVE R.    MOMENT  SECANT STIF    KB LOC RELATIVE R.     MOMENT      SECANT STIF
  5    3    1   0.13274E-02  0.82747E+02  0.62337E+05    4  1   0.16021E-01  0.29286E+03  0.18280E+05
  6    5    1   0.22701E-02  0.12545E+03  0.55261E+05    6  1   0.10863E-01  0.25904E+03  0.23846E+05

**** UNBALANCED FORCE ****    ( ) : PRESENT SURCHARGED NODAL FORCE (GRP) )
NOD NO       FX                       FY                          BM
  1    0.000000E+00 ( 0.000000E+00)  0.000000E+00 ( 0.000000E+00)  0.000000E+00 ( 0.000000E+00)
  2    0.000000E+00 ( 0.000000E+00)  0.000000E+00 ( 0.000000E+00)  0.000000E+00 ( 0.000000E+00)
  3    0.280233E-03 ( 0.128362E+02)  0.724769E-02 ( 0.262185E+02)  0.274510E-01 ( 0.402103E+03)
  4    0.550274E-03 (-0.553542E+01)  0.776829E-02 ( 0.240659E+02)  0.430558E-01 (-0.127762E+03)
  5    0.553720E-03 ( 0.513326E+01)  0.119023E-01 ( 0.201664E+02)  0.554760E-01 ( 0.189205E+03)
  6    0.188299E-03 (-0.148286E+01)  0.108670E-01 ( 0.195502E+02)  0.911275E-01 (-0.918895E+02)
*********************************************************************************************
```

FIGURE 2.18. *(continued)*

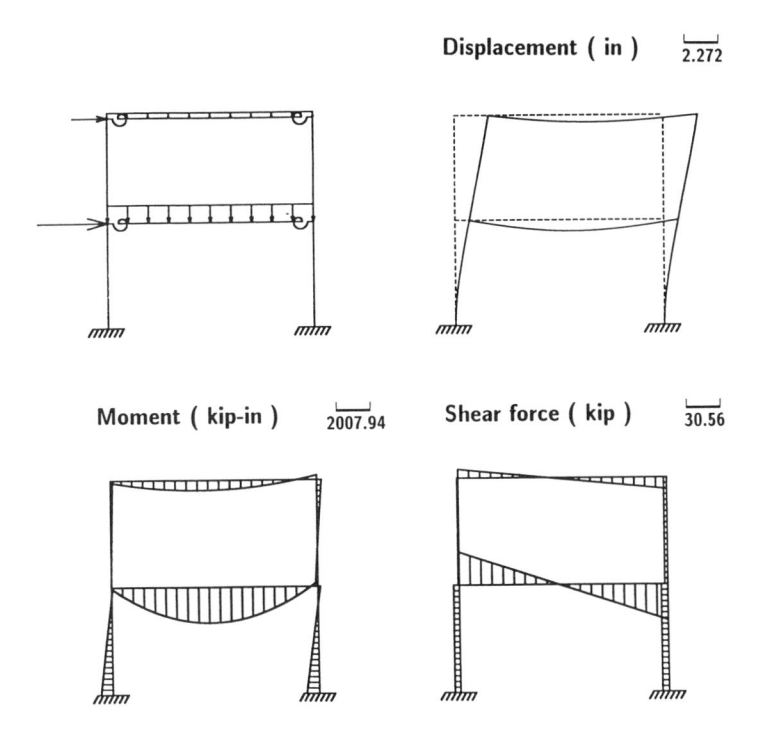

FIGURE 2.19. Computed results.

References

Byskov, E. (1989) Smooth post-buckling stress by a modified finite element method, *Int. J. Numerical Methods Eng.*, 28, 2877–2888.

Chen, W. F. and Saleeb, A. F. (1982) Uniaxial Behavior and Modeling in Plasticity, Structural Engineering Report No. CE-STR-82-35, School of Civil Engineering, Purdue University, West Lafayette, IN.

Connor, J. J., Logcher, R. D., and Chan, S. C. (1968) Nonlinear analysis of elastic framed structures, Proc. ASCE, *J. Struct. Div.*, 94(ST6), 1525–1547.

Dafalias, Y. E. and Popov, E. P. (1976) Plastic internal variables formalism of cyclic plasticity, *J. Appl. Mech.*, 43, 645–651.

Galambos, T. V., Ed. (1987) Finite element analysis of stability problems, in *Guide to Structural Design Criteria for Metal Structures*, 4th ed., Wiley-Interscience, New York, 673–686.

Goto, Y., Matsuura, S., Hasegawa, A., and Nishino, F. (1985) A new formulation of finite displacement theory of curved and twisted rods, *Struct. Eng./Earthquake Eng.*, 2(2), 151s–160s.

Goto, Y. and Chen, W. F. (1986) Documentation of Program FLFRM for Design Analysis of Flexibly Jointed Frames, Structural Engineering Report No. CE-STR-86-34, School of Civil Engineering, Purdue University, West Lafayette, IN.

Goto, Y. and Chen, W. F. (1987a) Second-order elastic analysis for frame design, *J. Struct. Eng.*, 113(7), 1501–1519.

Goto, Y. and Chen, W. F. (1987b) On the computer-based design analysis for the flexibly jointed frames, *J. Constr. Steel Res.*, 8, 203–231.

Goto, Y., Yamashita, T., and Matsuura, S. (1987c) Elliptic integral solutions for extensional elastica with constant initial curvature, *Struct. Eng. Earthquake Eng.*, 4(2), 299s–309s.

Goto, Y. and Chen, W.F. (1989) Closure to the discussions by Wan T. Tsai and by McGuire, M. and Ziemian, R. D., *J. Struct. Eng.*, 115(2), 500–506.

Goto, Y., Yoshimitsu, T., and Obata, M. (1990) Elliptic integral solutions of plane elastica with axial and shear deformations, *Int. J. Solids Struct.*, 26(4), 375–390.

Goto, Y., Suzuki, S., and Chen, W. F. (1991a) Analysis of critical behavior of semi-rigid frames with or without load history in connections, *Int. J. Solids Struct.*, 27(4), 467–483.

Goto, Y., Suzuki, S., and Chen, W. F. (1991b) Bowing effect on elastic stability of frames under primary bending moments, *J. Struct. Eng.*, 117(1), 111–127.

Hechtman, R. A. and Johnson, B. G. (1947) Riveted semi-rigid beam-to-column building connections, Progress Report, No. 1, AISC publication (Appendix B).

Hill, R. (1958) A general theory of uniqueness and stability in elastic-plastic solids, *J. Mech. Phys. Solids*, 6, 236–249.

Kishi, N. and Chen, W. F. (1986) Data Base of Steel Beam-to-Column Connections, Structural Engineering Report No. CE-STR-86-26, School of Civil Engineering, Purdue University, West Lafayette, IN, 653 pp.

Kishi, N., Chen, W. F., Goto, Y., and Matsuoka, K. (1991) Analysis Program for Design of Flexibly Jointed Frames, Structural Engineering Report No. CE-STR-91-26, School of Civil Engineering, Purdue University, West Lafayette, IN, 26 pp.

Kishi, N. and Chen, W. F. (1990) Moment-rotation of semi-rigid connections with angles, *J. Struct. Eng.*, 116(7), 1813–1834.

Korn, A. (1981) Effect of bowing on rectangular plane frames, *J. Struct. Div.*, 107(ST3), 569–574.

Lindsey, S. D., Ioannides, S. A., and Goverdhan, A. V. (1985) The effect of connection flexibility on steel members and frame stability, in *Connection Flexibility on Steel Members and Frame Stability*, Chen W. F., Ed., Proceeding of a session sponsored by ST division of ASCE, Detroit, MI, 6–12.

Mallet, R. H. and Marcal, P. V. (1968) Finite element analysis of nonlinear structures, Proc. ASCE, *J. Struct. Div.*, 94(ST9), 2081–2103.

Moncarz, P. D. and Gerstle, K. H. (1981) Steel frames with nonlinear connections, Proc. ASCE, *J. Struct. Div.*, 107(ST8), 1427–1441.

Nishino, F., Kasemset, C., and Lee, S. L. (1973) Variational formulation of stability problem for thin-walled members, *Ingen. Arch.*, 43, 58–68.

Nishino, F., Kurakata, Y., and Goto, Y. (1975) Finite displacement beam theory, *Proc. JSCE*, 237, 11–25.

Nishino, F. and Hasegawa, A. (1979) Thin-walled beam theory, *J. Fac. Eng. U. Tokyo*, B-35, 109–190.

Oran, C. (1973) Tangent stiffness in plane frames, Proc. ASCE, *J. Struct. Div.*, 99(ST6), 973–985.

Rathbun, J. C. (1936) Elastic properties of riveted connections, *Trans. ASCE*, 101.

Reissner, E. (1972) On one-dimensional finite-strain beam theory, *J. Appl. Math. Phys.*, 23, 795–804.

Riks, E. (1979) An incremental approach to the solutions of snapping and buckling problem, *Int. J. Solids Struct.*, 15(7), 529–551.

Saafan, S. A. (1963) Nonlinear behavior of structural plane frames, Proc. ASCE, *J. Struct. Div.*, 89(ST4), 557–579.

Timoshenko, S. P. and Gere, J. M. (1961) *Theory of Elastic Stability*, 2nd ed., McGraw-Hill, New York, 76–82.

Thompson, L. E., McKee, R. J., and Visintainer, D. A. (1970) An Investigation of Rotation Characteristics of Web Shear Framed Connections Using A-36 and A-441 Steels, Dept. of Civil Engineering, University of Missouri-Rolla.

Washizu, K. (1968) *Variational Method in Elasticity & Plasticity*, Pergammon Press, New York, 412 pp.

3: Semi-Rigid Connections

Norimitsu Kishi, *Department of Civil Engineering, Muroran Institute of Technology, Muroran, Japan*

3.1 Introduction

Conventional analysis and design of steel framework are performed using the assumption that the connections are either fully rigid or ideally pinned. The assumption of fully rigid connection implies that no relative rotation of connection occurs and the end moment of the beam is completely transferred to the column. On the other hand, the pinned connection implies that no restraint for rotation of connection exists and the connection moment is always zero. However, it is recognized that actual beam-to-column connections always provide some rigidity.

The primary distortion of a steel beam-to-column connection is its rotational deformation, θ_r, caused by the in-plane bending moment, M (Figure 3.1). The effect of this connection deformation has a destabilizing effect on frame stability since additional drift will occur as a result of the decrease in effective stiffness of the members to which the connections are attached. An increased frame drift will intensify the P-Δ effect and hence the overall stability of the frame will be affected.

Thus, the nonlinear characteristics of beam-to-column connection play a very important role for frame stability. New design method has shifted from Allowable Stress Design method (ASD) to limit state design method such as AISC Load and Resistance Factor Design (LRFD) specifications (AISC, 1986). The limit state design specifications refer to more sophisticated methods of analysis than allowable stress design specifications. For example, AISC-LRFD specifications require the limit state of structural stability taking into account the second-order P-Δ, P-δ effects, and give guidelines for the use of a first-order elastic analysis with B_1/B_2 amplification factors for considering these second-order effects. While two types of construction, fully restrained (FR) construction and partially restrained (PR) construction, are defined in the specifications, these guidelines for the design of PR construction are not given, since it is very difficult to evaluate the actual restraint of semi-rigid connections used in engineering practice.

To establish the guidelines for design of semi-rigid frames, it is necessary to know the $M - \theta_r$ behavior of actual beam-to-column connections and to formulate the appropriate $M - \theta_r$ model for use in the design and analysis of semi-rigid frames. In recent years, several researchers have published papers discussing the influences of connection rigidity on steel frame structures for all connection types. For example, Goverdhan (1983) collected much of the available test data on moment-rotation

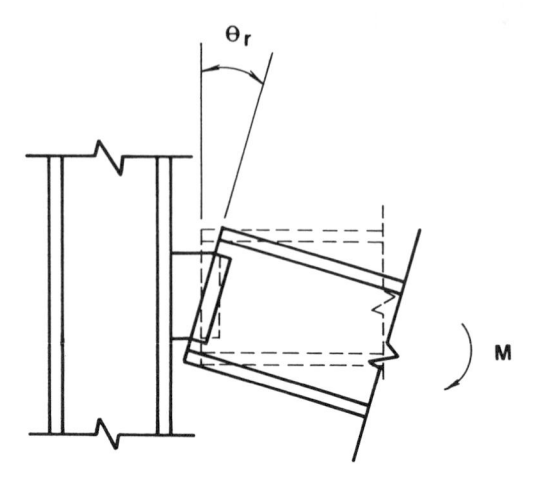

FIGURE 3.1. Rotational deformation of a connection.

characteristics and tried to formulate a prediction equation for each type of connections. Nethercot (1985) conducted an extensive literature survey for steel beam-to-column connection test data and their corresponding $M - \theta_r$ curve representations for the period 1915 to 1985. Kishi and Chen (1986b) and Chen and Kishi (1989) also collected experimental test data published from 1936 to 1985 and compiled a data base.

The aim of this chapter is to provide moment-rotation characteristics and corresponding parameters of semi-rigid beam-to-column connections used frequently in steel connection. Based on experimental test data collected by Kishi and Chen, a data bank on steel beam-to-column connections was developed at Purdue University Computer Center. To control this data base systematically, the Steel Connection Data Bank program (SCDB) has been developed (Kishi and Chen, 1986a). The SCDB program provides seven main functions including routines for tabulating the experimental test data and some proposed prediction equations. Three prediction equations, among many others, are chosen and installed in the program.

3.2 Types of Semi-Rigid Connections

3.2.1 Single Web-Angle Connections/Single Plate Connections

Single web-angle connections consist of an angle either bolted or welded to both the column and the beam web, as shown in Figure 3.2a. On the other hand, single plate connections use the plate instead of the angle. This connection type requires less materials than a single web-angle connection (Figure 3.2b). Generally, in designing these connections the single web-angle connections have moment rigidity equal to about one-half of the double web-angle connections, and the single plate connections have rigidity equal to or greater than the single web-angle connections since one side of the plate is fully welded with the column flange.

3.2.2 Double Web-Angle Connections

Double web-angle connections consist of two angles either bolted or riveted to both the column and the beam web as shown in Figure 3.2c. Rivets were used as fasteners in the earliest tests on double web-angle connections conducted by Rathbun (1936). In the 1950s, most specifications for the design of steel structures allowed the use of high-strength bolts in lieu of rivets. To clarify the effect of high-strength bolts on connection behavior when used in conjunction with rivets, Bell et al. (1958) and Lewitt et al. (1966) conducted experiments on riveted and bolted beam-to-column connections. Today, high strength bolts are used popularly as fasteners for this type of connection. Though the connection rigidity of this type of connections is stiffer than those of single web-angle and single plate connections, the AISC-ASD Specifications (1989) consider this connection type as a Type 2 construction (simple connection or shear connection).

3.2.3 Top- and Seat-Angle Connections with Double Web Angle

This type of connection is a combination of top- and seat-angle connections and double web-angle connections. A typical top- and seat-angle connection with double web-angle is shown in Figure 3.2d. Double web angles are used to improve the connection restraint characteristics of top- and seat-angle connections, and for shear transfer. This type of connections is considered as Type 3 framing of the AISC-ASD Specifications (1989), that is to say, a semi-rigid connection.

3.2.4 Top- and Seat-Angle Connections

A typical top- and seat-angle connection is shown in Figure 3.2e. The AISC-ASD Specifications (1989) described this type of connection as follows: (1) the top angle is used to provide lateral support of the compression flange of the beam and (2) the seat angle is to transfer only the vertical reaction of beam to column and should not give significant restraining moment on the end of the beam. However, according to the experimental results, these connections will be able to transfer not only the vertical reaction but also some end moment of the beam to the column.

3.2.5 Extended End-Plate Connections/ Flush End-Plate Connections

In general, end-plate connections are welded to the beam end along both the flanges and web in the fabricator's shop and bolted to the column in the field. The end-plate connection has been used extensively since the 1960s. The extended end-plate connections are classified into two types as end-plate either extended on the tension side only or on both the tension and compression sides as shown in Figures 3.2f and 3.2g. A typical flush end-plate connection is shown in Figure 3.2h. Since some end-plate connections are considered as Type FR construction rather than Type PR

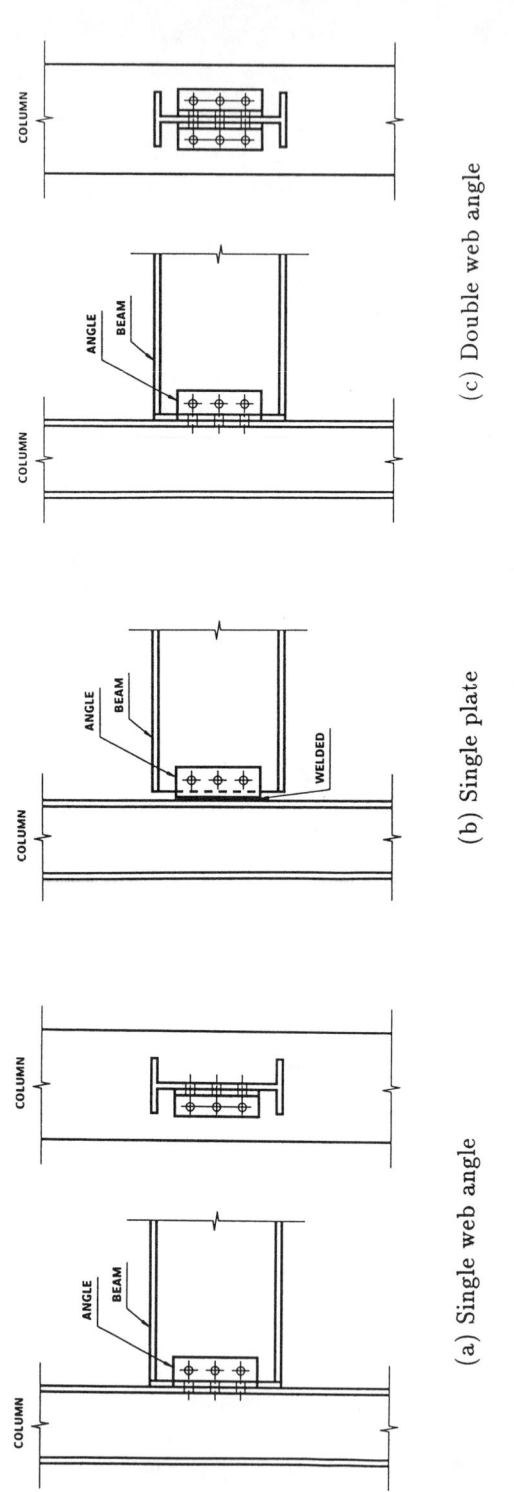

FIGURE 3.2. Typical types of beam-column connections.

(a) Single web angle

(b) Single plate

(c) Double web angle

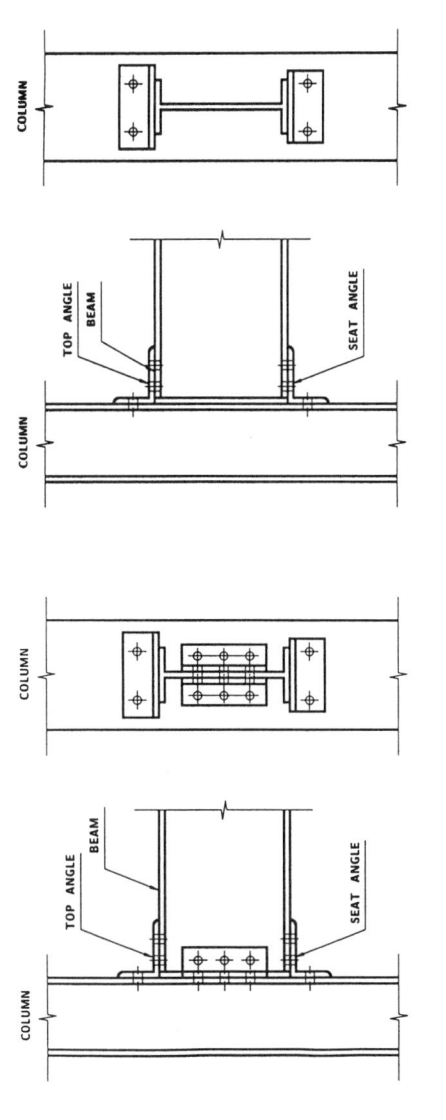

(d) Top and seat angles with double web angle

(e) Top and seat angles

FIGURE 3.2. (*continued*)

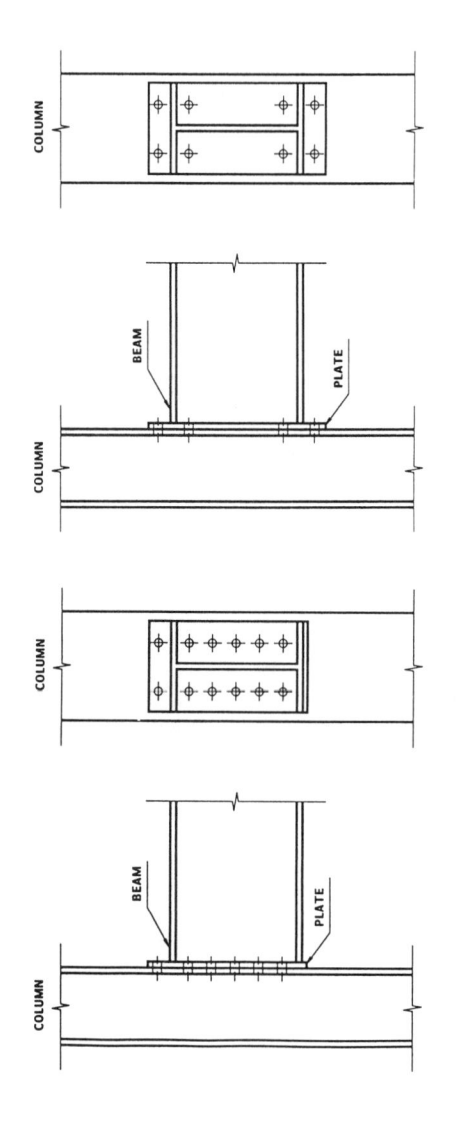

(f) Extended end plate on the tension side only (g) Extended end plate on both sides

FIGURE 3.2. (*continued*)

(h) Flush end plate (i) Header plate

FIGURE 3.2. (*continued*)

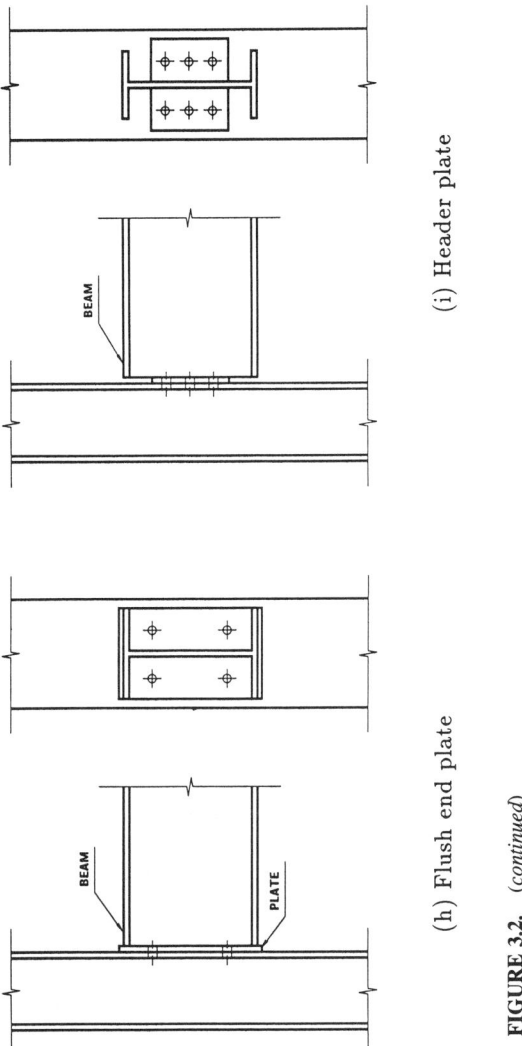

construction in AISC-LRFD Specifications (1986), they have often been used as means of transferring beam end moment to the column. The extended end-plate connection on both sides is preferred when the connection is subjected to moment reversal such as during severe earthquake loading. While the flush end-plate connection is weaker than the extended end-plate connection, this connection type is often used in roof details. The behavior of end-plate connection depends on whether the column flange near the connection is stiffened or not. The stiffeners of the column flanges act to prevent flexural deformation of the column flange, thereby influencing the behavior of the plate and fasteners.

3.2.6 Header-Plate Connections

A header-plate connection consists of an end plate, whose length is less than the depth of the beam, welded to the beam web, and bolted to the column as shown in Figure 3.2i. The moment-rotation characteristics of these connections are similar to those of double web-angle connections. Accordingly, a header-plate connection is used mainly to transfer the reaction of the beam to the column and classified in Type 2 framing of the AISC-ASD Specifications (1989).

3.3 Modeling of Connections

3.3.1 General Remarks

A number of analytical studies of connection behavior using the finite element method have been reported so far (Krishnamurthy et al., 1979; Patel and Chen, 1984; Driscoll, 1987; Kukreti et al., 1987). However, these approaches are unacceptable for practical use because cumbersome calculations are required to consider the material and geometrical nonlinearities. At the present, the most commonly used approach to describe the $M - \theta_r$ curve is to curve-fit the experimental data with simple expressions. Several analytical models have been developed to represent connection flexibility using available experimental test data.

Early models used the initial connection stiffness as the key parameter in a linear $M - \theta_r$ model (Rathbun, 1936; Moforton and Wu, 1963; Lightfoot and LeMessurier, 1974). Although the linear model is very easy to use, it has a serious disadvantage. It is suitable for only a small range of the initial relative rotation. A closer approximation of true connection behavior can be obtained by using either a bilinear model (Tarpy and Cardinal, 1981; Lui and Chen, 1983) and a piecewise linear model. The piecewise linear model is composed of a series of straight line segments. In these models, the abrupt changes in connection stiffness at the transition points make their practical use difficult. Jones et al. (1980, 1981) proposed a cubic-B-spline model to obtain a more suitable function. However, this model requires a large number of sampling data during the formulation process. Frye and Morris (1975) have reported a polynomial model to evaluate the behavior of several types of connections. In this model, the $M - \theta_r$ behavior is represented by an odd power polynomial.

Lui and Chen (1986) used an exponential function to curve-fit the experimental $M - \theta_r$ data. This model is a good representation of the monotonic nonlinear connection behavior. However, if there are some sharp changes in slope in the $M - \theta_r$ curve, this model cannot adequately represent it (Wu, 1989). Kishi and Chen (1986) refined the Lui-Chen exponential model to accommodate any sharp changes in slope in the $M - \theta_r$ (modified exponential model). The Kishi-Chen model can be really used in lieu of the $M - \theta_r$ of experimental data. Other exponential models (Yee-Melchers model [1986] and Wu-Chen model [1990]) have been reported. The former is a four-parameter model using the initial connection stiffness, the strain-hardening connection stiffness, plastic moment capacity, and constant. The later is the three-parameter model which is composed of initial connection stiffness, ultimate moment capacity, and shape parameter.

Models using power function, namely, power models, have also been reported. Colson and Louveau (1983) and Kishi and Chen (1990) proposed similar models independently using three parameters: initial connection stiffness, ultimate moment capacity, and shape parameter. Since these models use only three parameters, they are not as accurate as the cubic B-spline and/or the modified exponential models. And, they cannot estimate the $M - \theta_r$ curve which does not flatten out near the final loading. However, the number of data required for this model is drastically reduced. Ang and Morris (1984) used a standardized Ramberg-Osgood function (Ramberg and Osgood, 1943) in the power form. The Ang-Morris model is composed of four parameters. Richard et al. (1980) analytically described the $M - \theta_r$ curve of single plate connections with a nondimensional power model. The procedure used to establish this model was based upon a nonlinear finite element method in which nonlinear behavior of the bolts and connected plates was modeled by using a force-deformation relationship obtained from single shear bolt tests.

Some of the importance of connection models will be discussed in more detail in the following section.

3.3.2 Frye-Morris Polynomial Model

The most popular model for structural analysis is the polynomial function proposed by Frye and Morris (1975). The Frye-Morris model was developed based on a procedure formulated by Sommer (1969). They used the method of least square to determine the constants of the polynomial. This model has the general form shown in Eq. (3.1).

$$\theta_r = C_1(KM)^1 + C_2(KM)^3 + C_3(KM)^5 \tag{3.1}$$

where K is a standardization parameter dependent upon the connection type and geometry, and C_1, C_2, and C_3 are curve-fitting constants.

This model represents the $M - \theta_r$ behavior reasonably well. The main drawback is that the nature of a polynomial is to peak and trough within a certain range. Then, the first derivative of this function, which indicated tangent and connection stiffness, may become negative at some value of connection moment M. This is physically

Table 3.1 Curve-Fitting Constants and Standardization Constants for the Frye-Morris Polynomial Model[a]

Connection types	Curve-fitting constants	Standardization constants
Single web-angle connection	$C_1 = 4.28 \times 10^{-3}$ $C_2 = 1.45 \times 10^{-9}$ $C_3 = 1.51 \times 10^{-16}$	$K = d_a^{-2.4} t_a^{-1.81} g^{0.15}$
Double web-angle connection	$C_1 = 3.66 \times 10^{-4}$ $C_2 = 1.15 \times 10^{-6}$ $C_3 = 4.57 \times 10^{-8}$	$K = d_a^{-2.4} t_a^{-1.81} g^{0.15}$
Top- and seat-angle with double web-angle connection	$C_1 = 2.23 \times 10^{-5}$ $C_2 = 1.85 \times 10^{-8}$ $C_3 = 3.19 \times 10^{-12}$	$K = d^{-1.287} t^{-1.128} t_c^{-0.415} l_a^{-0.694} g^{1.350}$
Top- and seat-angle without double web-angle connection	$C_1 = 8.46 \times 10^{-4}$ $C_2 = 1.01 \times 10^{-4}$ $C_3 = 1.24 \times 10^{-8}$	$K = d^{-1.5} t^{-0.5} l_a^{-0.7} d_b^{-1.1}$
End-plate connection without column stiffeners	$C_1 = 1.83 \times 10^{-3}$ $C_2 = -1.04 \times 10^{-4}$ $C_3 = 6.38 \times 10^{-6}$	$K = d_g^{-2.4} t_p^{-0.4} t_f^{-1.5}$
End-plate connection with column stiffeners	$C_1 = 1.79 \times 10^{-3}$ $C_2 = -1.76 \times 10^{-4}$ $C_3 = 2.04 \times 10^{-4}$	$K = d_g^{-2.4} t_p^{-0.6}$
T-stub connection	$C_1 = 2.1 \times 10^{-4}$ $C_2 = 6.2 \times 10^{-6}$ $C_3 = -7.6 \times 10^{-9}$	$K = d^{-1.5} t^{-0.5} l_t^{-0.7} d_b^{-1.1}$
Header-plate connection	$C_1 = 5.10 \times 10^{-5}$ $C_2 = 6.20 \times 10^{-10}$ $C_3 = 2.40 \times 10^{-13}$	$K = t_p^{-1.6} g^{1.6} d_p^{-2.3} t_w^{-0.5}$

[a] All size parameters are in inches.

unacceptable. This negative stiffness makes structural analysis difficult if the analysis scheme with tangent connection stiffness is used.

Following the procedure of Frye-Morris, Picard et al. (1976) and Altman et al. (1982) developed prediction equations to describe the $M - \theta_r$ behavior for strap angle connections and top- and seat-angle connections with double web angles, respectively. Goverdhan (1983) reestimated a size parameter in the standardization constant K for flush end-plate connections to get a good agreement with the $M - \theta_r$ curve obtained from experimental results. The curve-fitting constants C_1, C_2, and C_3 and the standardization constant K for each connection type are summarized in Table 3.1, while the size parameters for each type of connections are shown schematically in Figure 3.3.

3.3.3 Modified Exponential Model

The Chen-Lui exponential model has been refined to accommodate linear components by Kishi and Chen (1986) and is referred to herein as the modified exponential model.

This model is represented by a function of the following form

$$M = M_0 + \sum_{m}^{j=1} C_j \left\{ 1 - \exp\left(-\frac{|\theta_r|}{2j\alpha}\right) \right\} + \sum_{n}^{k=1} D_k \left(|\theta_r| - |\theta_k|\right) H\left[|\theta_r| - |\theta_k|\right] \quad (3.2)$$

where M_0 is initial connection moment, α is a scaling factor for the purpose of numerical stability, C_j and D_k are curve-fitting parameters, θ_k is the starting rotation of the kth linear component given from the experimental $M - \theta_r$ curve, and $H[\theta]$ is Heaviside's step function (unity for $\theta \geq 0$, zero for $\theta < 0$).

Using the linear interpolation technique for the original $M - \theta_r$ data, the weight function for each $M - \theta_r$ datum is nearly equal. The constants C_j and D_k for the exponential and linear terms of the function are determined by a linear regression analysis.

The instantaneous connection stiffness R_k at an arbitrary relative rotation $|\theta_r|$ can be evaluated by differentiating Eq. (3.2) with respect to $|\theta_r|$.

When the connection is loaded, we have

$$R_k = R_{kt} = \left.\frac{dM}{d|\theta_r|}\right|_{|\theta_r| = |\theta_r|} = \sum_{m}^{j=1} \frac{C_j}{2j\alpha} \exp\left(-\frac{|\theta_r|}{2j\alpha}\right) + \sum_{n}^{k=1} D_k H\left[|\theta_r| - |\theta_k|\right] \quad (3.3)$$

When the connection is unloaded, we have

$$R_k = R_{ki} = \left.\frac{dM}{d|\theta_r|}\right|_{|\theta_r|=0} = \sum_{m}^{j=1} \frac{C_j}{2j\alpha} + D_k H\left[|\theta_k|\right]\bigg|_{k=1} \quad (3.4)$$

This model has the following merits:

(1) The formulation is relatively simple and straightforward.
(2) It can deal with connection loading and unloading for the full range of relative rotation in a second-order structural analysis with secant connection stiffness.
(3) The abrupt changing of the connection stiffness among the sampling data is only generated from inherent experimental characteristics.

The curve-fitting and tangent connection stiffness values from the experimental data examined are calculated with $m = 6$ in Eqs. (3.3) and (3.4). The comparison between the Chen-Lui exponential model (1985) and the modified exponential model for numerical example of the test data including a linear component is shown in Figure 3.4.

3.3.4 Three-Parameter Power Model

The modified exponential model mentioned above is a curve-fitting equation obtained by using the least-mean-square technique for the experimental test data. From

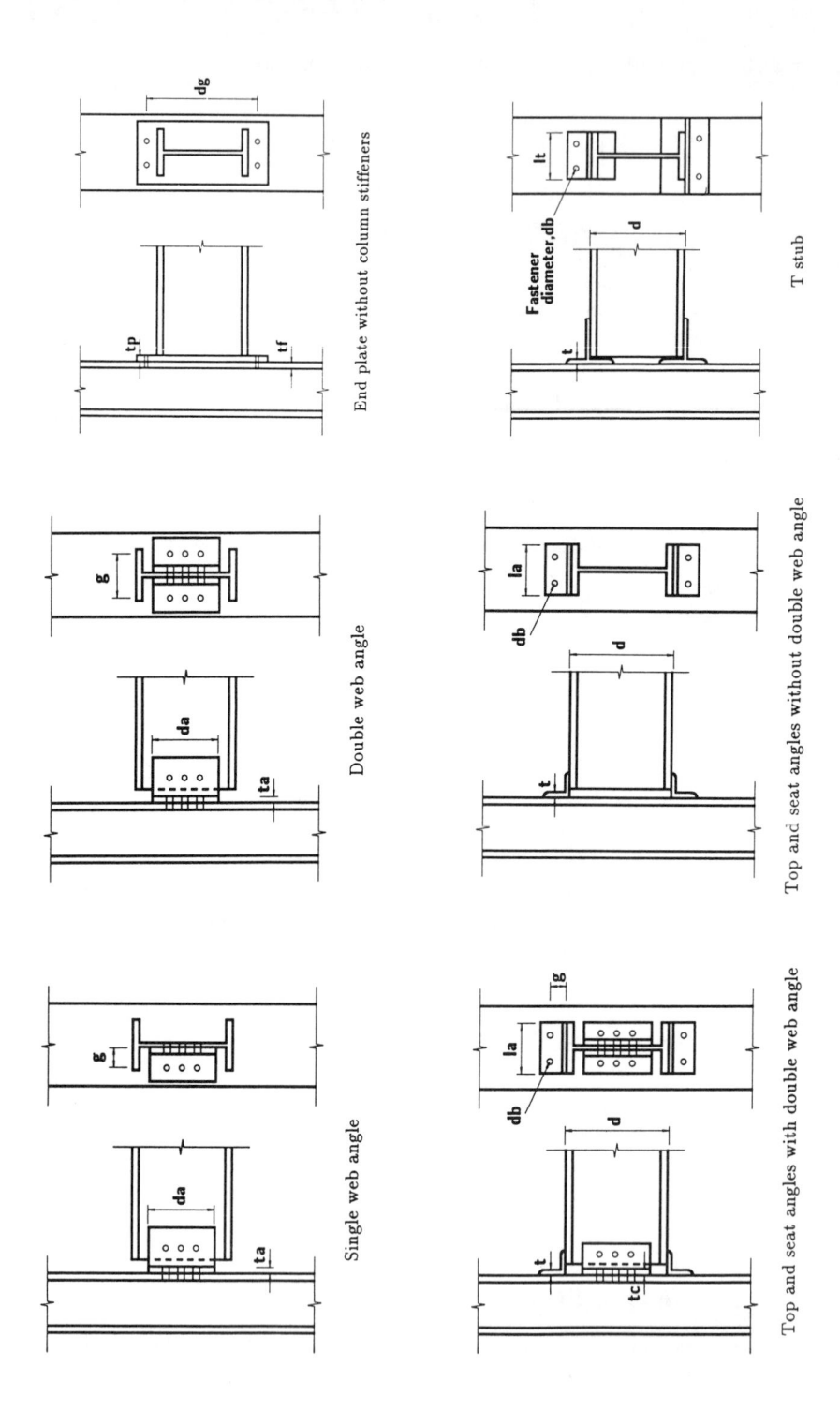

FIGURE 3.3. Size parameters for various connection types of Frye-Morris polynomial model.

End plate with column stiffeners

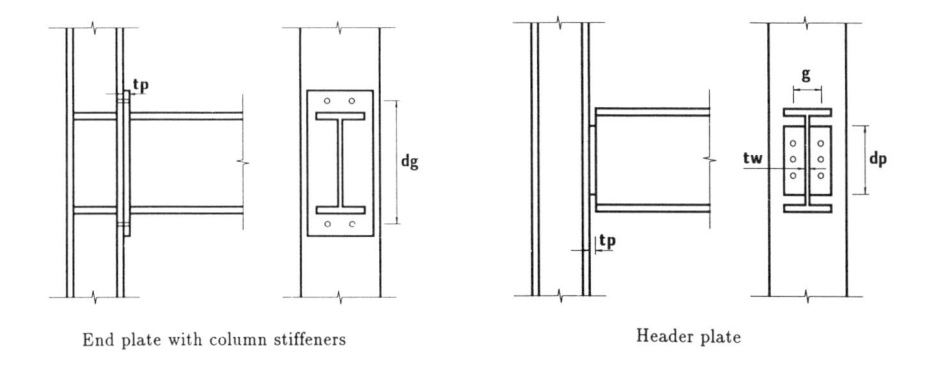

Header plate

FIGURE 3.3. (*continued*)

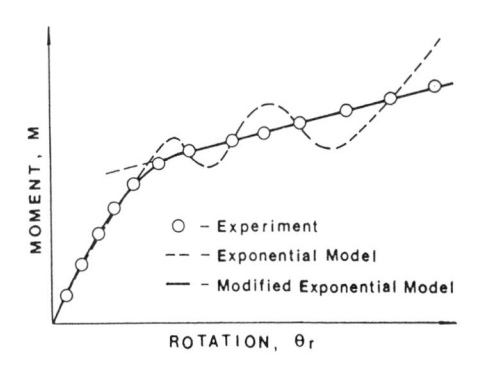

FIGURE 3.4. Comparison between results by exponential and modified exponential models for $M - \theta_r$ data including a linear term.

a different viewpoint, Chen and Kishi (1987) and Kishi et al. (1987a,b) developed another procedure to predict the moment-rotation characteristics of steel beam-to-column connections. In this procedure, the initial connection stiffness and ultimate moment capacity of the connection are determined by a simple analytical model. Using those values so obtained, a three-parameter power model given by Richard and Abbott (1975) was adopted to represent the moment-rotation relationship of the connection.

The generalized form of this model is

$$M = \frac{R_{ki}\theta_r}{\left\{1 + \left(\frac{\theta_r}{\theta_0}\right)^n\right\}^{1/n}} \tag{3.5}$$

where R_{ki} = initial connection stiffness; M_u = ultimate moment capacity; θ_0 = a reference plastic rotation = M_u/R_{ki}; and n = shape parameter. Eq (3.5) has the shape

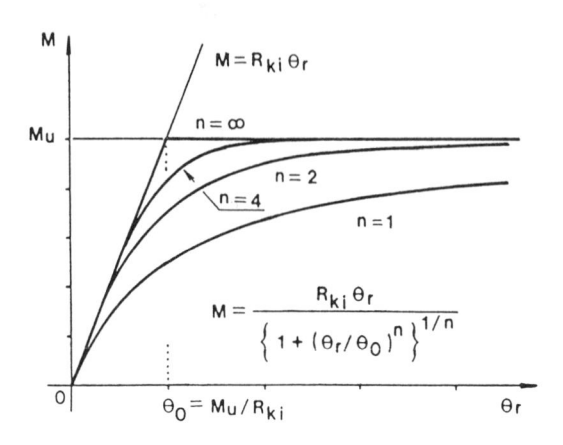

FIGURE 3.5. Three-parameter power model.

shown in Figure 3.5. From this figure, it is recognized that the larger the power index n, the steeper the curve. The shape parameter n can be determined by using the method of least squares for the differences between the predicted moments and the experimental test data.

This power model is an effective tool for designers to execute the second-order nonlinear structural analysis quickly and accurately. This is because the tangent connection stiffness R_k and relative rotation θ_r can be determined directly from Eq. (3.5) without iteration, in which the tangent connection stiffness R_k in Eq. (3.5) is

$$R_k = \frac{dM}{d\theta_r} = \frac{R_{ki}}{\left\{1 + \left(\frac{\theta_r}{\theta_0}\right)^n\right\}^{(n+1)/n}} \tag{3.6}$$

and the rotation θ_r is:

$$\theta_r = \frac{M}{R_{ki}\left\{1 - \left(\frac{M}{M_u}\right)^n\right\}^{1/n}} \tag{3.7}$$

3.4 Connection Data Base

The literature survey encompassed experimental data published from 1936 (Rathbun) to 1985, on riveted, bolted, and welded connections. All of 303 tests collected so far have been classified into seven types, as shown in Table 3.2. The references and number of experimental curves of each connection type are listed in Table 3.3. These test data are stored in the computer as the steel beam-to-column connection data base

SEMI-RIGID CONNECTIONS **105**

Table 3.2 Connection Types

Type number (1)	Connection type (2)
1	Single web-angle connections
2	Double web-angle connections
3	Top- and seat-angle connections with double web angle
4	Top- and seat-angle connections
5	Extended end-plate connections
6	Flush end-plate connections
7	Header-plate connections

at Purdue University Computer Center. The items from each test as registered in the data base are

1. Connection type and fastening mode.
2. Author, test I.D., and tested country.
3. The material properties and size of fasteners.
4. The material strength of the angles designed for connection elements.
5. All parameters used in beam-to-column connections.
6. Moment-rotation test data.

Numerical values registered in the data base are standardized in U.S. customary units by executing a subprogram as the preprocessor. The connection type is classified as shown in Table 3.2 and the fastening mode as shown in Table 3.4. The notations of major parameters for each connection type are identified in the next section.

3.5 Parameter Definition for Connection Type
3.5.1 Single Web-Angle Connections/Single Plate Connections

Collected experimental moment-rotation data for a single web-angle with bolted beam-to-column connections are tabulated in Table 3.5. Single plate and single web-angle with bolted to the beam and welded to the column connections are tabulated in Table 3.6. The notations for major parameters of these types of connections as shown in Figures 3.6 and 3.7 are identified as follows:

lp Angle height.
lu Distance from the top edge of the angle to the tension flange of the beam.
ll Distance from the bottom edge of the angle to the compression flange of the beam.
ta Angle thickness.
gb Distance from the angle's heel to the center of bolt holes in leg adjacent to the beam web face.
gc Distance from the angle's heel to the center of bolt holes in leg adjacent to the column face.

Table 3.3 Authors and Numbers of Experimental $M - \theta_r$ Curves for Each Connection Type

References for experimental curves (1)	Number of tests (2)
(a) Single web-angle connections, single plate connections	
S. L. Lipson (1968)	30
L. E. Thompson et al. (1970)	12
S. L. Lipson (1977)	8
R. M. Richard et al. (1982)	4
(b) Double web-angle connections	
J. C. Rathbun (1936)	7
W. C. Bell et al. (1958)	4
C. W. Lewitt et al. (1966)	6
W. H. Sommer (1969)	4
L. .E. Thompson et al. (1970)	48
B. Bose (1981)	1
(c) Top- and seat-angle connections with double web angle	
J. C. Rathbun (1936)	3
A. Azizinamini et al. (1985)	20
(d) Top- and seat-angle connections without double web angle	
J. C. Rathbun (1936)	3
R. A. Hechtman et al. (1947)	12
S. M. Maxwell et al. (1981)	12
M. J. Marley (1982)	26
(e) Extended end-plate connections	
L. G. Johnson et al. (1960)	1
A. N. Sherbourne (1961)	5
J. R. Bailey (1970)	26
J. O. Surtees et al. (1970)	6
J. A. Packer et al. (1977)	3
S. A. Ioannides (1978)	6
R. J. Dews (1979)	3
P. Grundy et al. (1980)	2
N. D. Johnstone et al. (1981)	8
(f) Flush end-plate connections	
J. R. Ostrander (1970)	24
(g) Header-plate connections	
W. H. Sommer (1969)	20

cu   Distance from the center of upper fastener holes to the top edge of the angle in the beam web.
cl   Distance from the center of lower fastener holes to the bottom edge of the angle in the beam web.
pb   Distance between two rows of fasteners in the beam web.
pc   Distance between two rows of fasteners in the column.
qb   Distance between two lines of fasteners in the beam web.

Table 3.4 Fastening Mode Patterns

Pattern number, N (1)	Fastening mode pattern (2)
1	All riveted
2	All bolted
3	Riveted-to-beam and bolted-to-column
4	Bolted-to-beam and riveted-to-column
5	Riveted-to-beam and welded-to-column
6	Bolted-to-beam and welded-to-column
7	Welded-to-beam and riveted-to-column
8	Welded-to-beam and bolted-to-column
9	All riveted without column stiffeners
10	All riveted with column stiffeners
11	All bolted without column stiffeners
12	All bolted with column stiffeners

qc Distance between two lines of fasteners in the column.
nb Total number of fasteners in the beam.
nc Total number of fasteners in the column.

3.5.2 Double Web-Angle Connections

Collected experimental moment-rotation data for double web-angle connections are tabulated in Table 3.7. The notations for major parameters of this connection type as shown in Figure 3.8 are identified as follows:

lp Angle height.
lu Distance from the top edge of the angle to the tension flange of the beam.
ll Distance from the bottom edge of the angle to the compression flange of the beam.
ta Angle thickness.
gb Distance from the angle's heel to the center of bolt holes in leg adjacent to the beam web face.
gc Distance from the angle's heel to the center of bolt holes in leg adjacent to the column face.
cu Distance from the center of upper fastener holes to the top edge of the angle in the beam web.
cl Distance from the center of lower fastener holes to the bottom edge of the angle in the beam web.
pb Distance between two rows of fasteners in the beam web.
pc Distance between two rows of fasteners in the column.
qb Distance between two lines of fasteners in the beam web.
qc Distance between two lines of fasteners in the column.
nb Total number of fasteners in the beam.
nc Total number of fasteners in the column.

3.5.3 Top- and Seat-Angle Connections with Double Web Angle

The experimental studies on this connection type were mainly performed by Altman et al. (1982) and Azizinamini et al. (1985) at the University of South Carolina. The

Table 3.5 Collected Test Data for Single Web-Angle Bolted Beam-to-Column Connections

No	Author	Test Id.	Angle	l (in)	gc (in)	pc (in)	gb (in)	db	nc	nb
1	S.L.Lipson (1968)	AA-2/1	4×3.5×1/4	5.50	2.56	2.25	3.00	3/4	1×2	1×2
2		AA-2/2	4×3.5×1/4	5.50	2.56	2.25	3.00	3/4	1×2	1×2
3		AA-3/1	4×3.5×1/4	8.50	2.56	2.25	3.00	3/4	1×3	1×3
4		AA-3/2	4×3.5×1/4	8.50	2.56	2.25	3.00	3/4	1×3	1×3
5		AA-4/1	4×3.5×1/4	11.50	2.56	2.25	3.00	3/4	1×4	1×4
6		AA-4/2	4×3.5×1/4	11.50	2.56	2.25	3.00	3/4	1×4	1×4
7		AA-5/1	4×3.5×1/4	14.50	2.56	2.25	3.00	3/4	1×6	1×6
8		AA-5/2	4×3.5×1/4	14.50	2.56	2.25	3.00	3/4	1×6	1×6
9		AA-6/1	4×3.5×1/4	17.50	2.56	2.25	3.00	3/4	1×6	1×6
10		AA-6/2	4×3.5×1/4	17.50	2.56	2.25	3.00	3/4	1×6	1×6
11		BB-4/1	4×3.5×5/16	11.50	2.56	2.25	3.00	3/4	1×4	1×4
12		BB-4/2	4×3.5×5/16	11.50	2.56	2.25	3.00	3/4	1×4	1×4
13		B2-4	4×3.5×5/16	11.50	2.56	2.25	3.00	3/4	1×4	1×4
14		B3-4	4×3.5×5/16	11.50	2.56	2.25	3.00	3/4	1×4	1×4
15		C-4	5×3.5×5/16	11.50	1.94	2.25	3.00	3/4	1×4	1×4
16		D-4	4×3.5×5/16	11.50	1.94	3.25	3.00	3/4	1×4	1×4

Table 3.6 Collected Test Data for Single Web-Angle Bolted Beam and Welded to Column

No	Author	Test Id.	Angle	l (in)	pc (in)	db	nb
1	S.L.Lipson (1968)	P1-2/1	--	5.50	2.50	3/4	1×2
2		P1-2/2	--	5.50	2.50	3/4	1×2
3		P1-3/1	--	8.50	2.50	3/4	1×3
4		P1-3/2	--	8.50	2.50	3/4	1×3
5		P1-4/1	--	11.50	2.50	3/4	1×4
6		P1-4/2	--	11.50	2.50	3/4	1×4
7		P1-5/1	--	14.50	2.50	3/4	1×5
8		P1-5/2	--	14.50	2.50	3/4	1×5
9		P1-6/1	--	17.50	2.50	3/4	1×6
10		P1-6/2	--	17.50	2.50	3/4	1×6
11		Q-2/1	4×3.5×5/16	4.75	2.50	3/4	1×2
12		Q-3/1	4×3.5×5/16	8.00	2.50	3/4	1×3
13		Q-4/1	4×3.5×5/16	12.00	2.50	3/4	1×4
14		Q-5/1	4×3.5×5/16	14.50	2.50	3/4	1×5
15	S.L.Lipson et al. (1977)	A1-5	4×3×3/8	14.50	2.50	3/4	1×5
16		A1-6	4×3×3/8	17.50	2.50	3/4	1×6
17		A1-7	4×3×3/8	20.50	2.50	3/4	1×7
18		A1-8	4×3×3/8	23.50	2.50	3/4	1×8
19		A1-9	4×3×3/8	26.50	2.50	3/4	1×9
20		A1-10	4×3×3/8	29.50	2.50	3/4	1×10
21		A1-11	4×3×3/8	32.50	2.50	3/4	1×11
22		A1-12	4×3×3/8	35.50	2.50	3/4	1×12
23	L.E.Thompson et al. (1970)	F-1 alt	--	11.50	2.25	3/4	2× -
24		F-1 alt	--	11.50	2.25	3/4	2× -
25		G-1 alt	--	11.50	2.25	3/4	2× -
26		G-1 alt	--	11.50	2.25	3/4	2× -
27		H-1 alt	--	11.50	2.25	3/4	2× -
28		H-1 alt	--	11.50	2.25	3/4	2× -
29		F-2-1 alt	--	11.50	2.25	3/4	2× -
30		F-2-1 alt	--	11.50	2.25	3/4	2× -
31		F-2-2 alt	--	11.50	2.25	3/4	2× -
32		F-2-2 alt	--	11.50	2.25	3/4	2× -
33		F-2-3 alt	--	11.50	2.25	3/4	2× -
34		F-2-3 alt	--	11.50	2.25	3/4	2× -
35	R.M.Richard et al. (1982)	RGKL2	--	6.00	--	3/4	1× 2
36		RGKL3	--	9.00	--	3/4	1× 3
37		RGKL5	--	15.00	--	3/4	1× 5
38		RGKL7	--	21.50	--	7/8	1× 7

test program included specimen subject to both cyclic loading as well as static loading. However cyclic loading test data are excluded from the data base, as listed in Table 3.8. The notations for major parameters of this connection type as shown in Figure 3.9 are as follows:

On the Web Angle:

lp   Angle height.
lu   Distance from the top edge of the angle to the tension flange of the beam.
ll   Distance from the bottom edge of the angle to the compression flange of the beam.

110 ADVANCED ANALYSIS OF STEEL FRAMES

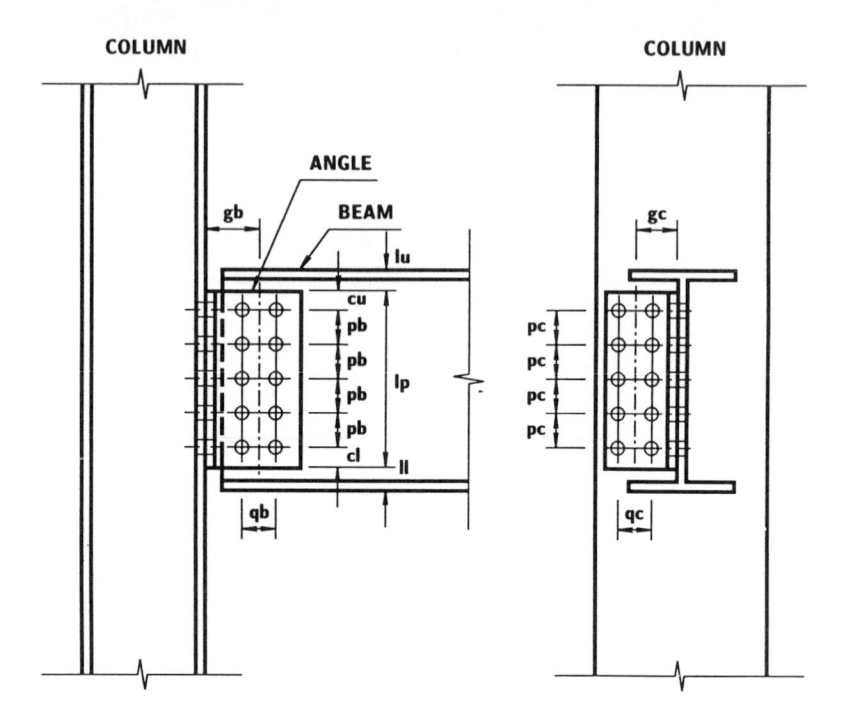

FIGURE 3.6. Size parameters for single web-angle connection.

- ta   Web-angle thickness.
- gb   Distance from the angle's heel to the center of bolt holes in leg adjacent to the beam web face.
- gc   Distance from the angle's heel to the center of bolt holes in leg adjacent to the column face.
- cu   Distance from the center of upper fastener holes to the top edge of the angle in the beam web.
- cl   Distance from the center of lower fastener holes to the bottom edge of the angle in the beam web.
- pb   Distance between two rows of fasteners in the beam web.
- pc   Distance between two rows of fasteners in the column.
- qb   Distance between two lines of fasteners in the beam web.
- qc   Distance between two lines of fasteners in the column.
- nb   Total number of fasteners in the beam.
- nc   Total number of fasteners in the column.

On the Top and Seat Angles:

- lt   Width of the top angle along the column.
- tt   Top-angle thickness.
- ls   Width of the seat angle along the column.
- ts   Seat-angle thickness.

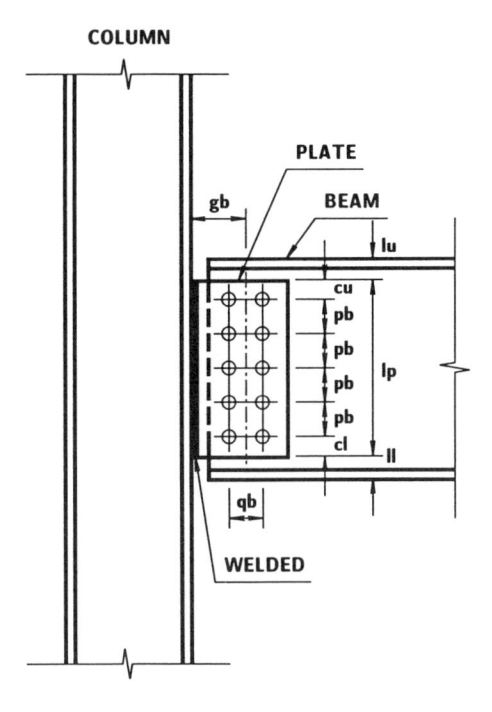

FIGURE 3.7. Size parameters for single plate connection.

gt   Gage from the top angle's heel to the center of fastener holes in leg seating on the tension-beam flange.

gt'   Gage from the top angle's heel to the center of fastener holes in leg adjacent to the column face.

gs   Gage from the seat angle's heel to the center of fastener holes in leg under the compression-beam flange.

gs'   Gage from the seat angle's heel to the center of fastener holes in leg adjacent to the column face.

qt   Distance between two inner lines of fasteners in leg seating on the tension flange.

qt2   Distance between outer and inner line of fasteners in leg seating on the tension-beam flange where the fastener line number exceeds three.

rt   Distance between two inner lines of fasteners in leg adjacent to the column face in tension side.

rt2   Distance between outer and inner line of fasteners in leg adjacent to the column face in tension side where the fastener line exceeds three.

qs   Distance between two inner lines of fasteners in leg under the compression-beam flange.

qs2   Distance between outer and inner line of fasteners in leg under the compression-beam flange where the fastener line number exceeds three.

rs   Distance between two inner lines of fasteners in leg adjacent to the column face in compression side.

rs2   Distance between outer and inner line of fasteners in leg adjacent to the column face in compression side where the fastener line number exceeds three.

112 ADVANCED ANALYSIS OF STEEL FRAMES

Table 3.7 Collected Test Data for Double Web-Angle Connections

No	Author	Test Id.	Angle	l (in)	gc (in)	pc (in)	gb (in)	db	nc	nb
1	J.C.Rathbun (1936)	A-1	6×4×3/8	2.50	2.63	3.50	3.00	7/8	1×1	2×1
2		A-2	6×4×3/8	6.00	2.63	3.50	3.00	7/8	1×2	2×2
3		A-3	6×6×3/8	6.00	2.63	3.50	3.00	7/8	2×2	2×2
4		A-4	4×3.5×3/8	9.00	2.56	2.25	3.00	7/8	1×3	1×3
5		A-5	6×6×3/8	9.00	3.56	3.50	3.00	7/8	2×3	2×3
6		A-6	4×3.5×3/8	15.00	2.50	2.25	3.00	7/8	1×5	1×5
7		A-7	6×6×3/8	15.00	3.50	3.50	3.00	7/8	2×5	2×5
8	W.C.Bell et al. (1958)	FK-4A	6×4×3/8	11.50	2.57	4.50	3.00	M20	1×4	2×4
9		FK-4B	6×4×3/8	11.50	2.57	4.50	3.00	M20	1×4	2×4
10	W.C.Bell et al. (1958)	FK-4C	6×6×3/8	11.50	2.57	4.50	3.00	M20	1×4	2×4
11		FK-4R	6×4×3/8	11.50	2.57	4.50	3.00	M20	1×4	2×4
12	C.W.Lewitt et al. (1966)	FK-3	6×4×3/8	8.50	2.63	3.50	3.00	3/4	1×3	2×3
13	C.W.Lewitt et al. (1966)	FK-4AB-M	6×4×3/8	11.50	2.07	3.50	3.00	3/4	2×4	2×4
14	C.W.Lewitt et al. (1966)	FK-4P	6×4×3/8	11.50	2.07	3.50	3.00	3/4	2×4	2×4
15		WK-4	6×4×3/8	11.50	2.57	3.50	3.00	3/4	1×4	2×4
16		FB-4	6×3.5×3/8	11.50	2.57	2.25	3.00	3/4	1×4	1×4
17		FK-5	6×4×7/16	14.50	2.55	3.50	3.00	3/4	1×5	2×5
18	B.Bose (1981)	B-1	150×90×15	15.75	2.38	3.54	2.95	M20	1×5	2×-
19	L.E.Thompson et al. (1970)	A1-1 alt	4×3.5×5/16	11.50	2.55	2.25	--	3/4	-×-	-×2
20		A1-1 alt	4×3.5×5/16	11.50	2.55	2.25	--	3/4	-×-	-×2
21		A1-2 alt	4×3.5×5/16	11.50	2.55	2.25	--	3/4	-×-	-×2
22		A1-2 alt	4×3.5×5/16	11.50	2.55	2.25	--	3/4	-×-	-×2
23		A1-3 alt	4×3.5×5/16	11.50	2.55	2.25	--	3/4	-×-	-×2
24		A1-3 alt	4×3.5×5/16	11.50	2.55	2.25	--	3/4	-×-	-×2
25		B1-1 alt	4×3.5×5/16	11.50	2.55	2.25	--	3/4	-×-	-×2
26		B1-1 alt	4×3.5×5/16	11.50	2.55	2.25	--	3/4	-×-	-×2
27		B1-2 alt	4×3.5×1/2	11.50	2.55	2.25	--	3/4	-×-	-×2
28		B1-2 alt	4×3.5×1/2	11.50	2.55	2.25	--	3/4	-×-	-×2
29		B1-3 alt	4×3.5×1/2	11.50	2.55	2.25	--	3/4	-×-	-×2
30		B1-3 alt	4×3.5×1/2	11.50	2.55	2.25	--	3/4	-×-	-×2
31		D1-1 alt	4×3.5×5/16	11.50	1.80	2.25	--	3/4	-×-	-×2
32		D1-1 alt	4×3.5×5/16	11.50	1.80	2.25	--	3/4	-×-	-×2
33		D1-2 alt	4×3.5×5/16	11.50	1.80	2.25	--	3/4	-×-	-×2
34		D1-2 alt	4×3.5×5/16	11.50	1.80	2.25	--	3/4	-×-	-×2
35		D1-3 alt	4×3.5×5/16	11.50	1.80	2.25	--	3/4	-×-	-×2

Table 3.7 (continued)

36	D1-3 alt	4×3.5×5/16	11.50	1.80	2.25	---	3/4	-x-	-x2
37	E1-1 alt	4×3.5×1/2	11.50	1.80	2.25	---	3/4	-x-	-x2
38	E1-1 alt	4×3.5×1/2	11.50	1.80	2.25	---	3/4	-x-	-x2
39	E1-2 alt	4×3.5×1/2	11.50	1.80	2.25	---	3/4	-x-	-x2
40	E1-2 alt	4×3.5×1/2	11.50	1.80	2.25	---	3/4	-x-	-x2
41	E1-3 alt	4×3.5×1/2	11.50	1.80	2.25	---	3/4	-x-	-x2
42	E1-3 alt	4×3.5×1/2	11.50	1.80	2.25	---	3/4	-x-	-x2
43	A2-1 alt	4×3.5×5/16	11.50	2.55	2.25	---	3/4	-x-	-x2
44	A2-1 alt	4×3.5×5/16	11.50	2.55	2.25	---	3/4	-x-	-x2
45	A2-2 alt	4×3.5×5/16	11.50	2.55	2.25	---	3/4	-x-	-x2
46	A2-2 alt	4×3.5×5/16	11.50	2.55	2.25	---	3/4	-x-	-x2
47	A2-3 alt	4×3.5×5/16	11.50	2.55	2.25	---	3/4	-x-	-x2
48	A2-3 alt	4×3.5×5/16	11.50	2.55	2.25	---	3/4	-x-	-x2
49	B2-1 alt	4×3.5×5/16	11.50	1.80	2.25	---	3/4	-x-	-x2
50	B2-1 alt	4×3.5×5/16	11.50	1.80	2.25	---	3/4	-x-	-x2
51	B2-2 alt	4×3.5×5/16	11.50	1.80	2.25	---	3/4	-x-	-x2
52	B2-2 alt	4×3.5×5/16	11.50	1.80	2.25	---	3/4	-x-	-x2
53	B2-3 alt	4×3.5×5/16	11.50	1.80	2.25	---	3/4	-x-	-x2
54	B2-3 alt	4×3.5×5/16	11.50	1.80	2.25	---	3/4	-x-	-x2
55	C2-1 alt	4×3.5×1/2	11.50	2.55	2.25	---	3/4	-x-	-x2
56	C2-1 alt	4×3.5×1/2	11.50	2.55	2.25	---	3/4	-x-	-x2
57	C2-2 alt	4×3.5×1/2	11.50	2.55	2.25	---	3/4	-x-	-x2
58	C2-2 alt	4×3.5×1/2	11.50	2.55	2.25	---	3/4	-x-	-x2
59	C2-3 alt	4×3.5×1/2	11.50	2.55	2.25	---	3/4	-x-	-x2
60	C2-3 alt	4×3.5×1/2	11.50	2.55	2.25	---	3/4	-x-	-x2
61	D2-1 alt	4×3.5×1/2	11.50	1.80	2.25	---	3/4	-x-	-x2
62	D2-1 alt	4×3.5×1/2	11.50	1.80	2.25	---	3/4	-x-	-x2
63	D2-2 alt	4×3.5×1/2	11.50	1.80	2.25	---	3/4	-x-	-x2
64	D2-2 alt	4×3.5×1/2	11.50	1.80	2.25	---	3/4	-x-	-x2
65	D2-3 alt	4×3.5×1/2	11.50	1.80	2.25	---	3/4	-x-	-x2
66	D2-3 alt	4×3.5×1/2	11.50	1.80	2.25	---	3/4	-x-	-x2
W.H.Sommer (1969)									
67	TEST21	3.5×3×3/8	9.00	1.83	2.25	3.00	3/4	1×3	-x2
68	TEST22	3.5×3×3/8	12.00	1.83	2.25	3.00	3/4	1×4	-x2
69	TEST23	4×3×3/8	15.00	2.53	2.25	3.00	3/4	1×5	-x2
70	TEST24	4×3×3/8	18.00	2.53	2.25	3.00	3/4	1×6	-x2

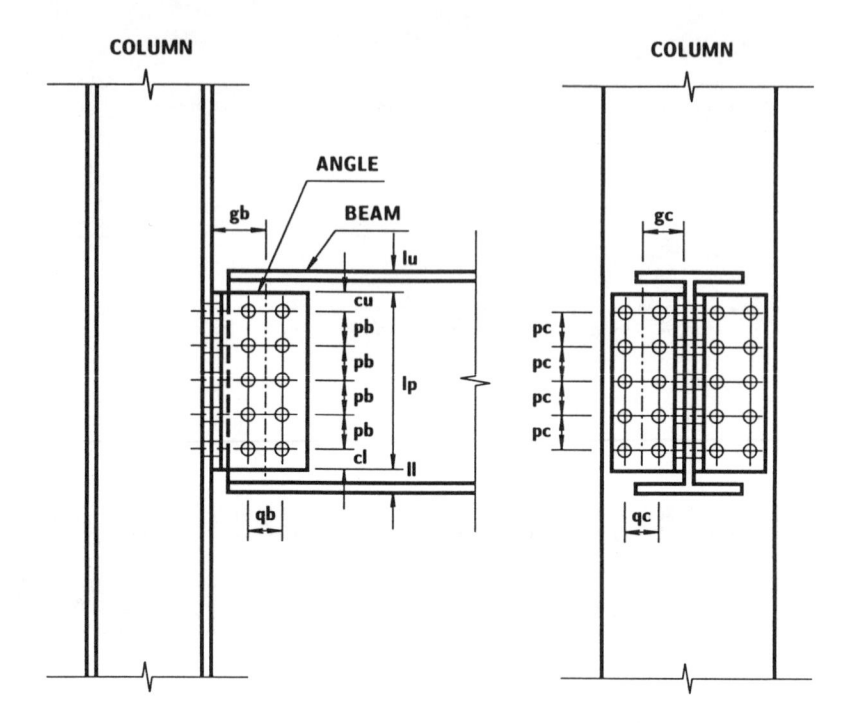

FIGURE 3.8. Size parameters for double web-angle connection.

pt   Distance between two rows of fasteners in leg seating on the tension-beam flange.
pt' Distance between two rows of fasteners in leg adjacent to the column face in tension side.
ps   Distance between two rows of fasteners in leg under the compression-beam flange.
ps' Distance between two rows of fasteners in leg adjacent to the column face in compression side.
nt   Total number of fasteners in the tension-beam flange.
nt' Total number of fasteners in the leg adjacent to the column face in tension side.
ns   Total number of fasteners in the compression-beam flange.
ns' Total number of fasteners in the leg adjacent to the column face in compression side.

3.5.4 Top- and Seat-Angle Connections

The collected test data for this connection type are listed in Table 3.9. The notations for major parameters of this connection type as shown in Figure 3.10 are as follows:

lt   Width of the top angle along the column.
tt   Top-angle thickness.
ls   Width of the seat angle along the column.
ts   Seat-angle thickness.
gt   Gage from the top angle's heel to the center of fastener holes in leg seating on the tension-beam flange.

SEMI-RIGID CONNECTIONS 115

Table 3.8 Collected Test Data for Top- and Seat-Angle with Double Web-Angle Connections

No	Author	Test Id.	Beam	Column	F.Angle	W.Angle
1	A.Azizinamini et al. (1985)	8S1	W8×21	W12×58	6×3.5×5/16×6.0	4×3.5×1/4×5.5
2		8S2	W8×21	W12×58	6×3.5×3/8×6.0	4×3.5×1/4×5.5
3		8S3	W8×21	W12×58	6×3.5×5/16×8.0	4×3.5×1/4×5.5
4		8S4	W8×21	W12×58	6×6.0×3/8×6.0	4×3.5×1/4×5.5
5		8S5	W8×21	W12×58	6×4.0×3/8×8.0	4×3.5×1/4×5.5
6		8S6	W8×21	W12×58	6×4.0×5/16×6.0	4×3.5×1/4×5.5
7		8S7	W8×21	W12×58	6×4.0×3/8×6.0	4×3.5×1/4×5.5
8		8S8	W8×21	W12×58	6×3.5×5/16×6.0	4×3.5×1/4×5.5
9		8S9	W8×21	W12×58	6×3.5×3/8×6.0	4×3.5×1/4×5.5
10		8S10	W8×21	W12×58	6×3.5×1/2×6.0	4×3.5×1/4×5.5
11		14S1	W14×38	W12×96	6×4.0×3/8×8.0	4×3.5×1/4×8.5
12		14S2	W14×38	W12×96	6×4.0×1/2×8.0	4×3.5×1/4×8.5
13		14S3	W14×38	W12×96	6×4.0×3/8×8.0	4×3.5×1/4×5.5
14		14S4	W14×38	W12×96	6×4.0×3/8×8.0	4×3.5×3/8×8.5
15		14S5	W14×38	W12×96	6×4.0×3/8×8.0	4×3.5×1/4×8.5
16		14S6	W14×38	W12×96	6×4.0×1/2×8.0	4×3.5×1/4×8.5
17		14S8	W14×38	W12×96	6×4.0×5/8×8.0	4×3.5×1/4×8.5
18		14S9	W14×38	W12×96	6×4.0×1/2×8.0	4×3.5×1/4×8.5
19	A.Azizinamini et al. (1985)	14WS1	W14×38	W12×96	6×4.0×3/8×8.0	4×3.5×1/4×8.5
20		14WS2	W14×38	W12×96	6×4.0×1/2×8.0	4×3.5×1/4×8.5
21	J.C.Rathbun (1936)	B-11	W12×31.8	--	6×6.0×3/8×9 (TOP)	4×3.5×3/8×9
22		B-12	W12×31.8	--	6×6.0×3/8×14 (TOP)	4×3.5×3/8×9

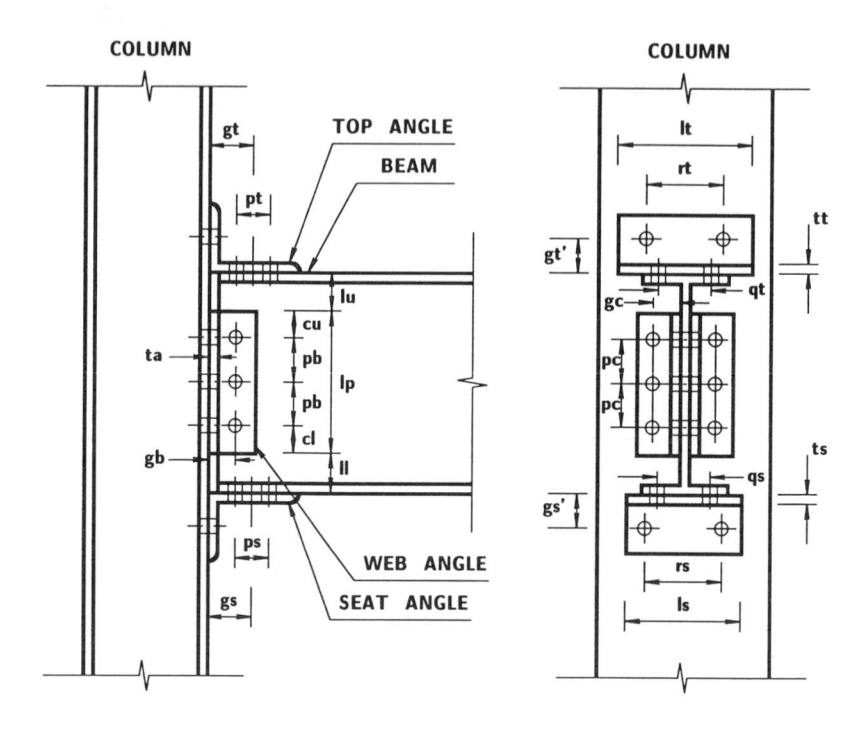

FIGURE 3.9. Size parameters for top- and seat-angle connections with double web angle.

gt′   Gage from the top angle's heel to the center of fastener holes in leg adjacent to the column face.
gs   Gage from the seat angle's heel to the center of fastener holes in leg under the compression-beam flange.
gs′   Gage from the seat angle's heel to the center of fastener holes in leg adjacent to the column.
qt   Distance between two inner lines of fasteners in leg seating on the tension-beam flange where the fastener line number exceeds three.
qt2   Distance between outer and inner line of fasteners in leg seating on the tension-beam flange where the fastener line number exceeds three.
rt   Distance between two inner lines of fasteners in leg adjacent to the column face in tension side.
rt2   Distance between outer and inner line of fasteners in leg adjacent to the column face in tension side where the fastener line exceeds three.
qs   Distance between two inner lines of fasteners in leg under the compression-beam flange.
qs2   Distance between outer and inner line of fasteners in leg under the compression-beam flange where the fastener line number exceeds three.
rs   Distance between two inner lines of fasteners in leg adjacent to the column face in compression side.
rs2   Distance between outer and inner line of fasteners in leg adjacent to the column face in compression side where the fastener line number exceeds three.
pt   Distance between two rows of fasteners in leg seating on the tension-beam flange.

SEMI-RIGID CONNECTIONS 117

Table 3.9 Collected Test Data for Top- and Seat-Angle Connections

No	Author	Test Id.	Beam	Column	Angle	lt (in)	gt (in)	tt (in)
1	J.C.Rathbun (1936)	B-8	W12×31.8	--	6×6×3/8×6 (TOP)	6.00	3.50	0.38
2		B-9	W12×31.8	--	6×6×3/8×8 (TOP)	6.00	3.50	0.38
3		B-10	W12×31.8	--	6×6×3/8×14 (TOP)	6.00	3.50	0.38
4	R.A.Hechtmann et al. (1947)	NO 2	W12×25	W10×49	6×6×1/2×6.75 (TOP)	4.00	3.50	0.63
5		NO 5	W18×47	W12×65	6×6×7/8×7.5 (TOP)	4.00	3.50	0.50
6		NO 9	W18×47	W12×65	6×6×7/8×7.5 (TOP)	4.00	3.50	0.63
7		NO 10	W18×47	W12×65	6×6×7/8×7.5 (TOP)	4.00	3.50	0.75
8		NO 11	W18×47	W14×58	6×6×7/8×7.5 (TOP)	4.00	3.50	0.50
9		NO 16	W12×25	W10×49	6×6×1/2×6.75 (TOP)	4.00	4.75	0.50
10		NO 17	W12×25	W10×49	6×6×1/2×6.75 (TOP)	4.00	3.50	0.50
11		NO 18	W12×50	W10×49	6×6×1/2×8 (TOP)	4.00	3.50	0.50
12		NO 20	W14×34	W12×65	6×6×5/8×7.25 (TOP)	4.00	3.50	0.63
13		NO 22	W16×40	W12×65	6×6×3/4×7.25 (TOP)	4.00	3.50	0.63
14		NO 23	W18×40	W14×58	6×6×3/4×7.25 (TOP)	4.00	3.50	0.63
15		NO 24	W18×47	W12×65	6×6×7/8×7.5 (TOP)	4.00	3.50	0.63
16	M.J.Marley (1982)	A-1/4-1	W5×16	W5×16	4×4×1/4×5.0	4.00	--	0.25
17		A-1/4-2	W5×16	W5×16	4×4×1/4×5.0	4.00	--	0.25
18		B-1/4-1	W5×16	W5×16	4×4×1/4×5.0	4.00	--	0.25
19		B-1/4-2	W5×16	W5×16	4×4×1/4×5.0	4.00	--	0.25
20		C1-1/4-2	W5×16	W5×16	4×4×1/4×5.0	4.00	--	0.25
21		C2-1/4-1	W5×16	W5×16	4×4×1/4×5.0	4.00	--	0.25
22		C2-1/4-2	W5×16	W5×16	4×4×1/4×5.0	4.00	--	0.25
23		D1-1/4-1	W5×16	W5×16	4×4×1/4×5.0	4.00	--	0.25
24		D1-1/4-2	W5×16	W5×16	4×4×1/4×5.0	4.00	--	0.25
25		D2-1/4-1	W5×16	W5×16	4×4×1/4×5.0	4.00	--	0.25
26		D2-1/4-2	W5×16	W5×16	4×4×1/4×5.0	4.00	--	0.25

118 ADVANCED ANALYSIS OF STEEL FRAMES

Table 3.9 (continued)

27		D3-1/4-1	W5×16	W5×16	4×4×1/4×5.0	4.00	---	0.25
28		D3-1/4-2	W5×16	W5×16	4×4×1/4×5.0	4.00	---	0.25
29		A-1/2-1	W5×16	W5×16	4×4×1/4×5.0	4.00	---	0.25
30		B-1/2-1	W5×16	W5×16	4×4×1/4×5.0	4.00	---	0.25
31		B-1/2-2	W5×16	W5×16	4×4×1/4×5.0	4.00	---	0.25
32		C1-1/2-1	W5×16	W5×16	4×4×1/4×5.0	4.00	---	0.25
33		C1-1/2-2	W5×16	W5×16	4×4×1/4×5.0	4.00	---	0.25
34		C2-1/2-1	W5×16	W5×16	4×4×1/4×5.0	4.00	---	0.25
35		C2-1/2-2	W5×16	W5×16	4×4×1/4×5.0	4.00	---	0.25
36		D1-1/2-1	W5×16	W5×16	4×4×1/4×5.0	4.00	---	0.25
37		D1-1/2-2	W5×16	W5×16	4×4×1/4×5.0	4.00	---	0.25
38		D2-1/2-1	W5×16	W5×16	4×4×1/4×5.0	4.00	---	0.25
39		D2-1/2-2	W5×16	W5×16	4×4×1/4×5.0	4.00	---	0.25
40		D3-1/2-1	W5×16	W5×16	4×4×1/4×5.0	4.00	---	0.25
41		D3-1/2-2	W5×16	W5×16	4×4×1/4×5.0	4.00	---	0.25
42	S.M.Maxwell et al. (1981)	A1	457×191×UB	305×305×UC	150×90×10×150mm	3.54	---	0.39
43		A2	457×191×UB	305×305×UC	150×90×10×200mm	3.54	---	0.39
44		A3	457×191×UB	305×305×UC	150×90×10×150mm	3.54	---	0.39
45		A4	457×191×UB	305×305×UC	150×90×10×200mm	3.54	---	0.39
46		B1	457×191×UB	305×305×UC	150×90×12×150mm	3.54	---	0.47
47		B2	457×191×UB	305×305×UC	150×90×12×200mm	3.54	---	0.47
48		B3	457×191×UB	305×305×UC	150×90×12×150mm	3.54	---	0.47
49		B4	457×191×UB	305×305×UC	150×90×12×200mm	3.54	---	0.47
50		C1	457×191×UB	305×305×UC	150×90×15×150mm	3.54	---	0.59
51		C2	457×191×UB	305×305×UC	150×90×15×200mm	3.54	---	0.59
52		C3	457×191×UB	305×305×UC	150×90×15×150mm	3.54	---	0.59
53		C4	457×191×UB	305×305×UC	150×90×15×200mm	3.54	---	0.59

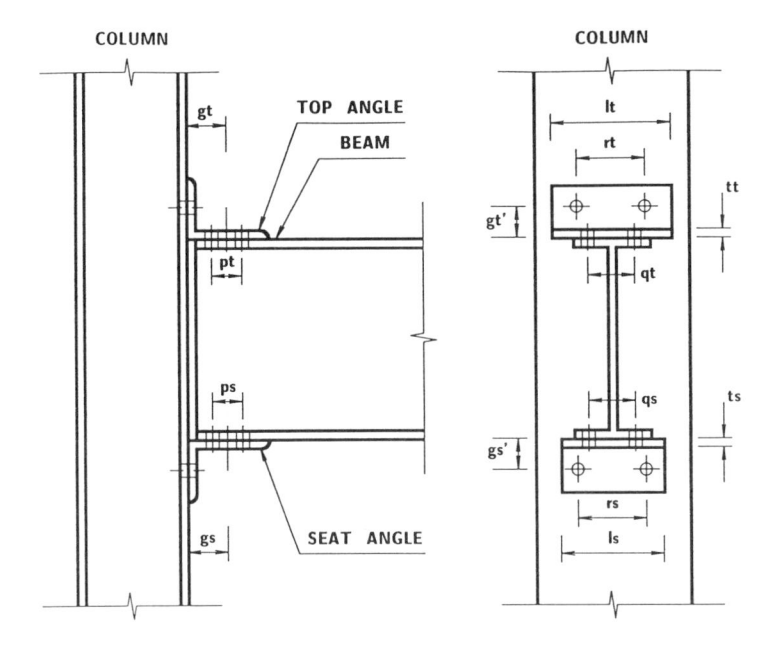

FIGURE 3.10. Size parameters for top- and seat-angle connections without double web angle.

pt′   Distance between two rows in leg adjacent to the column face in tension side.
ps   Distance between two rows of fasteners in leg under the compression-beam flange.
ps′   Distance between two rows of fasteners in leg adjacent to the column face in compression side.
nt   Total number of fasteners in the tension-beam flange.
nt′   Total number of fasteners in leg adjacent to the column face in tension side.
ns   Total number of fasteners in the compression-beam flange.
ns′   Total number of fasteners in leg adjacent to the column face in compression side.

3.5.5 Extended End-Plate Connections

The type of extended end-plate connections includes both cases of extended on tension and compression sides and extended on tension side only. Collected experimental moment-rotation data for extended end-plate connections are tabulated in Table 3.10. The notations for major parameters of this connection type shown in Figure 3.11 are identified as follows:

ct   Distance from the tension-side edge of plate to the center of the outer tension-side fastener holes.
pt   Distance between two rows of tension-side fasteners.
li   Distance from the center of the lower tension side fastener holes to the center of upper compression-side fasteners holes.

Table 3.10 Collected Test Data for Extended End-Plate Connections

No	Author	Test Id.	Beam	Column	lp (in)	tp (in)	g (in)	p (in)	db
1	S.A.Ioannides (1978)	TEST 1	W14×22	W8×35	17.25	0.63	5.50	3.50	3/4"
2		TEST 2	W18×35	W10×49	21.50	0.75	5.50	3.75	7/8"
3		TEST 3	W24×55	W14×48	27.25	0.88	5.50	4.00	1"
4		TEST 4	W14×22	W8×35	17.25	0.88	5.50	3.50	3/4"
5		TEST 5	W18×35	W10×49	21.50	1.25	5.50	3.75	7/8"
6		TEST 6	W24×55	W14×48	27.25	1.25	5.50	4.00	1"
7	R.J.Dews (1979)	TEST 1	W14×22	W8×31	16.81	0.97	5.50	3.50	3/4"
8		TEST 2	W14×22	W8×31	16.92	0.94	5.53	3.50	3/4"
9		TEST 3	W16×26	W10×33	19.14	1.56	5.38	3.75	1"
10	A.N.Sherbourne (1961)	TEST A1	15×5 RSJ 42	8×8 UC 35	14.50	1.25	4.00	3.50	3/4"
11	J.R.Bailey (1970)	TEST A1-L	12×5 RSJ 32	8×8 UC 48	--	1.00	4.00	5.50	7/8"
12		TEST A1-R	12×5 RSJ 32	8×8 UC 48	--	1.00	4.00	4.50	7/8"
13		TEST A2-L	10×4 UB 17	8×8 UC 40	--	0.75	4.00	4.38	5/8"
14		TEST A2-R	10×4 UB 17	8×8 UC 40	--	0.75	4.00	4.38	5/8"
15		TEST A3-L	10×4 UB 19	8×8 UC 40	--	0.75	4.00	4.38	3/4"
16		TEST A3-R	10×4 UB 19	8×8 UC 40	--	0.75	4.00	4.38	3/4"
17	J.O.Surtees et al. (1970)	TEST C1	12×5 UB 25	8×8 UC 48	--	0.75	3.50	4.50	3/4"
18		TEST C2	15×6 UB 40	10×10 UC 60	--	0.75	3.50	4.50	3/4"
19		TEST C3	15×6 UB 40	10×10 UC 60	--	0.75	3.50	4.50	1"
20		TEST C4	15×6 UB 40	10×10 UC 60	--	1.00	3.50	4.50	1"
21		TEST C5	15×6 UB 40	10×10 UC 60	--	1.00	3.50	4.50	1"
22	J.A.Packer et al. (1977)	TEST J1	254×102 UB 2	152×152 UC 3	9.76	0.59	3.39	3.82	M16
23		TEST J2	254×102 UB 2	152×152 UC 3	9.75	0.59	3.39	3.81	M16
24		TEST J3	254×102 UB 2	152×152 UC 2	9.76	0.59	3.39	3.82	M16
25	P.Grundy et al. (1980)	T1	610 UB 113	310 UC 240	--	1.00	3.50	3.75	7/8"
26		T2	610 UB 113	310 UC 240	--	1.25	3.50	3.75	7/8"
27	A.N.Sherbourne (1961)	TEST A2	15×5 RSJ 42	8×8 UC 35	14.50	1.25	4.00	3.50	3/4"
28		TEST A3	15×5 RSJ 42	8×8 UC 35	14.50	0.75	4.00	3.50	3/4"
29		TEST B1	15×5 RSJ 42	8×8 UC 35	17.00	1.00	4.00	5.00	7/8"

Table 3.10 (continued)

#	Reference	Test	Beam	Column					Bolt
30		TEST B2	15×5 RSJ 42	8×8 UC 35	17.00	0.75	4.00	5.00	7/8"
31	L.G.Johnson et al. (1960)	TEST 5	10×4.5 RSJ	8×8 UC 45	--	0.50	3.50	3.75	3/4"
32	J.R.Bailey (1970)	B4-L	12×5 UB 32	8×8 UC 40	--	1.25	4.00	4.50	7/8"
33		B4-R	12×5 UB 32	8×8 UC 40	--	1.25	4.00	4.50	7/8"
34		B5-L	14×5 UB 26	8×8 UC 40	--	1.00	4.00	4.25	3/4"
35		B5-R	14×5 UB 26	8×8 UC 40	--	1.00	4.00	4.25	3/4"
36		B6-L	14×6.75 UB	8×8 UC 40	--	1.38	4.00	4.31	1"
37		B6-R	14×6.75 UB	8×8 UC 40	--	1.38	4.00	4.31	1"
38		B7-L	12×6.5 UB	8×8 UC 40	--	1.25	4.00	4.50	1"
39		B7-R	12×6.5 UB	8×8 UC 40	--	1.25	4.00	4.50	1"
40		B8-L	8×5.25 UB 2	8×8 UC 40	--	0.75	4.00	4.50	3/4"
41		B8-R	8×5.25 UB 2	8×8 UC 40	--	0.75	4.00	4.50	3/4"
42		C9-L	8×5.25 UB 2	8×8 UC 40	--	0.75	4.00	4.50	5/8"
43		C9-R	8×5.25 UB 2	8×8 UC 40	--	0.75	4.00	4.50	5/8"
44		C10-L	12×5 UB 32	8×8 UC 40	--	0.81	4.00	4.50	3/4"
45		C10-R	12×5 UB 32	8×8 UC 40	--	0.81	4.00	4.50	3/4"
46		C11-L	14×5 UB 26	8×8 UC 40	--	0.88	4.00	4.25	5/8"
47		C11-R	14×5 UB 26	8×8 UC 40	--	0.88	4.00	4.25	5/8"
48		C12-L	12×6.5 UB	8×8 UC 40	--	1.06	4.00	4.38	7/8"
49		C12-R	12×6.5 UB	8×8 UC 40	--	1.06	4.00	4.38	7/8"
50		C13-L	14×6.75 UB	8×8 UC 40	--	1.25	4.00	4.63	1"
51		C13-R	14×6.75 UB	8×8 UC 40	--	1.25	4.00	4.63	1"
52	J.O.Surtees et al. (1970)	TEST C6	18×6 UB 55	10×10 UC 89	--	1.13	6.00	6.00	1 1/8"
53	N.D.Johnstone et al. (1981)	TEST 1-L	310 UB 46	250 UC 89	--	1.26	5.51	7.01	M30
54		TEST 1-R	310 UB 46	250 UC 89	--	1.26	5.51	7.01	M30
55		TEST 2-L	310 UB 46	250 UC 89	--	1.26	5.51	7.01	M30
56		TEST 2-R	310 UB 46	250 UC 89	--	1.26	5.51	7.01	M30
57		TEST 3-L	310 UB 46	250 UC 89	--	0.94	5.12	4.02	M24
58		TEST 3-R	310 UB 46	250 UC 89	--	0.94	5.12	4.02	M24
59		TEST 4-L	310 UB 46	250 UC 89	--	0.63	5.12	4.53	M24
60		TEST 4-R	310 UB 46	250 UC 89	--	0.63	5.12	4.53	M24

122 ADVANCED ANALYSIS OF STEEL FRAMES

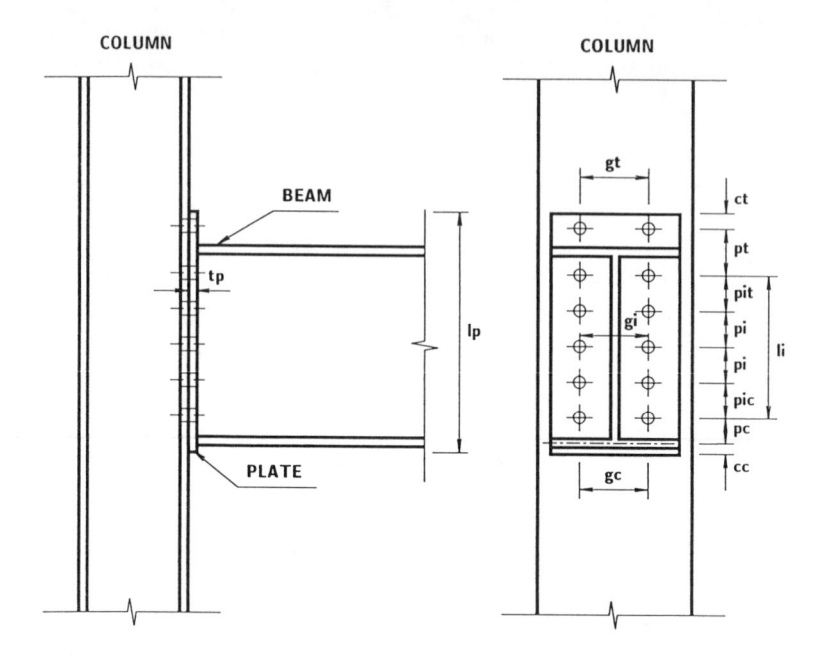

FIGURE 3.11. Size parameters for extended end-plate connections.

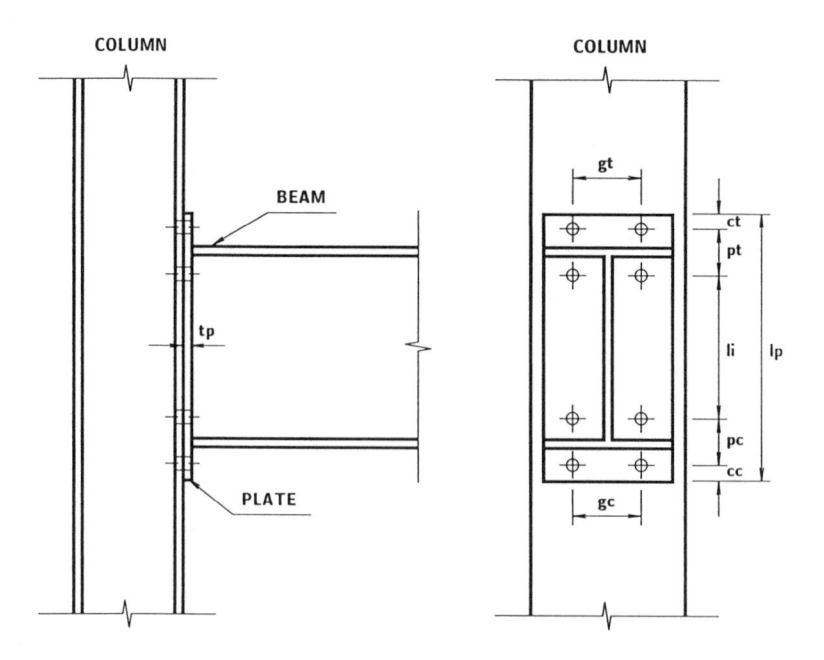

FIGURE 3.11. (*continued*)

The case of end-plate connections extended only on the tension side.

pc Distance from the inner row of fasteners on the compression side to the center of compression-side flange of the beam.
cc Distance from the center of the compression-side flange of the beam to the compression-side edge of the plate.

The case of end-plate connections extended on both the tension and the compression sides.

pc   Distance between two rows of compression-side fasteners.
cc   Distance from the outer row of fasteners on the compression side to the compression-side edge of plate.
pit Distance from the inner row of fasteners on the tension side to the upper row of the central fasteners.
pi   Distance between the two inner rows of the central fasteners.
pic Distance from the inner row of fasteners on the compression side to the lower row of the central fasteners.
gt   Distance between two inner lines of tension side fasteners.
gt2 Distance from the outer line to the inner line of tension-side fasteners where the fastener line number exceeds three.
gi   Distance between two inner lines of central fasteners.
gi2 Distance from the outer line to the inner line of central fasteners where the fastener line number exceeds three.
gc   Distance between two inner lines of compression-side fasteners.
gc2 Distance from outer line to inner line of compression-side fasteners where the fastener line number exceeds three.
lp   Plate length.
tp   Plate thickness.

3.5.6 Flush End-Plate Connections

Collected experimental moment-rotation data for flush-end connections are tabulated in Table 3.11. The notations for major parameters of this connection type as shown in Figure 3.12 are identified as follows:

ct   Distance from the tension-side edge of plate to the outer surface of the tension-side flange of the beam.
pt   Distance from the outer surface of the tension-side flange to the row of tension-side fasteners.
li   Distance from the center of tension-side fastener holes to the center of the compression-side fastener holes.
pc   Distance from the row of compression-side fasteners to the outer surface of the compression-side flange of the beam.
cc   Distance from the outer surface of the compression-side flange of the beam to the compression-side edge of the plate.
pit Distance from the row of tension-side fasteners to the upper row of central fasteners.

Table 3.11 Collected Test Data for Flush End-Plate Connections

No	Author	Test Id.	Beam	Column	li (in)	g (in)	tp (in)	db
1	J.R.Ostrander (1970)	TEST 1	W10×21	W8×28	5.00	3.50	0.50	3/4"
2		TEST 3	W10×21	W8×28	5.00	3.50	0.38	3/4"
3		TEST 4	W10×21	W8×28	5.00	3.50	0.25	3/4"
4		TEST 9	W10×21	W8×28	5.00	3.50	0.75	3/4"
5		TEST 11	W12×27	W8×40	7.00	4.00	0.38	3/4"
6		TEST 12	W12×27	W8×40	7.00	4.00	0.50	3/4"
7		TEST 13	W12×27	W8×40	7.00	4.00	0.63	3/4"
8		TEST 17	W12×27	W8×24	7.00	4.00	0.38	3/4"
9		TEST 18	W12×27	W8×24	7.00	4.00	0.50	3/4"
10		TEST 19	W12×27	W8×24	7.00	4.00	0.63	3/4"
11		TEST 23	W12×27	W8×48	7.00	4.00	0.63	3/4"
12	J.R.Ostrander (1970)	TEST 2	W10×21	W8×28	5.00	3.50	0.50	3/4"
13		TEST 5	W10×21	W8×28	5.00	3.50	0.50	3/4"
14		TEST 6	W10×21	W8×28	5.00	3.50	0.38	3/4"
15		TEST 7	W10×21	W8×28	5.00	3.50	0.25	3/4"
16		TEST 8	W10×21	W8×28	5.00	3.50	0.75	3/4"
17		TEST 10	W10×21	W8×28	5.00	3.50	0.38	3/4"
18		TEST 14	W12×27	W8×40	7.00	4.00	0.50	3/4"
19		TEST 15	W12×27	W8×40	7.00	4.00	0.63	3/4"
20		TEST 16	W12×27	W8×40	7.00	4.00	0.38	3/4"
21		TEST 20	W12×27	W8×24	7.00	4.00	0.50	3/4"
22		TEST 21	W12×27	W8×24	7.00	4.00	0.63	3/4"
23		TEST 22	W12×27	W8×24	7.00	4.00	0.50	3/4"
24		TEST 24	W12×27	W8×48	7.00	4.00	0.63	3/4"

SEMI-RIGID CONNECTIONS **125**

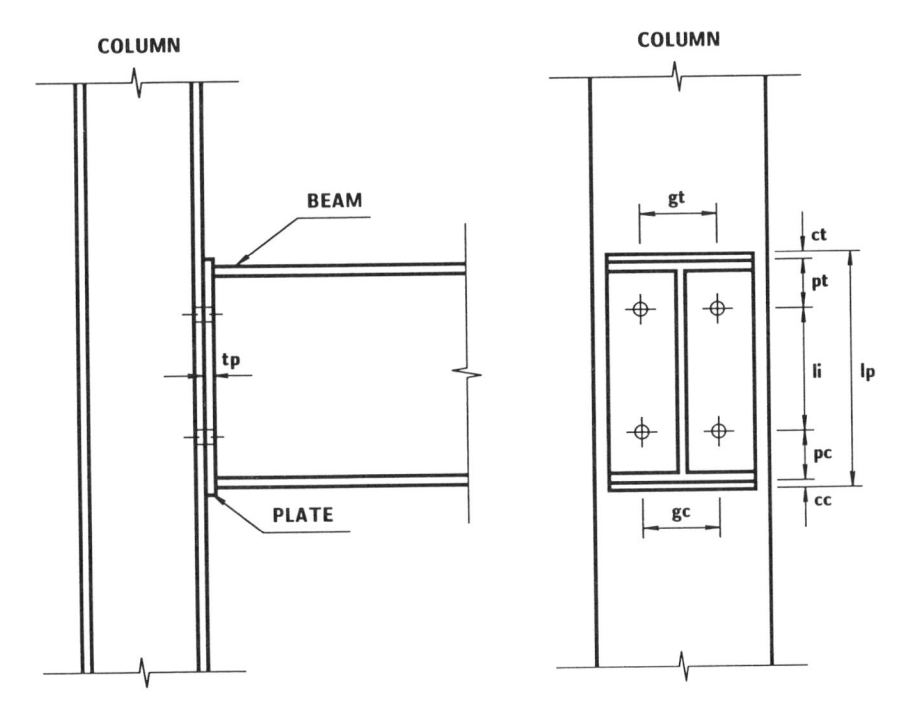

FIGURE 3.12. Size parameters for flush end-plate connections.

pi   Distance between two inner rows of central fasteners.
pic   Distance from the row of compression-side fasteners to the row of central fasteners.
gt   Distance between two inner lines of tension-side fasteners.
gt2   Distance from the outer line to the inner line of tension-side fasteners where the fastener line number exceeds three.
gi   Distance between two inner lines of central fasteners.
gi2   Distance from the outer line tot he inner line of central fasteners where the fastener line exceeds three.
gc   Distance between two inner lines of compression-side fasteners.
gc2   Distance from the outer line to the inner line of compression-side fasteners where the fastener line number exceeds three.
lp   Plate length.
tp   Plate thickness.

3.5.7 Header-Plate Connections

All of collected moment-rotation tests on header-plate connections were performed by Sommer at the University of Toronto in 1969. He tested 20 specimens listed in Table 3.12. The notations for major parameters of this connection type as shown in Figure 3.13 are identified as follows:

lt   Distance from the tension-beam flange to the top edge of the plate.
ct   Distance from the center of the upper fastener holes to the top edge of the plate.

Table 3.12 Collected Test Data for Header-Plate Connections

No	Author	Test Id.	Beam	Column	l (in)	g (in)	tp (in)	db
1	W.H.Sommer (1969)	TEST 5	W18×45	--	15.00	4.00	0.25	3/4"
2		TEST 6	W24×76	--	9.00	4.00	0.25	3/4"
3		TEST 7	W24×76	--	12.00	4.00	0.25	3/4"
4		TEST 8	W24×76	--	15.00	4.00	0.25	3/4"
5		TEST 9	W24×76	--	18.00	4.00	0.25	3/4"
6		TEST 10	W18×45	--	9.00	4.00	0.38	3/4"
7		TEST 11	W18×45	--	12.00	4.00	0.38	3/4"
8		TEST 12	W18×45	--	15.00	4.00	0.38	3/4"
9		TEST 13	W24×76	--	9.00	4.00	0.38	3/4"
10		TEST 14	W24×76	--	12.00	4.00	0.38	3/4"
11		TEST 15	W24×76	--	15.00	5.50	0.38	3/4"
12		TEST 16	W24×76	--	18.00	5.50	0.38	3/4"
13		TEST 17	W24×76	--	18.00	5.50	0.38	3/4"
14		TEST 18	W24×76	--	15.00	5.50	0.25	3/4"
15		TEST 19	W24×76	--	12.00	5.50	0.50	3/4"
16		TEST 20	W24×76	--	15.00	5.50	0.50	3/4"
17		TEST 25	W18×45	--	12.00	4.00	0.25	3/4"
18		TEST 26	W18×45	--	9.00	4.00	0.25	3/4"
19		TEST 27	W12×27	--	9.00	4.00	0.25	3/4"
20		TEST 28	W12×27	--	6.00	4.00	0.25	3/4"

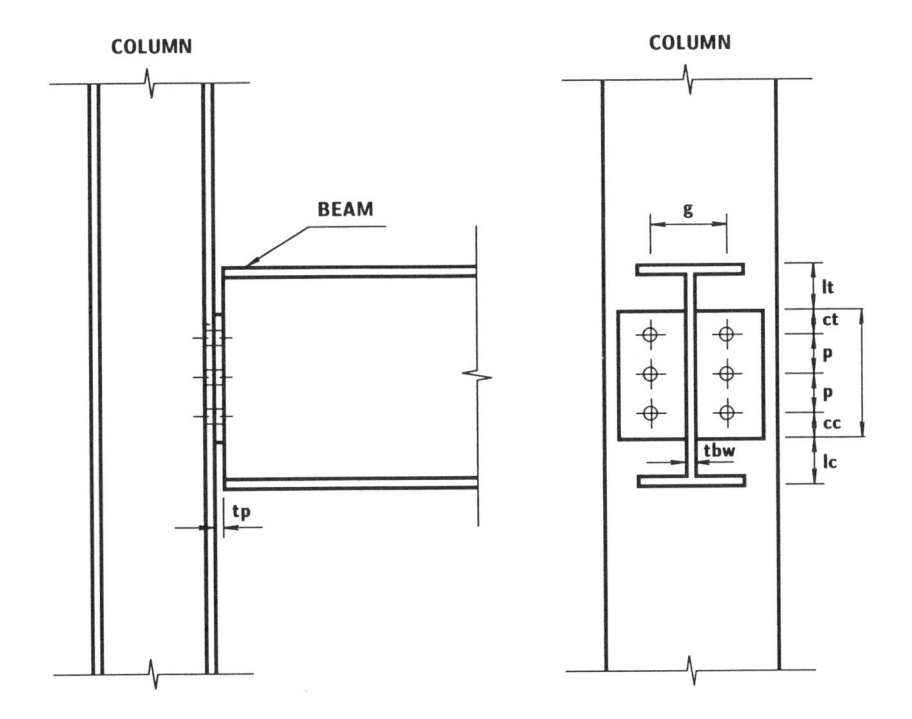

FIGURE 3.13. Size parameters for header-plate connections.

p   Distance between two rows of fasteners.
cc   Distance from the center of the lower fastener holes to the bottom edge of the plate.
lc   Distance from the compression-beam flange to the bottom edge of the plate.
g   Distance between two lines of fasteners.
lp   Plate depth.
tp   Plate thickness.
tbw   Beam-web thickness.
n   Total number of fasteners.

3.6 Steel Connection Data Bank Program
3.6.1 Outline of SCDB

The experimental connection test data collected in the previous section can be controlled by the Steel Connection Data Bank program (SCDB). The first version of SCDB program has seven main routines. In order to be controlled by any workstation with a FORTRAN compiler, the plotter routines have been removed. A new version of the SCDB program for this book consists of the following main routines:

(1) Transformation of the output unit system from original U.S. customary units to the MKS system.
(2) Set and print the selected test data.

Table 3.13 List of Files Used in SCDB Program

File name	Contents
scdb.d	input data file
BANK/no1 - BANK/no7	original data base of each connection type
bank.d	output file
temp	working file

Table 3.14 Fastening Mode Patterns for Making Table

Pattern number, N (1)	Fastening mode pattern (2)
1	All riveted
2	All bolted
3	Riveted-to-beam and bolted-to-column
4	Bolted-to-beam and riveted-to-column
5	Riveted-to-beam and welded-to-column
6	Bolted-to-beam and welded-to-column
7	Welded-to-beam and riveted-to-column
8	Welded-to-beam and bolted-to-column
9	All riveted without column stiffeners
10	All riveted with column stiffeners
11	All bolted without column stiffeners
12	All bolted with column stiffeners
13	Riveted/bolted-to-beam and column (1, 2, 3, 4)
14	Riveted/bolted-to-beam and welded-to-column (5, 6)
15	Riveted/bolted-to-column and welded-to-beam (7, 8)
16	Riveted/bolted without column stiffener (9, 11)
17	Riveted/bolted with column stiffener (10, 12)
20	All modes included

(3) Set and print general tables of test data concerning connection type and mode.
(4) Determine and print the values of three prediction curves for selected test data.

In this program, three prediction equations are prepared: the Frye-Morris polynomial model, the modified exponential model, and the three-parameter power model. The prediction curves obtained from the three-parameter power model are limited to the angle types of connections. In the three-parameter power model, initial connection stiffness R_{ki} and ultimate connection moment M_u are based on the values determined from the simple mechanics and mechanism formulated by Kishi and Chen (1990). The shape parameter n is determined numerically by applying the least-mean-square technique for the connection moments between experimental test data and Eq. (3.5). The validity of using either the three-parameter power model or the modified experimental model to fit the experimental curves for practical use is discussed by Kishi and Chen et al. (1991). It was concluded that the three-parameter power model can replace the experimental test data in describing adequately the connection $M - \theta_r$ behavior for practical use.

Depending on the user's demand, it is possible to execute the SCDB program together with the above-mentioned routines. This program enables the user to make comparative studies of the role of different joint parameters on $M - \theta_r$ behavior.

3.6.2 User's Manual for Program SCDB

The program SCDB is written in FORTRAN language and controlled by a UNIX system which is composed of four files (**scdb1.f-scdb4.f**). These files are provided in the source list of the attached diskette. A **Makefile** is also included to provide a specification on how an executable program can be built using the make utility of the UNIX system. The user can compile and make the executable file by himself. This program has been tested in computing environments by the author: a Sun SPARCstation IPX using a Sun FORTRAN 77 compiler and a Sony RISC NEWS using a RISC NEWS FORTRAN. A hard disk is generally required. To execute the program SCDB, the files shown in Table 3.13 must to be prepared. It should be noted that **scdb.d** and **BANK/no1-BANK/no7** in Table 3.13 are files for input, while **bank.d** is for output. The files of **BANK/no1-BANK/no7** are the files for the original data base of each type of connections and are also provided on the attached diskette. The file **scdb.d** should be defined in the same directory as an executable program, and input data should be created according to the user's purpose as follows:

Line 1 to 7 (7a50)
    File names with full pass of files **BANK/no1** to **BANK/no7**.

Line 8 (4i5)
- idsi: Index for unit name of output data
  - 0: U.S.A. customary units
  - 1: MKS units, except dimensions (cm) of connection
- icnct: Identification number of the connection type referred to Table 3.14.
  - In this table, the cases of types of end-plate connections are taken from number 9 through 12 or numbers 16 and 17.
- ilist: Flag for output.
  - 0: Print data bank list.
  - n: Print the table concerning n-th fastening mode referred to Table 3.13.
- nterm: Number of terms considered in the modified exponential curve-fitting equation. The case of nterm = 6 will give a good result.
  - 0: Default set to 6.

Line 9 (i5)
- ngroup: Number of calculating group of experimental test data for the type of connection given by **icnct** in Line 8.

Line 10 (10i5)
- istart(i), i=1,ngroup:
  - Starting the sequential number of test data in i-th group specified by the above variable **ngroup**.

130   ADVANCED ANALYSIS OF STEEL FRAMES

Table 3.15   Content of **scdb.d** for Example 1

BANK/no1			
BANK/no2			
BANK/no3			
BANK/no4			
BANK/no5			
BANK/no6			
BANK/no7			
0	2	0	0
1			
1			
1			

Line 11 (10i5)
iend(i), i=1,ngroup:
> Finishing the sequential number of test data in i-th group specified by the above variable **ngroup**.

3.6.3 Examples

EXAMPLE 1: PRINTING OUT OF BANK LIST

The first example is for making a bank list: connection parameters, experimental $M - \theta_r$ data, parameters for each prediction equation, and the comparison of the values obtained from experimental test data with the values of prediction equations. The input file named **scdb.d** for this example is given in Table 3.15 and may be used for making a bank list on the experimental test data of sequential number 1 for the type of double-web angle connections. The content of the output file **bank.d** is shown in Table 3.16 and can be used to verify program SCDB upon installation on a computer system. On the attached diskette, input and output files for this example are prepared in files of **scdb_ex1.d** and **bank_ex1.d**, respectively.

EXAMPLE 2: PRINTING OUT OF GENERAL TABLES OF TEST DATA

The second example is for making the general table of experimental test data concerning type and mode. In this example, an attempt is made to make a general table of test data concerning double-web angle connections with riveted-to-beam and bolted-to-column. The input file, **scdb.d**, for this example is given in Table 3.17. The output file, **scdb.d**, obtained from the program SCDB, is shown in Table 3.18. Those files are provided on the attached diskette in files named **scdb_ex2.d** and **bank_ex2.d**, respectively.

Table 3.16 Content of **bank.d** for Example 1

```
Connection type : Double web-angle connections
           Mode : All riveted

Tested by :  J. C. Rathbun (1936)            U. S. A.
Test  Id. :  A-1

Column :  --                    Fasteners:   -- -X-7/8"D
Beam   : W6X12.5                             15/16" Oversize holes
Angle  : 6 X 4 X 3/8            Material : G40.21
                                    Fy = 44.00 ksi
                                    Fu = 64.80 ksi

                        Major parameters

      lp = 2.5000"    lu = 1.7500"    ll = 1.7500"    ta = 0.3750"
      gb = 3.5000"    gc = 2.6250"    cu = 1.2500"    cl = 1.2500"
      qb = 2.5000"    nb = 2 X 1      nc = 1 X 1

Remark   1)  CONNECTION ANGLE RIVETED TO 1/2" MOUNING PLATE.
         2)
```

No	Moment (k-in)	Rotation (radians) X 1/1000	No	Moment (k-in)	Rotation (radians) X 1/1000
1	0.0	0.00	26	17.1	19.17
2	0.6	0.60	27	17.6	19.94
3	1.2	1.20	28	18.0	20.85
4	1.9	1.82	29	18.5	21.76
5	2.5	2.44	30	19.3	23.10
6	3.2	3.12	31	19.5	23.98
7	4.0	3.81	32	19.8	24.85
8	4.6	4.41	33	20.0	25.78
9	5.3	5.01	34	20.2	26.71
10	6.2	6.08	35	20.6	28.15
11	7.1	7.02	36	20.7	29.00
12	7.9	7.79	37	20.9	29.84
13	8.7	8.55			
14	9.3	9.31			
15	10.0	10.06			
16	10.8	10.86			
17	11.7	11.80			
18	12.4	12.67			
19	13.2	13.54			
20	13.7	14.21			
21	14.2	14.88			
22	14.9	15.85			
23	15.6	16.82			
24	16.1	17.61			
25	16.6	18.40			

Table 3.16 (continued)

Moment-rotation prediction equations (R : X 1/1000 radians)

Frye and Morris polynominal model : R = Sum (Ai X (K*Bm)**Pi X 10**Qi)
xd = 2.500000"   g = 5.500000"   t = 0.375000"
A1 = 3.660000   A2 = 1.150000   A3 = 4.570000   K = 0.845282
P1 = 1   P2 = 3   P3 = 5   Q1 = -1   Q2 = -3   Q3 = -5

Modified exponential model :
Bm = Sum (Ai X (1 - exp(- R/(2*i*C)))) + Sum (RKj X (R - Rj0)) + Bm0
C = 0.39904000E+00   Bmo= 0.00000000E+00   Nexp= 6   Nliner= 0
A1 = -0.56289585E+01   0.10550337E+03   -0.69927898E+03   0.19892244E+04   -0.24849136E+04   0.11693B1E+04

Power model : Bm = (Rki X R) / (1 + (R/R0)**rn)**(1/rn)
rn = 1.807   rki = 0.13188761E+01   rmu = 0.24538170E+02

No	Rotation (radians) X 1/1000	Moment (kip-inch)			Connection stiffness (kip-inch) X 1000		
	Expri.	Expri.	Poly.	M.Expo.	P.Model.	Expri.	Poly.	M.Expo.	P.Model.
1	0.00	0.0	0.0	0.0	0.0	0.1019E+01	0.3232E+01	0.6440E+00	0.1319E+01
3	1.20	1.2	3.7	1.2	1.6	0.1024E+01	0.2799E+01	0.1109E+01	0.1304E+01
5	2.44	2.5	6.5	2.5	3.2	0.1047E+01	0.1736E+01	0.1017E+01	0.1268E+01
7	3.81	4.0	8.4	3.9	4.9	0.1073E+01	0.1064E+01	0.1027E+01	0.1210E+01
9	5.01	5.3	9.5	5.2	6.3	0.1023E+01	0.7775E+00	0.1034E+01	0.1148E+01
11	7.02	7.1	10.7	7.2	8.5	0.9703E+00	0.5358E+00	0.9863E+00	0.1031E+01
13	8.55	8.7	11.5	8.7	10.0	0.9527E+00	0.4344E+00	0.9390E+00	0.9378E+00
15	10.06	10.1	12.1	10.1	11.3	0.9321E+00	0.3676E+00	0.9037E+00	0.8478E+00
17	11.80	11.7	12.7	11.6	12.7	0.9088E+00	0.3133E+00	0.8700E+00	0.7494E+00
19	13.54	13.2	13.2	13.1	13.9	0.8136E+00	0.2737E+00	0.8293E+00	0.6591E+00
21	14.88	14.2	13.5	14.2	14.8	0.7470E+00	0.2498E+00	0.7872E+00	0.5959E+00
23	16.82	15.6	14.0	15.6	15.9	0.6690E+00	0.2222E+00	0.7081E+00	0.5143E+00
25	18.40	16.6	14.3	16.7	16.6	0.6361E+00	0.2040E+00	0.6325E+00	0.4564E+00
27	19.94	17.6	14.6	17.6	17.3	0.5832E+00	0.1893E+00	0.5538E+00	0.4064E+00
29	21.76	18.5	15.0	18.5	18.0	0.5405E+00	0.1747E+00	0.4619E+00	0.3552E+00
31	23.98	19.5	15.3	19.4	18.7	0.2999E+00	0.1598E+00	0.3580E+00	0.3023E+00
33	25.78	20.0	15.6	20.0	19.2	0.2456E+00	0.1496E+00	0.2845E+00	0.2660E+00
35	28.15	20.6	16.0	20.6	19.8	0.2234E+00	0.1383E+00	0.2048E+00	0.2259E+00
37	29.84	20.9	16.2	20.9	20.2	0.2279E+00	0.1312E+00	0.1597E+00	0.2018E+00

Table 3.17 Content of **scdb.d** for Example 2

BANK/no1			
BANK/no2			
BANK/no3			
BANK/no4			
BANK/no5			
BANK/no6			
BANK/no7			
0	2	3	0
1			
1			
70			

Table 3.18 Content of **bank.d** for Example 2

```
                        II.  Double web-angle connections
                             Riveted-to-beam and bolted-to-column

 No   Author                 Test Id.   Angle           lp(in)    gc(in)   gb(in)   pc(in)   db      nc    nb
  1   W.C.Bell et al. (1958) FK-4A      6 X 4 X 3/8    11.5000   2.5730   4.5000   3.0000   3/4"   1X 4   2X 4
  2                          FK-4B      6 X 4 X 3/8    11.5000   2.5730   4.5000   3.0000   3/4"   1X 4   2X 4
  3                          FK-4C      6 X 4 X 3/8    11.5000   2.5730   4.5000   3.0000   3/4"   1X 4   2X 4
  4   C.W.Lewitt et al.(1966)FK-3       6 X 4 X 3/8     8.5000   2.6320   3.5000   3.0000   3/4"   1X 3   2X 3
  5                          FK-4P      6 X 4 X 3/8    11.5000   2.0710   3.5000   3.0000   3/4"   2X 4   2X 4
  6                          WK-4       6 X 4 X 3/8    11.5000   2.5710   3.5000   3.0000   3/4"   1X 4   2X 4
  7                          FB-4       6 X 3.5 X 3/8  11.5000   2.5710   2.2500   3.0000   3/4"   1X 4   1X 4
  8                          FK-5       6 X 4 X 7/16   14.5000   2.5500   3.5000   3.0000   3/4"   1X 5   2X 5
```

References

AISC (1989) Allowable Stress Design and Plastic Design Specifications for Structural Steel Buildings, American Institute of Steel Construction, Chicago, 329 pp.

AISC (1986) Load and Resistance Factor Design Specifications for Structural Steel Buildings, American Institute of Steel Construction, Chicago, 313 pp.

Altman, W. G., Jr., Azizinamini, A., Bradburn, J. H., and Radziminski, J. B. (1982) Moment-Rotation Characteristics of Semi-Rigid Steel Beam-To-Column Connection, Department of Civil Engineering, University of South Carolina, Columbia.

Ang, K. M. and Morris, G. A. (1984) Analysis of three-dimensional frame with flexible beam-column connections, *Can. J. Civ. Eng.*, 11, 245–254.

Azizinamini, A., Bradburn, J. H., and Radziminski, J. B. (1985) Static and Cyclic Behavior of Semi-Rigid Steel Beam-Column Connections, Department of Civil Engineering, University of South Carolina, Columbia.

Bailey, J. R. (1970) Strength and Rigidity of Bolted Beam-To-Column Connections, Conf. on Joints in Structures, University of Sheffield, England, Vol. 1, Paper 4.

Bell, W. G., Chesson, E., Jr., and Munse, W. H. (1958) Static Tests of Standard Riveted and Bolted Beam-To-Column Connections, University of Illinois Engineering Experiment Station, Urbana, IL.

Bose, B. (1981) Moment-rotation characteristics of semi-rigid joints in steel structures, *J. Inst. Eng. (India)*, 62(2), 128–132.

Chen, W. F. and Kishi, N. (1987) Moment-Rotation Relation of Top- and Seat-Angle Connections, CE-STR-87-4, School of Civil Engineering, Purdue University, West Lafayette, IN.

Chen, W. F. and Kishi, N. (1989) Semirigid steel beam-to-column connections: data base and modeling, *ASCE*, 115(ST1), 105–119.

Chen, W. F. and Lui, E. M. (1985) Column with end restraint and bending in load and resistance factor design, *AISC J.*, 22(4), 105–132.

Colson, A. and Louveau, J. M. (1983) Connections incidence on the inelastic behavior of steel structures, *Euromech Coll.*, 174, Oct.

Dews, R. J. (1979) Experimental Test Results on Experimental End-Plate Moment Connections, M.S. thesis, Vanderbilt University, Nashville, TN.

Driscoll, G. C. (1987) Elastic-plastic analysis of top- and seat-angle connections, *J. Constr. Steel Res.*, 8, 119–136.

Frye, M. J. and Morris, G. A. (1975) Analysis of flexibly connected steel frames, *Can. J. Civ. Eng.*, 2, 280–291.

Goverdhan, A. V. (1983) A Collection of Experimental Moment-Rotation Curves and Evaluation of Prediction Equations for Semi-Rigid Connections, M.S. thesis, Vanderbilt University, Nashville, TN.

Grundy, P., Thomas, I. R., and Bennetts, I. D. (1980) Beam-to-column moment connections, *ASCE*, 106(ST1), 313–330.

Hechtman, R. A. and Johnston, B. G. (1947) Riveted Semi-Rigid Beam-To-Column Building Connections, Progress Report No. 1, AISC Research at Lehigh University, Bethlehem, PA.

Ioannides, S. A. (1978) Flange Behavior in Bolted End-Plate Moment Connections, Ph.D. thesis, Vanderbilt University, Nashville, TN.

Johnson, L. G., Cannon, J. C., and Spooner, L. A. (1960) High tensile preloaded bolted joints for development of full plastic moments, *Br. Welding J.*, 7, 560–569.

Johnstone, N. D. and Walpole, W. R. (1981) Bolted end-plate beam to column connections under earthquake type loading, Research Report 81-7, Department of Civil Engineering, University of Canterbury, Christchurch, New-Zealand.

Jones, S. W., Kirby, P. A., and Nethercot, D. A. (1980) Effect of semi-rigid connections on steel column strength, *J. Const. Steel Res.*, 1(1), 38–46.

Jones, S. W., Kirby, P. A., and Nethercot, D. A. (1981) Modeling of semi-rigid connection behaviour and its influence on steel column behaviour, in *Joints in Structural Steelwork*, Howlett, J. H., Jenkins, W. M., and Stainsby, R., Eds., Pentech Press, United Kingdom, 5.73–5.78.

Kishi, N. and Chen, W. F. (1986a) Steel Construction Data Bank Program, CE-STR-86-18, School of Civil Engineering, Purdue University, West Lafayette, IN.

Kishi, N. and Chen, W. F. (1986b) Data Base of Steel Beam-To-Column Connections, CE-STR-86-26, School of Civil Engineering, Purdue University, West Lafayette, IN.
Kishi, N., Chen, W. F., Matsuoka, K. G., and Nomachi, S. G. (1987a) Moment-rotation relation of top- and seat-angle with double web-angle connections, *Proc. of the State-of-the-Art Workshop on Connections and the Behavior, Strength and Design of Steel Structures*, Bjorhovde, R., Brozzetti, J., and Colson, A., Eds., Ecole Normale Superieure de Cachan, France, May 25–27, 1987.
Kishi, N., Chen, W. F., Matsuoka, K. G., and Nomachi, S. G. (1987b) Moment-rotation relation of single/double web-angle connections, *Proc. of the State-of-the-Art Workshop on Connections and the Behavior, Strength and Design of Steel Structures*, Bjorhovde, R., Brozzetti, J., and Colson, A., Eds., Ecole Normale Superieure de Cachan, France, May 25–27, 1987.
Kishi, N. and Chen, W. F. (1990) Moment-rotation relations of semi-rigid connections with angles, *ASCE*, 116(ST7), 1813–1834.
Kishi, N., Chen, W. F., Goto, Y., and Matsuoka K. G. (1991) Applicability of three-parameter power model to structural analysis of flexibly jointed frames, Proc. of Mechanics Computing in 1990's and Beyond, Columbus, OH, May 20–22, 233–237.
Krishnamurthy, N., Huang, H. T., Jefferey, P. K., and Avery, L. K. (1979) Analytical M-θ curves for end-plate connections, *ASCE*, 105(ST1), 133–145.
Kukreti, A. R., Murray, T. M., and Abolmaali, A. (1987) End-plate connection moment-rotation relationship, *J. Constr. Steel Res.*, 8, 137–157.
Lewitt, C. W., Chesson, E., Jr., and Munse, W. H. (1966) Restraint characteristics of flexible riveted and bolted beam-to-column connections, Department of Civil Engineering, University of Illinois, Urbana, IL.
Lightfoot, E. and LeMessurier, A. P. (1974) Elastic analysis of frameworks with elastic connections, *ASCE*, 100(ST6), 1297–1309.
Lipson, S. L. (1968) Single-Angle and Single-Plate Beam Framing Connections, Canadian Structural Engineering Conference, Toronto, Ontario, 141–162.
Lipson, S. L. (1977) Single angle welded-bolted connections, *ASCE*, 103(ST3), 559–571.
Lui, E. M. and Chen, W. F. (1986) Analysis and behavior of flexibly jointed frames, *Eng. Struct.*, 8(2), 107–118.
Marley, M. J. and Gerstle, K. H. (1982) Analysis and Tests of Flexibly-Connected Steel Frames, Report to AISC under Project 199, American Institute of Steel Construction, Chicago, IL.
Maxwell, S. M. et al. (1981) A realistic approach to the performance and application of semi-rigid jointed structures, in *Joints in Structural Steelwork*, Howlett, J. H., Jenkins, W. M., and Stainsby, R., Eds., Pentech Press, United Kingdom, 2.71–2.98.
Monforton, A. R. and Wu, T. S. (1963) Matrix analysis of semi-rigidly connected frames, *ASCE*, 87(ST6), 13–42.
Nethercot, D. A. (1985) Steel beam-to-column connections — A review list of test data. CIRIA, London.
Ostrander, J. R. (1970) An Experimental Investigation of End-Plate Connections, M.S. thesis, University of Saskatchewan, Saskatchewan, Canada.
Packer, J. A. and Morris, L. J. (1977) A limit state design method for the tension region of bolted beam-column connections, *Struct. Eng.*, 5(10), 446–458.
Patel, K. V. and Chen, W. F. (1984) Nonlinear analysis of steel moment connections, ASCE, 110(ST8), 1861–1874.
Picard, A., Giroux, Y.-M., and Brun, P. (1976) Discussion of analysis of flexibly connected steel frames, *Can. J. Civ. Eng.*, 3(2), 350–352.
Ramberg, W. and Osgood, W. R. (1943) Description of Stress-Strain Curves by Three Parameters, Technical Note, 902, National Advisory Committee for Aeronautics, Washington, D.C.
Rathbun, J. C. (1936) Elastic properties of riveted connections, *Trans. ASCE*, Paper No. 1933, 101, 524–563.
Richard, R. M., Gillet, P. E., Kriegh, J. D., and Lewis, B. A. (1980) The analysis and design of single plate framing connections, *AISC Eng. J.*, 2nd Quarter, 38–52.
Richard, R. M., Kriegh, J. D., and Hormby, D. E. (1982) Design of single plate framing connections with A307 bolts, *AISC Eng. J.*, 4th Quarter, 209–213.

Sherbourne, A. N. (1961) Bolted beam-to-column connections, *Struct. Eng.*, 39(Jun), 203–210.

Simitses, G. J. and Vlahinos, A. S. (1984) Elastic stability of rigidly and semi-rigidly connected unbraced frames, in *Steel Frame Structures*, Narayanan, R., Ed., Elsevier, London, 115–152.

Sommer, W. H. (1969) Behavior of Welded-Header-Plate Connections, M.S. thesis, University of Toronto, Toronto, Canada.

Surtees, J. O. and Mann, A. J. (1970) End Plate Connection in Plastically Designed Structures, Cong. on Joints in Structures, 1(5), University of Sheffield, Sheffield, England.

Tarpy, T. S. and Cardinal, J. W. (1981) Behavior of semi-rigid beam-to-column end plate connections, in *Joints in Structural Steelwork*, Howlett, J. H., Jenkins, W. M., and Stainsby, R., Eds., Pentech Press, United Kingdom, 2.3–2.25.

Thompson, L. E., McKee, R. J., and Visintainer, D. A. (1970) An Investigation of Rotation Characteristic of Web Shear Framed Connections Using A-36 and A-441 Steel, Department of Civil Engineering, University of Missouri, Rolla, MO.

Wu, F. S. (1989) Semi-Rigid Connections in Steel Frames, Ph.D. thesis, Purdue University, West Lafayette, IN.

Wu, F. S. and Chen, W. F. (1990) A design model for semi-rigid connections, *Eng. Struct.*, 12(2), 88–97.

Yee, Y. L. and Melchers, R. E. (1986) Moment-rotation curves for bolted connections, *ASCE*, 112(ST3), 615–635.

4: Second-Order Plastic Hinge Analysis of Frames

J. Y. Richard Liew, *Department of Civil Engineering, National University of Singapore, Republic of Singapore*

W. F. Chen, *School of Civil Engineering, Purdue University, West Lafayette, Indiana*

Notations

The following symbols are used in this chapter:

A	=	Cross-sectional area
b	=	Cross-section width
d	=	Cross-section depth
e	=	Axial displacement
E	=	Modulus of elasticity
E_t	=	Column tangent modulus
f	=	Element force vector
f_c	=	Element force vector in corotational coordinates
f_g	=	Structural force vector in global coordinates
F_{rc}	=	Maximum compressive residual stress
F_y	=	Material yield stress
G, G_A, G_B	=	Ratio of the bending stiffness of the columns versus that of the beam at a beam-column joint (subscripts apply to the respective ends of the column)
I, I_b, I_c	=	Moment of inertia of cross-section, b and c denoting beam and column
I_x, I_y	=	Moment of inertia about the strong- and weak-axis
k	=	Distance from the section heel to the toe of the fillet
k_c	=	Element corotational stiffness matrix
k_g	=	Element global stiffness matrix
K, K_i	=	Effective length factor, subscript i refers to i^{th} column
K_g	=	Structural global stiffness matrix
L, L_c	=	Length of member or story height, subscript c refers to column
L_i	=	Length of ith column
L_f	=	Length of a deformed element

M	=	Bending moment
M_p	=	Plastic bending moment
n_c	=	Number of columns in the plane of the story
n_s	=	Number of storys in the plane of the frame
P	=	Axial force
P_e	=	Euler buckling load, $\pi^2 EI/L^2$
P_n	=	Nominal axial strength
P_u	=	Required axial strength of a member; factored vertical load
P_y	=	Axial load at full-yield condition or squash load
r, r_x, r_y	=	Radius of gyration (subscripts apply to the respective bending axis)
S_1, S_2	=	Stability functions
t_f	=	Thickness of column flange
t_w	=	Web thickness
T_{cg}	=	Corotational global transformation matrix
T_i	=	Initial stress transformation matrix
x_o	=	Horizontal projection of the undeformed member
x_f	=	Horizontal projection of the deformed member
X_c	=	Location of the maximum moment along the length of a member
y_o	=	Vertical projection of the undeformed member
y_f	=	Vertical projection of the deformed member
Z	=	Plastic section modulus
α	=	Force-state parameter defined by Eq. (4.41)
α_o	=	The force-state parameter associated with the initial yield surface
β	=	Ratio of the load on the beam to the total load on the frame;
δ_o	=	Maximum magnitude of member initial out-of-straightness
Δ	=	Relative lateral translation between the column ends; lateral story or frame drift; an increment
Δ_i	=	Out-of-plumbness of ith column
ε	=	Normal strain, e/L
ε_y	=	Normal yield strain, F_y/E
λ_c	=	Column slenderness parameter, $(KL/(\pi r))\sqrt{F_y/E}$
ϕ_b, ϕ_c	=	Resistance factors for beam and column in AISC-LRFD
ϕ, ϕ_A, ϕ_B	=	Stiffness reduction parameter (subscripts refer to the respective ends of the element)
ψ_o	=	Sway angle associated with column out-of-plumbness
ρ	=	Normalized axial force, $P/P_e = PL^2/(\pi^2 EI)$
$\theta, \theta_A, \theta_B$	=	Element end rotation (subscripts refer to the respective ends of the element)

4.1 Introduction

In elastic-plastic hinge analysis, inelasticity in frame elements is assumed to concentrate at "zero-length" plastic hinges. Regions in the frame elements other than at the plastic hinges are assumed to behave elastically. If the cross-section forces at any particular locations in an element are less than the cross-section plastic capacity, elastic behavior is assumed. If the section plastic capacity is reached, a plastic hinge is formed and the element stiffness matrix is adjusted to account for the presence of a plastic hinge. The cross-section response after the formation of a plastic hinge is usually assumed to be perfectly plastic with no strain hardening. Although the effects of biaxial bending, shear, and bimoment may be included in the modeling of the cross-section plastic strength, shear and bimoment effects are often neglected in most studies (Orbison, 1982; Duan and Chen, 1990).

There are two common approaches for elastic-plastic hinge analysis. The first approach is called the beam-column stability function approach in which the plastic hinge can undergo plastic rotation only. The change in the axial force in a frame element is based solely on the element axial force-displacement relationship, with no effect from the inelasticity at the plastic hinges. The second approach is based on a force-space plasticity formulation in which an associated flow rule is used to describe the relationship between the axial and rotation plastic deformations at a fully plastified cross-section. For most practical cases, the differences in results predicted by these two plastic hinge approaches are small. When the axial force is zero, the two approaches are identical. Also, both approaches can account for force-point movement on the plastic strength surface (Riahi et al., 1978; Orbison, 1982; Lui, 1985; Ziemian, 1990; Deierlein et al., 1991; Liew, 1992; Liew et al., 1992a-c). In other words, if the axial force is increased on a fully plastified cross-section, the bending moment would need to be decreased so that the cross-section plastic capacity is not violated.

Elastic-plastic hinge analyses can be further classified into two categories; first order and second order, depending on whether geometric second-order effects are accounted for. For *first-order elastic-plastic hinge analysis*, equilibrium is formulated based on an undeformed geometry. Thus, only the inelasticity effects that influence the strength of the structure are included; the geometric nonlinear effects on the equilibrium of the structure are not considered. First-order elastic-plastic hinge analysis predicts the maximum load of the structure corresponding to the formation of a plastic collapse mechanism. This analysis approach essentially predicts the same maximum load as the conventional rigid-plastic analysis approach (Beedle, 1958; Baker and Heyman, 1969, Heyman, 1969).

If equilibrium is formulated based on deformed structural geometry, the plastic-hinge analysis is called *second-order elastic-plastic hinge analysis*. For frame members subjected to end forces only, the second-order elastic-plastic hinge approach typically employs only one beam-column element per member for global frame analysis. Therefore, it is computationally more efficient and economical than second-order plastic-zone analyses (see Chapters 5 and 6 for plastic-zone analyses). However, elastic-plastic hinge analysis only models yielding at zero-length plastic hinges, it does not accurately represent the distributed plasticity and the associated P-δ effects within individual members of the frame. For slender structures in which elastic instability is the predominant mode of failure, both the plastic hinge and "plastic-zone" methods lead to almost identical results. However, for structures that exhibit significant yielding in the members, the elastic-plastic hinge method often over-predicts the actual stiffness and strength of the structure. Therefore, some refinements to the basic elastic-plastic hinge theory are necessary to generalize its application for the analysis of a wider range of structural systems (Liew et al., 1992a-c).

In this chapter, a method called the *refined plastic hinge* approach is developed. This method is comparable to the elastic-plastic hinge analysis in efficiency and simplicity, but it does not possess its limitations. The refined plastic hinge analysis utilizes a two-surface yield model and an effective tangent modulus to account for stiffness degradation due to distributed plasticity in frame members. The member stiffness is assumed to degrade gradually as the second-order forces at critical locations approach the cross-section plastic strength. Also, a column tangent modulus is used to represent the effective stiffness of the member when it is loaded with high axial load. Thus, the refined plastic hinge model approximates the effects of distributed plasticity along the element length caused by initial imperfections and large bending and axial force actions. In fact, research by Liew et al. (1992b and c) has shown that the refined plastic hinge analysis is capable of capturing the interaction effects of strength and stability of structural systems and that of their component elements. This type of analysis method may, therefore, be classified as an *advanced analysis* for which separate specification member capacity checks are not required.

Section 4.2 of this chapter defines the scope and the assumptions involved in the development of plastic hinge based elements suitable for inelastic analysis of planar steel frames. Sections 4.3 and 4.4 deal with the modeling of frame and truss elements that can undergo large rigid-body displacements. An updated Lagrangian approach proposed by Lui (1985) is employed in which a moving corotational coordinate system attached to the element is related to a fixed global coordinate system at each step of load increments. Section 4.5 focuses on the formulation of the elastic-plastic hinge model, and provides the framework for the development of the refined plastic hinge method. Section 4.6 outlines the development of the refined plastic hinge model. This analysis model provides essentially the same predictions of member and system performance as plastic-zone methods. Section 4.7 presents a procedure that incorporates connection nonlinear effects in the analysis of two-dimensional steel frames. Section 4.8 discusses the numerical procedures, and provides some thoughts for implementing the second-order plastic hinge based analyses using a tangent-stiffness iteration technique. Sections 4.9 describes the user's manual of a second-order plastic hinge based analysis program developed by the authors. Examples illustrating the use of the program are also provided for several frames.

4.2 Assumptions and Scope

The general assumptions used in modeling the beam-column elements are:

(1) All elements are initially straight and prismatic. Plane cross-sections remain plane after deformation.
(2) Local buckling and lateral-torsional buckling are not considered. Therefore, all members are assumed to be fully compact and adequately braced to preclude out-of-plane deformations.

(3) Large rigid-body displacements are allowed, but the member deformations and strains are small.
(4) The element stiffness formulation is based on the conventional beam-column stability functions. It includes axial and bending deformations, but not shear. Element bowing effects are neglected.
(5) The formulation is limited by its ability to model plastic hinges only at the element ends.
(6) Plastic hinges can sustain inelastic rotations only. Strain hardening is not considered.
(7) Connection moment-rotation behavior is modeled by nonlinear rotational springs attached at the element ends.

The assumptions that the member is prismatic and member distortions are small are reasonable for ordinary steel frame structures. Although the steel frame may undergo large rigid-body displacements at collapse, the distortion of each member with respect to its chord length in the displaced configuration will remain small since steel members with compact cross sections usually exhibit high bending rigidity. Large deformation theory is useful to model the full post-collapse behavior of members in the structure. However, for many types of steel structures, large strains usually do not occur until the members are loaded far into the post-collapse region.

Element bowing effects are not considered in the present work because many practical frame members usually have slenderness ratios in the range for which the axial shortening is often dominated by inelastic axial deformation. However, for very slender beam-column members, the bowing effects may need to be considered in the element stiffness formulation. Alternatively, the member may be discretized into several elements to approximate the member bowing effects.

Inelastic behavior in the member is assumed to be contained within a zero-length plastic hinge. The reduction of the plastic moment capacity at the plastic hinge due to the presence of axial force is considered. Once a plastic hinge is formed, the cross-section forces are allowed to move on the plastic strength surface. However, the possible benefits of strain hardening at the plastic hinges are not considered. In fact, the ability of a beam-column to develop significant strain hardening is dependent on many factors, such as moment gradient and interaction of local and lateral-torsional buckling effects, and distributed yielding along the member length. Presently, the precise effects of yielding, strain hardening, and possible local and lateral torsional buckling on the full moment-rotation characteristics of beam-column members are still not well enough quantified for direct implementation in a practical inelastic analysis.

Although the numerical formulation is limited to model plastic hinges only at the element ends, maximum second-order elastic moment that occurs within the element length can be detected using the analytical expressions given in Eqs. (1.17-1.19) in Chapter 1. Usually, the analysis of frame members with maximum moment within their span lengths, and the analysis of beam-columns with in-span loading and inclined members subjected to gravity loads require more than one beam-column element per member to capture the plastic hinge formation in the member. However,

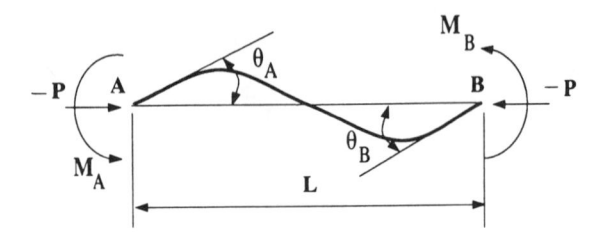

FIGURE 4.1. Force-displacement relationships of a frame element.

research by Chen and Atsuta (1976) shows that the insertion of a plastic hinge at the maximum moment point in a frame member is not required for an accurate estimate of the maximum strength of the member. As long as the "exact" location of the plastic hinge in a member is not more than L/6 from the assumed position, the strength prediction is not more than 5% in error compared with the "exact" solutions. The above observation is true for beam-columns subjected to various in-span loading cases (Chen and Atsuta, 1976).

In the proposed analysis approach, loads can be applied proportionally or nonproportionally. However, proportionally loaded frames are generally weaker than frames in which the gravity loads are applied first, followed by lateral loading. This is because of the elastic unloading that occurs at the inelastically loaded connections and at the plastic hinges for the nonproportional load case. In the present work, elastic unloading at the fully plastified cross-section are neglected. The analytical results obtained based on this assumption are generally conservative for statically loaded frames. The numerical implementation following this assumption is also straightforward in concept, and it does not possess the problem of being trapped into a recurring process of loading, unloading, and reloading of plastic hinges (Orbison, 1982).

4.3 Modeling of Elastic Frame Elements

Consider a prismatic frame element of length L and moment of inertia I with modulus of elasticity E shown in Figure 4.1. The incremental force-displacement relationship of this element may be written as:

$$\begin{Bmatrix} \dot{M}_A \\ \dot{M}_B \\ \dot{P} \end{Bmatrix} = \frac{EI}{L} \begin{bmatrix} S_1 & S_2 & 0 \\ S_2 & S_1 & 0 \\ 0 & 0 & A/I \end{bmatrix} \begin{Bmatrix} \dot{\theta}_A \\ \dot{\theta}_B \\ \dot{e} \end{Bmatrix} \tag{4.1}$$

in which $\dot{M}_A$, $\dot{M}_B$, $\dot{\theta}_A$, and $\dot{\theta}_B$ are the incremental end moments and the corresponding joint rotations at element end A and B, respectively. $\dot{P}$ and $\dot{e}$ (positive in tension) are the incremental axial force and displacement in the longitudinal direction of the element. S_1 and S_2 are the stability functions that account for the effect of the axial force on the bending stiffness of the member. The conventional stability functions may be written as:

$$S_1 = \begin{cases} \dfrac{\pi\sqrt{\rho}\sin(\pi\sqrt{\rho}) - \pi^2\rho\cos(\pi\sqrt{\rho})}{2 - 2\cos(\pi\sqrt{\rho}) - \pi\sqrt{\rho}\sin(\pi\sqrt{\rho})} & \text{if } P < 0 \\[2ex] \dfrac{\pi^2\rho\cosh(\pi\sqrt{\rho}) - \pi\sqrt{\rho}\sinh(\pi\sqrt{\rho})}{2 - 2\cosh(\pi\sqrt{\rho}) + \pi\sqrt{\rho}\sinh(\pi\sqrt{\rho})} & \text{if } P > 0 \end{cases} \quad (4.2)$$

$$S_2 = \begin{cases} \dfrac{\pi^2\rho - \pi\sqrt{\rho}\sin\rho(\pi\sqrt{\rho})}{2 - 2\cos(\pi\sqrt{\rho}) - \pi\sqrt{\rho}\sin(\pi\sqrt{\rho})} & \text{if } P < 0 \\[2ex] \dfrac{\pi\sqrt{\rho}\sinh(\pi\sqrt{\rho}) - \pi^2\rho}{2 - 2\cosh(\pi\sqrt{\rho}) + \pi\sqrt{\rho}\sinh(\pi\sqrt{\rho})} & \text{if } P > 0 \end{cases} \quad (4.3)$$

where $\rho = P/(\pi^2 EI/L^2)$, and P is taken as positive in tension.

The numerical solutions obtained from Eqs. (4.2) and (4.3) are indeterminate when the axial force is equal to zero. To circumvent this problem and to avoid the use of different expressions of S_1 and S_2 for a different sign of axial forces, Goto and Chen (1987) have proposed a set of expressions that make use of power-series expansions to approximate the stability functions. The power-series expressions have been shown to converge to a high degree of accuracy within the first ten terms of the series expansions. Alternatively, if the axial force in the member falls within the range $-2.0 \leq \rho \leq 2.0$, the following simplified expressions may be used to closely approximate the stability functions (Lui, 1985):

$$S_1 = 4 + \frac{2\pi^2\rho}{15} - \frac{(0.01\rho + 0.543)\rho^2}{4 + \rho} - \frac{(0.004\rho + 0.285)\rho^2}{8.183 + \rho} \quad (4.4a)$$

$$S_2 = 2 - \frac{\pi^2\rho}{30} + \frac{(0.01\rho + 0.543)\rho^2}{4 + \rho} - \frac{(0.004\rho + 0.285)\rho^2}{8.183 + \rho} \quad (4.4b)$$

Equations (4.4a) and (4.4b) are applicable for members in tension (positive P) and compression (negative P). For most practical applications, Eqs. (4.4a-b) give an excellent correlation to the "exact" expressions given by Eqs. (4.2) and (4.3). However, for ρ other than the range of $-2.0 \leq \rho \leq 2.0$, the conventional stability functions should be used instead. The stability function approach enables the use of only one element for each frame member and still maintains good accuracy in the element stiffness terms and in the recovery of element end forces for all ranges of axial load. The finite element geometric stiffness approach may lead to some error in the stiffness terms and in the force recovery process; therefore, in using the geometric stiffness approach, its range of applicability must be understood (White and Hajjar, 1991).

The element tangent stiffness relationship from Eq. (4.1) may be written symbolically as

$$\dot{f}_c = k_c \dot{d}_c \quad (4.5)$$

in which $\dot{f}_c$ and $\dot{d}_c$ are the incremental element end forces and displacements, respectively, and k_c is the element basic tangent stiffness matrix. For a plane frame member, three additional degrees of freedom are required to describe the total displacements of the member. If d_{g1}, d_{g2} ... and d_{g6} are defined as the global translational and rotational degrees of freedom of a frame member (Figure 4.2), it can be shown that the local displacements are related to the global displacements by

$$d_{c1} = \theta_A = \theta_o + d_{g3} - \tan^{-1}\frac{y_o + d_{g5} - d_{g2}}{x_o + d_{g4} - d_{g1}} \qquad (4.6a)$$

$$d_{c2} = \theta_B = \theta_o + d_{g6} - \tan^{-1}\frac{y_o + d_{g5} - d_{g2}}{x_o + d_{g4} - d_{g1}} \qquad (4.6b)$$

$$d_{c3} = \frac{(2x_o + d_{g4} - d_{g1})(d_{g4} - d_{g1}) + (2y_o + d_{g5} - d_{g2})(d_{g5} - d_{g2})}{L_f + L} \qquad (4.6c)$$

The expression for d_{c3} in Eq. (4.6c) is more accurate than the value calculated from $L_f - L$. This is because Eq. (4.6c) avoids finding the small difference between large member lengths (Cook et al., 1989). Equation (4.6c) is obtained by writing $d_{c3} = (L_f^2 - L^2)/(L_f + L)$, and than solving for d_{c3}. In the denominator, $L_f + L \approx 2L$ may be used, since small displacement theory is presumed for the corotational chord element (Belytschko and Hsieh, 1973).

Upon differentiation of Eqs. (4.6a–c) with respect to time (or pseudo-time), the incremental kinematic relationship relating the two sets of displacement vectors may be written as

$$\dot{d}_c = \frac{\partial d_{cj}}{\partial d_{gk}}\dot{d}_g = T_{cg}\dot{d}_g \qquad (4.7)$$

where $j = 1, 2, 3$, $k = 1, 2 \ldots 6$, and

$$\dot{d}_c = [\dot{\theta}_A \dot{\theta}_B \dot{e}]^T$$

$$\dot{d}_g = [\dot{d}_{g1} \dot{d}_{g2} \dot{d}_{g3} \dot{d}_{g4} \dot{d}_{g5} \dot{d}_{g6}]^T$$

$$T_{cg} = \begin{bmatrix} -s/L & c/L & 1 & s/L & -c/L & 0 \\ -s/L & c/L & 0 & s/L & -c/L & 1 \\ -c & -s & 0 & c & s & 0 \end{bmatrix} \qquad (4.8)$$

in which $c = \cos\theta$, $s = \sin\theta$, and θ is the angle of inclination of the chord of the deformed member.

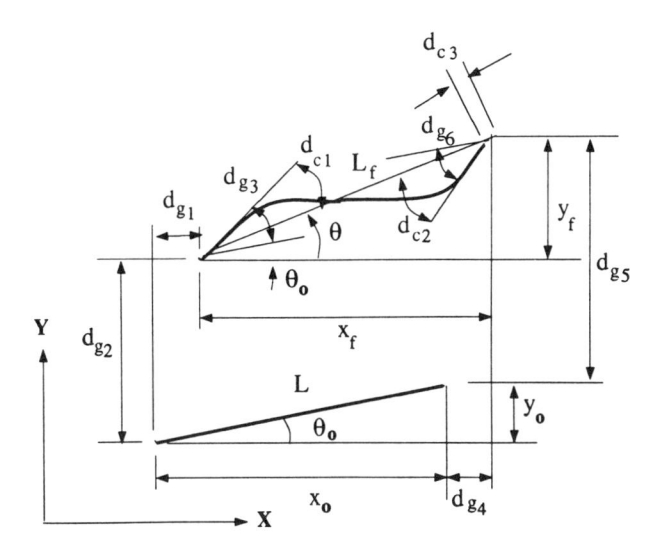

FIGURE 4.2. Kinematic relationships between local and global displacements of a frame element.

Based on the principle of equilibrium, the forces in the two systems are related by

$$f_g = T_{cg}^T f_c \tag{4.9}$$

Taking derivatives on both sides of Eq. (4.9) gives

$$\dot{f}_g = T_{cg}^T \dot{f}_c + \dot{T}_{cg}^T f_c \tag{4.10}$$

In view of Eqs. (4.5) and (4.7), Eq. (4.10) may be further written as

$$\dot{f}_g = T_{cg}^T k_c T_{cg} \dot{d}_g + \dot{T}_{cg}^T f_c \tag{4.11}$$

The matrix $\dot{T}_{cg}^T$ in Eq. (4.11) can be evaluated by taking derivative of T_{cg}^T with respect to the pseudo time as

$$\dot{T}_{cg}^T = \frac{\partial T_{cg}^T}{\partial d_{gk}} \dot{d}_{gk} = \left[\frac{\partial^2 d_{ci}}{\partial d_{gj} \partial d_{gk}}\right]^T \dot{d}_{gk} \tag{4.12}$$

Substituting $\dot{T}_{cg}^T$ from Eq. (4.12) into Eq. (4.11) gives the following expression for $\dot{f}_g$

$$\dot{f}_g = \left(T_{cg}^T k_c T_{cg} + T_1 M_A + T_2 M_B + T_3 P\right) \dot{d}_g \tag{4.13}$$

where

$$T_1 = T_2 = \frac{1}{L^2} \begin{bmatrix} -2sc & c^2-s^2 & 0 & 2sc & -(c^2-s^2) & 0 \\ & 2cs & 0 & -(c^2-s^2) & -2sc & 0 \\ & & 0 & 0 & 0 & 0 \\ \text{sym.} & & & -2sc & c^2-s^2 & 0 \\ & & & & 2sc & 0 \\ & & & & & 0 \end{bmatrix} \quad (4.14)$$

$$T_3 = \frac{1}{L} \begin{bmatrix} s^2 & -sc & 0 & -s^2 & sc & 0 \\ & c^2 & 0 & sc & -c^2 & 0 \\ & & 0 & 0 & 0 & 0 \\ \text{sym.} & & & s^2 & -sc & 0 \\ & & & & c^2 & 0 \\ & & & & & 0 \end{bmatrix} \quad (4.15)$$

Equation (4.13) is the incremental force-displacement relationships of a frame element in the global coordinate system, and it may be written symbolically as

$$\dot{f}_g = k_g \dot{d}_g \quad (4.16)$$

where k_g represents the tangent stiffness matrix of a frame element. It should be noted that in the derivation of the tangent stiffness matrix, k_g, the joints are assumed to be rigid. If plastic hinges or connections are presented at the element ends, the tangent stiffness matrix needs to be modified. These modifications are discussed in Sections 4.5 and 4.7, respectively.

4.4 Modeling of Elastic Truss Elements

For structures subjected to lateral wind or earthquake loadings, truss diagonal bracings may be used to reduce the frame drifts and to enhance the lateral-load resistance of the structure. In design, these braces are usually assumed to carry axial force only. Therefore, it is justifiable to use truss elements to model the bracing members. The truss elements also may be used for modeling gravity columns that do not participate in the lateral-force resisting system. These gravity columns (leaned columns), which are widely used in many types of low-rise industrial buildings and tall office building frames (Springfield, 1991), are usually designed to carry only gravity loads. Therefore, they can be modeled by the truss element described in this section.

The tangent stiffness relationship for a bracing element can be obtained from the tangent stiffness relationship of a frame element by deleting the appropriate rows and columns in Eq. (4.13) that correspond to the rotational degrees of freedom of the element. The resulting tangent stiffness relationship of a truss element is

$$\dot{f}_g = \left(T_{cg}^T k_c T_{cg} + TP\right)\dot{d}_g = k_g \dot{d}_g \quad (4.17)$$

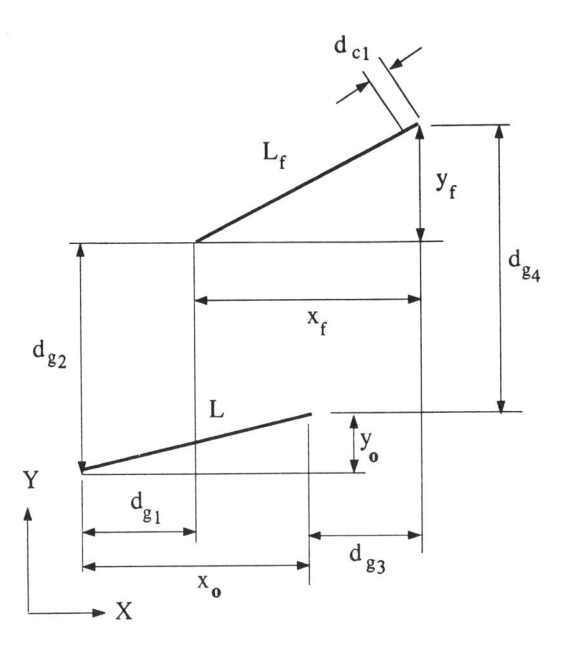

FIGURE 4.3. Kinematic relationships between local and global displacements of a truss element.

where (refer to Figure 4.3)

$$\dot{f}_g = \begin{bmatrix} \dot{f}_{g1} \dot{f}_{g2} \dot{f}_{g3} \dot{f}_{g4} \end{bmatrix}^T \tag{4.18}$$

$$\dot{d}_g = \begin{bmatrix} \dot{d}_{g1} \dot{d}_{g2} \dot{d}_{g3} \dot{d}_{g4} \end{bmatrix}^T \tag{4.19}$$

$$T_{cg} = [-c \ -s \ c \ s] \tag{4.20}$$

$$k_c = \frac{EA}{L} \tag{4.21}$$

$$T = \frac{1}{L} \begin{bmatrix} s^2 & -sc & -s^2 & sc \\ & c^2 & sc & -c^2 \\ \text{sym.} & & s^2 & -sc \\ & & & c^2 \end{bmatrix} \tag{4.22}$$

in which $s = \sin\theta$ and $c = \cos\theta$. θ is the inclination angle of the displaced element chord.

4.5 Second-Order Elastic-Plastic Hinge Analysis

In a second-order elastic-plastic hinge analysis, the frame element is assumed to remain elastic until the second-order forces at the critical location in the element

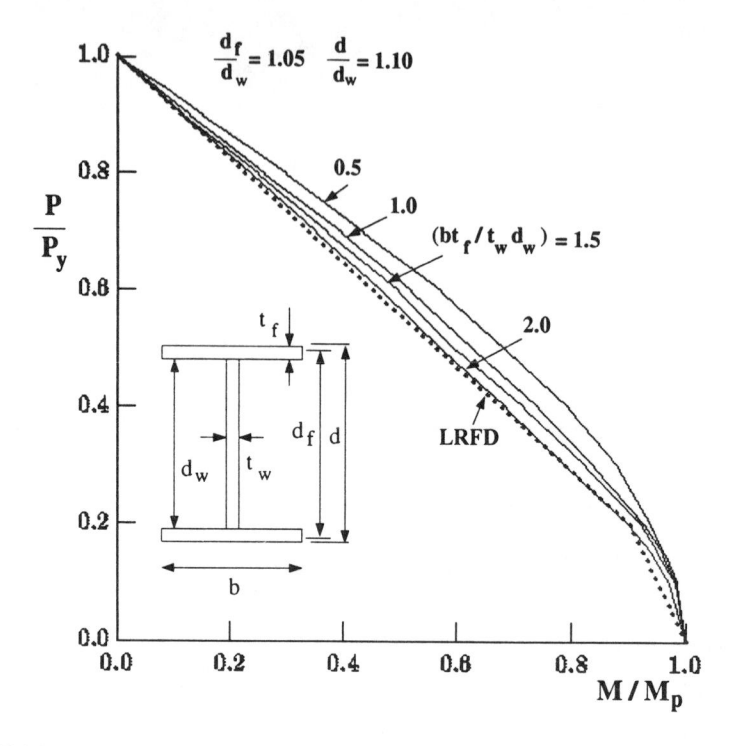

FIGURE 4.4. Strength interaction curves for wide-flange sections bending about the strong axis.

reach the cross-section plastic strength. Once the plastic strength is reached, a plastic hinge is formed and the cross-section behavior is assumed to be perfectly plastic with no strain hardening.

4.5.1 Cross-Section Plastic Strength

The AISC-LRFD bilinear interaction equations (AISC-LRFD, 1986) for a member of compact cross-section and of zero length are used in the present formulation for elastic-plastic hinge analysis of members subjected to strong- or weak-axis bending. These equations are written as:

$$\frac{P}{P_y} + \frac{8}{9}\frac{M}{M_p} = 1.0 \quad for \quad \frac{P}{P_y} \geq 0.2 \tag{4.23a}$$

$$\frac{P}{2P_y} + \frac{M}{M_p} = 1.0 \quad for \quad \frac{P}{P_y} < 0.2 \tag{4.23b}$$

where P_y is the squash load, M_p is the plastic moment capacity for member under pure bending action, and P and M are the second-order axial force and bending moment at the cross-section being considered. Equations (4.23a and b) are plotted in Figures 4.4 and 4.5 for comparison with the more "exact" analytical solutions. It can be

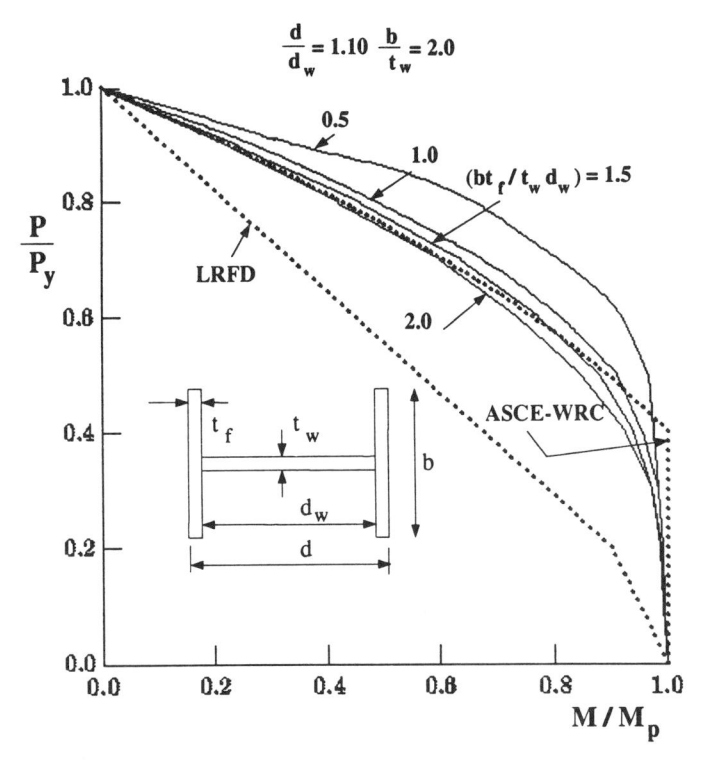

FIGURE 4.5. Strength interaction curves for wide-flange sections bending about the weak axis.

observed that Eqs. (4.23a and b), which assume the same functional relationships for both strong- and weak-axis strengths, provide a reasonable lower-bound fit to most of the strong-axis strengths, but they are rather conservative for the weak-axis strength when the bending moment and axial force are dominant. A simplified weak-axis cross-section strength equation proposed by ASCE-WRC (1971), as shown in Figure 4.5, gives a better fit to the weak-axis analytical strength results.

It should be noted that the resistance factors ϕ_b and ϕ_c have been omitted in Eqs. (4.23a and b), but they may be incorporated for use in Load and Resistance Factor Design (LRFD). Detailed discussions on use of second-order elastic-plastic hinge analysis for direct frame design without the need of specification member capacity checks are reported in Liew et al. (1992a).

4.5.2 Modification of Element Stiffness for the Presence of Plastic Hinges

If the state of forces at any cross-section equals or exceeds the plastic section capacity, a plastic hinge is formed, and slope discontinuity at the plastic-hinge location occurs. Therefore, the element force-displacement relationships need to be modified to reflect the change in element behavior due to formation of plastic hinges at the element ends.

If a plastic hinge is formed at element end A, the incremental force-displacement relationship from Eq. (4.1) may be written as

$$\begin{Bmatrix} \Delta \dot{M}_{pcA} \\ \dot{M}_B \\ \dot{P} \end{Bmatrix} = \frac{EI}{L} \begin{bmatrix} S_1 & S_2 & 0 \\ S_2 & S_1 & 0 \\ 0 & 0 & A/I \end{bmatrix} \begin{Bmatrix} \dot{\theta}_A \\ \dot{\theta}_B \\ \dot{e} \end{Bmatrix} \quad (4.24)$$

where ΔM_{pcA} is the change in plastic moment capacity at end A as P changes. $\dot{\theta}_A$ can be solved from the first row of Eq. (4.24) as

$$\dot{\theta}_A = \frac{L \Delta \dot{M}_{pcA}/EI - S_2 \dot{\theta}_B}{S_1} \quad (4.25)$$

Back-substituting Eq. (4.25) into the second and third row of Eq. (4.24), the modified element force-displacement relationship can be obtained

$$\begin{Bmatrix} \dot{M}_A \\ \dot{M}_B \\ \dot{P} \end{Bmatrix} = \frac{EI}{L} \begin{bmatrix} 0 & 0 & 0 \\ 0 & (S_1 - S_2^2/S_1) & 0 \\ 0 & 0 & A/I \end{bmatrix} \begin{Bmatrix} \dot{\theta}_A \\ \dot{\theta}_B \\ \dot{e} \end{Bmatrix} + \begin{Bmatrix} 1 \\ S_2/S_1 \\ 0 \end{Bmatrix} \Delta \dot{M}_{pcA} \quad (4.26)$$

Equation (4.26) represents the modified incremental force-displacement relationship of a frame element with a plastic hinge formed at end A. If a plastic hinge is formed at the end B, a similar approach can be followed and the corresponding basic incremental force-displacement relationship is

$$\begin{Bmatrix} \dot{M}_A \\ \dot{M}_B \\ \dot{e} \end{Bmatrix} = \frac{EI}{L} \begin{bmatrix} (S_1 - S_2^2/S_1) & 0 & 0 \\ 0 & 0 & 0 \\ 0 & 0 & A/I \end{bmatrix} \begin{Bmatrix} \dot{\theta}_A \\ \dot{\theta}_B \\ \dot{P} \end{Bmatrix} + \begin{Bmatrix} S_2/S_1 \\ 1 \\ 0 \end{Bmatrix} \Delta \dot{M}_{pcB} \quad (4.27)$$

If plastic hinges are formed at both ends of the element, θ_A and θ_B can be written in terms of the change in moment at the respective end of the element. The resulting matrix equation is

$$\begin{Bmatrix} \dot{M}_A \\ \dot{M}_B \\ \dot{P} \end{Bmatrix} = \frac{EI}{L} \begin{bmatrix} 0 & 0 & 0 \\ 0 & 0 & 0 \\ 0 & 0 & A/I \end{bmatrix} \begin{Bmatrix} \dot{\theta}_A \\ \dot{\theta}_B \\ \dot{e} \end{Bmatrix} + \begin{Bmatrix} \Delta \dot{M}_{pcA} \\ \Delta \dot{M}_{pcB} \\ 0 \end{Bmatrix} \quad (4.28)$$

where ΔM_{pcA} and ΔM_{pcB} are the change in the plastic moment capacity at the respective end of the member as P changes.

Equations (4.26–4.28) account for the presence of plastic hinge(s) at the element end(s). They may be written symbolically as

$$\dot{f}_c = k_{ch} \dot{d}_c + \dot{f}_{cp} \quad (4.29)$$

where k_{ch} is the modified basic tangent stiffness matrix due to the presence of plastic hinge(s). $\dot{f}_{cp}$ is an equilibrium force correction vector that results from the change in moment capacity as P changes.

If Eq. (4.5) is replaced by Eq. (4.29) and the procedure in Section 4.3 is followed, the modified element force-displacement relationship in global coordinates may be written as

$$\dot{f}_g = \left(T_{cg}^T k_{ch} T_{cg} + T_1 M_A + T_2 M_B + T_3 P\right) \dot{d}_g + T_{cg}^T \dot{f}_{cp} \tag{4.30}$$

or

$$\dot{f}_g = K_{gh} \dot{d}_g + \dot{f}_{gp} \tag{4.31}$$

where

$$K_{gh} = T_{cg}^T k_{ch} T_{cg} + T_1 M_A + T_2 M_B + T_3 P \tag{4.32}$$

is the modified element tangent stiffness matrix, and

$$\dot{f}_{gp} = T_{cg}^T \dot{f}_{cp} \tag{4.33}$$

is the global equilibrium force correction vector.

Once a plastic hinge is formed in a member, the subsequent change in the plastic moment capacity due to the change in axial force will affect the force-displacement relationship of the beam-column element. In other words, referring to Figure 4.6, once the plastic strength is reached at point Q, the state of moment is stationary at Q. However, because of the presence of axial force, the state of moment will change. If the axial force P is increased, the force point should move from Q to R. Under no circumstances should the forces be allowed to breach the strength surface associated with the full plastic cross-section. Several plastic-hinge models proposed in the past do not satisfy this requirement. Errors can result if the element is assumed to accept additional axial load at a constant plastic moment after hinges form.

4.5.3 Illustrative Example

Figure 4.7 compares the in-plane strength curve obtained by the elastic-plastic hinge analysis with the "exact" plastic-zone results generated by Kanchanalai (1977). The portal frame shown in the inset of Figure 4.7 is one of the series benchmark frames studied by Liew (1992) for the assessment of the elastic-plastic hinge method for representing in-plane beam-column strength performance. In Kanchanalai's plastic-zone analyses, both the in-plane strength curves for strong- and weak-axis bending are presented, whereas in the plastic-hinge analysis, only the strength curve for the strong-axis bending is shown. The weak-axis strength curve from this analysis is identical with the strong-axis strength curve, since the results are presented in a nondimensional form, and only one plastic-strength curve (Eqs. 4.23a and b) is used for both the strong- and weak-axis section strengths.

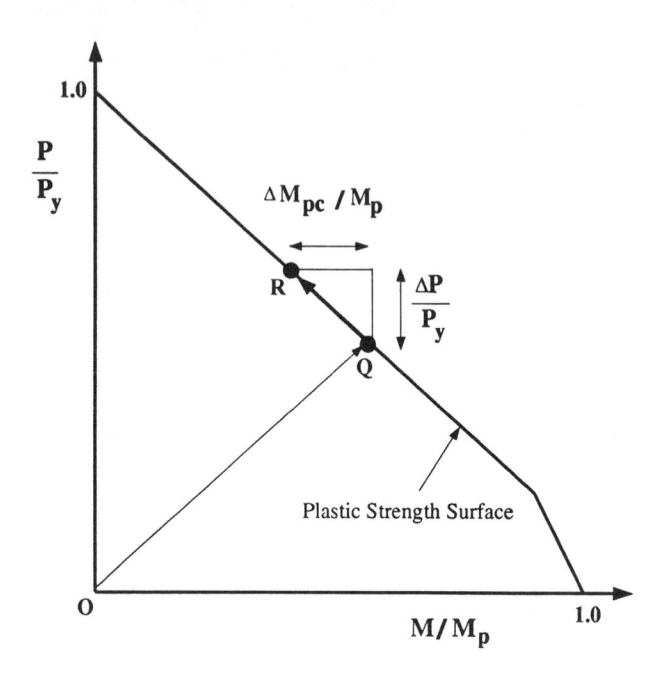

FIGURE 4.6. Equilibrium correction for force-point movement on the plastic strength surface.

The study shows that the elastic-plastic hinge analysis over predicts the maximum strength of the frame. The strong-axis strength from the plastic hinge model is approximately in error by a maximum of 21% for the case of $L_c/L = 40$ (the errors are measured radially from the origin of the plots). The weak-axis strength is also significantly over-predicted by the elastic-plastic hinge analysis, despite the fact that a more conservative weak-axis plastic strength curve is used (see Section 4.5.1) For beam-columns with $L_c/r = 40$, the maximum error in the weak-axis strength prediction based on the elastic-plastic hinge analysis is 30% unconservative. Since the errors cover quite a wide range of axial force and moment combinations, it can be concluded that some refinements to the conventional elastic-plastic hinge model are necessary to generalize its application for the analysis of a wider range of structural systems. These refinements are discussed in the next section.

4.6 Second-Order Refined Plastic Hinge Analysis

The elastic-plastic hinge model described in Section 4.5 can be refined by considering the actual distribution of plasticity in a beam-column element loaded by arbitrary end forces, and then attempting to model the effective stiffness of the element by approximating the effects of the distributed yielding. Since the distributed yielding is influenced largely by member initial imperfections (member out-of-straightness and residual stresses) as well as by axial force, these distributed plasticity effects can be modeled by an effective tangent-modulus approach for the calculation of column strength. However, for members subjected to significant in-plane moment, this

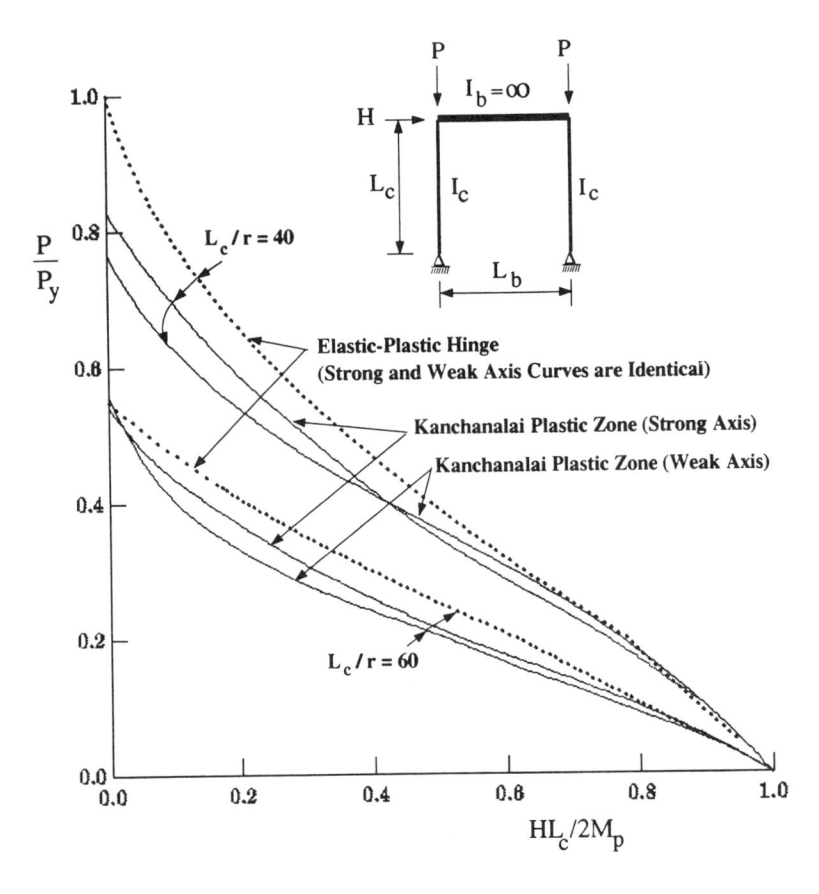

FIGURE 4.7. Comparisons of second-order elastic-plastic hinge strength curves with plastic-zone strengths curves from Kanchanalai (1977).

approach is not sufficient to represent the gradual stiffness degradation as yielding progress through the volume of the member (Liew et al., 1992b and c). Additional distributed plasticity effects along the member length are associated with bending actions. These effects may be represented by modifying the elastic-plastic hinge model such that the member stiffness degrades gradually from the elastic stiffness at the onset of yielding to that associated with the cross-section plastic strength when a plastic hinge is developed at the element end. The detailed formulation of the proposed refined plastic hinge model is described below.

4.6.1 Tangent Modulus Approach

An approximate scheme based on a tangent-modulus approach is employed to reduce the modulus of elasticity in the element stiffness calculation. The tangent modulus may be evaluated from the column equations of the specification used by the designer.

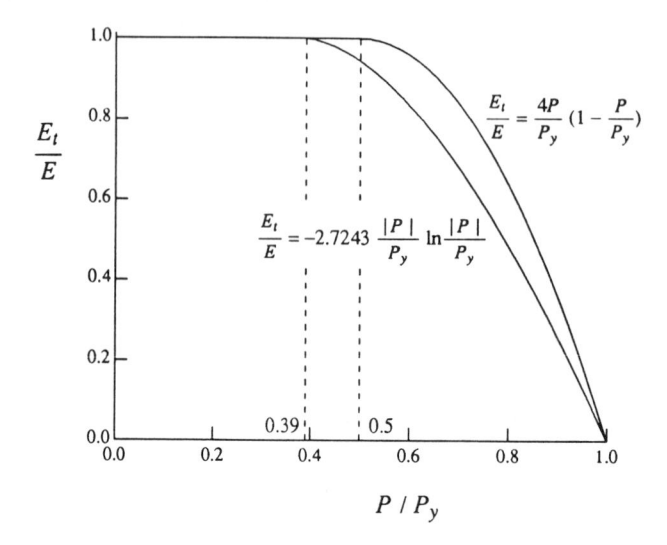

Figure shows E_t/E vs P/P_y curves with two expressions: $\dfrac{E_t}{E} = \dfrac{4P}{P_y}\left(1 - \dfrac{P}{P_y}\right)$ and $\dfrac{E_t}{E} = -2.7243 \dfrac{|P|}{P_y} \ln\dfrac{|P|}{P_y}$, with transition at 0.39 and 0.5.

FIGURE 4.8. Inelastic stiffness reductions for axial force effect.

In AISC-LRFD, this modulus is directly related to the inelastic stiffness reduction factors for calculation of column strength in the inelastic range (AISC-LRFD, 1986). The ratio of the tangent modulus to the elastic modulus, E_t/E, obtained from this procedure may be written for members subjected to axial compression as (Liew et al., 1991 and 1992b):

$$\frac{E_t}{E} = -2.7243 \frac{P}{P_y} \ln\left[\frac{P}{P_y}\right] \quad \text{for } P > 0.39 P_y \tag{4.34}$$

Since this E_t model is derived from the LRFD column strength formula, it implicitly includes the effects of residual stresses as well as the member initial out-of-straightness in modeling the member effective stiffness (Liew, 1992). Equation 4.34 is plotted in Figure 4.8. The tangent modulus is reduced from the elastic value for $P > 0.39 P_y$.

The compressive axial load-displacement relationship for an element can be derived based on the following equations:

$$e = \frac{0.39 P_y L}{EA} + \int_{0.39 P_y}^{P} \frac{L}{AE_t} dP \quad \text{for } P > 0.39 P_y \tag{4.35}$$

Upon substituting from Eq. (4.34) for E_t and carrying out the integration in Eq. (4.35), the normalized axial force-deformation relationship can be written as:

$$\frac{P}{P_y} = \exp\left[-0.9416 \exp\left(2.7243\left[0.39 - \frac{\varepsilon}{\varepsilon_y}\right]\right)\right] \quad \text{for } P > 0.39 P_y \tag{4.36}$$

where $\varepsilon/\varepsilon_y = Ee/(F_y L)$ is the total normalized axial strain in the member.

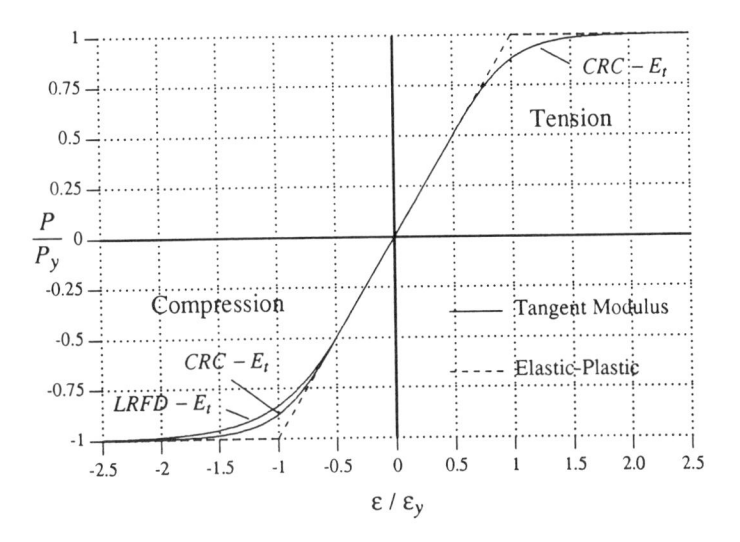

FIGURE 4.9. Normalized axial force-strain relationship.

The CRC column strength equations (Galambos, 1988) also may be employed in deriving the tangent modulus. The E_t/E expressions can be written as (Chen and Lui, 1992):

$$\frac{E_t}{E} = \frac{4P}{P_y}\left(1 - \frac{P}{P_y}\right) \quad \text{for } P > 0.5P_y \tag{4.37}$$

Equation (4.37) is also plotted in Figure 4.8. The tangent-modulus stiffness reduction becomes effective when the axial force reaches a value of $P = 0.5P_y$. The total axial displacement based on the E_t model in Eq. (4.37) may be derived as:

$$e = \frac{0.5P_y L}{EA} + \int_{0.5P_y}^{P} \frac{L}{AE_t} dP \quad \text{for } P > 0.5P_y \tag{4.38}$$

Upon substituting from Eq. (4.37) for E_t and carrying out the integration, the normalized axial force-deformation relationship is:

$$\frac{P}{P_y} = \frac{1}{1 + \exp\left(2 - 4\varepsilon/\varepsilon_y\right)} \quad \text{for } P > 0.5P_y \tag{4.39}$$

The normalized axial force-strain relationships described by Eqs. (4.36) and (4.39) are shown in Figure 4.9. It should be noted that the normalized axial stress-strain relationships of an element described by the tangent moduli in Eqs. (4.34) and (4.37) are nonlinear in nature, whereas for the conventional elastic-plastic hinge model, a linear elastic, perfectly plastic normalized axial stress-strain relationship is tacitly assumed. Also, for a member subjected to tensile axial force, only the residual

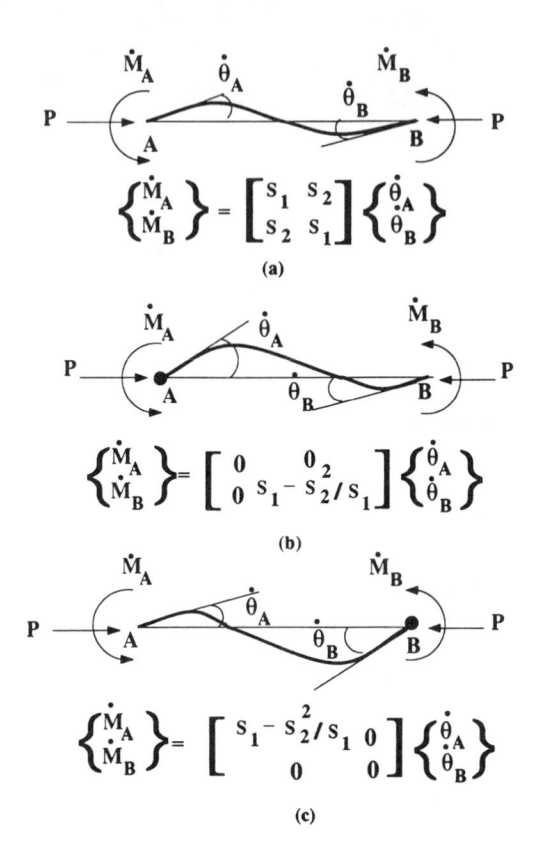

FIGURE 4.10. Incremental force displacement relationships of frame elements. (a) No plastic hinge. (b) Plastic hinge at end A. (c) Plastic hinge at end B.

stress effects should be represented in the member stiffness degradation. In this case, the CRC tangent modulus would be more appropriate. The main difference between the CRC-E_t and the LRFD-E_t is that the former considers only the residual stress effects in modeling the column effective stiffness, whereas the later is based on LRFD column strength equations that account for both the effects of geometric imperfections and residual stresses.

4.6.2 Two-Surface Stiffness Degradation Model

In the refined plastic hinge approach, the element stiffness is assumed to degrade parabolically after the element end forces exceed a predefined initial yield function. When the cross-section plastic strength is reached, the section at the plastic hinge is modeled as a real hinge with an applied moment, M_{pc}. The element tangent stiffness is then adjusted to account for the presence of a plastic hinge.

The force-displacement relationships, with and without plastic hinges at their ends, have been derived in Section 4.5, and they are summarized in Figures 4.10a to c. M_{pcA} and M_{pcB} in Figures 4.10b and c are the plastic moment capacity at element ends

Effect of Plastification at End A Only

Consider the frame elements shown in Figures 4.10a and b. To represent a gradual transition from the elastic stiffness at the onset of yielding (Figure 4.10a) to the stiffness associated with a full plastic hinge at end A (Figure 4.10b), the element incremental force-displacement relationship may be written as

$$\begin{Bmatrix} \dot{M}_A \\ \dot{M}_B \end{Bmatrix} = \frac{E_t I}{L} \begin{bmatrix} \phi_A S_1 & \phi_A S_2 \\ \phi_A S_2 & \left[S_1 - \frac{S_2^2}{S_1}(1-\phi_A) \right] \end{bmatrix} \begin{Bmatrix} \dot{\theta}_A \\ \dot{\theta}_B \end{Bmatrix} \quad (4.40)$$

The terms S_1 and S_2 in this equation are the conventional beam-column stability functions, but with E_t used in place of the elastic modulus. The term ϕ_A is a scalar parameter that allows for gradual inelastic stiffness reduction of the element associated with the effect of plastification at end A. This term is equal to one when the element is elastic (Figure 4.10a), and it is zero when a plastic hinge is formed at end A (Figure 4.10b). The parameter ϕ is assumed to vary according to a prescribed function

$$\phi = 4\alpha(1-\alpha) \quad for\ \alpha > 0.5 \quad (4.41)$$

where α is a force-state parameter that measures the magnitude of axial force (P) and bending moment (M) at the element end. The term α is expressed as:

$$\alpha = \frac{P}{P_y} + \frac{8}{9}\frac{M}{M_p} \quad for\ \frac{P}{P_y} \geq \frac{2}{9}\frac{M}{M_p} \quad (4.42a)$$

$$\alpha = \frac{P}{2P_y} + \frac{M}{M_p} \quad for\ \frac{P}{P_y} < \frac{2}{9}\frac{M}{M_p} \quad (4.42b)$$

Initial yielding is assumed to occur based on a yield surface that has the same shape as the plastic-strength surface and with the force-state parameter denoted as α_o. The plastic-strength surface and initial-yield surface with $\alpha_o = 0.5$ are shown in Figure 4.11. If the state of forces is changed in such a manner that the force point moves inside or along the initial yield surface, the element is assumed to remain fully elastic with no stiffness reduction. If the force point moves beyond the initial yield surface, the element stiffness is reduced to account for the effect of plastification at the element end.

Figure 4.12 shows the relationship between the stiffness reduction factor, ϕ, and the force state parameter, α, modeled by Eq. (4.41). A gradual stiffness degradation is observed when the force state moves from the initial yield surface (α_o) toward the

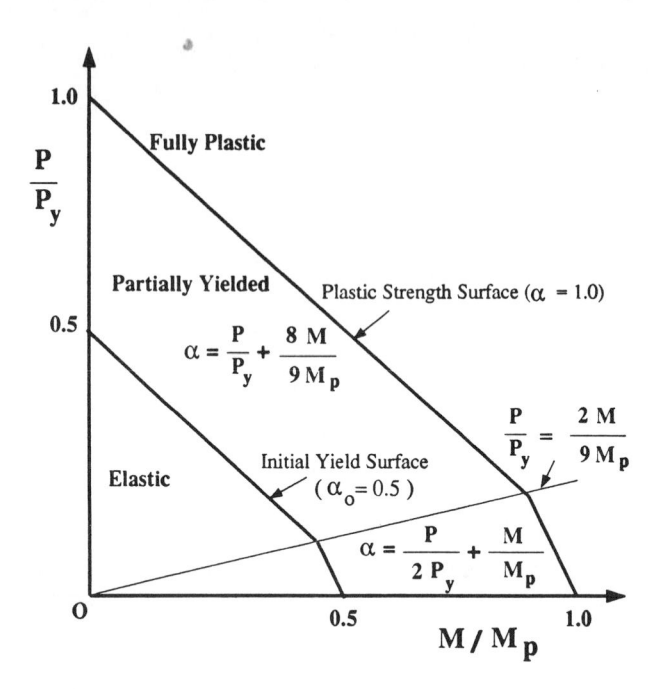

FIGURE 4.11. Two-surface stiffness degradation model for refined plastic-hinge analysis.

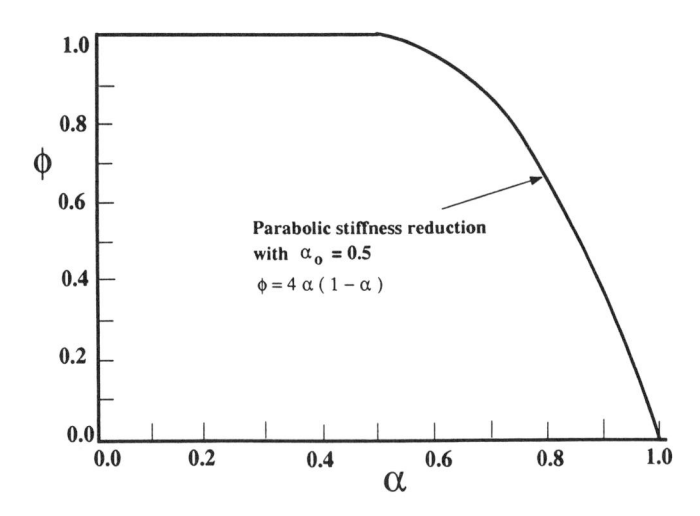

FIGURE 4.12. Element stiffness degradation function.

plastic strength surface ($\alpha = 1.0$). The expression of ϕ in Eq. (4.41) is obtained based on calibration with the "exact" plastic-zone solutions of simple portal frames and beam-column subassemblages (Liew, 1992; Liew et al. 1992c). It is important to note that only a simple relationship for ϕ is required to "adequately" describe the degradation in stiffness associated with distributed plasticity effects. Although more

Effect of Plastification at End B Only

To represent a gradual transition from the elastic beam-column stiffness (Figure 4.10a) to the stiffness associated with a plastic hinge at end B (Figure 4.10c), the incremental stiffness equations may be written as

$$\begin{Bmatrix} \dot{M}_A \\ \dot{M}_B \end{Bmatrix} = \frac{E_t I}{L} \begin{bmatrix} \left[S_1 - \dfrac{S_2^2}{S_1}(1-\phi_B) \right] & \phi_B S_2 \\ \phi_B S_2 & \phi_B S_1 \end{bmatrix} \begin{Bmatrix} \dot{\theta}_A \\ \dot{\theta}_B \end{Bmatrix} \qquad (4.43)$$

Equations (4.40) and (4.43) represent the gradual plastification effects at one end of the beam-column element only.

Effect of Plastification at Both Ends

For the general case where plastic hinges may form at both ends of an element, the incremental force-displacement relationships may be written as

$$\begin{Bmatrix} \dot{M}_A \\ \dot{M}_B \end{Bmatrix} = \frac{E_t I}{L} \begin{bmatrix} \phi_A \left[S_1 - \dfrac{S_2^2}{S_1}(1-\phi_B) \right] & \phi_A \phi_B S_2 \\ \phi_A \phi_B S_2 & \phi_B \left[S_1 - \dfrac{S_2^2}{S_1}(1-\phi_A) \right] \end{bmatrix} \begin{Bmatrix} \dot{\theta}_A \\ \dot{\theta}_B \end{Bmatrix} \qquad (4.44)$$

It should be noted that:

(1) when $1 > \phi_A > 0$ and $1 > \phi_B > 0$, Eq. (4.44) accounts for the effects of partial plastification at both ends of the element.
(2) when $\phi_A = \phi_B = 1$, both ends of the beam-column element are fully elastic. Equation (4.44) reduces to the stiffness equations shown in Figure 4.10a.
(3) when $\phi_A = 1$ and $1 > \phi_B > 0$, end A is elastic and end B is partially yielded. Equation (4.44) then reduces to the stiffness equations shown in Figure 4.10c.
(4) when $\phi_B = 1$ and $1 > \phi_A > 0$, end B is elastic and end A is partially yielded. Equation (4.44) then reduces to the form shown in Figure 4.10b.

In a similar formulation proposed by King et al. (1991), it is assumed that once a plastic hinge formed at the element end, the moment at the plastic hinge will remain unchange as the axial force is increased. However, as discussed in Section 4.5.2, neglecting force point movement on the plastic strength surface may violate the cross-section plastic strength, which is not acceptable for maximum strength analysis

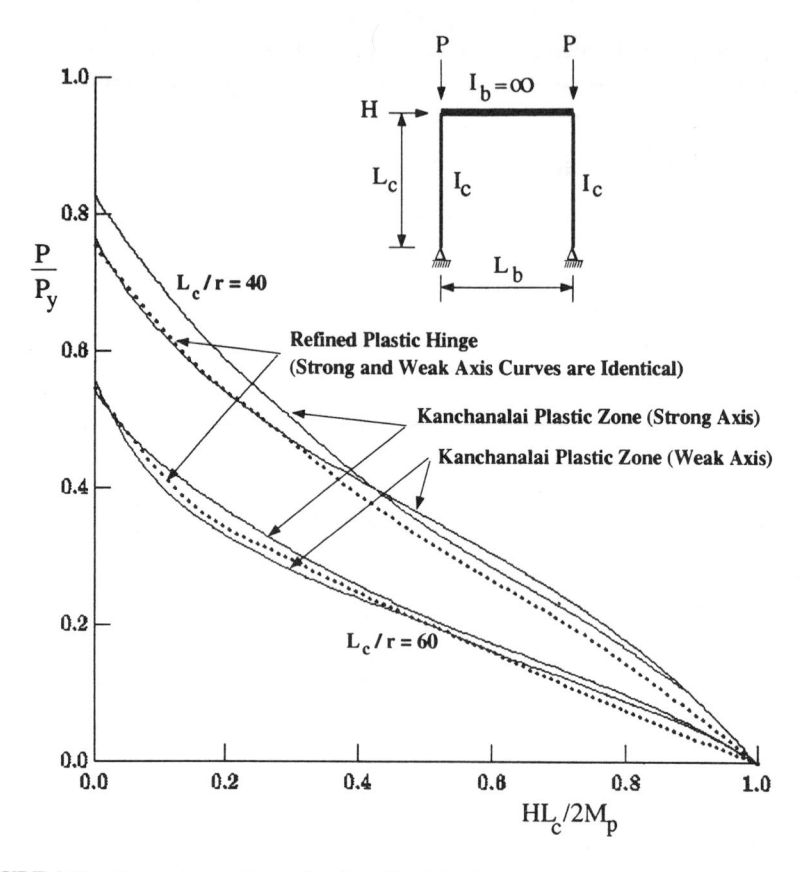

FIGURE 4.13. Comparisons of second-order refined plastic hinge strength curves with plastic-zone strengths curves from Kanchanalai (1977).

of members. In the present refined plastic hinge formulation, the force-point movement on the plastic strength surface is handled in the same way as in the elastic-plastic hinge formulation in Section 4.5.2. The reader is referred to Liew (1992) for a detailed formulation pertinent to this aspect.

4.6.3 Illustrative Example

Figure 4.13 compares the in-plane strength curves generated by the second-order refined plastic hinge analysis with the "exact" plastic-zone results generated by Kanchanalai (1977). The plastic-zone strength curves were generated for frames with no initial geometric imperfections, but the residual stress effects have been included. For comparison of these solutions, the refined plastic hinge analysis based on the CRC tangent modulus, which represents the effects of residual stress only, is the appropriate model.

The refined plastic hinge method gives excellent strength predictions with errors not more than 1% unconservative. It should be noted that when the axial force in the frame member is large, the use of the tangent modulus alone in the elastic-plastic hinge analyses gives good estimate of the ultimate strengths. However, at the intermediate axial load range, this model somewhat over-estimates the "exact" strength. In this load range, additional distributed plasticity effects in the member are associated with bending actions. Therefore, the use of E_t alone in the elastic-plastic hinge analyses is not sufficient to represent the member stiffness degradation, particularly when the bending moment in the member is not insignificant. Hence, if distributed plasticity effects associated with the bending actions are dominant, the use of the refined plastic hinge model, which accounts for distributed plasticity effects, provides an overall improvement in the inelastic strength predictions.

Recent research by Liew et al. (1992a) proposes an alternative approach which does not involve any modifications to the basic elastic-plastic hinge theory, and which promises to satisfy the in-plane strength requirements for inelastic analysis of steel frames. This method of analysis involves the use of notional lateral loads to account approximately for the influence of member imperfections and distributed plasticity effects on the overall system strength. Interested readers are referred to this reference for details.

4.7 Analysis of Semi-Rigid Frames
4.7.1 Modeling of Connections

The three-parameter connection model proposed by Kishi and Chen (1990) is adopted in the present work to model the moment-rotation behavioral response of semi-rigid connections, which are assumed to attach at the element ends. The generalized equation of the three-parameter model (Kishi and Chen, 1990) has the form:

$$m = \frac{\theta}{\left(1+\theta^n\right)^{1/n}} \quad \text{for } \theta > 0 \text{ and } m > 0 \tag{4.45}$$

or equivalently,

$$\theta = \frac{m}{\left(1-m^n\right)^{1/n}} \quad \text{for } \theta > 0 \text{ and } m > 0 \tag{4.46}$$

The parameters in these equations are defined as:

m = M/M_u
θ = θ_r/θ_o
θ_o = M_u/R_{ki}, the reference plastic rotation
M_u = ultimate moment capacity of the connection
R_{ki} = initial connection stiffness
n = shape parameter

164 ADVANCED ANALYSIS OF STEEL FRAMES

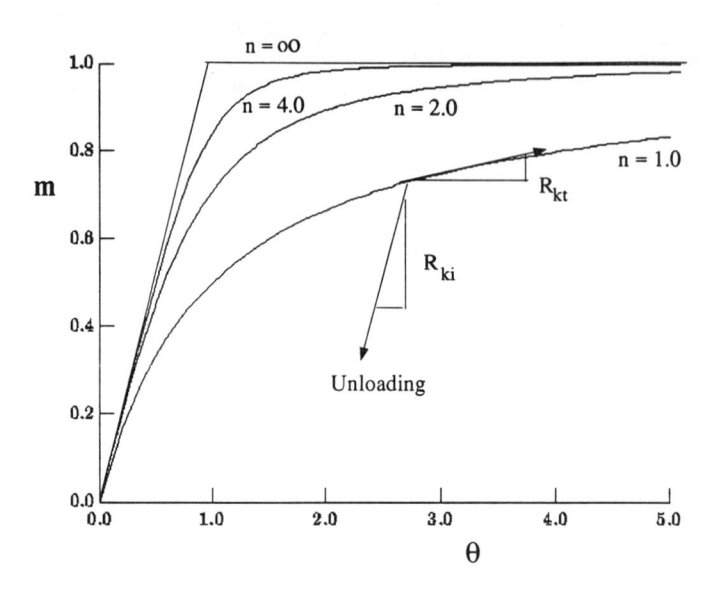

FIGURE 4.14. Moment-rotation behavior of connections.

The connection tangent stiffness R_{kt} at an arbitrary rotation $|\theta_r|$ can be evaluated by differentiating M with respect to $|\theta_r|$, and it is expressed as

$$R_{kt} = \frac{dM}{d|\theta_r|}\bigg|_{|\theta_r|} = \frac{M_u}{\theta_o(1-\theta^n)^{1+1/n}} \qquad (4.47)$$

when the connection is loaded, and it is

$$R_{kt} = \frac{dM}{d|\theta_r|}\bigg|_{|\theta_r|=0} = \frac{M_u}{\theta_o} = R_{ki} \qquad (4.48)$$

when the connection is unloaded.

Equation (4.45) or (4.46) has the shape shown in Figure 4.14. The principal merit of this model is that it allows the designer to execute the nonlinear structural analysis quickly and accurately. This is because the connection stiffness and rotation can be determined directly from Eqs. (4.46 to 4.48) without iteration. Also, this model is based on connection parameters that can be determined analytically based on the connection configuration, thus making it more appropriate for practical use (Liew, 1992).

4.7.2 Modification of Element Stiffness to Account for End Connections

To incorporate the effect of connection flexibility into the member stiffness relationships, the connection is modeled as a rotational spring with the moment-rotation

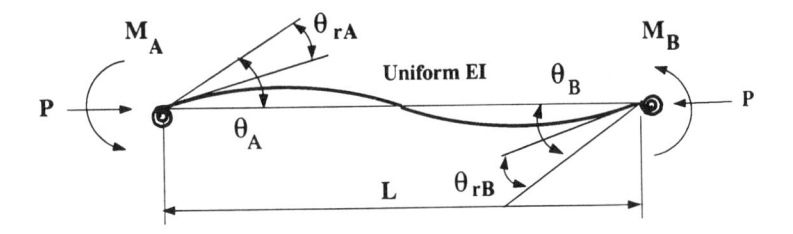

SECOND-ORDER PLASTIC HINGE ANALYSIS OF FRAMES 165

FIGURE 4.15. Frame element with end connections.

relationship described by Eq. (4.45) or (4.46). Consider a beam-column element subjected to end moments (M_A and M_B) and axial force (P) with connections attached at both ends as shown in Figure 4.15. The presence of the connections introduces relative incremental rotations of θ_{rA} and θ_{rB} at the A-th and B-th end of the member, respectively. Denoting R_{ktA} and R_{ktB} as the tangent stiffnesses of connections A and B, respectively, the relative incremental rotation between the joint and the beam end (i.e., rotational deformation of the connection) can be expressed as

$$\dot{\theta}_{rA} = \frac{\dot{M}_A}{R_{ktA}} \qquad \dot{\theta}_{rB} = \frac{\dot{M}_B}{R_{ktB}} \tag{4.49}$$

If $\dot{\theta}_A$ and $\dot{\theta}_B$ are denoted as the incremental joint rotations at A-th and B-th end of the beam-column, the incremental force-displacement equations for the beam-column element modified for the presence of end connections can be expressed as (Chen and Lui, 1991):

$$\dot{M}_A = \frac{EI}{L}\left[S_{ii}\left(\dot{\theta}_A - \frac{\dot{M}_A}{R_{ktA}}\right) + S_{ij}\left(\dot{\theta}_B - \frac{\dot{M}_B}{R_{ktB}}\right)\right] \tag{4.50a}$$

$$\dot{M}_B = \frac{EI}{L}\left[S_{ij}\left(\dot{\theta}_A - \frac{\dot{M}_A}{R_{ktA}}\right) + S_{jj}\left(\dot{\theta}_B - \frac{\dot{M}_B}{R_{ktB}}\right)\right] \tag{4.50b}$$

S_{ii}, S_{ij}, and S_{jj} are defined in the following manner.

(1) If elastic-plastic hinge assumption is assumed, then

$$S_{ii} = S_{jj} = S_1 \quad \text{and} \quad S_{ij} = S_2 \tag{4.51}$$

where S_1 and S_2 are the stability functions as defined in Section 4.3.

(2) If gradual stiffness degradation due to plasticity effects in the element is considered, then

$$S_{ii} = \left[S_1 - \frac{S_2^2}{S_1}(1-\phi_B)\right]\phi_A \tag{4.52a}$$

$$S_{jj} = \left[S_1 - \frac{S_2^2}{S_1}(1-\phi_A)\right]\phi_B \tag{4.52b}$$

$$S_{ij} = \phi_A \phi_B S_2 \tag{4.52c}$$

and the elastic of modulus, E, in Eq. (4.50a) and (4.50b) is replaced by the tangent modulus E_t.

Solving Eqs. (4.50a) and (4.50b) for $\dot{M}_A$ and $\dot{M}_B$ gives

$$\dot{M}_A = \frac{E_t I}{L}\left[S_{ii}^* \dot{\theta}_A + S_{ij}^* \dot{\theta}_B\right] \tag{4.53a}$$

$$\dot{M}_B = \frac{E_t I}{L}\left[S_{ij}^* \dot{\theta}_A + S_{jj}^* \dot{\theta}_B\right] \tag{4.53b}$$

where

$$S_{ii}^* = \left(S_{ii} + \frac{E_t I S_{ii} S_{jj}}{L R_{ktB}} - \frac{E_t I S_{ij}^2}{L R_{ktB}}\right)\bigg/ R^* \tag{4.54a}$$

$$S_{jj}^* = \left(S_{jj} + \frac{E_t I S_{ii} S_{jj}}{L R_{ktA}} - \frac{E_t I S_{ij}^2}{L R_{ktA}}\right)\bigg/ R^* \tag{4.54b}$$

$$S_{ij}^* = S_{ij}/R^* \tag{4.54c}$$

and

$$R^* = \left(1 + \frac{E_t I S_{ii}}{L R_{ktA}}\right)\left(1 + \frac{E_t I S_{jj}}{L R_{ktB}}\right) - \left(\frac{E_t I}{L}\right)^2 \frac{S_{ij}^2}{R_{ktA} R_{ktB}} \tag{4.55}$$

The incremental axial force-displacement equation is

$$\dot{P} = \frac{E_t A}{L}\dot{e} \tag{4.56}$$

Finally, the incremental element force-displacement relationship can be written as:

$$\begin{Bmatrix} \dot{M}_A \\ \dot{M}_B \\ \dot{P} \end{Bmatrix} = \frac{E_t I}{L} \begin{bmatrix} S_{ii}^* & S_{ij}^* & 0 \\ S_{ij}^* & S_{jj}^* & 0 \\ 0 & 0 & A/I \end{bmatrix} \begin{Bmatrix} \dot{\theta}_A \\ \dot{\theta}_B \\ \dot{e} \end{Bmatrix} \tag{4.57}$$

The element tangent stiffness matrix expressed in Eq. (4.57) accounts for both the effects of distributed plasticity and connection flexibility on the element. In addition, the connection stiffness at both ends of the element need not be equal. In other words, this formulation allows the consideration of cases where the axial force, P, in the element is appreciable and the connections at the ends of the element may behave differently. For members subjected to end loading only, Eq. (4.57), which is formulated based on the use of one frame element per member, is sufficiently accurate in capturing the inelasticity and connection nonlinear effects on member performance.

4.8 Numerical Implementation

The simple incremental solution method is the simplest and most direct nonlinear global solution technique. This numerical procedure is straightforward in concept and implementation, and the numerical algorithm is generally well behaved and often exhibits good computational efficiency. This is especially true when the structure is loaded into the inelastic region by which a trace of hinge-by-hinge formation is required in the element stiffness formulations. However, for a finite increment size, this approach only approximates the nonlinear structural response, and equilibrium between the external applied loads and the internal element forces is not satisfied.

Iterative methods, such as the Newton-Raphson method, satisfy the equilibrium equations at a specific external load magnitude. In this method, the equilibrium out-of-balance that is present following an initial linear load step is eliminated within a given tolerance by applying additional corrective steps. The Newton-Raphson method possesses the advantage of providing results that lie upon the "exact" load-displacement trace. However, the requirement to trace the hinge-by-hinge formation in the structure may render the plastic hinge analysis slightly inefficient in the numerical iteration process.

In the analysis programs under discussion, both the elastic-plastic hinge and refined plastic hinge analyses are implemented using an automatic load-increment procedure. To prevent plastic hinges from forming within a constant-stiffness load increment, load step sizes less than or equal to the specified increment magnitude are internally computed such that plastic hinges form only after the load increment. Thus, the subsequent element stiffness formulation will account for the stiffness reduction due to the presence of plastic hinges. For elements that are partially yielded at their ends, a limit is placed on the magnitude of the increment in the element end forces. Increments of these force components that are relatively large, compared to their respective plastic strengths, may indicate that the path taken from the initial yield surface to the plastic strength surface may have deviated excessively from the "exact" path, upon which the calculation of element effective stiffness may lead to some error. The analysis program automatically scales an attempted load increment when the change in the element stiffness parameter ($\Delta\phi$) exceeds a predefined tolerance. A default tolerance value of $\Delta\phi = 0.1$ has been specified in the solution scheme to suppress the error that may be incurred during the computation.

168   ADVANCED ANALYSIS OF STEEL FRAMES

The applied load increment in the above solution procedure may be reduced for any of the following reasons:

(1) Formation of new plastic hinge(s) prior to the full application of incremental loads.
(2) The increment in the element nodal forces at plastic hinges is excessive.
(3) Nonpositive definiteness of the structural stiffness matrix.

As the stability limit point is approached in the analysis, large step increments may overstep a limit point. Therefore, a smaller step size is used near the limit point to obtain accurate collapse displacements and second-order forces.

4.9 PHINGE — A Second-Order Plastic Hinge Based Analysis Program

4.9.1 Program Overall View

This section describes PHINGE, a FORTRAN program for second-order static plastic hinge analysis of two-dimensional steel frames. This program has been written to be an integral constitute of the materials covered in this chapter. Methods and techniques developed throughout the chapter are employed in PHINGE. The names of variables and arrays are kept as close as possible to those used in the development of the theory.

The computer program is divided into two parts: the first part consists of a FORTRAN program, INPUT, which reads an input file called INFILE and generates three data files, DATA0, DATA1, and DATA2. The second program, PHINGE, reads the input data files DATA0, DATA1, and DATA2 and generates an output file called OUTPUT. After the OUTPUT file is generated, the user can review this file on the screen or print it out for detailed information on force distribution and load deformation characteristic of the structure. The schematic diagram shown in Figure 4.16 explains the operation procedure of the computer program.

The computer program has been tested in two computing environments. The first is in an IBM or equivalent personal computer system using a MicroSoft FORTRAN 77 compiler, version 5.0. The second is in a Sun 3 or Sun 4 machine using a Sun FORTRAN 77 compiler. The memory required to run the program depends on the size of the problem. A hard disk is generally required. Two executable program files named INPUT and PHINGE are given. These programs are executed by issuing the commands "INPUT" and "PHINGE" separately at the terminals.

The complete source code listing for the computer program is provided in the attached diskette; the user is encouraged to modify or improve the program for his or her use. Program PHINGE is in its preliminary stage of development, therefore many improvements are still required to enhance its capability for practical analysis/design use.

The next section presents the input instructions necessary for the user to execute PHINGE. The new user is advised to read all the instructions, paying particular attention to the appended notes, to gain an initial overall view of the data required

SECOND-ORDER PLASTIC HINGE ANALYSIS OF FRAMES 169

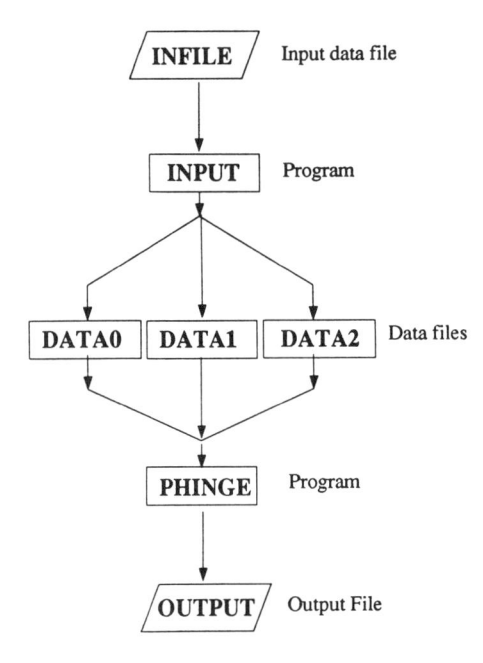

FIGURE 4.16. Operating procedure of the plastic hinge-based program.

to specify an analysis task to PHINGE. Readers without prior knowledge or experience on nonlinear analysis will benefit from careful study of the example problems given in Section 4.9.3. They are intended to help clarify the data generation process.

Upon examing the input instructions, the reader should recognize that no system of units is assumed by the program. It is the user's responsibility to specify data in the consistent units of his or her choice. Overlooking this requirement is a common source of erroneous results.

4.9.2 Input Instructions

The input sequence and data structure for INFILE are described below. The executive program can analyze any frame structures with maximum degrees of freedom not more than 165. However, it is possible to recompile the source code to accommodate more degrees of freedom by changing the size of the arrays in the PARAMETER statements. This problem can be overcome by using dynamic storage allocation. This procedure is rather common in finite element programs (Hughes, 1987; Cook, 1989), and it will be used in the next release of the program.

Line 1: General Comment Line

Note	Columns	Variable	Description
(1)	1–8	—	A comment card

Note: The comment line is skipped, without interpretation, at the execution of INPUT. This line is usually required to identify one input file from the other.

170 ADVANCED ANALYSIS OF STEEL FRAMES

Line 2 (I5): Analysis Methods Control Card

Note	Columns	Variable	Description
(1)	1–5	ISOLVE	Specify the analysis method ISOLVE = 0, elastic-plastic hinge analysis ISOLVE = 1, refined plastic hinge analysis

Notes:
1. IF ISOLVE = 0 is selected, the second-order elastic-plastic method, as described in Sections 4.3 through 4.5, is launched. If ISOLVE = 1 is specified, the second-order refined plastic hinge method described in Section 4.6 is initiated.

Line 3 (2I5): Analysis Attributes Card — This card is skipped if ISOLVE = 0

Note	Columns	Variable	Description
(1)	1–5	IET	Tangent modulus model IET = 1, CRC tangent modulus IET = 2, LRFD tangent modulus
(2)	6–10	IPHI	Element stiffness degradation model (IPHI = 1) IPHI = 1, parabolic stiffness reduction

Notes:
1. If IET = 1 is selected, the CRC tangent modulus model given by Eq. (4.37) is adopted in the refined plastic hinge analysis for members subjected to compression or tension. If IET = 2 is specified, the LRFD tangent modulus from Eq. (4.34) is effected for members subjected to axial compression only. For members subjected to tension, the CRC tangent modulus model is automatically selected, as discussed in Section 4.6.1.
2. A parabolic stiffness reduction function (Eq. 4.41) with initial yield surface equal to one half the plastic strength surface is adopted in the refined plastic hinge analysis if IPHI = 1 is selected. Other possible stiffness reduction schemes used in calibrating the refined plastic hinge model to best fit the "exact" plastic zone solutions are given in the subroutine ESTIFF. The reader is referred to Liew (1992) for information pertinent to this aspect.

Line 4 (4I5): Job Control Card

Note	Columns	Variable	Description
(1)	1–5	NDOFS	Number of degrees of freedom of the structure (≤ 165)
(2)	6–10	NINCRE	Number of load increments
(3)	11–15	NSEQN	Number of load sequences (≤ 2)
(4)	16–20	IPRTFC	Flag for printing member end forces IPRTFC = 0, do not print element end forces IPRTFC = 1, print element end forces

Notes:
1. The structure degrees of freedom is required. This value should not be larger that the default limit of 165. If this value is exceeded, the user needs to recompile the source code by setting a larger value for "MAXDOF" in the PARAMETER statements.
2. The total load incremental steps are required. The analysis is terminated when the number of the total load incremental steps is reached.
3. Only two load sequences are allowed in the present version of the program.

4. Member end forces are not printed if IPRTFC = 0. These forces are printed at each step of load increment if IPRTFC = 1; this means a larger storage capacity is required for OUTPUT.

Line 5 (3I5): Element Types Control Card

Note	Columns	Variable	Description
(1)	1–5	NCTYPE	Number of connection types (≤10)
(1)	6–10	NFTYPE	Number of frame types (≤20)
(1)	11–15	NTTYPE	Number of truss types (≤10)

Note:
1. If the limits specified in the above are exceeded, the user needs to recompile the program by specifying suitable values for the variables in the PARAMETER statements of the source code.

Line 6 (3I5): Element Group Control Card

Note	Columns	Variable	Description
(1)	1–5	NUMCNT	Number of connection elements (≤10)
(1)	6–10	NUMFRM	Number of frame elements (≤200)
(1)	11–15	NUMTRS	Number of truss elements (≤50)

Note:
1. If the limits specified in the above are exceeded, the user needs to recompile the program by specifying suitable values for the variables in the PARAMETER statements of the source code.

Line 7 (I5, 3D10.0): Connection Properties Cards — This line should be omitted if NCTYPE = 0. Otherwise, this line must be supplied NCTYPE times.

Note	Columns	Variable	Description
(1)	1–5	ICTYPE	Connection type identifier (an integer)
(2)	6–15	M_u	Connection ultimate moment capacity
(2)	16–25	R_{ki}	Initial connection stiffness
(2)	26–35	n	Connection shape parameter

Notes:
1. Connection property sets must be specified in the order as shown. Users should refer to Section 4.7 for the definition of the connection parameters.
2. The connection property set is stored in an array CTYPE(ICTYPE, J=1,3). Step-by-step procedures for calculating M_u and R_{ki} for semi-rigid connections composed of angles are given in Liew (1992).

Line 8 (I5, 5D10.0): Frame Element Properties Cards — If NFTYPE ≠ 0, this line must be supplied NFTYPE times. Otherwise, this line should be omitted.

Note	Columns	Variable	Description
(1)	1–5	IFTYPE	Frame type identifier (an integer)
(1)	6–15	A	Cross sectional area
(1)	16–25	I	Moment of inertia
(1)	26–35	Z	Plastic section modulus
(1)	36–45	E	Modulus of elasticity
(1)	46–55	F_y	Yield stress

172 ADVANCED ANALYSIS OF STEEL FRAMES

Notes:
1. Material property sets for the frame elements must be specified in the order as shown. The material set is stored in an array FTYPE(IFTYPE, J=1,5).

Line 9 (I5, 4D10.0): Truss Element Properties Cards — If NTTYPE ≠ 0, this line must be supplied NTTYPE times. Otherwise, this line should be omitted.

Note	Columns	Variable	Description
	1–5	ITTYPE	Truss type identifier (an integer)
(1)	6–15	A	Cross sectional area
(1)	16–25	I	Moment of inertia
(1)	26–35	E	Modulus of elasticity
(1)	36–45	F_y	Yield stress

Notes:
1. Material property sets for the truss elements must be specified in the order as shown. For a truss element under axial compression, the analysis is terminated once the axial force in the truss element exceeds the column strength implied by the LRFD column strength equations (AISC-LRFD, 1986). Therefore, the moment of inertial for weak-axis bending should be specified for a pinned-end truss element. For element subjected to tensile force, the cross-section effective area (with allowance for bolt holes) may be specified. The material set is stored in an array TTYPE(ITTYPE, J=1,4).

Line 10 (4I5): Connection Data Cards — This line should be omitted if NUMCNT = 0. Otherwise, this line must be supplied NUMCNT times.

Note	Columns	Variable	Description
	1–5	LCNT	Connection element number
	6–10	IFMCNT(LCNT)	Frame element number to which the connection is attached
(1)	11–15	IEND(LCNT)	Frame element end to which the connection is attached. IEND = 1 for connection at end A IEND = 2 for connection at end B (Figure 4.17)
	16–20	JDCNT(LCNT)	Connection type number

Notes:
1. See Figure 4.17 for the respective end of the frame element.

Line 11 (I5, 2D10.0, 7I5): Frame Element Data Cards — This line should be omitted if NUMFRM = 0. Otherwise, this line must be supplied NUMFRM times.

Note	Columns	Variable	Description
	1–5	LFRM	Frame element number
(1)	6–15	FXO(LFRM)	Horizontal projected length
(1)	16–25	FYO(LFRM)	Vertical projected length
	26–30	JDFRM(LFRM)	Frame type number
(2)	31–35	NFRMCO(JDFRM,1)	Node number for degree of freedom 1
(2)	36–40	NFRMCO(JDFRM,2)	Node number for degree of freedom 2
(2)	41–45	NFRMCO(JDFRM,3)	Node number for degree of freedom 3
(2)	46–50	NFRMCO(JDFRM,4)	Node number for degree of freedom 4

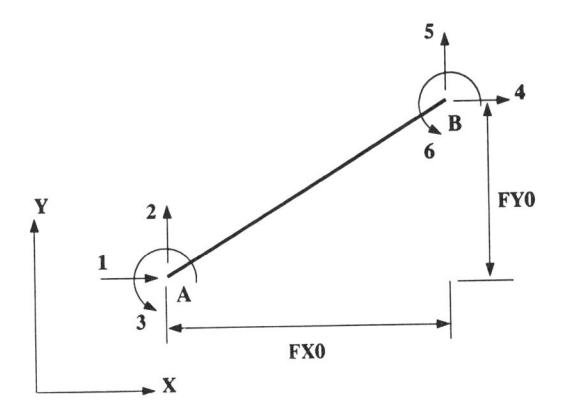

SECOND-ORDER PLASTIC HINGE ANALYSIS OF FRAMES 173

FIGURE 4.17. Degrees of freedom numbering for the frame element.

Note	Columns	Variable	Description
(2)	51–55	NFRMCO(JDFRM,5)	Node number for degree of freedom 5
(2)	56–60	NFRMCO(JDFRM,6)	Node number for degree of freedom 6

Notes:
1. The projected lengths for a frame element are shown in Figure 4.17.
2. The nodal numbering sequence must follow the order of the degrees of freedom for the frame element shown in Figure 4.17.

Line 12 (I5, 2D10.0, 5I5): Truss Element Data Card — This line should be omitted if NUMTRS = 0. Otherwise, this line must be supplied NUMTRS times.

Note	Columns	Variable	Description
	1–5	LTRS	Truss element number
(1)	6–15	TXO(LTRS)	Horizontal projected length (in inches)
(1)	16–25	TYO(LTRS)	Vertical projected length (in inches)
	26–30	JDTRS(LTRS)	Truss type number
(2)	31–35	NTRSCO(JTRS,1)	Node number for degree of freedom 1
(2)	36–40	NTRSCO(JTRS,2)	Node number for degree of freedom 2
(2)	41–45	NTRSCO(JTRS,3)	Node number for degree of freedom 3
(2)	46–50	NTRSCO(JTRS,4)	Node number for degree of freedom 4

Notes:
1. The projected lengths for a truss element are shown in Figure 4.18.
2. The nodal numbering sequence must follow the order of the degrees of freedom for the truss element shown in Figure 4.18.

Line 13 (I5, 3D10.0): Nodal Load Increments Card

Note	Columns	Variable	Description
(1)	1–5	NLDOF	Degree of freedom at which a load is applied
(2)	6–15	FAMG(I,1)	Magnitude of load increment of load sequence 1
(3)	16–25	FAMG(I,2)	Magnitude of load at which load sequence 1 end
(3)	26–35	FAMG(I,3)	Magnitude of load increment of load sequence 2

174 ADVANCED ANALYSIS OF STEEL FRAMES

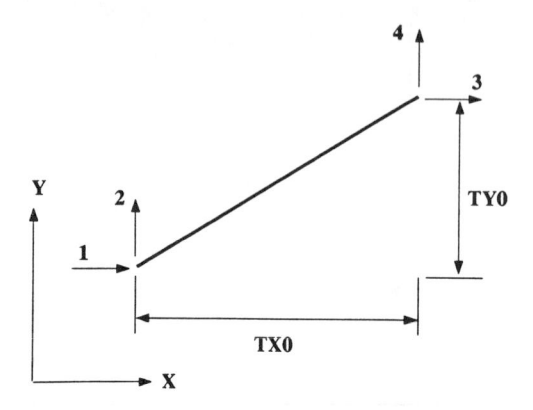

FIGURE 4.18. Degrees of freedom numbering for the truss element.

Notes:
1. All concentrated forces must be applied at the nodal point in the X and Y directions of the global coordinates by specifying the translational degrees of freedom of the node. Concentrated moment may be applied at the nodal point by specifying the rotational degree of freedom of the node.
2. If the number of load increments NINCRE = 1, values assigned to FAMG(I,2) and FAMG(I,3) will not affect the analysis results. In this case, the analysis is terminated when the number of load increment NINCRE is reached or when the structural stiffness matrix is nonpositive definite after attempting several load reduction steps near the frame's limit point.
3. If NINCRE = 2, the user must specify values for FAMG(I,2) and FAMG(I,3) so that the analysis knows when to apply load sequence two once the total load parameter for load sequence one is reached.

4.9.3 Examples

Example 1: Factored load analysis of semi-rigid braced frame using second-order elastic-plastic hinge method.

The first example is a semi-rigid braced frame, shown schematically with structural degrees of freedom in Figure 4.19 and subjected to factored static loadings in Figure 4.20. The input file for this problem is given in Table 4.1. After executing the program INPUT, three data files named DATA0, DATA1, and DATA2 are generated as shown in Table 4.2. Finally, the output file OUTPUT is generated by executing the program PHINGE. Part of the output file is shown in Table 4.3, and they can be used to verify PHINGE upon installation on a computer system.

Example 2: Limit load analysis of semi-rigid braced frame using second-order refined plastic hinge method.

The frame shown in Figure 4.19 is loaded to its limit of resistance, and the analysis is carried out using the refined plastic hinge analysis. The input file, INFILE, for this problem is given Table 4.4. It should be noted that a larger number of load increments should be specified when a limit load analysis is performed. Three data files are generated after executing the program INPUT, and they are shown in Table 4.5. Part of the OUTPUT file obtained from the analysis program PHINGE is shown in Table 4.6.

SECOND-ORDER PLASTIC HINGE ANALYSIS OF FRAMES 175

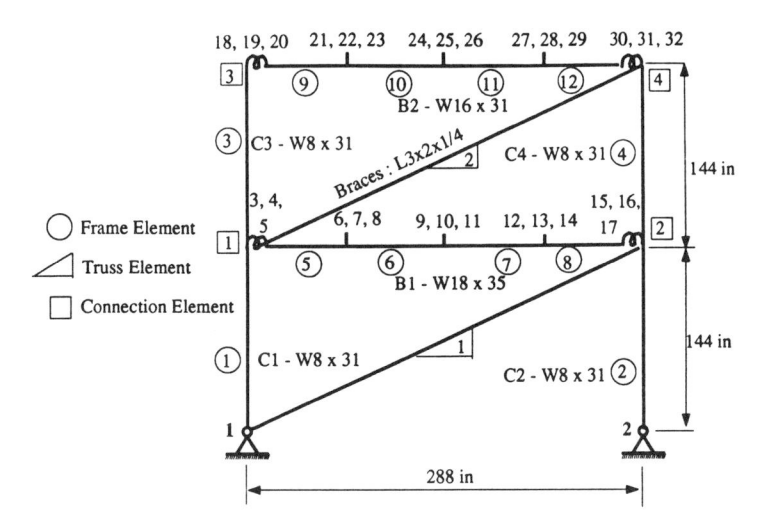

FIGURE 4.19. Structural modeling of a semi-rigid braced frame.

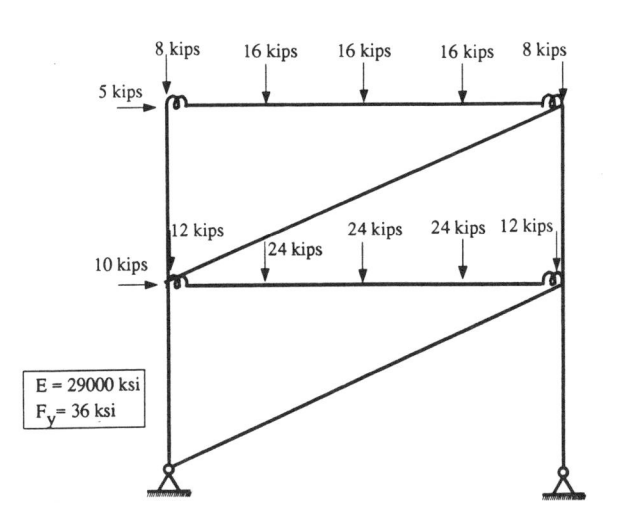

FIGURE 4.20. Factored loading on the frame.

Table 4.1 Content of INFILE for Example 1

```
/* ELASTIC-PLASTIC HINGE ANALYSIS AT FACTORED LOAD LEVEL
    0
    32      30       1       1
     2       3       1
     4      12       2
     1    1773.0 954013.0            0.80
     2     814.0 205924.0            1.57
     1       9.13      110.         30.4       29000.          36.0
     2      10.3       510.         66.5       29000.          36.0
     3       9.12      375.         54.0       29000.          36.0
     1       1.19     0.392      29000.0          36.0
     1       5       1       1
     2       8       2       1
     3       9       1       2
     4      12       2       2
     1       0.0      144.      1       0       0       1       3       4       5
     2       0.0      144.      1       0       0       2      15      16      17
     3       0.0      144.      1       3       4       5      18      19      20
     4       0.0      144.      1      15      16      17      30      31      32
     5      72.0        0.     2       3       4       5       6       7       8
     6      72.0        0.     2       6       7       8       9      10      11
     7      72.0        0.     2       9      10      11      12      13      14
     8      72.0        0.     2      12      13      14      15      16      17
     9      72.0        0.     3      18      19      20      21      22      23
    10      72.0        0.     3      21      22      23      24      25      26
    11      72.0        0.     3      24      25      26      27      28      29
    12      72.0        0.     3      27      28      29      30      31      32
     1     288.0      144.0    1       0       0      15      16
     2     288.0      144.0    1       3       4      30      31
     3       0.50       10.0       0.
    18       0.25        5.0       0.
     4      -0.600      -12.0       0.
     7      -1.200      -24.0       0.
    10      -1.200      -24.0       0.
    13      -1.200      -24.0       0.
    16      -0.200      -12.0       0.
    19      -0.4000     -8.00       0.
    22      -0.8000    -16.0        0.
    25      -0.8000    -16.0        0.
    28      -0.8000    -16.0        0.
    31      -0.4000     -8.00       0.
```

Table 4.2 Content of DATA0, DATA1, and DATA2 for Example 1

     0

----
    32   30   1   1
     2   3   1
     4   12   2
     1 0.17730D+04 0.95401D+06 0.80000D+00
     2 0.81400D+03 0.20592D+06 0.15700D+01
 0.00000D+00   0.00000D+00   0.50000D+00-0.60000D+00   0.00000D+00   0.00000D+00-
0.12000D+01 0.00000D+00
 0.00000D+00-0.12000D+01   0.00000D+00   0.00000D+00-0.12000D+01   0.00000D+00
 0.00000D+00-0.20000D+00
 0.00000D+00   0.25000D+00-0.40000D+00   0.00000D+00   0.00000D+00-0.80000D+00
 0.00000D+00   0.00000D+00   -0.80000D+00   0.00000D+00   0.00000D+00-0.80000D+00
 0.00000D+00 0.00000D+00-0.40000D+00 0.00000D+00

----
     1 0.00000D+00 0.14400D+03 0.91300D+01 0.11000D+03 0.30400D+02 0.29000D+05
0.36000D+02
       0   0   0
       0   0   1   3   4   5
     2 0.00000D+00 0.14400D+03 0.91300D+01 0.11000D+03 0.30400D+02 0.29000D+05
0.36000D+02
       0   0   0
       0   0   2   15   16   17
     3 0.00000D+00 0.14400D+03 0.91300D+01 0.11000D+03 0.30400D+02 0.29000D+05
0.36000D+02
       0   0   0
       3   4   5   18   19   20
     4 0.00000D+00 0.14400D+03 0.91300D+01 0.11000D+03 0.30400D+02 0.29000D+05
0.36000D+02
       0   0   0
      15   16   17   30   31   32
     5 0.72000D+02 0.00000D+00 0.10300D+02 0.51000D+03 0.66500D+02 0.29000D+05
0.36000D+02
       1   1   1
       3   4   5   6   7   8
     6 0.72000D+02 0.00000D+00 0.10300D+02 0.51000D+03 0.66500D+02 0.29000D+05
0.36000D+02
       0   0   0
       6   7   8   9   10   11
     7 0.72000D+02 0.00000D+00 0.10300D+02 0.51000D+03 0.66500D+02 0.29000D+05
0.36000D+02
       0   0   0
       9   10   11   12   13   14
     8 0.72000D+02 0.00000D+00 0.10300D+02 0.51000D+03 0.66500D+02 0.29000D+05
0.36000D+02
       1   2   1
      12   13   14   15   16   17
     9 0.72000D+02 0.00000D+00 0.91200D+01 0.37500D+03 0.54000D+02 0.29000D+05
0.36000D+02
       1   1   2
      18   19   20   21   22   23
    10 0.72000D+02 0.00000D+00 0.91200D+01 0.37500D+03 0.54000D+02 0.29000D+05
0.36000D+02
       0   0   0
      21   22   23   24   25   26
    11 0.72000D+02 0.00000D+00 0.91200D+01 0.37500D+03 0.54000D+02 0.29000D+05
0.36000D+02
       0   0   0
      24   25   26   27   28   29
    12 0.72000D+02 0.00000D+00 0.91200D+01 0.37500D+03 0.54000D+02 0.29000D+05
0.36000D+02
       1   2   2
      27   28   29   30   31   32
     1 0.28800D+03 0.14400D+03 0.11900D+01 0.39200D+00 0.29000D+05 0.36000D+02
       0   0   15   16
     2 0.28800D+03 0.14400D+03 0.11900D+01 0.39200D+00 0.29000D+05 0.36000D+02
       3   4   30   31

------------------------------------   ------------------

178   ADVANCED ANALYSIS OF STEEL FRAMES

Table 4.3   Content of OUTPUT for Example 1

```
--------------------------------------------------------------------------
****************************************************
* SECOND-ORDER ELASTIC-PLASTIC HINGE ANALYSIS *
****************************************************

==================
= STRUCTURE DATA =
==================

    NUMBER OF DEGREES OF FREEDOM OF THE STRUCTURE = 32
    NUMBER OF LOAD INCREMENTS                     = 30
    NUMBER OF LOAD SEQUENCES                      =  1

    NUMBER OF CONNECTION TYPES                    =  2
    NUMBER OF FRAME TYPES                         =  3
    NUMBER OF TRUSS TYPES                         =  1

    NUMBER OF CONNECTION ELEMENTS                 =  4
    NUMBER OF FRAME ELEMENTS                      = 12
    NUMBER OF TRUSS ELEMENTS                      =  2


===================
= CONNECTION DATA =
===================
    ---------------------
    - CONNECTION TYPE  1 -
    ---------------------
        MU = 0.17730D+04       RKI =  0.95401D+06       N =  0.80000D+00
    ---------------------
    - CONNECTION TYPE  2 -
    ---------------------
        MU = 0.81400D+03       RKI =  0.20592D+06       N =  0.15700D+01

======================
= FRAME ELEMENT DATA =
======================
    ---------------------
    - ELEMENT NUMBER  1 -
    ---------------------
        XO    = 0.00000D+00       YO  = 0.14400D+03       LO  = 0.14400D+03
        A     = 0.91300D+01       I   = 0.11000D+03       ZP  = 0.30400D+02
        E     = 0.29000D+05       FY  = 0.36000D+02
        ICNT  =        0          IEND =        0         IDCNT=       0
        ELEMENT   DOF :    1     2    3    4    5    6
        ---------------------------------------------
        STRUCTURE DOF :    0     0    1    3    4    5

    ---------------------
    - ELEMENT NUMBER  2 -
    ---------------------
        XO    = 0.00000D+00       YO  = 0.14400D+03       LO  = 0.14400D+03
        A     = 0.91300D+01       I   = 0.11000D+03       ZP  = 0.30400D+02
        E     = 0.29000D+05       FY  = 0.36000D+02
        ICNT  =        0          IEND =        0         IDCNT=       0
        ELEMENT   DOF :    1     2    3    4    5    6
        ---------------------------------------------
        STRUCTURE DOF :    0     0    2   15   16   17

    ---------------------
    - ELEMENT NUMBER  3 -
    ---------------------
        XO    = 0.00000D+00       YO  = 0.14400D+03       LO  = 0.14400D+03
        A     = 0.91300D+01       I   = 0.11000D+03       ZP  = 0.30400D+02
        E     = 0.29000D+05       FY  = 0.36000D+02
        ICNT  =        0          IEND =        0         IDCNT=       0
        ELEMENT   DOF :    1     2    3    4    5    6
        ---------------------------------------------
        STRUCTURE DOF :    3     4    5   18   19   20
```

```
----------------------
- ELEMENT NUMBER  4 -
----------------------
    XO    = 0.00000D+00      YO    = 0.14400D+03      LO  = 0.14400D+03
    A     = 0.91300D+01      I     = 0.11000D+03      ZP  = 0.30400D+02
    E     = 0.29000D+05      FY    = 0.36000D+02
    ICNT  =           0      IEND  =           0      IDCNT=          0
    ELEMENT   DOF :   1   2   3   4   5   6
    ---------------------------------------------
    STRUCTURE DOF :  15  16  17  30  31  32

----------------------
- ELEMENT NUMBER  5 -
----------------------
    XO    = 0.72000D+02      YO    = 0.00000D+00      LO  = 0.72000D+02
    A     = 0.10300D+02      I     = 0.51000D+03      ZP  = 0.66500D+02
    E     = 0.29000D+05      FY    = 0.36000D+02
    ICNT  =          '1      IEND  =           1      IDCNT=          1
    ELEMENT   DOF :   1   2   3   4   5   6
    ---------------------------------------------
    STRUCTURE DOF :   3   4   5   6   7   8

----------------------
- ELEMENT NUMBER  6 -
----------------------
    XO    = 0.72000D+02      YO    = 0.00000D+00      LO  = 0.72000D+02
    A     = 0.10300D+02      I     = 0.51000D+03      ZP  = 0.66500D+02
    E     = 0.29000D+05      FY    = 0.36000D+02
    ICNT  =           0      IEND  =           0      IDCNT=          0
    ELEMENT   DOF :   1   2   3   4   5   6
    ---------------------------------------------
    STRUCTURE DOF :   6   7   8   9  10  11

----------------------
- ELEMENT NUMBER  7 -
----------------------
    XO    = 0.72000D+02      YO    = 0.00000D+00      LO  = 0.72000D+02
    A     = 0.10300D+02      I     = 0.51000D+03      ZP  = 0.66500D+02
    E     = 0.29000D+05      FY    = 0.36000D+02
    ICNT  =           0      IEND  =           0      IDCNT=          0
    ELEMENT   DOF :   1   2   3   4   5   6
    ---------------------------------------------
    STRUCTURE DOF :   9  10  11  12  13  14

----------------------
- ELEMENT NUMBER  8 -
----------------------
    XO    = 0.72000D+02      YO    = 0.00000D+00      LO  = 0.72000D+02
    A     = 0.10300D+02      I     = 0.51000D+03      ZP  = 0.66500D+02
    E     = 0.29000D+05      FY    = 0.36000D+02
    ICNT  =           1      IEND  =           2      IDCNT=          1
    ELEMENT   DOF :   1   2   3   4   5   6
    ---------------------------------------------
    STRUCTURE DOF :  12  13  14  15  16  17

----------------------
- ELEMENT NUMBER  9 -
----------------------
    XO    = 0.72000D+02      YO    = 0.00000D+00      LO  = 0.72000D+02
    A     = 0.91200D+01      I     = 0.37500D+03      ZP  = 0.54000D+02
    E     = 0.29000D+05      FY    = 0.36000D+02
    ICNT  =           1      IEND  =           1      IDCNT=          2
    ELEMENT   DOF :   1   2   3   4   5   6
    ---------------------------------------------
    STRUCTURE DOF :  18  19  20  21  22  23

----------------------
- ELEMENT NUMBER 10 -
----------------------
    XO    = 0.72000D+02      YO    = 0.00000D+00      LO  = 0.72000D+02
    A     = 0.91200D+01      I     = 0.37500D+03      ZP  = 0.54000D+02
    E     = 0.29000D+05      FY    = 0.36000D+02
    ICNT  =           0      IEND  =           0      IDCNT=          0
    ELEMENT   DOF :   1   2   3   4   5   6
    ---------------------------------------------
    STRUCTURE DOF :  21  22  23  24  25  26
```

Table 4.3 *(continued)*

```
----------------------
- ELEMENT NUMBER 11 -
----------------------
    XO     = 0.72000D+02      YO   = 0.00000D+00       LO   = 0.72000D+02
    A      = 0.91200D+01      I    = 0.37500D+03       ZP   = 0.54000D+02
    E      = 0.29000D+05      FY   = 0.36000D+02
    ICNT   =        0         IEND =        0          IDCNT=         0
    ELEMENT    DOF :    1     2     3     4     5     6
    ---------------------------------------------------
    STRUCTURE DOF :    24    25    26    27    28    29

----------------------
- ELEMENT NUMBER 12 -
----------------------
    XO     = 0.72000D+02      YO   = 0.00000D+00       LO   = 0.72000D+02
    A      = 0.91200D+01      I    = 0.37500D+03       ZP   = 0.54000D+02
    E      = 0.29000D+05      FY   = 0.36000D+02
    ICNT   =        1         IEND =        2          IDCNT=         2
    ELEMENT    DOF :    1     2     3     4     5     6
    ---------------------------------------------------
    STRUCTURE DOF :    27    28    29    30    31    32

======================
= TRUSS ELEMENT DATA =
======================
----------------------
- ELEMENT NUMBER  1 -
----------------------
    XO= 0.28800D+03       YO= 0.14400D+03       LO= 0.32199D+03
    A = 0.11900D+01       I = 0.39200D+00       E = 0.29000D+05
    FY= 0.36000D+02
    ELEMENT    DOF :    1     2     3     4
    ----------------------------------------
    STRUCTURE DOF :     0     0    15    16

----------------------
- ELEMENT NUMBER  2 -
----------------------
    XO= 0.28800D+03       YO= 0.14400D+03       LO= 0.32199D+03
    A = 0.11900D+01       I = 0.39200D+00       E = 0.29000D+05
    FY= 0.36000D+02
    ELEMENT    DOF :    1     2     3     4
    ----------------------------------------
    STRUCTURE DOF :     3     4    30    31


=================================
= STRUCTURE DEGREES OF FREEDOM =
=================================
    NUMBER OF TRANSLATIONAL DOF =  20
    DOF NUMBERING :   3  4 15 16 18 19 30 31  6  7  9 10 12 13 21 22 24 25 27 28
    NUMBER OF ROTATIONAL DOF    =  12
    DOF NUMBERING :   1  5  2 17 20 32  8 11 14 23 26 29
```

```
==============================
= LOAD-DEFLECTION BEHAVIOR =
==============================
    -----------------
    - LOAD STEP   1 -
    -----------------
        LOAD SEQUENCE  1
        DETERMINANT OF STIFFNESS MATRIX = 0.67687 TIMES 10 TO THE POWER 128
        TRACE OF STIFFNESS MATRIX       = 0.10028 TIMES 10 TO THE POWER   8

              DOF         NODAL FORCE     NODAL DISPLACEMENT

               1          0.00000D+00        0.82881D-04
               2          0.00000D+00       -0.26216D-03
               3          0.50000D+00        0.96548D-02
               4         -0.60000D+00       -0.20956D-02
               5          0.00000D+00       -0.36690D-03
               6          0.00000D+00        0.95794D-02
               7         -0.12000D+01       -0.36352D-01
               8          0.00000D+00       -0.41732D-03
               9          0.00000D+00        0.95040D-02
              10         -0.12000D+01       -0.52343D-01
              11          0.00000D+00        0.64318D-05
              12          0.00000D+00        0.94286D-02
              13         -0.12000D+01       -0.35674D-01
              14          0.00000D+00        0.41984D-03
              15          0.00000D+00        0.93532D-02
              16         -0.20000D+00       -0.22299D-02
              17          0.00000D+00        0.32947D-03
              18          0.25000D+00        0.13568D-01
              19         -0.40000D+00       -0.29607D-02
              20          0.00000D+00       -0.27397D-03
              21          0.00000D+00        0.13353D-01
              22         -0.80000D+00       -0.37585D-01
              23          0.00000D+00       -0.41394D-03
              24          0.00000D+00        0.13137D-01
              25         -0.80000D+00       -0.53522D-01
              26          0.00000D+00        0.22773D-05
              27          0.00000D+00        0.12922D-01
              28         -0.80000D+00       -0.37365D-01
              29          0.00000D+00        0.41402D-03
              30          0.00000D+00        0.12706D-01
              31         -0.40000D+00       -0.31638D-02
              32          0.00000D+00        0.24751D-03
```

Table 4.3 *(continued)*

```
ELEMENT FORCES IN GLOBAL COORDINATE
------------------------------------
FRAME ELEMENT :
  ELEMENT NUMBER                    ELEMENT FORCES
       1           0.13865D+00       0.38532D+01       0.00000D+00
                  -0.13865D+00      -0.38532D+01      -0.19928D+02
       2          -0.18177D+00       0.41000D+01      -0.35527D-14
                   0.18177D+00      -0.41000D+01       0.26213D+02
       3           0.54143D+00       0.15906D+01      -0.41038D+02
                  -0.54143D+00      -0.15906D+01      -0.36921D+02
       4          -0.57551D+00       0.17173D+01       0.43255D+02
                   0.57551D+00      -0.17173D+01       0.39624D+02
       5           0.31284D+00       0.17703D+01       0.60966D+02
                  -0.31284D+00      -0.17703D+01       0.66508D+02
       6           0.31284D+00       0.57041D+00      -0.66508D+02
                  -0.31284D+00      -0.57041D+00       0.10758D+03
       7           0.31284D+00      -0.62945D+00      -0.10758D+03
                  -0.31284D+00       0.62945D+00       0.62257D+02
       8           0.31284D+00      -0.18294D+01      -0.62257D+02
                  -0.31284D+00       0.18294D+01      -0.69468D+02
       9           0.79138D+00       0.11902D+01       0.36921D+02
                  -0.79138D+00      -0.11902D+01       0.48803D+02
      10           0.79138D+00       0.39044D+00      -0.48803D+02
                  -0.79138D+00      -0.39044D+00       0.76928D+02
      11           0.79138D+00      -0.40921D+00      -0.76928D+02
                  -0.79138D+00       0.40921D+00       0.47452D+02
      12           0.79138D+00      -0.12090D+01      -0.47452D+02
                  -0.79138D+00       0.12090D+01      -0.39624D+02

TRUSS ELEMENT :
  ELEMENT NUMBER                    ELEMENT FORCES
       1          -0.70636D+00      -0.35316D+00       0.70636D+00       0.35316D+00
       2          -0.21583D+00      -0.10791D+00       0.21583D+00       0.10791D+00

                               .
                               .
                               .
                               .
                               .
```

```
------------------
- LOAD STEP  30 -
------------------
    LOAD SEQUENCE  1
    DETERMINANT OF STIFFNESS MATRIX = 0.11992 TIMES 10 TO THE POWER 124
    TRACE OF STIFFNESS MATRIX       = 0.76609 TIMES 10 TO THE POWER   7

         DOF         NODAL FORCE        NODAL DISPLACEMENT

          1          0.00000D+00          0.18821D-02
          2          0.00000D+00         -0.59462D-02
          3          0.11000D+02          0.29643D+00
          4         -0.13200D+02         -0.46521D-01
          5          0.00000D+00         -0.98125D-02
          6          0.00000D+00          0.27180D+00
          7         -0.26400D+02         -0.18987D+01
          8          0.00000D+00         -0.24609D-01
          9          0.00000D+00          0.25575D+00
         10         -0.26400D+02         -0.33671D+01
         11          0.00000D+00          0.15439D-01
         12          0.00000D+00          0.23967D+00
         13         -0.26400D+02         -0.18974D+01
         14          0.00000D+00          0.24602D-01
         15          0.00000D+00          0.21515D+00
         16         -0.44000D+01         -0.49302D-01
         17          0.00000D+00          0.72691D-02
         18          0.55000D+01          0.40453D+00
         19         -0.88000D+01         -0.65669D-01
         20          0.00000D+00         -0.39618D-02
         21          0.00000D+00          0.39495D+00
         22         -0.17600D+02         -0.93935D+00
         23          0.00000D+00         -0.10193D-01
         24          0.00000D+00          0.38936D+00
         25         -0.17600D+02         -0.13312D+01
         26          0.00000D+00          0.78245D-05
         27          0.00000D+00          0.38377D+00
         28         -0.17600D+02         -0.93906D+00
         29          0.00000D+00          0.10174D-01
         30          0.00000D+00          0.37424D+00
         31         -0.88000D+01         -0.70103D-01
         32          0.00000D+00          0.26906D-02
```

Table 4.3 *(continued)*

```
ELEMENT FORCES IN GLOBAL COORDINATE
-----------------------------------
FRAME ELEMENT :
  ELEMENT NUMBER            ELEMENT FORCES
         1      0.36756D+01    0.84992D+02    0.99476D-13
               -0.36756D+01   -0.84992D+02   -0.50408D+03
         2     -0.38239D+01    0.90370D+02   -0.72831D-13
                0.38239D+01   -0.90370D+02    0.57009D+03
         3      0.11333D+02    0.35131D+02   -0.94174D+03
               -0.11333D+02   -0.35131D+02   -0.68647D+03
         4     -0.11169D+02    0.38094D+02    0.90693D+03
                0.11169D+02   -0.38094D+02    0.70745D+03
         5      0.82427D+01    0.39498D+02    0.14458D+04
               -0.82427D+01   -0.39498D+02    0.14124D+04
         6      0.87768D+01    0.13076D+02   -0.14124D+04
               -0.87768D+01   -0.13076D+02    0.23665D+04
         7      0.87787D+01   -0.12366D+02   -0.23665D+04
               -0.87787D+01    0.12366D+02    0.14635D+04
         8      0.82307D+01   -0.39715D+02   -0.13967D+04
               -0.82307D+01    0.39715D+02   -0.14770D+04
         9      0.16670D+02    0.26319D+02    0.68647D+03
               -0.16670D+02   -0.26319D+02    0.12230D+04
        10      0.16809D+02    0.87231D+01   -0.12230D+04
               -0.16809D+02   -0.87231D+01    0.18576D+04
        11      0.16808D+02   -0.88693D+01   -0.18576D+04
               -0.16808D+02    0.88693D+01    0.12124D+04
        12      0.16669D+02   -0.26466D+02   -0.12124D+04
               -0.16669D+02    0.26466D+02   -0.70745D+03

TRUSS ELEMENT :
  ELEMENT NUMBER                  ELEMENT FORCES
         1     -0.16340D+02   -0.81611D+01    0.16340D+02    0.81611D+01
         2     -0.56621D+01   -0.28298D+01    0.56621D+01    0.28298D+01
```

==
= SEQUENCE OF PLASTIC HINGE FORMATION =
==

```
    PLASTIC HINGE              LOCATION
          1            B-TH END OF FRAME ELEMENT   6
```

Table 4.4 Content of INFILE for Example 2

```
/* REFINED PLASTIC HINGE LIMIT-LOAD ANALYSIS
    1
    1    1
   32   80      1      1
    2    3      1
    4   12      2
    1     1773.0  954013.0          0.80
    2      814.0  205924.0          1.57
    1         9.13      110.       30.4      29000.        36.0
    2        10.3       510.       66.5      29000.        36.0
    3         9.12      375.       54.0      29000.        36.0
    1         1.19     0.392    29000.0         36.0
    1    5      1      1
    2    8      2      1
    3    9      1      2
    4   12      2      2
    1      0.0       144.     1     0      0     1     3     4     5
    2      0.0       144.     1     0      0     2    15    16    17
    3      0.0       144.     1     3      4     5    18    19    20
    4      0.0       144.     1    15     16    17    30    31    32
    5     72.0         0.     2     3      4     5     6     7     8
    6     72.0         0.     2     6      7     8     9    10    11
    7     72.0         0.     2     9     10    11    12    13    14
    8     72.0         0.     2    12     13    14    15    16    17
    9     72.0         0.     3    18     19    20    21    22    23
   10     72.0         0.     3    21     22    23    24    25    26
   11     72.0         0.     3    24     25    26    27    28    29
   12     72.0         0.     3    27     28    29    30    31    32
    1    288.0       144.0    1     0      0    15    16
    2    288.0       144.0    1     3      4    30    31
    3      0.50       10.0     0.
   18      0.25        5.0     0.
    4     -0.600     -12.0     0.
    7     -1.200     -24.0     0.
   10     -1.200     -24.0     0.
   13     -1.200     -24.0     0.
   16     -0.200     -12.0     0.
   19     -0.4000    -8.00     0.
   22     -0.8000   -16.0      0.
   25     -0.8000   -16.0      0.
   28     -0.8000   -16.0      0.
   31     -0.4000    -8.00     0.
```

Table 4.5 Content of DATA0, DATA1, and DATA2 for Example 2

```
      1
      1    1
---------------------------------------------------------------------------

     32     80    1    1
      2      3    1
      4     12    2
      1  0.17730D+04 0.95401D+06 0.80000D+00
      2  0.81400D+03 0.20592D+06 0.15700D+01
 0.00000D+00    0.00000D+00    0.50000D+00-0.60000D+00   0.00000D+00   0.00000D+00-
0.12000D+01 0.00000D+00
 0.00000D+00-0.12000D+01    0.00000D+00    0.00000D+00-0.12000D+01   0.00000D+00
 0.00000D+00-0.20000D+00
 0.00000D+00    0.25000D+00-0.40000D+00   0.00000D+00   0.00000D+00-0.80000D+00
 0.00000D+00    0.00000D+00   -0.80000D+00   0.00000D+00   0.00000D+00-0.80000D+00
 0.00000D+00 0.00000D+00-0.40000D+00 0.00000D+00

---------------------------------------------------------------------------

---------------------------------------------------------------------------
      1  0.00000D+00  0.14400D+03  0.91300D+01  0.11000D+03  0.30400D+02  0.29000D+05
 0.36000D+02
      0      0      0
      0      0      1    3    4    5
      2  0.00000D+00  0.14400D+03  0.91300D+01  0.11000D+03  0.30400D+02  0.29000D+05
 0.36000D+02
      0      0      0
      0      0      2   15   16   17
      3  0.00000D+00  0.14400D+03  0.91300D+01  0.11000D+03  0.30400D+02  0.29000D+05
 0.36000D+02
      0      0      0
      3      4      5   18   19   20
      4  0.00000D+00  0.14400D+03  0.91300D+01  0.11000D+03  0.30400D+02  0.29000D+05
 0.36000D+02
      0      0      0
     15     16     17   30   31   32
      5  0.72000D+02  0.00000D+00  0.10300D+02  0.51000D+03  0.66500D+02  0.29000D+05
 0.36000D+02
      1      1      1
      3      4      5    6    7    8
      6  0.72000D+02  0.00000D+00  0.10300D+02  0.51000D+03  0.66500D+02  0.29000D+05
 0.36000D+02
      0      0      0
      6      7      8    9   10   11
      7  0.72000D+02  0.00000D+00  0.10300D+02  0.51000D+03  0.66500D+02  0.29000D+05
 0.36000D+02
      0      0      0
      9     10     11   12   13   14
      8  0.72000D+02  0.00000D+00  0.10300D+02  0.51000D+03  0.66500D+02  0.29000D+05
 0.36000D+02
      1      2      1
     12     13     14   15   16   17
      9  0.72000D+02  0.00000D+00  0.91200D+01  0.37500D+03  0.54000D+02  0.29000D+05
 0.36000D+02
      1      1      2
     18     19     20   21   22   23
     10  0.72000D+02  0.00000D+00  0.91200D+01  0.37500D+03  0.54000D+02  0.29000D+05
 0.36000D+02
      0      0      0
     21     22     23   24 • 25   26
     11  0.72000D+02  0.00000D+00  0.91200D+01  0.37500D+03  0.54000D+02  0.29000D+05
 0.36000D+02
      0      0      0
     24     25     26   27   28   29
     12  0.72000D+02  0.00000D+00  0.91200D+01  0.37500D+03  0.54000D+02  0.29000D+05
 0.36000D+02
      1      2      2
     27     28     29   30   31   32
      1  0.28800D+03 0.14400D+03 0.11900D+01 0.39200D+00 0.29000D+05 0.36000D+02
      0      0     15   16
      2  0.28800D+03 0.14400D+03 0.11900D+01 0.39200D+00 0.29000D+05 0.36000D+02
      3      4     30   31

---------------------------------------------------------------------------
```

SECOND-ORDER PLASTIC HINGE ANALYSIS OF FRAMES 187

Table 4.6 Content of OUTPUT for Example 2

```
****************************************
* SECOND-ORDER REFINED PLASTIC HINGE ANALYSIS *
****************************************
===================================
= CRC Tangent Modulus model       =
= Parabolic stiffness reduction   =
===================================

==================
= STRUCTURE DATA =
==================
      NUMBER OF DEGREES OF FREEDOM OF THE STRUCTURE =   32
      NUMBER OF LOAD INCREMENTS                     =   80
      NUMBER OF LOAD SEQUENCES                      =    1

      NUMBER OF CONNECTION TYPES                    =    2
      NUMBER OF FRAME TYPES                         =    3
      NUMBER OF TRUSS TYPES                         =    1

      NUMBER OF CONNECTION ELEMENTS                 =    4
      NUMBER OF FRAME ELEMENTS                      =   12
      NUMBER OF TRUSS ELEMENTS                      =    2

===================
= CONNECTION DATA =
===================
   ---------------------
   - CONNECTION TYPE  1 -
   ---------------------
      MU = 0.17730D+04      RKI = 0.95401D+06      N = 0.80000D+00
   ---------------------
   - CONNECTION TYPE  2 -
   ---------------------
      MU = 0.81400D+03      RKI = 0.20592D+06      N = 0.15700D+01

======================
= FRAME ELEMENT DATA =
======================
   ---------------------
   - ELEMENT NUMBER  1 -
   ---------------------
      XO    = 0.00000D+00     YO    = 0.14400D+03     LO  = 0.14400D+03
      A     = 0.91300D+01     I     = 0.11000D+03     ZP  = 0.30400D+02
      E     = 0.29000D+05     FY    = 0.36000D+02
      ICNT  =      0          IEND  =      0          IDCNT=      0
      ELEMENT   DOF :    1    2    3    4    5    6
      ---------------------------------------------
      STRUCTURE DOF :    0    0    1    3    4    5

   ---------------------
   - ELEMENT NUMBER  2 -
   ---------------------
      XO    = 0.00000D+00     YO    = 0.14400D+03     LO  = 0.14400D+03
      A     = 0.91300D+01     I     = 0.11000D+03     ZP  = 0.30400D+02
      E     = 0.29000D+05     FY    = 0.36000D+02
      ICNT  =      0          IEND  =      0          IDCNT=      0
      ELEMENT   DOF :    1    2    3    4    5    6
      ---------------------------------------------
      STRUCTURE DOF :    0    0    2   15   16   17
```

Table 4.6 *(continued)*

```
---------------------
- ELEMENT NUMBER  3 -
---------------------
    XO    = 0.00000D+00       YO   = 0.14400D+03       LO   = 0.14400D+03
    A     = 0.91300D+01       I    = 0.11000D+03       ZP   = 0.30400D+02
    E     = 0.29000D+05       FY   = 0.36000D+02
    ICNT  =        0          IEND =        0          IDCNT=        0
    ELEMENT    DOF :    1    2    3    4    5    6
    ------------------------------------------------
    STRUCTURE DOF  :    3    4    5   18   19   20

---------------------
- ELEMENT NUMBER  4 -
---------------------
    XO    = 0.00000D+00       YO   = 0.14400D+03       LO   = 0.14400D+03
    A     = 0.91300D+01       I    = 0.11000D+03       ZP   = 0.30400D+02
    E     = 0.29000D+05       FY   = 0.36000D+02
    ICNT  =        0          IEND =        0          IDCNT=        0
    ELEMENT    DOF :    1    2    3    4    5    6
    ------------------------------------------------
    STRUCTURE DOF  :   15   16   17   30   31   32

---------------------
- ELEMENT NUMBER  5 -
---------------------
    XO    = 0.72000D+02       YO   = 0.00000D+00       LO   = 0.72000D+02
    A     = 0.10300D+02       I    = 0.51000D+03       ZP   = 0.66500D+02
    E     = 0.29000D+05       FY   = 0.36000D+02
    ICNT  =        1          IEND =        1          IDCNT=        1
    ELEMENT    DOF :    1    2    3    4    5    6
    ------------------------------------------------
    STRUCTURE DOF  :    3    4    5    6    7    8

---------------------
- ELEMENT NUMBER  6 -
---------------------
    XO    = 0.72000D+02       YO   = 0.00000D+00       LO   = 0.72000D+02
    A     = 0.10300D+02       I    = 0.51000D+03       ZP   = 0.66500D+02
    E     = 0.29000D+05       FY   = 0.36000D+02
    ICNT  =        0          IEND =        0          IDCNT=        0
    ELEMENT    DOF :    1    2    3    4    5    6
    ------------------------------------------------
    STRUCTURE DOF  :    6    7    8    9   10   11

---------------------
- ELEMENT NUMBER  7 -
---------------------
    XO    = 0.72000D+02       YO   = 0.00000D+00       LO   = 0.72000D+02
    A     = 0.10300D+02       I    = 0.51000D+03       ZP   = 0.66500D+02
    E     = 0.29000D+05       FY   = 0.36000D+02
    ICNT  =        0          IEND =        0          IDCNT=        0
    ELEMENT    DOF :    1    2    3    4    5    6
    ------------------------------------------------
    STRUCTURE DOF  :    9   10   11   12   13   14
---------------------
- ELEMENT NUMBER  8 -
---------------------
    XO    = 0.72000D+02       YO   = 0.00000D+00       LO   = 0.72000D+02
    A     = 0.10300D+02       I    = 0.51000D+03       ZP   = 0.66500D+02
    E     = 0.29000D+05       FY   = 0.36000D+02
    ICNT  =        1          IEND =        2          IDCNT=        1
    ELEMENT    DOF :    1    2    3    4    5    6
    ------------------------------------------------
    STRUCTURE DOF  :   12   13   14   15   16   17

---------------------
- ELEMENT NUMBER  9 -
---------------------
    XO    = 0.72000D+02       YO   = 0.00000D+00       LO   = 0.72000D+02
    A     = 0.91200D+01       I    = 0.37500D+03       ZP   = 0.54000D+02
```

```
    E       = 0.29000D+05     FY    = 0.36000D+02
    ICNT  =      1            IEND  =       1           IDCNT=        2
    ELEMENT   DOF :    1     2     3     4     5     6
    -------------------------------------------------
    STRUCTURE DOF :   18    19    20    21    22    23

---------------------
- ELEMENT NUMBER 10 -
---------------------
    XO      = 0.72000D+02     YO    = 0.00000D+00     LO   = 0.72000D+02
    A       = 0.91200D+01     I     = 0.37500D+03     ZP   = 0.54000D+02
    E       = 0.29000D+05     FY    = 0.36000D+02
    ICNT  =      0            IEND  =       0           IDCNT=        0

    ELEMENT   DOF :    1     2     3     4     5     6
    -------------------------------------------------
    STRUCTURE DOF :   21    22    23    24    25    26

---------------------
- ELEMENT NUMBER 11 -
---------------------
    XO      = 0.72000D+02     YO    = 0.00000D+00     LO   = 0.72000D+02
    A       = 0.91200D+01     I     = 0.37500D+03     ZP   = 0.54000D+02
    E       = 0.29000D+05     FY    = 0.36000D+02
    ICNT  =      0            IEND  =       0           IDCNT=        0
    ELEMENT   DOF :    1     2     3     4     5     6
    -------------------------------------------------
    STRUCTURE DOF :   24    25    26    27    28    29

---------------------
- ELEMENT NUMBER 12 -
---------------------
    XO      = 0.72000D+02     YO    = 0.00000D+00     LO   = 0.72000D+02
    A       = 0.91200D+01     I     = 0.37500D+03     ZP   = 0.54000D+02
    E       = 0.29000D+05     FY    = 0.36000D+02
    ICNT  =      1            IEND  =       2           IDCNT=        2
    ELEMENT   DOF :    1     2     3     4     5     6
    -------------------------------------------------
    STRUCTURE DOF :   27    28    29    30    31    32

======================
= TRUSS ELEMENT DATA =
======================
---------------------
- ELEMENT NUMBER  1 -
---------------------
    XO= 0.28800D+03     YO= 0.14400D+03       LO= 0.32199D+03
    A = 0.11900D+01     I = 0.39200D+00       E = 0.29000D+05
    FY= 0.36000D+02
    ELEMENT   DOF :    1     2     3     4
    ---------------------------------------
    STRUCTURE DOF :    0     0    15    16

---------------------
- ELEMENT NUMBER  2 -
---------------------
    XO= 0.28800D+03     YO= 0.14400D+03       LO= 0.32199D+03
    A = 0.11900D+01     I = 0.39200D+00       E = 0.29000D+05
    FY= 0.36000D+02
    ELEMENT   DOF :    1     2     3     4
    ---------------------------------------
    STRUCTURE DOF :    3     4    30    31

==================================
= STRUCTURE DEGREES OF FREEDOM =
==================================
    NUMBER OF TRANSLATIONAL DOF =   20
    DOF NUMBERING :   3   4 15 16 18 19 30 31  6  7  9 10 12 13 21 22 24 25 27 28
    NUMBER OF ROTATIONAL DOF     =   12
```

190 ADVANCED ANALYSIS OF STEEL FRAMES

Table 4.6 *(continued)*

```
    DOF NUMBERING :   1   5   2  17  20  32   8  11  14  23  26  29

=============================
= LOAD-DEFLECTION BEHAVIOR =
=============================
-----------------
 - LOAD STEP    1 -
-----------------
    LOAD SEQUENCE   1
    DETERMINANT OF STIFFNESS MATRIX  = 0.67687 TIMES 10 TO THE POWER 128
    TRACE OF STIFFNESS MATRIX        = 0.10028 TIMES 10 TO THE POWER   8

         DOF           NODAL FORCE         NODAL DISPLACEMENT

          1             0.00000D+00           0.82881D-04
          2             0.00000D+00          -0.26216D-03
          3             0.50000D+00           0.96548D-02
          4            -0.60000D+00          -0.20956D-02
          5             0.00000D+00          -0.36690D-03
          6             0.00000D+00           0.95794D-02
          7            -0.12000D+01          -0.36352D-01
          8             0.00000D+00          -0.41732D-03
          9             0.00000D+00           0.95040D-02
         10            -0.12000D+01          -0.52343D-01
         11             0.00000D+00           0.64318D-05
         12             0.00000D+00           0.94286D-02
         13            -0.12000D+01          -0.35674D-01
         14             0.00000D+00           0.41984D-03
         15             0.00000D+00           0.93532D-02
         16            -0.20000D+00          -0.22299D-02
         17             0.00000D+00           0.32947D-03
         18             0.25000D+00           0.13568D-01
         19            -0.40000D+00          -0.29607D-02
         20             0.00000D+00          -0.27397D-03
         21             0.00000D+00           0.13353D-01
         22            -0.80000D+00          -0.37585D-01
         23             0.00000D+00          -0.41394D-03
         24             0.00000D+00           0.13137D-01
         25            -0.80000D+00          -0.53522D-01
         26             0.00000D+00           0.22773D-05
         27             0.00000D+00           0.12922D-01
         28             0.80000D+00          -0.37365D-01
         29             0.00000D+00           0.41402D-03
         30             0.00000D+00           0.12706D-01
         31            -0.40000D+00          -0.31638D-02
         32             0.00000D+00           0.24751D-03

    ELEMENT FORCES IN GLOBAL COORDINATE
    -----------------------------------
    FRAME ELEMENT :
    ELEMENT NUMBER                          ELEMENT FORCES
          1           0.1386470D+00    0.3853178D+01    0.0000000D+00
                     -0.1386470D+00   -0.3853178D+01   -0.1992796D+02
          2          -0.1817656D+00    0.4100000D+01   -0.3552714D-14
                      0.1817656D+00   -0.4100000D+01    0.2621259D+02
          3           0.5414269D+00    0.1590615D+01   -0.4103832D+02
                     -0.5414269D+00   -0.1590615D+01   -0.3692094D+02
          4          -0.5755107D+00    0.1717302D+01    0.4325537D+02
                      0.5755107D+00   -0.1717302D+01    0.3962393D+02
          5           0.3128379D+00    0.1770332D+01    0.6096628D+02
                     -0.3128379D+00   -0.1770332D+01    0.6650830D+02
          6           0.3128379D+00    0.5704108D+00   -0.6650830D+02
                     -0.3128379D+00   -0.5704108D+00    0.1075829D+03
          7           0.3128379D+00   -0.6294473D+00   -0.1075829D+03
                     -0.3128379D+00    0.6294473D+00    0.6225746D+02
          8           0.3128379D+00   -0.1829375D+01   -0.6225746D+02
                     -0.3128379D+00    0.1829375D+01   -0.6946796D+02
          9           0.7913836D+00    0.1190234D+01    0.3692094D+02
                     -0.7913836D+00   -0.1190234D+01    0.4880331D+02
         10           0.7913837D+00    0.3904395D+00   -0.4880331D+02
```

SECOND-ORDER PLASTIC HINGE ANALYSIS OF FRAMES   191

```
                        -0.7913837D+00   -0.3904395D+00    0.7692757D+02
              11         0.7913837D+00   -0.4092078D+00   -0.7692757D+02
                        -0.7913837D+00    0.4092078D+00    0.4745182D+02
              12         0.7913836D+00   -0.1209010D+01   -0.4745182D+02
                        -0.7913836D+00    0.1209010D+01   -0.3962393D+02

     TRUSS ELEMENT :
ELEMENT NO.                        ELEMENT FORCES

      1      -0.70636D+00  -0.35316D+00    0.70636D+00    0.35316D+00

      2      -0.21583D+00  -0.10791D+00    0.21583D+00    0.10791D+00

                              .
                              .
                              .
                              .
                              .

------------------
- LOAD STEP  44 -
------------------
     LOAD SEQUENCE   1
     DETERMINANT OF STIFFNESS MATRIX = 0.12224 TIMES 10 TO THE POWER 108
     TRACE OF STIFFNESS MATRIX       = 0.45705 TIMES 10 TO THE POWER   7

         DOF         NODAL FORCE       NODAL DISPLACEMENT

          1          0.00000D+00         0.69325D-03
          2          0.00000D+00        -0.66512D-02
          3          0.11342D+02         0.59764D+00
          4         -0.13611D+02        -0.48903D-01
          5          0.00000D+00        -0.14014D-01
          6          0.00000D+00         0.49587D+00
          7         -0.27222D+02        -0.39319D+01
          8          0.00000D+00        -0.53053D-01
          9          0.00000D+00         0.41234D+00
         10         -0.27222D+02        -0.74533D+01
         11          0.00000D+00         0.49915D-01
         12          0.00000D+00         0.32879D+00
         13         -0.27222D+02        -0.39317D+01
         14          0.00000D+00         0.53044D-01
         15          0.00000D+00         0.22716D+00
         16         -0.45369D+01        -0.50926D-01
         17          0.00000D+00         0.89554D-02
         18          0.56712D+01         0.75005D+00
         19         -0.90738D+01        -0.68705D-01
         20          0.00000D+00        -0.55122D-02
         21          0.00000D+00         0.72754D+00
         22         -0.18148D+02        -0.17010D+01
         23          0.00000D+00        -0.20829D-01
         24          0.00000D+00         0.71399D+00
         25         -0.18148D+02        -0.28564D+01
         26          0.00000D+00         0.32314D-01
         27          0.00000D+00         0.70045D+00
         28         -0.18148D+02        -0.17016D+01
         29          0.00000D+00         0.20806D-01
         30          0.00000D+00         0.67802D+00
         31         -0.90738D+01        -0.72922D-01
         32          0.00000D+00         0.16876D-02

     ELEMENT FORCES IN GLOBAL COORDINATE
     -----------------------------------

        FRAME ELEMENT :
        ELEMENT NUMBER              ELEMENT FORCES
              1         0.4645826D+01   0.8772992D+02  -0.1092850D-12
                       -0.4645826D+01  -0.8772992D+02  -0.6165452D+03
              2        -0.4347195D+01   0.9332283D+02   0.1679039D-12
```

Table 4.6 *(continued)*

	0.4347195D+01	-0.9332283D+02	0.6472017D+03
3	0.1226154D+02	0.3626211D+02	-0.9900048D+03
	-0.1226154D+02	-0.3626211D+02	-0.7701281D+03
4	-0.1206669D+02	0.3923494D+02	0.9733495D+03
	0.1206669D+02	-0.3923494D+02	0.7819378D+03
5	0.7689109D+01	0.4083871D+02	0.1606550D+04
	-0.7689109D+01	-0.4083871D+02	0.1359414D+04
6	0.8974052D+01	0.1357457D+02	-0.1359409D+04
	-0.8974052D+01	-0.1357457D+02	0.2367208D+04
7	0.8965608D+01	-0.1355455D+02	-0.2367208D+04
	-0.8965608D+01	0.1355455D+02	0.1360876D+04
8	0.7671789D+01	-0.4093686D+02	-0.1352390D+04
	-0.7671789D+01	0.4093686D+02	-0.1620551D+04
9	0.1748467D+02	0.2717707D+02	0.7701281D+03
	-0.1748467D+02	-0.2717707D+02	0.1214660D+04
10	0.1781266D+02	0.9118748D+01	-0.1214658D+04
	-0.1781266D+02	-0.9118748D+01	0.1891705D+04
11	0.1781088D+02	-0.9183675D+01	-0.1891705D+04
	-0.1781088D+02	0.9183675D+01	0.1209997D+04
12	0.1748160D+02	-0.2725967D+02	-0.1208731D+04
	-0.1748160D+02	0.2725967D+02	-0.7819378D+03

```
     TRUSS ELEMENT :
ELEMENT NO.                    ELEMENT FORCES

    1    -0.17300D+02 -0.86403D+01  0.17300D+02  0.86403D+01

    2    -0.58635D+01 -0.29305D+01  0.58635D+01  0.29305D+01

>> SOLUTION DOES NOT CONVERGE, A LIMIT STATE MAY HAVE BEEN REACHED <<

========================================
= SEQUENCE OF PLASTIC HINGE FORMATION =
========================================

       PLASTIC HINGE              LOCATION

            1          B-TH END OF FRAME ELEMENT  6
            2          B-TH END OF FRAME ELEMENT 10
```

References

AISC (1986) *Load and Resistance Factor Design Specification for Structural Steel Buildings*, 1st ed., American Institute of Steel Construction, Chicago.

ASCE-WRC (1971) Plastic Design in Steel — A Guide and Commentary, ASCE Manuals and Reports on Engineering Practice, No. 41, 2nd ed., Jointed Committee of the Welding Research Council and American Institute of Steel Contruction, Chicago.

Baker, L. and Heyman, J. (1969) *Plastic Design of Frames — Volume 1: Fundamentals*, Cambridge University Press, Cambridge, England, 228 pp.

Beedle, L. S. (1958), *Plastic Design of Steel Frames*, John Wiley & Sons, New York.

Belytschko, T. and Hsieh, B. J. (1973) Non-linear transient finite element analysis with convected coordinates, *Int. J. Num. Meth. Eng.*, 7(3), 255–271.

Brinstiel, C. and Iffland, J. S. B. (1980) Factors influencing frame instability, *J. Struct. Div.*, 106(2), 491–504.

Chen, W. F. and Atsuta, T. (1976) *Theory of Beam-Column, Vol. 1, In-Plane Behavior and Design*, McGraw-Hill, New York, 513 pp.

Chen, W. F., Duan, L., and Zhou, S. P. (1990) Second-order inelastic analysis of braced portal frames — evaluation of design formulae in LRFD and GBJ specification, *J. Singapore Struct. Steel Soc. Steel Struct.*, 1(1), 5–15.

Chen, W. F. and Lui, E. M. (1991) *Stability Design of Steel Frames*, CRC Press, Boca Raton, FL, 380 pp.

Clarke, M. J., Bridge, R. Q., Hancock, G. J., and Trahair, N. S., (1991), Design using advanced analysis, Annual Tech. Session Proc., SSRC, April 15–17, Chicago, IL, 27–40.

Cook, R. D., Malkus, D. S., and Plesha, M. E. (1989) *Concepts and Applications of Finite Element Analysis*, 3rd ed., John Wiley & Sons, New York, 630 pp.

Deierlein, G. G., Zhao, Y., and McGuire, W. (1991) A two-surface concentrated plasticity model for analysis of 3D framed structures, Annual Tech. Session Proc., SSRC, April 15–17, Chicago, IL, 423–432.

Duan, L. and Chen, W. F. (1990) A yield surface equation for doubly symmetrical section, *Eng. Struct.*, 12, 114–119.

Ekhande, S. G., Selvappalam, M., and Madugula, M. K. S. (1989) Stability functions for three-dimensional beam-columns, *J. Struct. Eng.*, 115(2), 467–479.

Galambos, T. V., Ed. (1988) *Guide to Stability Design Criteria for Metal Structures*, 4th ed., John Wiley & Sons, New York.

Heyman, J. (1969) *Plastic Design of Frames — Volume 2: Applications*, Cambridge University Press, Cambridge, England.

Hughes, T. J. R. (1987) *The Finite Element Method: Linear Static and Dynamic Finite Element Analysis*, Prentice-Hall, Englewood Cliffs, NJ, 803 pp.

Kanchanalai, T. (1977) The Design and Behavior of Beam-Columns in Unbraced Steel Frames, AISI Project No. 189, Report No. 2, Civil Engineering/Structures Research Lab., University of Texas at Austin, 300 pp.

King, W. S., White, D. W., and Chen, W. F. (1991) On Second-Order Inelastic Methods for Steel Frame Design, Structural Engineering Report, CE-STR-91-16, Purdue University, West Lafayette, IN, 36 pp.

Kishi, N. and Chen, W. F. (1990) Moment-rotation relations of semi-rigid connections with angles, *J. Struct. Eng.*, 116(7), 1813–1834.

Korn, A. and Galambos, T. V. (1967) Behavior of elastic-plastic frame, *J. Struct. Div.*, 94(ST5), 1119–1142.

Liew, J. Y. R., White, D. W., and Chen, W. F. (1991) Beam-column design in steel frameworks — insight on current methods and trends, *J. Constr. Steel Res.*, 18, 269–308.

Liew J. Y. R. (1992) Advanced Analysis for Frame Design, Ph.D dissertation, School of Civil Engineering, Purdue University, West Lafayette, IN, 393 pp.

Liew, J. Y. R., White, D. W., and Chen, W. F. (1992a) Notional Load Plastic Hinge Method for Frame Design, Structural Engineering Report, CE-STR-92-4, Purdue University, West Lafayette, IN, 42 pp.

Liew, J. Y. R., White, D. W., and Chen, W. F. (1992b) Second-Order Refined Plastic Hinge Analysis of Frames, Structural Engineering Report, CE-STR-92–12, Purdue University, West Lafayette, IN, 35 pp.

Liew, J. Y. R., White, D. W., and Chen, W. F. (1992c) Second-Order Refined Plastic Hinge Analysis for Frame Design, Structural Engineering Report, CE-STR-92–14, Purdue University, West Lafayette, IN, 36 pp.

Lui, E. M. (1985) Effect of Connection Flexibility and Panel Zone Deformation on the Behavior of Plane Steel Frames, Ph.D dissertation, School of Civil Engineering, Purdue University, West Lafayette, IN.

Orbison, J. G. (1982), Nonlinear Static Analysis of Three-Dimensional Steel Frames, Department of Structural Engineering, Report No. 82–6, Cornell University, Ithaca, NY, 243 pp.

Renton J. D. (1962) Stability of space frames by computer analysis, *J. Struct. Div.*, 88(4), 539–608.

Riahi, A., Row, D. G., and Powell, G. H. (1978) Three-Dimensional Inelastic Frame Elements for the ANSR-1 Program, Report No. UCB/EERC 78/06, Earthquake Engineering Research Center, University of California, Berkely, CA, August, 67 pp.

Springfield, J. (1991) Limits on second-order elastic analysis, SSRC Proceedings, 1991 Annual Technical Session on Inelastic Behavior and Design of Frames, SSRC, April 25–27, Chicago, IL, 89–99.

Williams, F. W. (1981) Stability functions and frame instability — a fresh approach, *Int. J. Mech. Sci.*, 23(12), 715–722.

White, D. W. and Hajjar, J. F. (1991) Application of second-order elastic analysis in LRFD: research to practice, *Eng. J.*, 28(4), 133–148.

Ziemian, R. D. (1990), Advanced Methods of Inelastic Analysis in the Limit States Design of Steel Structures, Ph.D dissertation, School of Civil and Environmental Engineering, Cornell University, Ithaca, NY, 265 pp.

5: Plastic-Zone Analysis of Beam-Columns and Portal Frames

S. P. Zhou, *Department of Structural Engineering, Chongqing College of Architecture, Chongqing, Peoples Republic of China*

W. F. Chen, *School of Civil Engineering, Purdue University, West Lafayette, Indiana*

5.1 Introduction

The second-order inelastic analysis on the basis of the plastic-zone theory for in-plane beam-columns and portal frames is described in this chapter. There are three programs presented:

(1) *Program BCIN* is for beam-columns with I- and box-sections. The loading types are (1) the unequal end moments vary, but the axial force remains constant; (2) the axial force varies, but the unequal end moments remain constant. Under the two types of loading, any combination of axial force and unequal end moments can be analyzed by BCIN.
(2) *Program FRAMP* is for portal braced frames with vertical loads acted on the beam and top of the columns (no horizontal load).
(3) *Program FRAMH* is for portal unbraced frames with both vertical and horizontal loads acted on the beam and top of the columns. In this loading, the vertical loads remain constant but the horizontal load varies.

The beam and the columns, whose sections are I-shape, are rigidly connected in the programs FRAMP and FRAMH, and bent about their strong axis. The load-deformation curves, including ascending and descending branches, can be traced by BCIN and FRAMH, and the strength of the beam-columns or the frames is given by the peak value of the curves. However, FRAMP can only trace the ascending branch of the load-deformation curve.

The computer programs BCIN, FRAMP, and FRAMH are written in FORTRAN Language. The files of these programs are stored in the diskette as shown in Table 5.1. In these three programs for the analyses based on the plastic-zone theory, the members of beam and columns are divided into a number of segments, and the

196 ADVANCED ANALYSIS OF STEEL FRAMES

Table 5.1 List of Computer Files

Program names	Source file		Run file		Input file (example)	Sample problem file	Usage
	Name	Bytes	Name	Bytes			
BCIN	BCIN.FOR	30454	BCIN.EXE	91390	INB	OBC	Beam-column
FRAMP	FRAMP.FOR	36897	FRAMP.EXE	101760	INP	OFP	Portal braced frame
FRAMH	FRAMH.FOR	37941	FRAMH.EXE	88682	INH	OFH	Portal unbraced frame

sections of these members are further divided into a number of strip elements parallel to the bending axis. The deflection at each division point of these members are obtained by Newmark's integration method. The in-plane bending stresses in each strip can be considered uniform if the strips are narrow enough. The resultant force of residual stress in each strip is constant. Then, the elastic range and resultant force in each strip can be obtained by simple mathematical procedures: the stiffness of the section is obtained by integrating the stiffness of the elastic range area of each strip, and the internal axial force and moment of the section are also obtained by integrating the resultant force of each strip.

In this way, the gradual spreading of plastic-zone in the sections and along the members can be explicitly traced. The residual stresses and the initial crookedness of the member can be considered. It is recognized that the analysis based on the plastic-zone theory is most accurate. It can most realistically reflect the behaviors and the ultimate load-carrying capacity of the members or structures. The analysis of the three programs is based on the following assumptions:

(1) plane sections remain plane after loading;
(2) deformation of members is small compared to their length;
(3) steel is assumed to be an elastic, perfectly plastic material; and
(4) the out-of-plane buckling and local buckling are not considered.

5.2 Analysis of In-Plane Beam-Columns[*]

5.2.1 Principle of Analysis

The computer program BCIN, which was originally developed for reinforced concrete columns by Cranston (1967), has been modified for the analysis of steel beam-columns. Analytical conditions and methods are described in the following.

Analytical Conditions (Figure 5.1)

The analysis for isolated beam-columns is made under the following conditions:

---
[*] Zhou and Chen (1985).

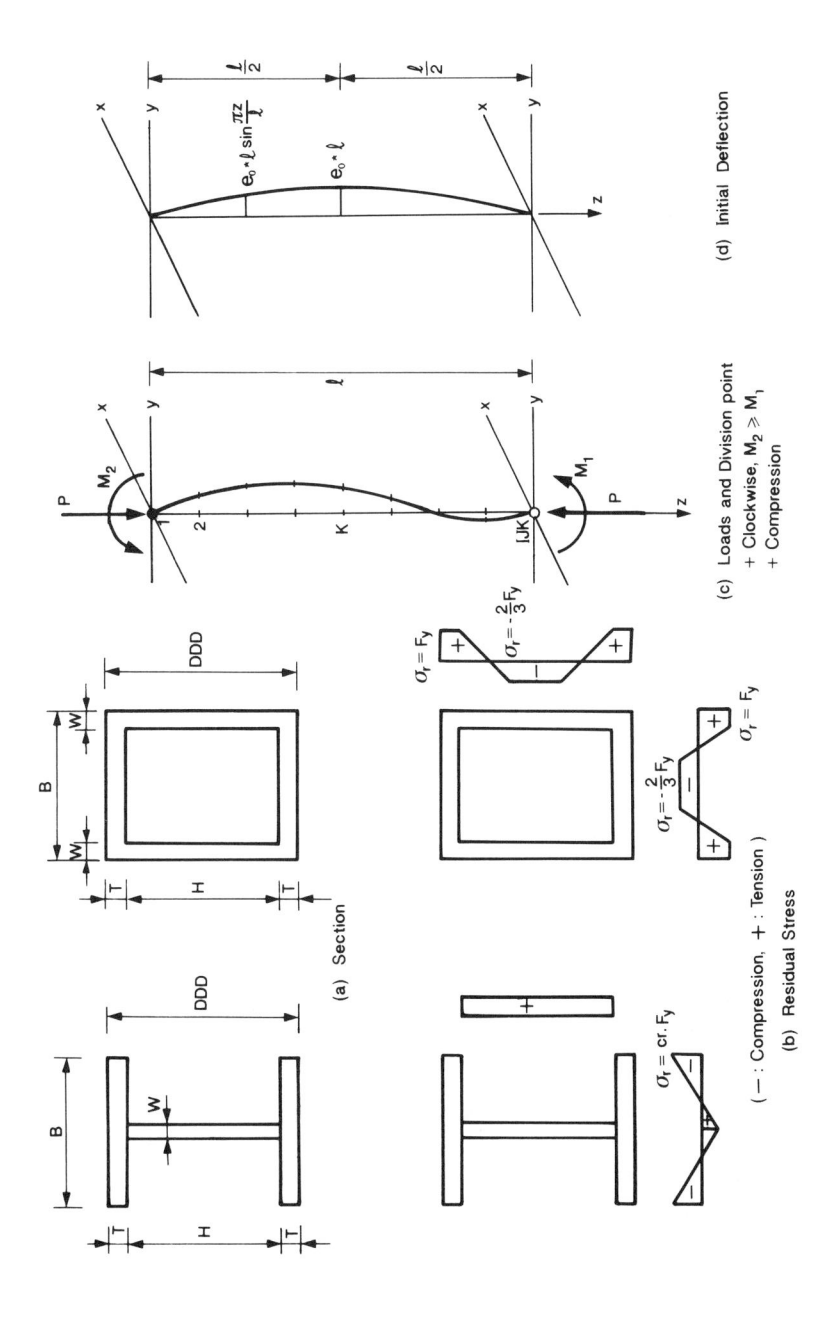

PLASTIC-ZONE ANALYSIS OF BEAM-COLUMNS AND PORTAL FRAMES   197

FIGURE 5.1. Steel beam-columns.

(1) section of beam-columns:
    I-section or box-section;
(2) bending axis:
    (a) bending about strong axis for I-section (the parameter $MJ = 0$ in BCIN);
    (b) bending about weak axis for I-section (the parameter $MJ = 1$ in BCIN);
    (c) bending about the axis perpendicular to the side plates DDD of box-section (the parameter $MJ = 2$ in BCIN);
(3) end condition: pinned;
(4) initial crookedness of beam-column:
    half-sine wave curve with the initial deflection at midspan e_o; when $e_o = 0$, the member is straight ($e_o = E0$ in BCIN);
(5) residual stress:
    shown in Figure 5.1; when the residual stress parameter cr is taken as zero ($cr = 0$), no residual stress is considered ($cr = RESID$ in BCIN);
(6) loads:
    (a) the axial load P remains constant while the unequal end moments M_1 and M_2 vary as the deformation of beam-column increases (the parameter $MP = 1$ in BCIN);
    (b) the unequal end moments M_1 and M_2 remain constant while the axial load P varies as the deformation of beam-column increases (the parameter $MP = 2$ in BCIN).

Analytical Method

For a beam-column subjected to the end moments shown in Figure 5.1, the equilibrium differential equation including the second-order effect is

$$y'' + k^2(y + y_o) = \frac{(M_1 + M_2)}{EI}\frac{z}{L} - \frac{M_2}{EI} \tag{5.1}$$

where

y = deflection of the member
y_o = initial deflection of the member
$k = \sqrt{\dfrac{P}{EI}}$
P = axial force
M_1, M_2 = smaller and larger end moments of the member
E = elastic modulus
I = moment of inertia about bending axis
L = length of the member
z = distance from the end with larger moment

In elastic range, the stiffness EI of the member remains constant and the closed-form solution of Eq. (5.1) can be obtained (see Eqs. (5.3) to (5.5)). Schematic load-

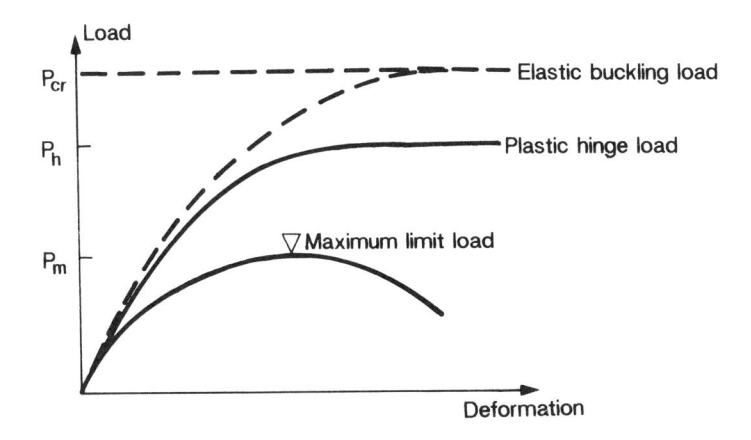

FIGURE 5.2. Load-deformation curves of curves of beam-columns.

deformation curves for beam-columns are shown in Figure 5.2. In the program BCIN, when the load is small and the member is in elastic state, the program follows Eqs. (5.3) to (5.5).

As the load increases, a part of the member goes into plastic range and the bending stiffness EI in Eq. (5.1) is not constant anymore. For inelastic members the behaviors of the load-deflection curves can be classified into two types. One type has an ascending branch and a descending branch. The member is in a stable equilibrium state in the ascending branch and in an unstable state in the descending branch. The failure in this type is due to buckling. The other type has an ascending branch and then approaches to a horizontal line (no descending branch). The failure in this type is due to a formation of plastic hinges (Figure 5.2). In order to trace these types of inelastic load-deformation curves, no closed-form solution is available and the numerical methods should be adopted.

In the program BCIN, when the load is large and the member is in an inelastic range, the analysis is programmed for computers by the following procedures. At the beginning, the member is divided into a number of segments and one of the divided points (usually the point at largest deflection) is chosen as the special point. A series of values of deflection for this special point is assigned. By the second-order plastic-zone analysis, a series of loads (M_1 and M_2 for $MP = 1$ or P for $MP = 2$) corresponding to the assigned deflection values can be obtained. For the first solution, the second-order elastic analysis is carried out by assuming small loads M_1 and M_2 (for $MP = 1$) or P (for $MP = 2$). Then, the second-order elastic deflection at the special point corresponding to the assumed small loads is taken as the assigned deflection of the special point. The entire curve of the second-order elastic deflection of the member and the assumed small loads are taken as a set of trial values.

Using the trial deflections and the loads, the external moment at each division point can be calculated. Now, the question is how to get the mean strain and curvature corresponding to the external moment at each division point. Here, we choose

$$EC = \varepsilon\, E \qquad (5.2\text{a})$$

$$CJ = \Phi\, E \qquad (5.2\text{b})$$

in which

ε = mean strain of section
Φ = curvature
EC = mean strain parameter
CJ = curvature parameter

The program BCIN adopts the tangent stiffness method (Chen and Atsuta, 1977) to obtain the parameters EC and CJ. In this way, with the calculated EC and CJ, the internal forces at each division point will satisfy the equilibrium condition. Here, a linear variation of curvature is assumed at each segment. Curvatures along the member are then integrated numerically to obtain deflections at all division points by Newmark's integration method. If the calculated deflection at the special division point does not meet with the initial assigned value, the trial set of loads must be modified and the procedures with the modified set of loads will be repeated until the difference is tolerated within an appropriate limit. Then, the deflections at all division points are checked one by one. If the differences between new and old calculated deflections at every division point are not within the specified limit, the new deflections at all division points will replace the old ones and the procedure will be repeated until the specified limit will be satisfied.

At this stage the compatibility as well as the equilibrium conditions at each division point are all satisfied and a solution is obtained. Then, a new value by adding an increment to the deflection of the last solution at the special point is assigned. By repeating the same procedures, a load-deformation curve of the steel beam-column can be traced.

As described above, there are three iteration processes in the program BCIN: the first is for the equilibrium of external and internal forces at each division point; the second is for the compatibility at the special point; and the third is for the compatibility at all division points. A success in the iterations depends on the appropriate trial values of deflections and loads to a larger extent. For the first solution, as stated before, the second-order elastic analysis is carried out by assuming small loads, and the results are used as the trial values. For other solutions, an extrapolation procedure is arranged to obtain the appropriate trial values.

5.2.2 Analytical Steps

The computer program takes the following analytical steps:

(1) Input the data from a screen and a file (see User's Manual).
(2) Calculate the geometric and the physical behaviors for sections such as sectional area A, moment of inertia I_x and I_y, radius of gyration r_x and r_y,

plastic section modulus Z_x and Z_y, plastic moment M_{px} and M_{py} etc., in which the subscripts x and y are the strong and weak axes for I-section or the axes perpendicular and parallel to the side plates DDD, respectively.

The inverses of I_x, I_y, r_x, r_y, Z_x, Z_y, M_{px}, and M_{py} are needed for the cases of $MJ = 0$ and 2 because the y-axis is assigned to the bending axis for $MJ = 0$ and 2 in BCIN.

(3) Divide the section into NF strips for each of the flanges and NW strips for the web: the strips are parallel to the bending axis. Calculate the coordinates along the axis perpendicular to the bending direction for the centroid of each strip. The ordinates are named by YI for $MJ = 0$ and 2 and XI for $MJ = 1$ in BCIN.

(4) Calculate the maximum compression and tension residual stress σ_{rc}(= SIC in BCIN) and σ_{rt}(= SIT in BCIN).

(5) Divide the member into $(IJK-1)$ segments. The division points are numbered 1, 2, 3, ... K, ... IJK. The point 1 is acted by larger end moment M_2, and the point IJK is acted by smaller end moment M_1. The point $IJ1$ is assigned as the special division point.

Calculate the initial deformations $YQ0$ at each division point by a half sine-wave curve.

(6) Calculate the loads M_1, M_2, and P from the input data $FF1$, $FF2$, and $AFA(\alpha)$. Carry out the second-order elastic analysis to obtain the deformation $YQ1$.

$$YQ1(K) = \frac{M_1}{P}\left(\frac{z}{L} - \frac{\sin kz}{\sin kL}\right) + \frac{M_2}{P}\left(\frac{\sin k(L-z)}{\sin kL} + \frac{z}{L} - 1\right) + \frac{YQ0(K)}{\left(\frac{\pi}{kL}\right)^2 - 1} \quad (5.3)$$

$$YQ1(IJK+1) = \frac{M_1}{P}\left(\frac{1}{L} - \frac{k}{\sin kL}\right) + \frac{M_2}{P}\left(\frac{1}{L} - \frac{k\cos kL}{\sin kL}\right) + \frac{YQ0(IJK+1)}{\left(\frac{\pi}{kL}\right)^2 - 1} \quad (5.4)$$

$$YQ1(IJK+2) = \frac{M_1}{P}\left(\frac{1}{L} - \frac{k\cos kL}{\sin kL}\right) + \frac{M_2}{P}\left(\frac{1}{L} - \frac{k}{\sin kL}\right) + \frac{YQ0(IJK+2)}{\left(\frac{\pi}{kL}\right)^2 - 1} \quad (5.5)$$

where

$YQ0(K)$ = initial deflections at division points, $K = 1$ to IJK
$YQ0(IJK + 1)$ = initial end rotation at Point 1
$YQ0(IJK + 2)$ = initial end rotation at Point IJK
$YQ1(K)$ = deflections of division points, $K = 1$ to IJK
$YQ1(IJK + 1)$ = end rotation at point 1
$YQ1(IJK + 2)$ = end rotation at point IJK

202 ADVANCED ANALYSIS OF STEEL FRAMES

$M_2 = F1*Mp$ = larger end moment acted on point 1 (= $F1*PMY = EE$ in BCIN)
$M_1 = aM_2$ = smaller end moment acted on point IJK (= $AFA*F1*PMY = F3*PMY = FF$ in BCIN)
$\alpha = M_1/M_2$ = ratio of end moments (= AFA in BCIN)
M_p = plastic bending moment (= PMY in BCIN)
$P = F2*P_y$ = axial force (= $F2*PY = GG$ in BCIN)
P_y = yield strength (= PY in BCIN)

Here, $F1 = FF1$ (for $MP = 1$) or $F2 = FF2$ (for $MP = 2$) is the input data, and $YQ1$ is obtained by the second-order elastic analysis. However, $F1$ or $F2$ will be modified and $YQ1$ will be replaced by the results of the second-order inelastic analysis in the later iterations.

Next, assign $YQ1(IJ1)$ as the deflection at the special point, i.e., $SDDY = YQ1(IJ1)$ in BCIN.

(7) Start the loop 100 in which a solution will be obtained by each time of the loop. It is numbered $N = 1, 2, 3, \ldots$

(8) Calculate the external moment including the second-order effect for each division point. As the first trial values in the iteration 200 the second-order elastic deflections $YQ1$ are taken for the first solution ($N = 1$). However, as the first trial value in the iteration, the extrapolated values $YQ1$ are taken for other solutions ($N > 1$). Of course, the $YQ1$ will be renewed during the 200 iterations.

Start the loop 300 to get the curvature at each division point by the tangent stiffness method. We know that for the elastic-perfect plastic material the normal stress on the section of the member is

$$\sigma = F_y \qquad \text{for} \qquad \sigma \geq F_y$$
$$\sigma = \sigma_r + E\varepsilon + E\Phi x \qquad \text{for} \quad -F_y > \sigma > F_y \qquad (5.6)$$
$$\sigma = -F_y \qquad \text{for} \qquad \sigma \leq -F_y$$

where

σ = normal stress
σ_r = residual stress
x = coordinate of strip (=YI for $MJ = 0$ and 2, =XI for $MJ = 1$ in BCIN)

The relationship between the deformation increment and the internal force increment has the usual form:

$$E\begin{bmatrix} A_e & S_e \\ S_e & I_e \end{bmatrix}\begin{Bmatrix} d\varepsilon \\ d\Phi \end{Bmatrix} = \begin{Bmatrix} dP \\ dM \end{Bmatrix} \qquad (5.7)$$

where

A_e, S_e, I_e = elastic area, static moment, moment of inertia of section
$d\varepsilon, d\Phi$ = strain and curvature increment
dP, dM = internal force and moment increment

In the elastic range, the terms A_e, S_e, and I_e are constant, but in the inelastic range they vary as the deformation increases. In BCIN, using Eq. (5.6), calculate the resultant force FFA and the elastic area EA of each strip of the section by σ_r, $E\varepsilon$, $E\Phi$, and x. Then, accumulate FFA and EA of each strip to obtain the internal axial force $P1$, the internal moment $R1$ and the tangent stiffness of the section at each division point. This process is arranged into the subroutines SECS, SECW, and SECB in BCIN.

Compare the external and the internal forces at each division point. If their differences are within the appropriate limit, the equilibrium condition of the external and the internal forces at the sections is satisfied and the curvatures are obtained. Then, step 9 can be followed. If not, solve Eq. (5.7) to obtain the deformation increments $d\varepsilon$ and $d\Phi$ which correspond to the differences of the external and the internal forces, then modify EC and CJ as follows:

$$EC = EC - E\,d\varepsilon \tag{5.8}$$

$$CJ = CJ - E\,d\Phi \tag{5.9}$$

After EC and CJ are modified, repeat the calculation of FFA, EA, $R1$, $P1$ ... until the equilibrium condition is reached. This is the iteration 400 in BCIN.

In calculating the first solution ($N = 1$), the first trial values of EC and CJ are taken from second-order elastic analysis. However, in calculating other solutions ($N > 1$), the extrapolated values of EC and CJ are used for the first trial values.

(9) The deflected shape of the member is shown in Figure 5.3. According to the assumption of linear variation of curvature in each segment and the assumption of small deformation, the deflection curve has the following geometric relationship:

$$y'_i = y'_{i-1} + Q_{i-1}\,dl + (2\Phi_{i-1} + \Phi_i)\,dl^2/6 \tag{5.10}$$

$$y_i = y'_n\,dl/l - y'_i \tag{5.11}$$

$$Q_i = Q_{i-1} + (\Phi_{i-1} + \Phi_i)\,dl/2 \tag{5.12}$$

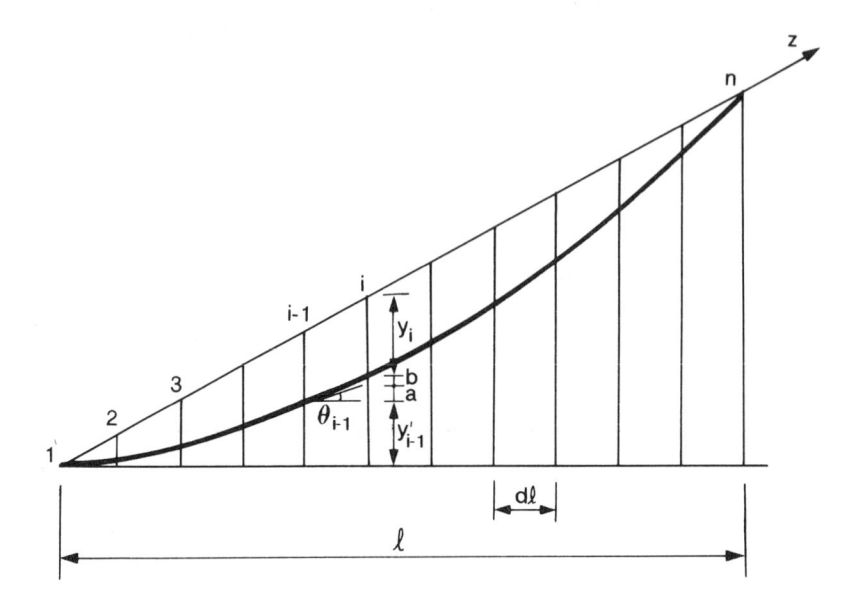

FIGURE 5.3. Deflection calculation by Newmark integration method.

where

y'_i	= a distance shown in Figure 5.3
y_i	= deflection of point i
Q_i	= angle between the tangent lines at point 1 and i shown in Figure 5.3, $Q_1 = 0$
Φ_i	= curvature of point i
dl	= length of segment
$Q_{i-1}dl$	= marked by "a" in Figure 5.3
$(2\Phi_{i-1} + \Phi_i)dl^2/6$	= marked by "b" in Figure 5.3

With Eqs. (5.10) and (5.11), the deflection can be successively calculated from Point 1 to n. This is known as Newmark's integration method. In BCIN, this calculation is programmed in loop 500 and 550. The term y_i is stored into the array YQ2(K) ($K = 1$ to *IJK*).

(10) Calculate DY = YQ1(IJ1)–YQ2(IJ1). If DY is within the acceptable limit, go to step 11; if not, a modification has taken place.

For *MP* = 1, the end moments, i.e., *F*1 and *F*3, need to be modified. Let us put a set of fictitious loads *Mp*, α*Mp*, and *P* on the beam-column as shown in Figure 5.4a and calculate the external moments at each division point, i.e., *PMY*1 in BCIN. Then, using the current tangent stiffness the corresponding curvatures are calculated by elastic analysis. From the curvatures, the deflection *F*1*Y* at the division point *IJ*1 is obtained by Newmark's integration method as before. Since *F*1*Y* is caused by *Mp*, α*Mp*, and *P*, we can infer that under the same current tangent stiffness the term *DF*1∗*Mp*, which causes the deflection *DY*, would have the following relationship:

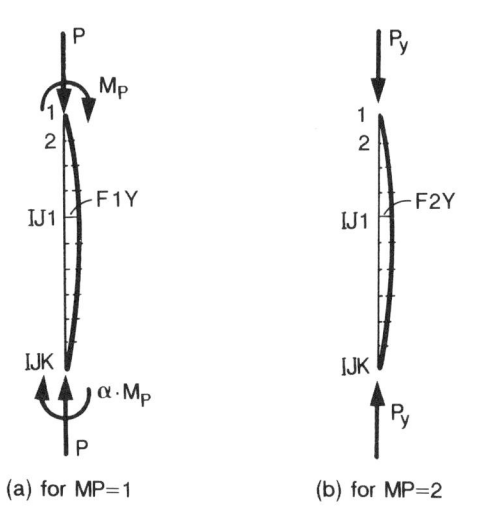

PLASTIC-ZONE ANALYSIS OF BEAM-COLUMNS AND PORTAL FRAMES 205

(a) for MP=1

(b) for MP=2

FIGURE 5.4. Fictitious loads on beam-columns.

$$DY/DF1 * M_p = F1Y/M_p$$

Hence,

$$DF1 = DY/F1Y \tag{5.13}$$

Now, we can see that $F1$ and $F3$ should be modified in order to eliminate DY as follows:

$$F1 = F1 - DF1 \tag{5.14}$$

$$F3 = \alpha F1 \tag{5.15}$$

Using the modified $F1$ and $F3$, repeat steps 8 to 10 until DY is tolerated within an appropriate limit.

For $MP = 2$, $F2$ needs to be modified and the fictitious load is P_y (Figure 5.4b). The corresponding deflection at the division point $IJ1$ is named by $F2Y$. $DF2*P_y$, which can cause DY, can be obtained by

$$DY/DF2 * P_y = F2Y/P_y$$

Hence,

$$DF2 = DY/F2Y \tag{5.16}$$

and the trial load $F2$ should be modified as

$$F2 = F2 - DF2 \tag{5.17}$$

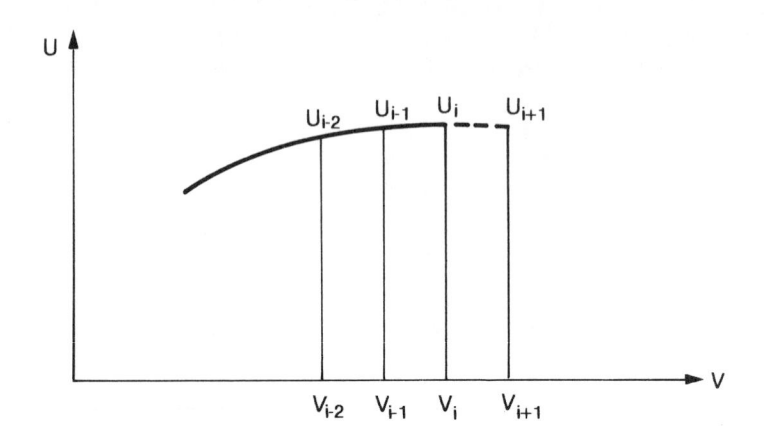

FIGURE 5.5. Extrapolation procedure in BCIN.

Using the modified $F2$, repeat steps 8 to 10 until DY is tolerated within an appropriate limit. This is the iteration 200–1 in BCIN (see Figure 5.6).

(11) Compare $YQ2$ with $YQ1$ at each division point. If they meet within the acceptable limit, a valid solution which satisfies both the equilibrium and compatibility conditions has been found; if not, let $YQ2$ equal the new $YQ1$ and repeat steps 8 to 11. This is the iteration 200–2 in BCIN (see Figure 5.6).

(12) Add a deflection increment DDY to $YQ1(IJ1)$ and use an extrapolation routine to obtain an estimated value for the next solution.

For a function $U = F(V)$, the second-order extrapolation routine adopted from Cranston's program can produce U_{i+1} corresponding to V_{i+1} from the last three solutions U_{i-2}, U_{i-1}, and U_i (Figure 5.5).

$$U_{i+1} = AU_{i-2} - BU_{i-1} + (1+C)U_i \tag{5.18}$$

where

$$A = \frac{(V_{i+1} - V_{i-1})(V_{i+1} - V_i)}{(V_{i-1} - V_{i-2})(V_i - V_{i-2})} \tag{5.18a}$$

$$B = \frac{(V_{i+1} - V_{i-2})(V_{i+1} - V_i)}{(V_{i-1} - V_{i-2})(V_i - V_{i-1})} \tag{5.18b}$$

$$C = \frac{(V_{i+1} - V_{i-2} + V_i - V_{i-1})(V_{i+1} - V_i)}{(V_i - V_{i-1})(V_i - V_{i-2})} \tag{5.18c}$$

In BCIN the term V is the deflection at Point $IJ1$. The terms V_{i-2}, V_{i-1}, and V_i are the deflections $YQ2(IJ1)$ for the last three solutions. The term V_{i+1} is the assigned value for the next step $(i + 1)$, which will be put into $YQ1(IJ1)$. In fact, the subtraction $(V_{i+1} - V_i)$ is the deflection increment DDY, but $(V_i - V_{i-1})$, $(V_{i-1} - V_{i-2})$, etc. are not exactly equal to DDY because of the iteration process.

The term U contains $YQ1, D(K,1), D(K,2)$, and $F1$ (for $MP = 1$) or $F2$ (for $MP = 2$). There are in total $3(IJK + 1)$ numbers stored in the array BB in BCIN. The term BB is divided into four sections, and each section has the length $L = 3(IJK + 1)$. Sections 1 to 3 store the solution values $YQ2$, $D(K,1)$, ... for $i - 2, i - 1, i$ steps and section 4 will store the forecasts from the extrapolation routine. A number of beam-column analyses have proved that the extrapolation routine is very successful.

(13) After the solutions are obtained for each deflection increasing, a load-deformation curve can be traced. The computation will automatically stop, and the maximum load $F1max$ (for $MP = 1$) or $F2max$ (for $MP = 2$) will be output in the following cases:
   (a) when the number of the solutions in the descending branch of the curve has reached NSTOP which is input data chosen by the user; or
   (b) when the ratio $T11$ of the elastic area to the whole area of the section becomes smaller than 0.05. In this case the curve is almost horizontal and the plastic hinge will be formed with a little more incremental deformation.

(14) The convergence may become difficult in the iteration due to the following cases:
   (a) the elastic area $T1$ or the determinant TZV of the tangent stiffness matrix of the section is close to zero;
   (b) in the iteration 400, the number $NO1$ of the iteration times is larger than $3*IJK$; in the iteration 200–1, the number $NA2$ of the iteration times is larger than 10; and in the iteration 200–2, the number $NB2$ of the iteration times is larger than 10.

For these cases, the procedure will jump to the statement labeled as 999, and the increment DDY will be reduced to one half. Using the reduced DDY, the extrapolation routine will again be carried out and the solution will be recalculated.

If the convergence is still not obtained after DDY is reduced four times, the message (as shown in Table 5.2) will be output and the calculation will be interrupted. At this time, if the maximum load is not obtained, the user can try again by adjusting the input data according to the output information.

5.2.3 Flow Chart of BCIN

The program BCIN consists of the main program and five subroutines (SECS, SECW, SECB, EXIN, and EXOUT). A flow chart of the program is shown in Figure 5.6.

208 ADVANCED ANALYSIS OF STEEL FRAMES

Table 5.2 Output for Interruption in BCIN

Condition			Output
Interrupt			INTERRUPT N K N01 NNK NA2 NB2 T1
NN ≥ 1			INTERRUPT AT DESCENDING BRANCH AT NN=~
		MP = 1	F1MAX = ~
		MP = 2	F2MAX = ~
	N = 1 {	MP = 1	SUGGESTION: CHANGE FF1 OR AEE AND TRY AGAIN!
		MP = 2	SUGGESTION: CHANGE FF2 OR AEE AND TRY AGAIN!
NN = 0	1 < N ≤ 10		SUGGESTION: CHANGE AEE OR ADDY OR IJ1 AND TRY AGAIN!
	N > 10		SUGGESTION: CHANGE ADDY OR IJ1 AND TRY AGAIN!

Note: (NN) Number of solution points in descending branch.

5.2.4 User's Manual of BCIN

In BCIN, a positive sign is taken as follows:

horizontal force: right direction
displacement: right direction
vertical load: downward direction
axial force and stress: compression
moment and rotation: clockwise

Input Data

There are two groups of input data in BCIN. One is input from the screen and the other is from a file.

Input from the screen:

D1   a character variable for the name of input file, not more than six characters
OUT a character variable for the name of output file, not more than six characters
ALD slenderness ratio about bending axis
FF1 larger end moment normalized by PMY; trial value for first solution when $MP = 1$; constant value when $MP = 2$
FF2 axial force normalized by PY; trial value for first solution when $MP = 2$; constant value when $MP = 1$
AFA ratio of smaller end moment to larger end moment, M_1/M_2

Note: PMY is the plastic moment without the effect of axial force, and PY is the yield strength.

The read sentences in BCIN are as follows:

  WRITE(*,'(A)') 'INPUT THE NAME OF INPUT FILE:'
  READ(*,'(A)') D1

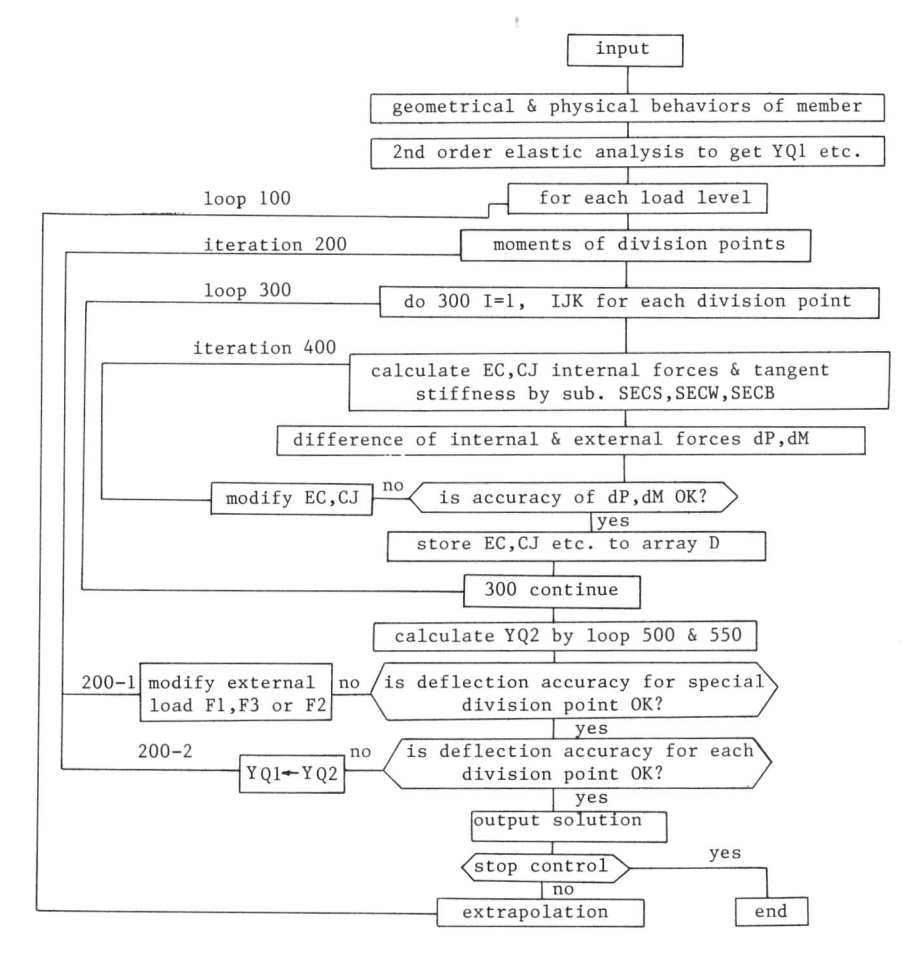

PLASTIC-ZONE ANALYSIS OF BEAM-COLUMNS AND PORTAL FRAMES

```
                          ┌─────────┐
                          │  input  │
                          └────┬────┘
              ┌────────────────┴────────────────────────┐
              │ geometrical & physical behaviors of member │
              └────────────────┬────────────────────────┘
              ┌────────────────┴────────────────────────┐
              │ 2nd order elastic analysis to get YQ1 etc. │
              └────────────────┬────────────────────────┘
   loop 100                    │
                 ┌─────────────┴──────────┐
                 │ for each load level    │
                 └─────────────┬──────────┘
   iteration 200               │
                 ┌─────────────┴──────────┐
                 │ moments of division points │
                 └─────────────┬──────────┘
   loop 300                    │
                 ┌─────────────┴────────────────────────┐
                 │ do 300 I=1, IJK for each division point │
                 └─────────────┬────────────────────────┘
   iteration 400               │
                 ┌─────────────┴────────────────────────┐
                 │ calculate EC,CJ internal forces & tangent │
                 │ stiffness by sub. SECS,SECW,SECB         │
                 └─────────────┬────────────────────────┘
                 ┌─────────────┴────────────────────────┐
                 │ difference of internal & external forces dP,dM │
                 └─────────────┬────────────────────────┘
                               │
      ┌──────────────┐   no   ╱                          ╲
      │ modify EC,CJ │◄──────< is accuracy of dP,dM OK?   >
      └──────────────┘         ╲                          ╱
                                    │ yes
                 ┌──────────────────┴──────────┐
                 │ store EC,CJ etc. to array D │
                 └──────────────────┬──────────┘
                           ┌────────┴──────┐
                           │  300 continue │
                           └────────┬──────┘
                 ┌──────────────────┴──────────┐
                 │ calculate YQ2 by loop 500 & 550 │
                 └──────────────────┬──────────┘
  200-1 ┌──────────────────┐  no   ╱                              ╲
        │ modify external  │◄─────< is deflection accuracy for special >
        │ load F1,F3 or F2 │       ╲  division point OK?           ╱
        └──────────────────┘            │ yes
  200-2                          no   ╱                            ╲
        ┌──────────┐                 ╱ is deflection accuracy for each ╲
        │ YQ1←YQ2  │◄───────────────<  division point OK?              >
        └──────────┘                 ╲                                  ╱
                                         │ yes
                               ┌─────────┴───────┐
                               │ output solution │
                               └─────────┬───────┘
                                   ╱           ╲      yes
                                  < stop control >─────────┐
                                   ╲           ╱           │
                                       │ no                ▼
                               ┌───────┴──────┐         ┌─────┐
                               │ extrapolation │        │ end │
                               └──────────────┘         └─────┘
```

FIGURE 5.6. Flow chart of BCIN.

```
WRITE(*,'(A)') 'INPUT THE FILE NAME TO STORE RESULTS:'
READ(*,'(A)') OUT
WRITE(*,'(A)') 'PLEASE PUT ALD FF1 FF2 AFA'
READ(*,*) ALD,FF1,FF2,AFA
```

Input from a file:

The data are stored in the file named by the user (in BCIN the file is characterized as "D1").

B	width of section flange
DDD	depth of section
T	thickness of section flange
W	thickness of section web
FY	yield stress

EEE   elastic modulus
EO   initial crookedness at midspan normalized by member length ($EO = e_0$ in Figure 5.1)
ADDY   increment of deflection at special division point
RESID residual stress parameter, (σ_r/F_y in Figure 5.1)
AEE   allowance in difference of deflections in iteration 200
MP   a flag for loading type
    =1: the end moments F1 and F3 change and the axial force F2 remains constant
    =2: the axial force F2 changes and the end moments F1 and F3 remain constant
MJ   a flag for member type
    =0: for bending about strong axis of I-section
    =1: for bending about weak axis of I-section
    =2: for bending about the axis perpendicular to the side plates DDD of box-section
NF   number of flange strips
NW   number of web strips
NSTOP assigned number of solution points of descending branch of load-deflection curve; when the solution point of descending branch reaches NSTOP, the calculation will stop automatically
IJK   number of division points of member (see Figure 5.1)
IJ1   location of special division point of member, i.e., number of division points from the end with larger moment to special division point

The read sentence in BCIN is as follows:

READ(9,*)   B,DDD,T,W,FY,EEE,EO,ADDY,RESID,AEE,MP,MJ,NF,NW, NSTOP,IJK,IJ1

Output Data

Output data are stored in a file named by the user (in BCIN the file is characterized as "OUT"). There are five groups of output data.

Head of output table:

B-DDD-T-W-ALD	see input data
FF1-FF2-AFA	"
ADDY-AEE	"
FY-EEE-RESID-EO	"
MP-MJ-NF-NW-NSTOP-IJK-IJ1	"

A-AIX-AIY	area of section, moment of inertia about x and y axes
ZX-ZY	plastic section moduli about x and y axes

PY-PMX-PMY   yield strength of section, plastic moments about x and y axes
ALDX-ALDY-CL   slenderness ratios about x and y axes, member length
Note: The axis y is a bending axis and x is perpendicular to the y axis. Notice that when $MJ = 0,2$ the axes are different from those in Figure 5.1.

Data about the load-deformation curve:
There is one line output for each solution.

N	sequence number of solution points of the curve
NO1	iteration times for tangent stiffness of section, from which the consumed calculation time can be estimated
F1	larger end moment normalized by PMY
Y	deflection at special division point (IJ1)
T11	ratio of minimum elastic area to whole area of section
IT	location of section with minimum elastic area

From IT and T11, the location of the section with the maximum plastic zone and the magnitude of the remainder elastic zone of the section can be known.

The maximum load capacity:
When the load-deformation curve reaches the descending branch or the plastic hinge is formed (i.e., $T11 < 0.05$), the message will be output as below:

F1MAX	maximum end moment capacity normalized by PMY for $MP = 1$
F2MAX	maximum axial force capacity normalized by PY for $MP = 2$

Information about reduction of deflection increment:
When the deflection increment DDY is reduced, the message will be output as below:

NNK	reduction times of deflection increment
DDY/2	reduced deflection increment

Interrupt information:
When the calculation is interrupted due to the convergence difficulty, the message will be output as in Table 5.2. We can know from N, K, and NNK, where the program stopped, and from NO1, NA2, NB2, T1, why it stopped.

K	sequence number of sections
NO1	number of iteration times (iteration 400) for tangent stiffness of section
NA2	number of iteration times (iteration 200–2) for special division point deflection
NB2	number of iteration times (iteration 200–1) for each division point deflection
T1	elastic area of section at Kth division point
NN	sequence number of solution points in descending branch of load-deformation curve

If the maximum load capacity is not obtained before the interruption (i.e., $NN = 0$), the user can try again following the output suggestion.

All the output data listed above are stored in the output file whose name is input by the user from the screen. The data of the groups (2), (3), (4), and (5) will also be displayed on the screen. In this way, the user can monitor the calculation process.

How to Determine the Input Data

(1) FF1 (for $MP = 1$) or FF2 (for $MP = 2$)

In general, choose $FF1 = 0.1$ or $FF2 = 0.1$. However, for larger axial force in case of $MP = 1$ or larger end moments in case of $MP = 2$, the value of $FF1$ or $FF2$ should be smaller, even be negative (see the sample problem 3 for FRAMP).

(2) ADDY

When the deflection increment ADDY is small, more computing time is required; the accuracy of the curve may be better. The convergence of the iterations in this case can be easily obtained since the forecast from the extrapolation is closer to the solution. In general, for the cases with larger slenderness or smaller constant axial force ($MP = 1$) or larger constant end moments ($MP = 2$), the value of $ADDY$ can be larger. In addition, it also relates to $IJ1$. If $IJ1$ is put at the point with larger deflection, $ADDY$ can also be larger. The specific value of $ADDY$ depends on the load and the length of the member and at the beginning users can try a few times to get the appropriate value of $ADDY$ for the problem.

(3) NF and NW

In general, choose $NF = 10$ to 20 and $NW = 60$ to 100 for $MJ = 0, 2$, and $NF = 60$ to 100 and $NW = 10$ to 20 for $MJ = 1$.

Note: $(2*NF + NW) \leq 120$ for $MJ = 0, 2$ and $(NF + NW) \leq 120$ for $MJ = 1$

(4) IJK

In general, choose $IJK = 11$ to 21.

Note: $(IJK + 2) \leq 30$

(5) IJ1

Any division point can be chosen as the point $IJ1$ except the points with very small deflection. Therefore, it would be better to choose the point with large deflection. For example, $IJ1 = IJK/2 + 1$ for $AFA = -1$ and $IJ1 = IJK/4 + 1$ for $AFA = 1$.

(6) AEE

In general, take $AEE = 0.001$. When the convergence is very difficult for the first solutions which often occur in the cases with larger axial force ($MP = 1$) or larger end moment ($MP = 2$), the value larger than 0.001 is recommended to initiate the process. The program will let $AEE = 0.001$ after the tenth solution ($N = 10$) is obtained. In this case, only the solutions after the tenth one are valid (see the sample problems of BCIN).

It would be difficult to get the solution for a large, specified $FF2$ ($MP = 1$) or for a large, specified end moment $FF1$ ($MP = 2$). In these cases, we can change $MP = 2$

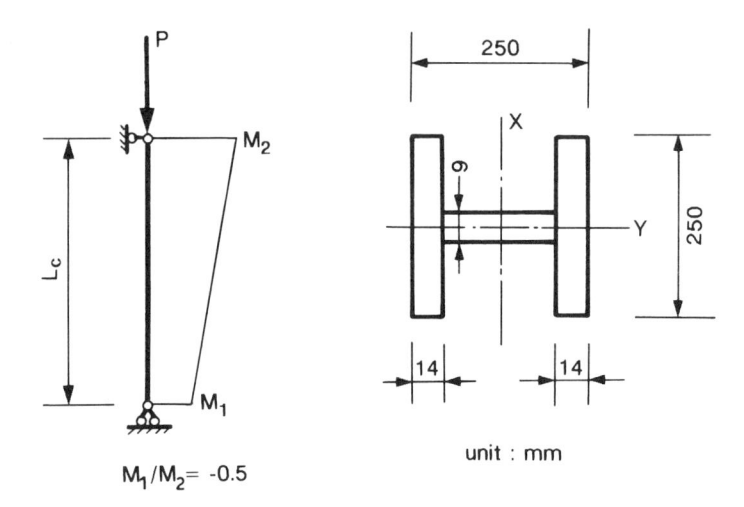

PLASTIC-ZONE ANALYSIS OF BEAM-COLUMNS AND PORTAL FRAMES   **213**

FIGURE 5.7. Sample problems of BCIN.

or 1 and get the solution for smaller constant $FF2$ or $FF1$, respectively, and try a few times to get the constant $FF2$ or $FF1$ with appropriate adjustment. The solution corresponding to the large specified load will be obtained. This is because the irreversibility of plastic deformation is not considered in BCIN and the solutions from $MP = 1$ and $MP = 2$ are the same. For example, with a beam-column with an I-section and $\lambda_x = 60$, bending about weak axis, the ratio of the end moments is $\alpha = -0.5$. The maximum value $F1MAX = 0.28238$ can be obtained from the constant $FF2 = 0.6$ and $MP = 1$ by the program BCIN. However, if we take $MP = 2$ and $FF1 = 0.28238$ constant, the maximum value F2MAX is 0.59988 by BCIN (see the sample problems of BCIN).

5.2.5 Sample Problems of BCIN

There are two sample problems for BCIN shown here. The first one is a steel beam-column with constant axial force $P = 0.6P_y$ and varying end moments M_2 and M_1 acted about the weak axis of the I-section of the member. The ratio of end moments is $\alpha = AFA = M_1/M_2 = -0.5$. The size of the section is shown in Figure 5.7 and its slenderness ratio about the weak axis is $\lambda_y = ALDY = 60$. The slenderness ratio about the strong axis is $\lambda_x = 35.23$ and the member length is $L_c = CL = 3820$ mm. The yield stress and the elastic modulus for steel are $F_y = 235$ N/mm² and $E = 206000$ N/mm², respectively. The residual stress and the initial crookedness of the member are the same as shown in Figure 5.1.

Accordingly, for this problem the input data from the screen are chosen as follows:

$ALD = \lambda_y = 60$
$FF1 = M_2/M_P = 0.1$ (a small value is recommended)
$FF2 = P/P_y = 0.6$
$AFA = \alpha = M_1/M_2 = -0.5$

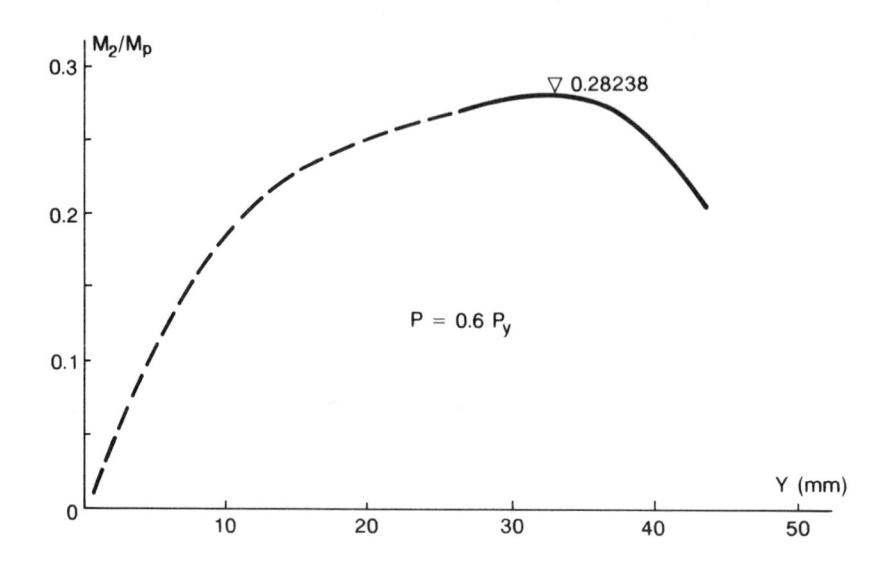

FIGURE 5.8. M_2/M_p~Y curve for problem 1 of BCIN.

The input data from the file are as follows:

B = DDD = 250 mm, T = 14 mm, W = 9 mm
FY = F_y = 235 N/mm², EEE = E = 206000 N/mm²
EO = 0.001
ADDY = 5 mm (selected after a few trials)
RESID = σ_r/F_y = 0.3
AEE = 0.005 (it will be changed to 0.001 after tenth solution)
MP = 1 (for constant axial force)
MJ = 1 (for bending about weak axis of I-section)
NF = 100, NW = 10 (for bending about weak axis of I-section)
NSTOP = 4
IJK = 11
IJ1 = 6 (at midspan of the member)

According to the output file by the program BCIN, the M_2/M_P~y curve (Figure 5.8) and A_e/A~y curve (Figure 5.9) for this member can be drawn. The abscissa y is the deflection at the midspan of the member and A_e/A is the ratio of the minimum elastic area to the whole area of the section. The M_2/M_P~y curve has ascending and descending branches with the maximum carrying capacity M_2/M_P = 0.28238. From the A_e/A~y curve it can be seen that the elastic area of the section is getting smaller as the deflection y decreases.

The sample problem 2 for BCIN is the same as the sample problem 1 except that the larger end moment M_2 = 0.28238M_P (equal to the maximum end moment of the

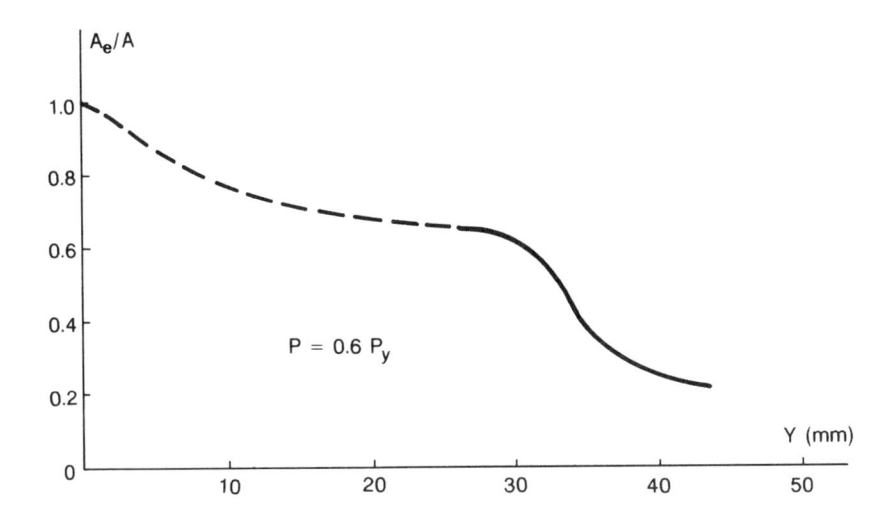

FIGURE 5.9. A_e/A~Y curve for problem 1 of BCIN.

problem 1) is kept constant while the axial force P/P_y changes. Accordingly, the input data are as follows:

$FF1 = 0.28238$, $FF2 = 0.1$ (a small value is recommended)
$MP = 2$ (for constant end moments)
The other input data are the same as those in the problem 1.

The M_2/M_P~y curve and the A_e/A~y curve for sample problem 2 are shown in Figures 5.10 and 5.11, respectively. In this case the maximum axial force $P/P_y = 0.59988 \approx 0.6$ is obtained. The maximum carrying capacities for these two sample problems are the same although their loading sequence is different. The input and output files for sample problems 1 and 2 are printed in Appendix A.

5.3 Analysis of Portal Braced Frames
5.3.1 Principle of Analysis

The program FRAMP is developed for the analysis of portal braced frames on the basis of the program BCIN (Chen and Zhou, 1987). Analytical concept is shown in Figure 5.12. In the analysis, the beam and the columns have the same residual stress as that of I-section in BCIN. The columns have the same initial crookedness as that in BCIN.

The total vertical load on the frame is $2P_t$.

$$2P_t = 2p_t P_y = qL_b + 2P_c \tag{5.19}$$

$$q = 2p_t P_y \beta / L_b \tag{5.20}$$

$$P_c = p_c P_y = p_t(1-\beta)P_y \tag{5.21}$$

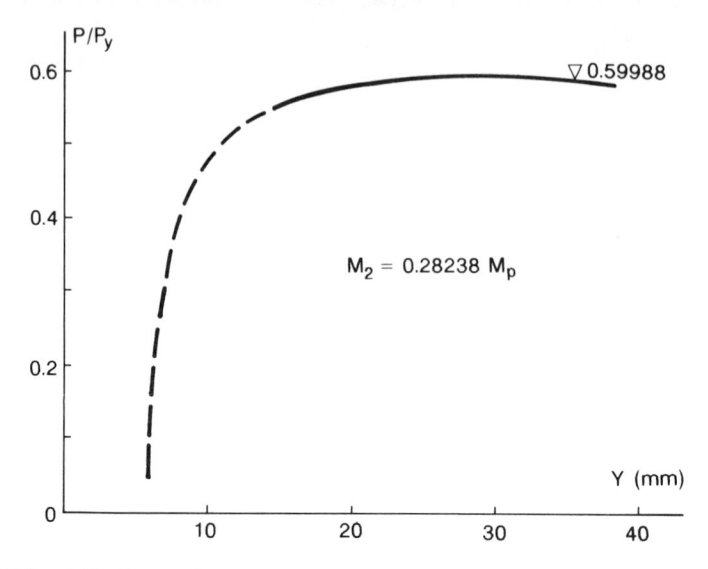

FIGURE 5.10. $P/P_y \sim Y$ curve for problem 2 of BCIN.

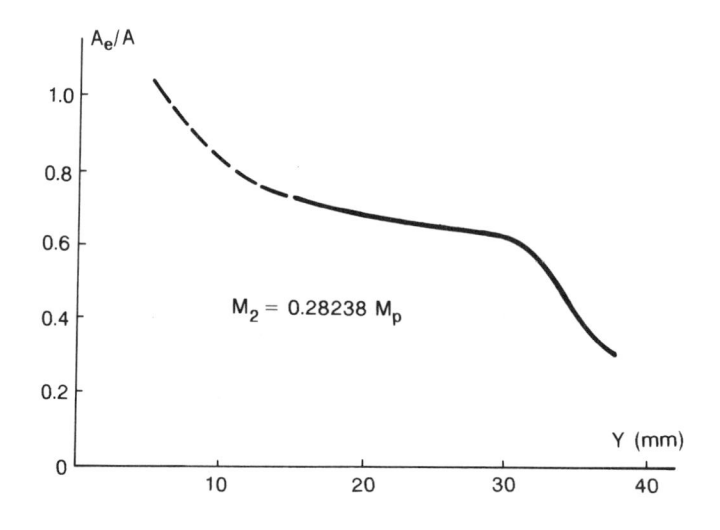

FIGURE 5.11. $A_e/A \sim Y$ curve for problem 2 of BCIN.

where

P_t = half total vertical load on frame
p_t = half total vertical load on frame normalized by P_y
q = uniform distributed load on beam
L_b = beam length
β = ratio of load on beam to total load on frame, $(qL_b/2P_t)$

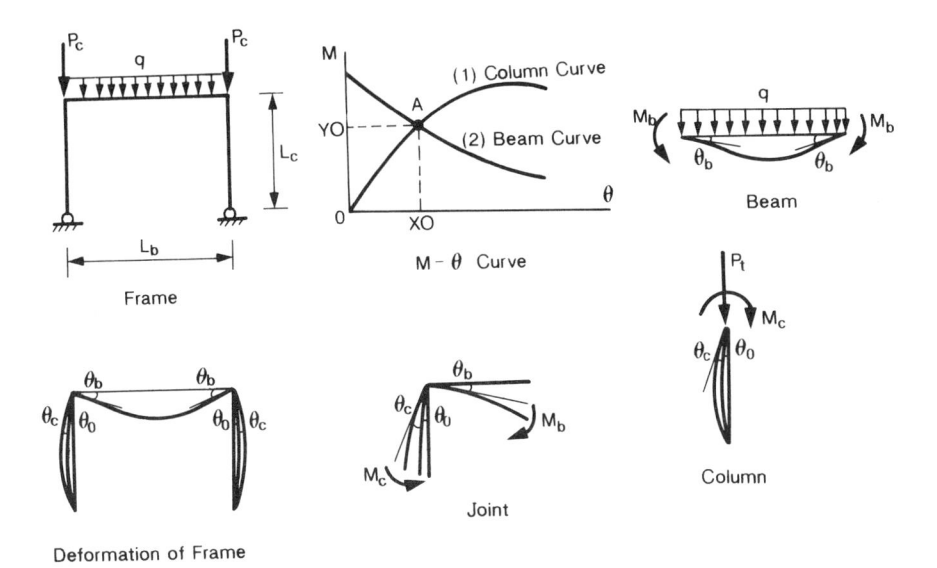

PLASTIC-ZONE ANALYSIS OF BEAM-COLUMNS AND PORTAL FRAMES   217

FIGURE 5.12. Braced portal frame.

P_c = load on top of each column
p_c = load on top of each column normalized by P_y
Note: The axial load of each column is P_t (not P_c).

In the analysis, the frame is considered as the assembly of three free bodies: two columns and one beam. From the symmetry in geometry and loading, half of the frame, i.e., one column and half of the beam, needs to be considered (see Figure 5.12).

At the beginning, the moment-rotation (M_c~θ_c) curve of the column is traced as shown by the curve (1) in Figure 5.12, using the second-order inelastic analysis. This procedure is arranged in the subroutine COLUMP. Since the axial force in the beam is very small and the second-order effect can be neglected, the moment-rotation curve (M_b~θ_b) of the beam is traced by the first-order inelastic analysis as shown by the curve (2) in Figure 5.12.

In order to satisfy the equilibrium and the compatibility, the relationships $M_c = M_b$ and $\theta_c = \theta_b$ must be in the frame. It follows that the intersection point A of the curves (1) and (2) is the solution point for the frame with the load p_t. In the program, the coordinates of A are stored in YO and XO. By varying the total load $2P_t$, a series of curves (1) and (2) can be produced. From these curves a series of the point A can be generated so that the complete response of an inelastic imperfect portal braced frame can be described.

The analytical principle of the subroutine COLUMP is the same as the case of $MJ = 0$, $MP = 1$, and $AFA = \alpha = 0$ in the program BCIN, i.e., it is the case for the beam-column with the I-section bent about strong axis, loaded by a constant axial force and varied moment on top of the column and no moment at bottom of the column. With making some modification on COLUMP, the subroutine BEAMP for beam analysis can be developed. In BEAMP, the second-order effect is not considered but the residual stress is included.

A half span of the beam is divided into a number of segments. For the given load q and the assumed end moment M_b, the curvature at each division point is calculated by the tangent stiffness method (the iteration 400 in BCIN). Assuming that the curvature in each segment varies linearly, the deflection and the end rotation θ_b can be calculated by Newmark integration method. By varying the end moment M_b, the $M_b - \theta_b$ curve can be traced. If the load q is small, the $M_b - \theta_b$ curve is a straight line, but as the load q is increased, the curve becomes nonlinear and approaches to a horizontal line. Eventually, the curve will be horizontal when the plastic hinge is formed.

5.3.2 Analytical Steps

According to the analysis principle described above, the computer program is arranged by the following steps:

(1) Input the data from a file and the screen (see User's Manual).
(2) Calculate the geometric and the physical behaviors of the section, the coordinates of the strips of the section, the residual stress parameters, and the initial deflection of the column.
(3) Input the load PT (P_t) on the frame and the data about the column from the screen.
(4) Calculate the $M_b - \theta_b$ curve for the beam by the subroutine BEAMP (see Figure 5.14). M_b and θ_b are stored in the arrays BM and CTAB. The curve will end when the plastic hinge is formed, i.e., the curve becomes a horizontal line, or the minimum M_b (=MB(NB), the last element in the array (BM) is smaller than −0.05PMC (PMC is the plastic moment of column section without the effect of axial force).
(5) Calculate the $M_c - \theta_c$ curve for the column and search the intersection point with the beam curve by the subroutine COLUMP. In fact, the $M_b - \theta_b$ curve and the $M_c - \theta_c$ curve are plotted by a number of discrete solution points. Assume that the line between the solution points is linear (see Figure 5.13) to obtain the intersection point. In the subroutine COLUMP, after each new segment of the column curve is obtained, the procedure for searching the intersection point of this segment with the beam curve is carried out. If the intersection point (YO, XO) is found (KK = 1 in the program, see Figure 5.13a and b), the calculation of the column curve for the current value of PT will end and the procedure will return to the main program. The user can choose a larger value of PT and repeat steps 3 to 5 when the procedure asks the user if the calculation should continue.
(6) If the intersection point (YO, XO) is not found (KK = 0 in the program), it may be caused by two cases:

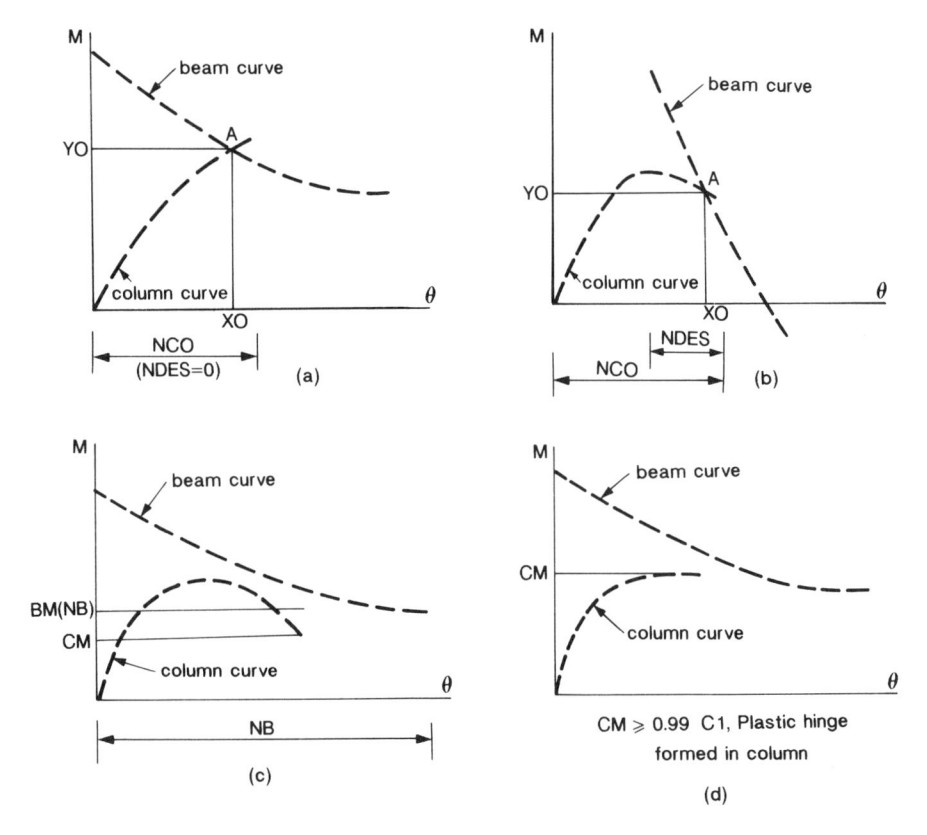

PLASTIC-ZONE ANALYSIS OF BEAM-COLUMNS AND PORTAL FRAMES   219

FIGURE 5.13. The intersection of column and beam curves.

(a) One cause is that the column load-deformation curve is not long enough to reach the intersection point and the procedure will continue to calculate a new segment and search for the intersection point again.

(b) The other cause is that the beam and the column load-deformation curves do not intersect. In this case, it is possible that the column curve has reached the descending branch and the CM of current solution point of the column curve is lower than the minimum value of BM (i.e., BM(NB)); or, the plastic hinge has been formed in the column (i.e., the column curve becomes a horizontal line, see Figures 5.13c and d). This means that the input load PT is too large for the frame. In this case, the calculation of the column curve is ended and a message to the user will be output shown as follows:

NO SOLUTION AT PT=........
SUGGESTION: CHOOSE SMALLER PT, CALCULATE AGAIN IF NEEDED.

Then, the procedure returns to the main program and asks the user if the calculation continues or not.

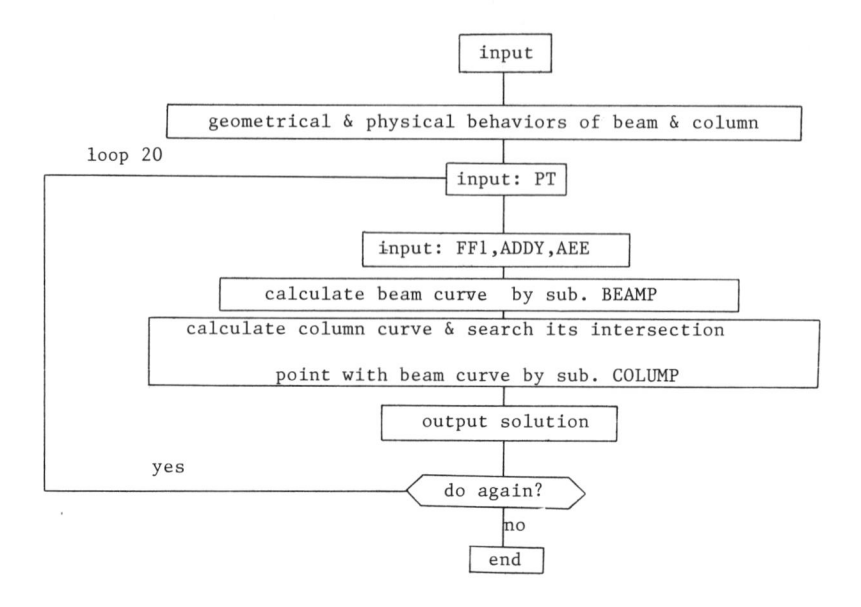

```
                    ┌─────────┐
                    │  input  │
                    └────┬────┘
        ┌───────────────────────────────────────────────────┐
        │ geometrical & physical behaviors of beam & column │
        └───────────────────────┬───────────────────────────┘
loop 20                  ┌──────┴──────┐
    ┌────────────────────┤ input: PT   │
    │                    └──────┬──────┘
    │              ┌────────────┴────────────┐
    │              │ input: FF1,ADDY,AEE     │
    │              └────────────┬────────────┘
    │       ┌───────────────────┴────────────────────┐
    │       │   calculate beam curve by sub. BEAMP   │
    │       └───────────────────┬────────────────────┘
    │  ┌────────────────────────┴─────────────────────────────┐
    │  │ calculate column curve & search its intersection     │
    │  │ point with beam curve by sub. COLUMP                 │
    │  └────────────────────────┬─────────────────────────────┘
    │                    ┌──────┴──────────┐
    │                    │ output solution │
    │                    └──────┬──────────┘
    │  yes                      │
    └───────────────────< do again? >
                                │ no
                            ┌───┴───┐
                            │  end  │
                            └───────┘
```

FIGURE 5.14. Flow chart of FRAMP.

(7) It is possible that the load on the beam is too large or the calculation of the column curve is interrupted due to the failure of convergence in iteration. In these cases, the program gives a message and asks the user if he would like to try again or not.

(8) It is considered that the maximum limit load for the frame is found when a slightly larger load than the maximum limit load can not be carried by the frame. For detail, see the user's manual and the sample problems.

5.3.3 Flow Chart of FRAMP

The program FRAMP consists of three parts: the column load-deformation curve; the beam load-deformation curve; and the intersection of the curves. Calculation procedures of the column curve is the same as in BCIN. Instead of getting a whole curve, once the intersection is obtained, the calculation will end. The procedure for the beam curve is developed by modifying the column curve case.

In FRAMP, there is a main program and five subroutines: BEAMP, COLUMP, SECS, EXIN, and EXOUT. The subroutines SECS, EXIN, and EXOUT are the same as those in BCIN. A flow chart of FRAMP is shown in Figure 5.14.

5.3.4 User's Manual of FRAMP

Positive directions are assigned as follows:

horizontal force: right direction
displacement: right direction

vertical load: downward
axial force and stress: compression
moment and rotation: clockwise

Input Data

There are two groups of input data for FRAMP: one is input from a file and the other is from the screen.

Input from a file:

A input file is named by the user and stored in the character "D2". The data are as follows:

FY	yield stress
EEE	elastic modulus
RESID	residual stress parameter, (σ_r/F_y in Figure 5.1)
BAT	ratio of the load on the beam to the total vertical load on the frame (β in Eq. 5.20), $0 < BAT \leq 1$, the case of $BAT = 0$ can not be treated by FRAMP.
DC	depth of column section
WC	web thickness of column section
BC	flange width of column section
TC	flange thickness of column section
CL	column length (L_c in Figure 5.12)
EO	initial crookedness at midspan normalized by column length
DB	depth of beam section
WB	web thickness of beam section
BB	flange width of beam section
TB	flange thickness of beam section
BL	beam length (L_b in Figure 5.12)
NW	number of strips in web of column and beam, take 60 to 100, (2∗NF + NW) $\leq$ 120
NF	number of strips in flange of column and beam, take 10 to 20
IJC	number of division points in column, take 11 to 21, $IJC \leq 28$
IHB	number of division points in a half span of beam, take 6 to 11, $IHB \leq 29$

The read sentences are as follows:

```
READ(7,*) FY,EEE,RESID,BAT
READ(7,*) DC,WC,BC,TC,CL,EO
READ(7,*) DB,WB,BB,TB,BL
READ(7,*) NW,NF,IJC,IHB
```

Input from the screen:

D2	a character variable for the name of the input file, not more than six characters

OUTF, OUTC, OUTB   character variables for the names of output files for frame, column, and beam, not more than six characters

PT   a half total load on frame normalized by PY, (p_t in Eq. 5.19)

FF1   first trial value of end moment at column top normalized by PMC

ADDY   increment of special division point deflection of column in frame; in FRAMP, the location of the special point is taken as $(IJC - 1)/3 + 1$

AEE   allowance for difference in column deflections in iteration, take AEE = 0.001, when the convergence is difficult at the first ten solutions of column, take AEE > 0.001

(PMC = plastic moment of column section without effect of axial force, PY = yield strength of column section)

The read sentences for file names are as follows:

```
WRITE(*,'(A)') 'INPUT THE NAME OF INPUT FILE:'
READ(*,'(A)' D2
WRITE(*,'(A)') 'INPUT THE FILE NAME TO STORE RESULTS OF FRAME:'
READ(*,'(A)') OUTF
WRITE(*,'(A)') 'INPUT THE FILE NAME TO STORE RESULTS OF
                COLUMN:'
READ(*,'(A)') OUTC
WRITE(*,'(A)') 'INPUT THE FILE NAME TO STORE RESULTS OF BEAM:'
READ(*,'(A)') OUTB
```

The read sentences for loading data are as follows:

```
WRITE(*,'(A)') 'INPUT HALF TOTAL VERTICAL LOAD PT FOR FRAME'
READ(*,*) PT
WRITE(*,'(A)') 'INPUT COLUMN DATA: FF1 ADDY AEE'
READ(*,*) FF1,ADDY,AEE
```

Output Data

There are three output files for load-deformation curves of the frame, the column, and the beam. The names of these files are input by the user and stored in the character variables: OUTF, OUTC, and OUTB.

Output for frame load-deformation curve (OUTF and unit 12):
Head of output table:

BAT	see input data
FY-E-RESID-EO	see input data

PLASTIC-ZONE ANALYSIS OF BEAM-COLUMNS AND PORTAL FRAMES **223**

BEAM-D-W-B-T-L   DC,WC,BC,TC,CL of input data
COLUMN-D-W-B-T-L DB,WB,BB,TB,BL of input data
NW-NF-IJC-IHB   see input data

Data of load-deformation curve of frame:

K	sequence number of solution points
PT	see input data, PT = 2∗PC + QB
PC	load on column top normalized by PY
QB	total load on beam normalized by PY
MC	end moment at column top normalized by PMC
QC	end rotation at column top, radian
MBC	beam moment at midspan normalized by PMC
MBB	beam moment at midspan normalized by PMB
PY	yield strength of column section
PMC, PMB	plastic moments of column and beam sections without effect of axial force
FF1	see input data
ADDY	see input data
AEE	see input data
NCO	sequence number of solution points after intersection point in column curve
NDES	sequence number of solution points after intersection point in descending branch of column curve

Note: If $NDES = 0$, the intersection point is in the ascending branch, and if $NDES > 0$, it is in descending branch. From NCO and NDES, the condition for the intersection point can be known. A selection of appropriate values of FF1, ADDY, and AEE is important to make the convergence successful. The output data of the previous calculation are good reference for the next calculation. Of course, the principle of selecting the values in BCIN is also valid in FRAMP. It should be advised that the solution for the frame curve is valid only with $NCO \geq 10$ if $AEE > 0.001$.

Interruption information:
    The calculation will be interrupted when the input data are not appropriate. In this case, the program will print out the messages as shown in Table 5.3 for the user's convenience.

Output for beam load-deformation curve (OUTB and unit 16):
Head of output table:

DB-BB-WB-TB-BL	see input data
PT-QB	"
FY-EEE-RESID	"
NF-NW-IHB	"
PY-PMC-PMB	see output data of frame curve

224 ADVANCED ANALYSIS OF STEEL FRAMES

Table 5.3 Output for Interruption in Frame Curve

Condition	Output
Curves can not intersect	NO SOLUTION AT PT = ~
	SUGGESTION: CHOOSE SMALLER PT, CALCULATE AGAIN, IF NEED.
Moment in beams is	PT = ~ IS TOO BIG FOR BEAM
too large	SUGGESTION: CHOOSE SMALLER PT AND TRY AGAIN!
Column curve interrupt	COLUMN CURVE INTERRUPT AT $\underset{\sim}{N}$ NDES $\underset{\sim}{AEE}$ $\underset{\sim}{ADDY}$
$N \ne 1$	SOLUTION CAN NOT BE OBTAINED AT PT =~
NDES = 0	SUGGESTION: CHANGE ADDY OR AEE AND TRY AGAIN!
N < 10	
NDES = 0	SUGGESTION: CHANGE ADDY AND TRY AGAIN!
N ≥ 10	SUGGESTION: CHECK BEAM & COLUMN CURVES TO JUDGE IF THEY CAN NOT CROSS
	IF SO, CHOOSE A SMALLER PT & CALCULATE AGAIN WHEN NEED
	IF NOT, CHANGE ADDY & TRY AGAIN!
$N = 1$	COLUMN CURVE STOP AT THE START FOR PT = ~
	SUGGESTION: CHANGE F11 OR AEE AND TRY AGAIN!

Note: (1) NDES in this table is the number of solution points in the descending branch of the column curve when calculation is interrupted.
(2) The N, PT, AEE, ADDY, and F11 are the same as before.
(3) After make some parameter changes, it would still be possible that when BAT is very small the solution can not be found due to iterative convergent difficulty in the numerical analysis.

Data of load-deformation curve:

N	sequence number of solution points
NO1	iteration times for tangent stiffness of beam section
MB	end moment normalized by PMC
QB	end rotation, radian
Y	deflection at midspan
AE	ratio of minimum elastic area to whole area of beam section

Output for column load-deformation curve (OUTC and unit 14):
Head of output table:

DC-BC-WC-TC-CL	see input data
FF1-PT	"
ADDY-AEE	"
FY-EEE-RESID-EO	"
NF-NW-IJC-IJ1	" , IJ1 = special division point, (IJC − 1)/3 + 1
AC-CI	area and moment of inertia of column section
PY-PMC	see output of frame curve

Data of load-deformation curve:

N	sequence number of solution points
NO1	iteration times for tangent stiffness of column section
MC	end moment at column top normalized by PMC
QC	end rotation at column top, radian
Y	deflection at special division point
AE	ratio of minimum elastic area to whole area of column section
NNK	reduction times of deflection increment, when NNK = 4, the calculation of the column curve will be interrupted
DDY/2	reduced deflection increment

Interruption information:

When the calculation of the column curve is interrupted, the message will appear not only in the output of frame curve as shown in Table 5.3 (unit 12) but also in the output of the column curve (unit 14). All the output data listed above are stored in the output files whose names are input by the user from the screen. Some of the data will be also displayed on screen in order for the user to monitor the calculation process.

Frame Failures

There are three types of frame failures possible depending on the situation of the intersection of the column and beam curves.

(1) The frame fails due to columns buckling. In this case, for the limit load (say P_{t1}) the corresponding column and beam curves are tangent to each other at a single point "A" as shown in Figure 5.15a. But for a slightly smaller load (say P_{t2}), the corresponding column and beam curves may have two intersection points "B" and "C", if the column curve is long enough. The point "A" for limit load lies somewhere between the "B" and "C". Point "B" represents a stable equilibrium state of the frame and lies on the ascending branch of the load-rotation curve of the frame. Point "C" represents an unstable equilibrium state and is on the descending branch. In FRAMP, the descending branch of frame curve is not calculated.

(2) The frame fails due to the formation of a plastic hinge at the midspan of the beam as well as the top of columns in the frame so the frame forms a plastic failure mechanism. In this case, both column and beam curves will overlap along a horizontal line at the maximum limit load of the frame as shown in Figure 5.15b.

(3) The frame fails due to the column buckling and the formation of a plastic hinge in the midspan of beam. In this case, the peak point of the column curve will be tangential to the horizontal branch of the beam curve at the maximum limit load of the frame (shown in Figure 5.15c).

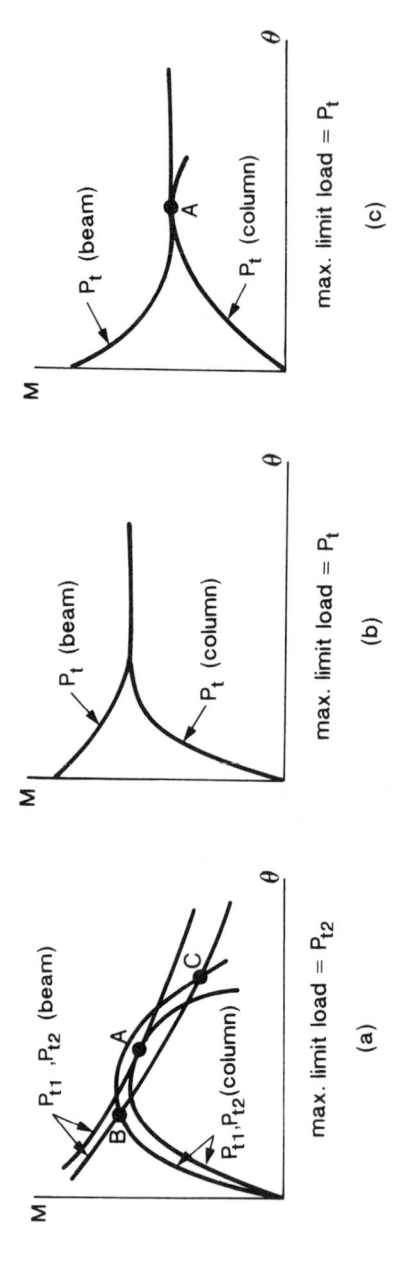

FIGURE 5.15. The maximum limit load of the braced frame.

When the user obtained the maximum limit load of the frame from program FRAMP, it would be better to draw the corresponding column and beam curves to check the correction of the solution according to the failure of the frame described above.

The Case with BAT = 0

The program FRAMP can not analyze the braced frames with all vertical load acting on the top of the columns, i.e., $\beta = 0$. Here, due to the limited space of this book, we can not discuss this case and the user can refer to Chen and Zhou (1987) and use the program BCIN with $MP = 2$ to analyze the information.

5.3.5 Sample Problems of FRAMP

There are three sample problems for FRAMP. The frame configuration is the same for these sample problems but the vertical load distributions differ. Sizes of the frame and the members are shown in Figure 5.16. The residual stress for the beam and columns and the initial crookedness for the columns are the same as shown in Figure 5.1. The yield stress and the elastic modulus are $F_y = 36$ *psi* and $E = 29000$ *psi*, respectively.

In sample problem 1, all the vertical load acts on the beam ($\beta = 1.0$, $q = 2P_t/L_b$, see Figure 5.17). In sample problem 2, one-third vertical load acts on the beam and one-third on each column ($\beta = 1/3$, $q = 2P_t/3L_b$, see Figure 5.18). In sample problem 3, one-tenth vertical load acts on the beam and 45% of the load acts on each column ($\beta = 0.1$, $q = 2P_t/10L_b$, see Figure 5.19).

Accordingly, the input data from the file are as follows:

FY = F_y = 36 psi, EEE = 29000 psi
RESID = σ_r/F_y = 0.3
DC = BC = 8 in, WC = 0.28 in, TC = 0.43 in, CL = 140 in
EO = 0.001
DB = 16.26 in, BB = 7.07 in, WB = 0.38 in, TB = 0.63 in, BL = 210 in
NW = 100, NF = 10 (for bending about strong axis)
IJC = 11,
IHB = 6
BAT = β = 1.0, 1/3, and 0.1, respectively for the problems 1, 2 and 3.

The input data PT, FF1, ADDY, and AEE from the screen are all shown in the output print of the frame curve.

In problem 1, the P_t~θ_c curve shown in Figure 5.17 approaches the maximum $P_t/P_y = 0.249$ as the rotation θ_c increases. The figure also shows the intersecting situation of the beam and column curves at $P_t/P_y = 0.249$ and 0.25; it indicates that the two curves overlap along a horizontal line at $P_t/P_y = 0.249$. Therefore, it is

228 ADVANCED ANALYSIS OF STEEL FRAMES

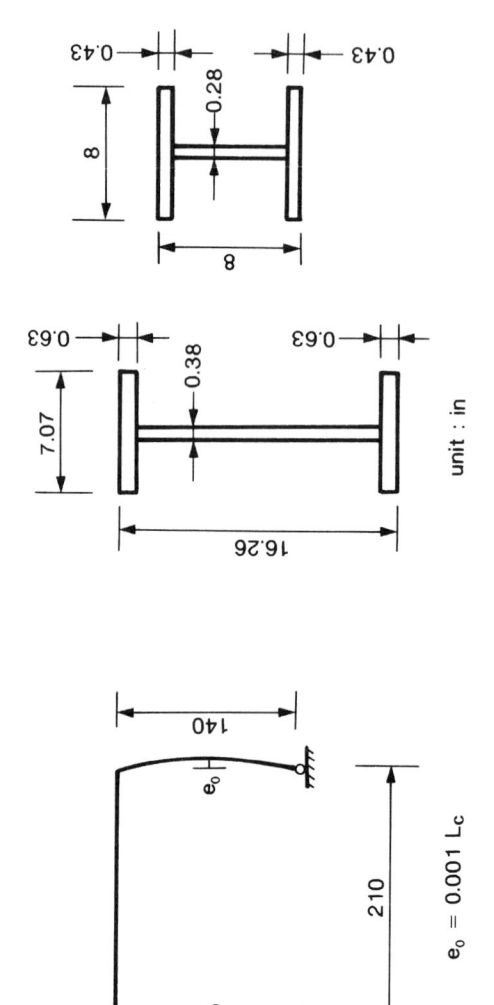

FIGURE 5.16. The frame for sample problems of FRAMP.

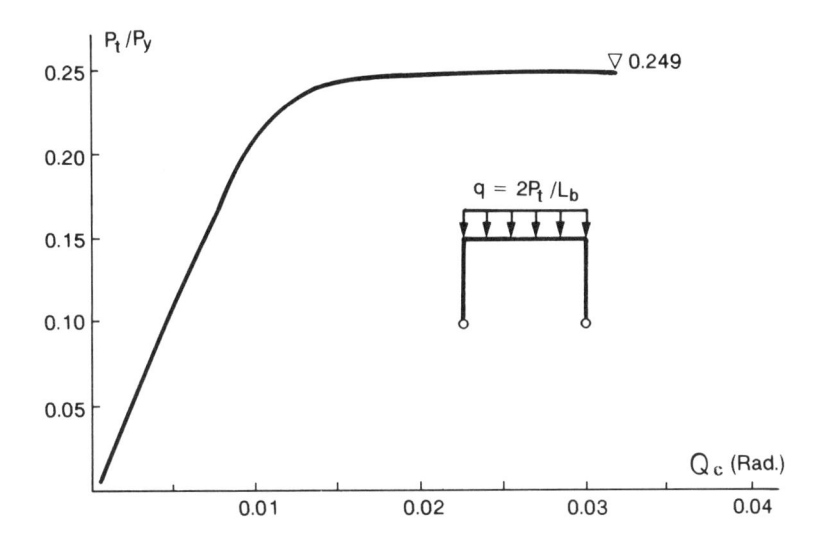

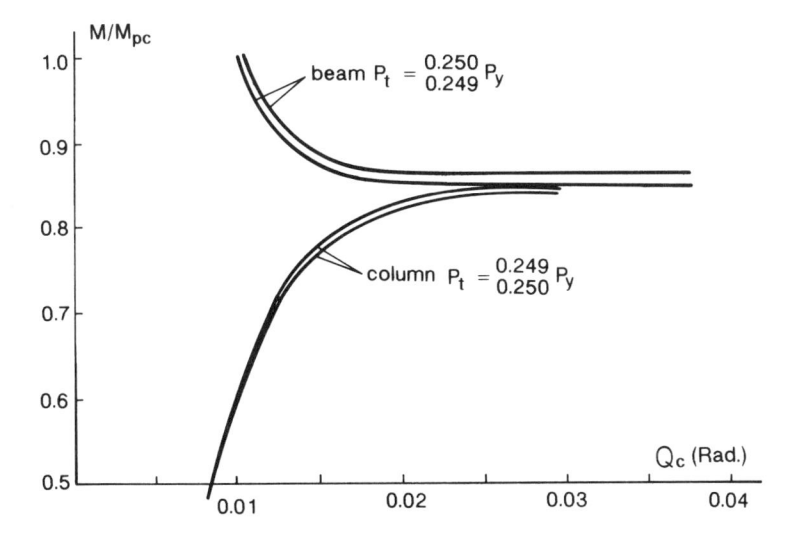

FIGURE 5.17. Behavior curves for problem 1 of FRAMP.

recognized that the failure of the frame is due to the plastic hinges formed in the beam and the columns. (For $P_t/P_y = 0.25$, the beam and column curves do not intersect.)

In sample problem 2, the P_t–θ_c curve and the intersecting situation of the beam and column curves is shown in Figure 5.18. At the load $P_t/P_y = 0.652$, the beam and column curves are tangent to each other; the tangent point is just at the peak of the column curve (NDES = 1 in the output) and on the horizontal branch of the beam curve. Therefore, it is recognized that the frame fails by the column buckling and the

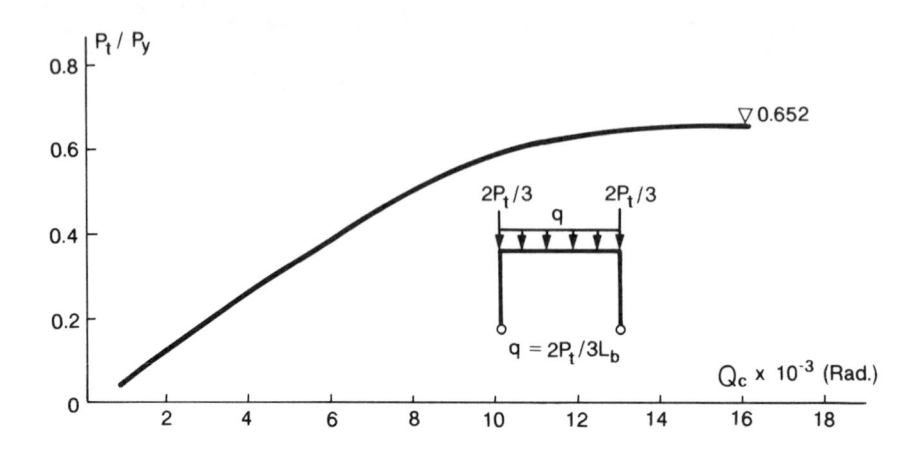

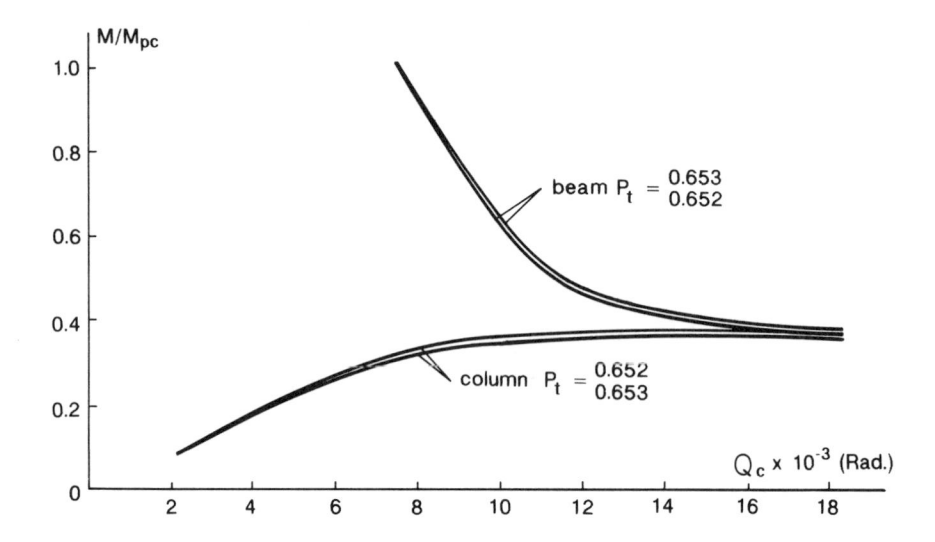

FIGURE 5.18. Behavior curves for problem 2 of FRAMP.

plastic hinge formed in the beam. (For $P_t/P_y = 0.653$, the beam and column curves do not intersect.)

In sample problem 3, the P_t–θ_c curve and the intersecting situation of the beam and column curves is shown in Figure 5.19. At the load $P_t/P_y = 0.896$, the beam and column curves are tangent to each other; the tangent point is on the descending branches of the beam and column curves (*NDES* > 1 in the output). Therefore, it is recognized that the frame fails by the column buckling. (For $P_t/P_y = 0.897$, the beam and column curves do not intersect.)

The input and output for the sample problems are printed in Appendix B.

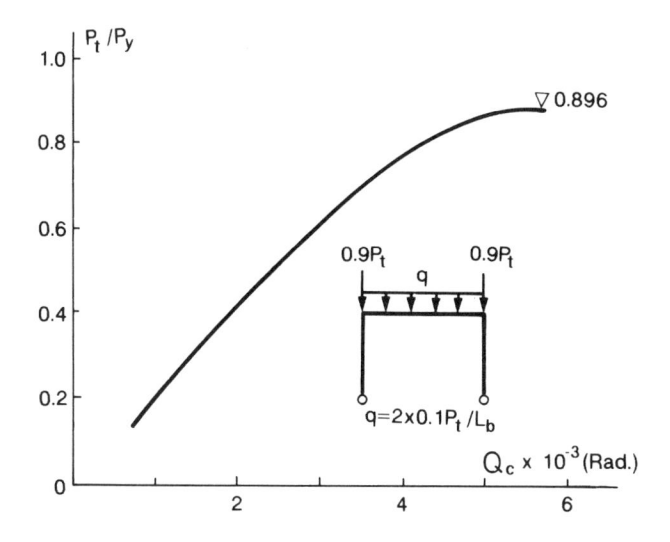

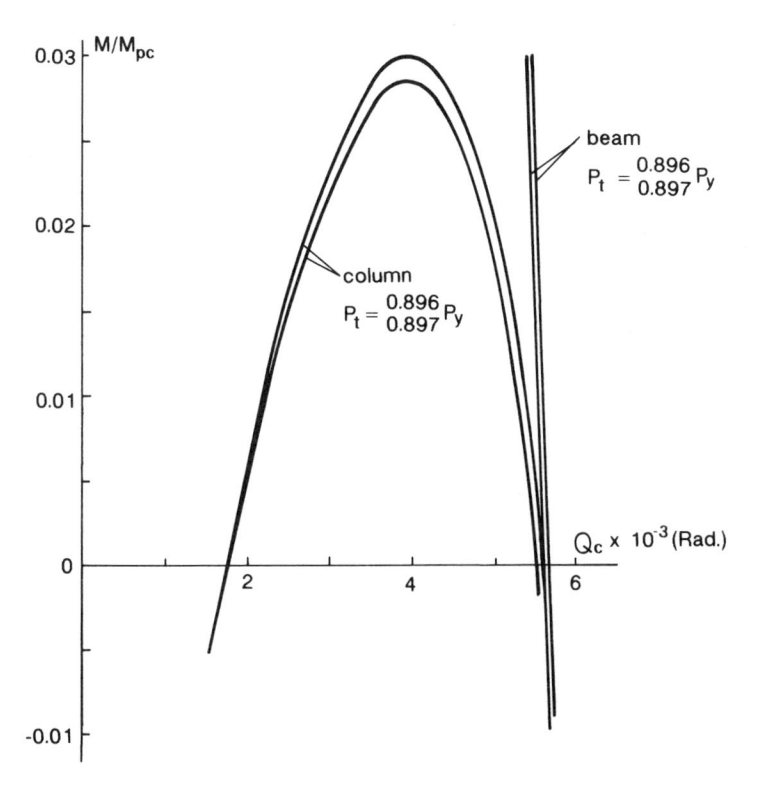

FIGURE 5.19. Behavior curves for problem 3 of FRAMP.

5.4. Analysis of Portal Unbraced Frame*
5.4.1 Principle of Analysis
Structure and Load of the Unbraced Frames

The program FRAMH is for the second-order inelastic analysis of a portal unbraced frame under combined horizontal and vertical load shown in Figure 5.20.

The residual stress for the beam and column and the initial crookedness for the columns are the same as that in the program FRAMP. However, depending on the e_{01} and e_{02}, the direction of initial crookedness for the two columns can have the following five types:

(a) Both columns are convex toward the windward direction, $e_{01} > 0$, $e_{02} > 0$
(b) Both columns are convex inward, $e_{01} > 0$, $e_{02} < 0$
(c) Both columns are straight, $e_{01} = e_{02} = 0$
(d) Both columns are convex outward, $e_{01} < 0$, $e_{02} > 0$
(e) Both columns are convex against the windward direction, $e_{01} < 0$, $e_{02} < 0$

The constant vertical loads are the same as that in the program FRAMP and the horizontal load H is first increased and then decreased as the lateral displacement Δ_1 at the top of left column increases.

Second-Order Elastic Analysis of the Unbraced Frames

In FRAMH, the inelastic analysis was developed based on the elastic analysis. Firstly, we divided the frame into three free bodies: two columns and a beam shown in Figure 5.20.

Here, the second-order effect of the beam is neglected and the axial deformation of the beam (i.e., $\Delta_1 - \Delta_2$) is determined approximately by elasticity. Then, using the equilibrium and the compatibility conditions for the three free bodies and a whole frame, we can calculate the elastic relations for the load-internal force-displacement of the unbraced frame as follows:

The relations of the force-deformation:
For left column

$$\theta_{f1} = \frac{M_{c1}}{P_1 L_c}\left(1 - \frac{k_1 L_c}{\sin k_1 L_c}\right) + \frac{\Delta_1}{L_c} + \frac{e_{01}}{\left(\dfrac{\pi}{k_1 L_c}\right)^2 - 1}\frac{\pi}{L_c} \tag{5.22}$$

$$\theta_{c1} = \frac{M_{c1}}{P_1 L_c}\left(1 - \frac{k_1 L_c \cos k_1 L_c}{\sin k_1 L_c}\right) + \frac{\Delta_1}{L_c} - \frac{e_{01}}{\left(\dfrac{\pi}{k_1 L_c}\right)^2 - 1}\frac{\pi}{L_c} \tag{5.23}$$

---
* Zhou et al., 1991.

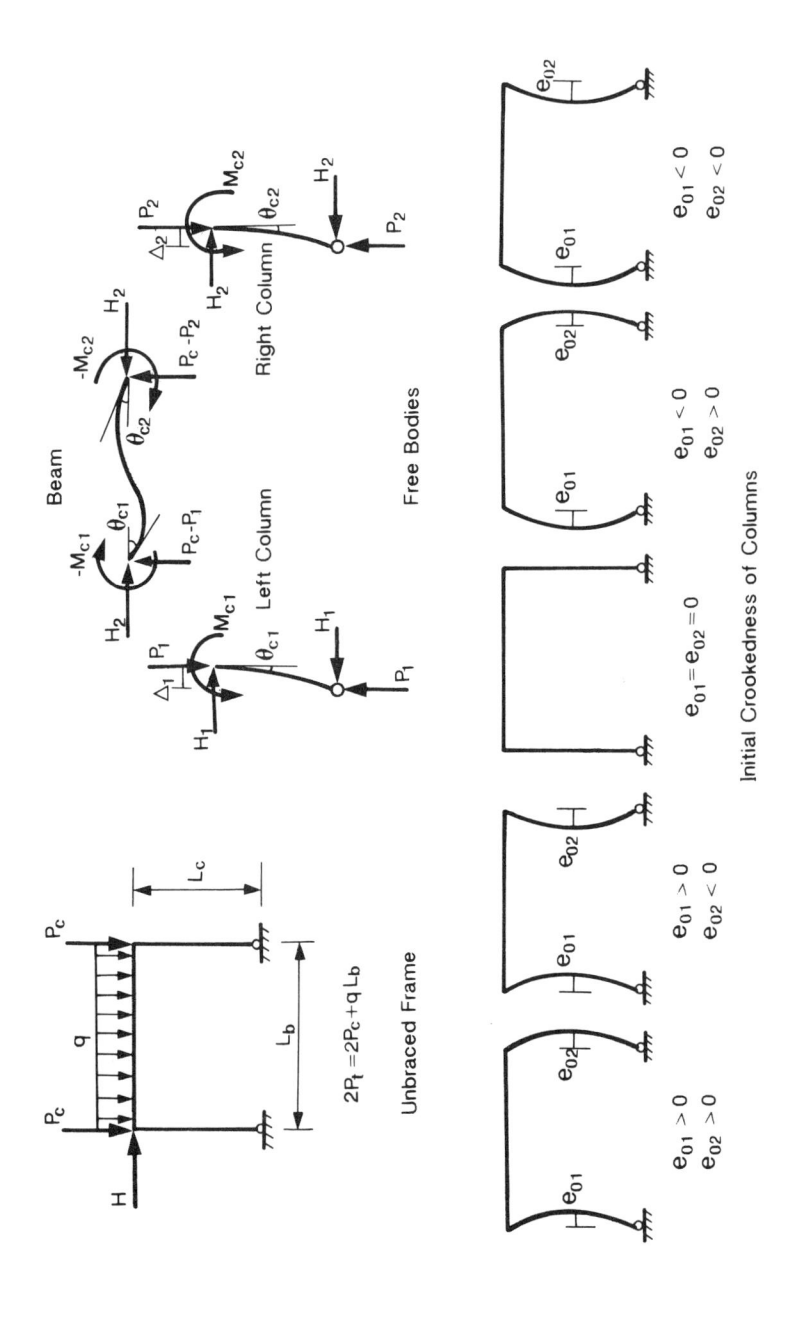

PLASTIC-ZONE ANALYSIS OF BEAM-COLUMNS AND PORTAL FRAMES **233**

FIGURE 5.20. Unbraced portal frame.

For right column

$$\theta_{f2} = \frac{M_{c2}}{P_2 L_c}\left(1 - \frac{k_2 L_c}{\sin k_2 L_c}\right) + \frac{\Delta_2}{L_c} + \frac{e_{02}}{\left(\frac{\pi}{k_2 L_c}\right)^2 - 1}\frac{\pi}{L_c} \quad (5.24)$$

$$\theta_{c2} = \frac{M_{c2}}{P_2 L_c}\left(1 - \frac{k_2 L_c \cos k_2 L_c}{\sin k_2 L_c}\right) + \frac{\Delta_2}{L_c} - \frac{e_{02}}{\left(\frac{\pi}{k_2 L_c}\right)^2 - 1}\frac{\pi}{L_c} \quad (5.25)$$

For beam

$$\theta_{c11} = \frac{b}{EI_b}\left(-\frac{M_{c1}}{3} + \frac{M_{c2}}{6}\right) + \frac{qL_b^3}{24EI_b} \quad (5.26)$$

$$\theta_{c22} = \frac{b}{EI_b}\left(\frac{M_{c1}}{6} - \frac{M_{c2}}{3}\right) - \frac{qL_b^3}{24EI_b} \quad (5.27)$$

$$\Delta_1 - \Delta_2 = \frac{H_2 L_b}{EA_b} \quad (5.28)$$

The equilibrium relations:
For left column:

$$P_1 \Delta_1 + H_1 L_c + M_{c1} = 0 \quad (5.29)$$

For right column

$$P_2 \Delta_2 + H_2 L_c + M_{c2} = 0 \quad (5.30)$$

For frame

$$P_1 = \left(P_c + \frac{1}{2}qL_b\right)\left(1 - \frac{\Delta_1 + \Delta_2}{L_b}\right) - H\frac{L_c}{L_b} \quad (5.31)$$

$$P_2 = \left(P_c + \frac{1}{2}qL_b\right)\left(1 + \frac{\Delta_1 + \Delta_2}{L_b}\right) + H\frac{L_c}{L_b} \quad (5.32)$$

in which the subscripts 1 and 2 are for left and right columns, respectively.

Δ_1, Δ_2 = lateral displacements at the top of columns
θ_{c1}, θ_{c2} = top end rotations of columns

θ_{f1}, θ_{f2} = bottom end rotations of columns
M_{c1}, M_{c2} = top end moments of columns
P_1, P_2 = axial forces of columns
H_1, H_2 = shear forces at ends of columns, H_2 is also axial force of beam
H = horizontal load acted on top of left column, (= $H_1 + H_2$)
$2P_t$ = total vertical load, (= $2P_c + qL_b$, $\beta = qL_b/2P_t$)

$$k_1 = \sqrt{\frac{P_1}{EI_c}}, \quad k_2 = \sqrt{\frac{P_2}{EI_c}}$$

I_c, I_b = moment of inertia for column and beam
L_c, L_b = length of column and beam
e_{01}, e_{02} = initial deflections of midspan for columns
$\theta_{c11}, \theta_{c22}$ = left and right end rotations of beam, $\theta_{c11} = \theta_{c1}, \theta_{c22} = \theta_{c2}$
A_b = section area of beam

The translation and force in the right direction are positive and the rotation and moment in the clockwise direction are positive.

In these 11 equations (Eq. 5.22 to 5.32), the P_c, q, and Δ_1 are known but there are 11 unknowns: $H_1, H_2, P_1, P_2, M_{c1}, M_{c2}, \theta_{c1}, \theta_{c2}, \theta_{f1}, \theta_{f2}$, and Δ_2. Using the Broyden method these 11 unknowns can be obtained.

(a) assign a value Δ_1
(b) assume the values of H_1 and H_2
(c) find Δ_2 by Eq. (5.28)
(d) find P_1 and P_2 by Eqs. (5.31) and (5.32)
(e) find M_{c1} and M_{c2} by Eqs. (5.29) and (5.30)
(f) find θ_{c1} and θ_{c2} by Eqs. (5.23) and (5.25)
(g) find θ_{c11} and θ_{c22} by Eqs. (5.26) and (5.27)
(h) $d\theta_{c1} = \theta_{c1} - \theta_{c11}$, $d\theta_{c2} = \theta_{c2} - \theta_{c22}$, If $d\theta_{c1}$ and $d\theta_{c2}$ are not in an appropriate limit, using the Broyden method, modify the assumed H_1 and H_2 and repeat the step (c) to (h) until satisfied. In this way, a solution corresponding to the assigned Δ_1 in step (a) is obtained. The terms θ_{f1} and θ_{f2} can also be obtained by Eqs. (5.22) and (5.24).
(i) Assign a new value of Δ_1 by adding a increment $d\Delta_1$ and take the last solution values of H_1 and H_2 as the trial values for next solution. Then, repeat steps (c) to (i). A new solution can then be obtained.

Second-Order Inelastic Analysis

Eqs. (5.29) to (5.32) represent the equilibrium relations, which can be used in both elastic and inelastic analysis. Eq. (5.28) represents the elastic axial force and axial deformation relation of the beam. Here, it is approximately used in the inelastic analysis. However, Eqs. (5.22) to (5.27) are the elastic rotation-moment equations for the columns and the beam. In order to get the second-order inelastic solution, we have developed two subroutines, COLUMH and BEAMH to replace Eqs. (5.22) to (5.27).

That means: in steps (f), (g), and (h), the terms θ_{c1}, θ_{c2}, θ_{c11}, and θ_{c22} as well as θ_{f1} and θ_{f2} are calculated by the subroutines COLUMH and BEAMH instead of Eqs. (5.22) to (5.27).

The principle of the subroutine COLUMH is basically the same as that in the program BCIN, but here both the P-δ effect of the member and the P-Δ effect of the frame are all considered, i.e., the deformation of the columns including bending deflection and a lateral displacement Δ_1 and Δ_2 of the top of the columns. In addition, the subroutine COLUMH is more simple than BCIN since only the end rotations θ_{c1}, θ_{f1} (θ_{c2}, θ_{f2}) are calculated under a set of given loads H_1, M_{c1}, P_1 (H_2, M_{c2}, P_2) and given lateral displacements Δ_1 (Δ_2).

The principle subroutine of BEAMH is basically the same as the subroutine BEAMP in the program FRAMP. However, since $M_{c1} = /M_{c2}$ and $P_1 = /P_2$, the beam in BEAMH is not symmetrical about the load and deformation. Therefore, in the analysis the whole span of the beam should be considered.

5.4.2 Analytical Steps

In FRAMH, the second-order elastic analysis carried out for the first three values of Δ_1 and their solution are taken as the trial values for the first three solutions of the inelastic analysis.

In order to get fast convergence, it is important to choose appropriate trial values of H_1 and H_2. In FRAMH, for the first solution of elastic analysis the trial values of H_1 and H_2 are chosen by the user since it is easy to get convergence. Then, the first (or second) solutions of H_1 and H_2 of elastic analysis are taken as the trial values to get the second (or third) solution of elastic analysis.

The first three elastic solutions H_1 and H_2 are also taken as the trial values for the first three solutions of the inelastic analysis, but in the later inelastic analysis the corresponding extrapolated values are taken as the trial values for next steps.

In this way, the analytical steps for FRAMH are arranged as follows:

(1) Input the data from a file and screen (see user's manual).
(2) Calculate the geometric and physical behaviors of the section, the coordinates of the strips of the section, etc.
(3) Take the input data H_1 and H_2 as the trial values and calculate the second-order elastic solutions H_1 and H_2. Then, take the current elastic solution H_1 and H_2 as the trial values and calculate the second-order inelastic solution.
(4) Add the increment $d\Delta_1$ to Δ_1. Taking the last elastic solution H_1 and H_2 as the trial values, calculate the second-order elastic solutions H_1 and H_2 corresponding to the new Δ_1. Then, taking the current elastic solutions H_1 and H_2 as the trial values, calculate the second-order inelastic solutions corresponding to the new Δ_1.
(5) Repeat step (4) until three inelastic solutions are obtained.
(6) Add the increment $d\Delta_1$ to Δ_1. Using the last three inelastic solutions, calculate the trial values H_1 and H_2 for the next solutions by an extrapolation routine.

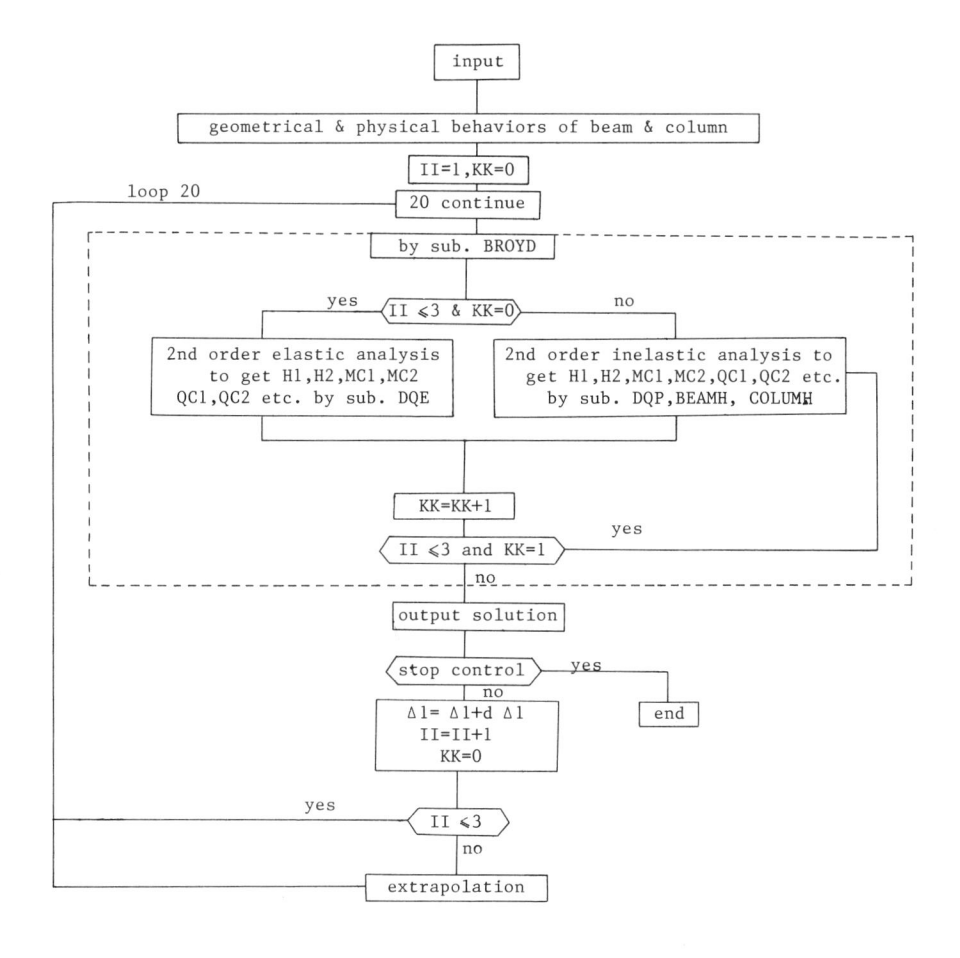

```
                    ┌───────┐
                    │ input │
                    └───┬───┘
        ┌───────────────┴──────────────────────────────┐
        │ geometrical & physical behaviors of beam & column │
        └───────────────┬──────────────────────────────┘
                    ┌───┴─────────┐
  loop 20           │ II=1, KK=0  │
  ┌─────────────────┤ 20 continue │
  │                 └──────┬──────┘
  │            ┌───────────┴──────────┐
  │            │    by sub. BROYD     │
  │            └───────────┬──────────┘
  │              yes  ┌────┴─────────┐  no
  │          ┌────────┤ II≤3 & KK=0  ├─────────┐
  │          │        └──────────────┘         │
  │  ┌───────┴──────────┐           ┌──────────┴─────────────┐
  │  │ 2nd order elastic │           │ 2nd order inelastic    │
  │  │ analysis to get   │           │ analysis to get        │
  │  │ H1,H2,MC1,MC2     │           │ H1,H2,MC1,MC2,QC1,QC2  │
  │  │ QC1,QC2 etc. by   │           │ etc. by sub. DQP,      │
  │  │ sub. DQE          │           │ BEAMH, COLUMH          │
  │  └───────┬───────────┘           └──────────┬─────────────┘
  │          └──────────────┬───────────────────┘
  │                   ┌─────┴─────┐
  │                   │  KK=KK+1  │
  │                   └─────┬─────┘              yes
  │                 ┌───────┴────────┐
  │                 │ II≤3 and KK=1  ├────────┐
  │                 └───────┬────────┘
  │                         │ no
  │                 ┌───────┴────────┐
  │                 │ output solution│
  │                 └───────┬────────┘
  │                  ┌──────┴──────┐   yes   ┌─────┐
  │                  │ stop control├─────────┤ end │
  │                  └──────┬──────┘         └─────┘
  │                         │ no
  │                 ┌───────┴────────┐
  │                 │ Δl= Δl+d Δl    │
  │                 │ II=II+1        │
  │                 │ KK=0           │
  │                 └───────┬────────┘
  │        yes      ┌───────┴───────┐
  └─────────────────┤     II≤3      │
                    └───────┬───────┘
                            │ no
                    ┌───────┴───────┐
                    │ extrapolation │
                    └───────────────┘
```

FIGURE 5.21. Flow chart of FRAMH.

Then, according to the extrapolated values the second-order inelastic analysis will take place.

(7) Repeat step (6) until the assigned solution points of the descending branch are obtained or a plastic failure mechanism is formed.

5.4.3 Flow Chart of FRAMH

The flow chart of FRAMH is shown in Figure 5.21. In the FRAMH, there is a main program and nine subroutines. The subroutines SECS, EXIN, and EXOUT are exactly the same as those in the program BCIN. Here, the incremental rotations $d\theta_{c1}$ and $d\theta_{c2}$ are calculated in the subroutine DQE by elastic analysis, and in the subroutine DQP by inelastic analysis. Then, in the subroutine BRYDEN, H_1 and H_2, which make $d\theta_{c1}$ and $d\theta_{c2}$ close to zero, will be found by iteration.

5.4.4 User's Manual of FRAMH

Positive directions are assigned as follows:

horizontal force: right direction
displacement: right direction
vertical load: downward
axial force and stress: compression
moment and rotation: clockwise

Input Data

There are two groups of input data for FRAMH: one is input from screen and the other is from a file.

Input from the screen:

D3	a character variable for the name of input file, not more than six characters
OUT	a character variable for the name of output file, not more than six characters
EO1, EO2	initial deflection at midspan of column normalized by column length, e_{01}/L_c and e_{02}/L_c (see Figure 5.20)
DTA1	assigned lateral displacement at left column top, Δ_1 in Figure 5.20
H11, H22	first trial values of shear forces in left and right columns, $H11 = H_1/P_y$ and $H22 = H_2/P_y$ in Figure 5.20
DDT	increment of DTA1
DDH	increments of H11 and H22 for iteration
NSTOP	assigned number of solution points in descending branch of load-deformation curve of frame, when it reaches to NSTOP, the calculation will end automatically

Note: In order to make the calculation successful, users may try a few times to get appropriate values of input data, and when the calculation is interrupted, the program will give a message for selection of the values (see Table 5.4). In general, take $H11 = H22 = 0.1$ to 0.01, $DDH = 0.002$ to 0.005. A smaller value of DTA1 would be better for iteration convergence since the frame will be in the elastic range for the value.

The read sentences are as follows:

> WRITE(*,'(A)') 'INPUT THE NAME OF INPUT FILE:'
> READ(*,'(A)' D3
> WRITE(*,'(A)') 'INPUT THE FILE NAME TO STORE RESULTS:'
> READ(*,'(A)') OUT
> WRITE(*,'(A)') 'PLEASE PUT IN EO1 EO2'
> READ(*,*) EO1,EO2

Table 5.4 Output for Interruption in FRAMH

Condition			Output
Interrupt in sub.BROYD			IT = ~
Interrupt in sub.BEAMH			BEAM STOP N01-T1 ~,~
Interrupt in sub.COLUMH			COLUMN STOP N02-N01-T1 ~,~,~
Interrupt when N = 1			SUGGESTION: CHANGE H11, H22, DTA1, DDH AND TRY AGAIN!
Interrupt when N ≠ 1			
NDES ≥ 1			INTERRUPT AT DESCENDING BRANCH NDES = ~
			HMAX = ~
NDES = 0	CM1 ≥ 0.99C1		INTERRUPT DUE TO PLASTIC HINGE FORMED AT THE TOP OF LEFT COLUMN
	CM2 ≥ 0.99C2		INTERRUPT DUE TO PLASTIC HINGE FORMED AT THE TOP OF RIGHT COLUMN
	BMAX ≥ 0.99		INTERRUPT DUE TO PLASTIC HINGE FORMED IN THE BEAM
			HMAX = ~
NDES = 0			SUGGESTION: CHANGE DDT OR DDH AND TRY AGAIN!
CM1 < 0.99C1			
CM2 < 0.99C2			
BMAX < 0.99			

```
WRITE(*,'(A)') 'PLEASE PUT IN DTA1 H11 H22 DDT DDH'
READ(*,*) DTA1,H11,H22,DDT,DDH
WRITE(*,'(A)') 'PLEASE PUT IN NSTOP'
READ(*,*) NSTOP
```

Input from a file:

A input file is named by the user and stored in the character "D3". The data are as follows:

FY	yield stress, F_y
EEE	elastic modulus, E
RESID	residual stress parameter, ($=\sigma_r/F_y$ in Figure 5.1)
BAT	ratio of the load on the beam to total vertical load on the frame, ($qL_b/2P_t$ in Figure 5.20)
PT	half total vertical load on frame normalized by PY, ($=P_t/P_y$ in Figure 5.20)
DC	depth of column section
WC	web thickness of column section
BC	flange width of column section
TC	flange thickness of column section
CL	column length (L_c in Figure 5.12)
DB	depth of beam section
WB	web thickness of beam section
BB	flange width of beam section
TB	flange thickness of beam section

240   ADVANCED ANALYSIS OF STEEL FRAMES

BL   beam length (L_b in Figure 5.20)
NW   number of strips in web of column and beam, take 60 to 100, (2*NF + NW) $\leq$ 120
NF   number of strips in flange of column and beam, take 10 to 20
IJC   number of division points in column, take 11 to 21, $IJC \leq 28$
IJB   number of division points in a half span of beam, take 6 to 11, $IHB \leq 28$

The read sentences are as follows:

READ(7,*) FY,EEE,RESID,BAT,PT
READ(7,*) DC,WC,BC,TC,CL
READ(7,*) DB,WB,BB,TB,BL
READ(7,*) NW,NF,IJC,IJB

Output Data

There are four groups of output in FRAMH.
Head of output table:

PT-BAT	see input data
FY-E-RESID-EO1-EO2	see input data
BEAM-D-W-B-T-L	DC,WC,BC,TC,CL of input data
COLUMN-D-W-B-T-L	DB,WB,BB,TB,BL of input data
NW-NF-IJC-IJB-NSTOP	see input data
H11-H22-DDH	"

Data of load-deformation curve of frame:

N	sequence number of solution points in frame curve
H1, H2	shear forces in left and right columns normalized by PY
H	horizontal load acted on left column top normalized by PY, (H1 + H2)
P1, P2	axial forces in left and right columns normalized by PY
MC1, MC2	end moments at top of left and right columns normalized by PMC
MB	maximum moment in beam normalized by PMB
C1, C2	plastic moment of left and right columns with effect of axial force normalized by PMC
PMC, PMB	plastic moment of column and beam without effect of axial force
PY	yield strength of column section
AE1, AE2	ratio of minimum elastic area to whole area of left and right columns
DTA1, DTA2	lateral displacements at top of left and right columns
QC1, QC2	rotations at top of left and right columns, radian
QF1, QF2	rotations at bottom of left and right columns, radian
IC	iteration times in BROYD, from which calculation time can be estimated

Load capacity of frame:
When the load-deformation curve of the frame reaches the descending branch or the plastic hinge is formed, HMAX will be printed out:

HMAX   horizontal load capacity normalized by PY

Interruption information:
When the calculation is interrupted due to the iteration failure in BROYD, BEAMH, or COLUMH, there will be a message as shown in Table 5.4. In the table, NO1 is iteration times for tangent stiffness. When NO1 $\geq$ (30∗IJC) for a column or NO1 $\geq$ (30∗IJB) for a beam, the calculation will be interrupted. NO2 is the iteration times for column deformation. When NO2 $\geq$ 10, the calculation will be interrupted. T1 is the ratio of the elastic area to the whole area at the interruption.

All the output data listed above are stored in the output file whose name is input by the user from the screen. Some of the data will also be displayed on the screen for the user to monitor the calculation process.

Frame Failures

There are three types of failures in unbraced frames.

(a) Frame fails due to column buckling, see Sample Problem 1
(b) Frame fails due to plastic hinge formation at the top of the columns, see sample problem 2
(c) Frame fails due to plastic hinge formation in the beam

In case (a), the load-deformation curve has both ascending and descending branches. In cases (b) and (c), the curve ascends first and becomes a horizontal line. When the moment reaches 99% of the plastic moment, which includes the effect of the axial force, the calculation will stop.

5.4.5 Sample Problems for FRAMH

There are three sample problems for FRAMH. The frame configuration of these sample problems is the same as that of FRAMP shown in Figure 5.16 except the initial crookedness which bends both columns convexly toward the windward direction, i.e., $e_{01} = e_{02} = +L_c/1000$. Accordingly, the input data are the same as those of FRAMP, but EO1 = EO2 = 0.001 and IJB = 11.

Figure 5.22 shows the results of sample problem 1 in which the vertical loads $P_t/P_y = 0.5$ act on the top of the columns ($P_c = P_t$). Accordingly, the input data are BAT=0 and PT=0.5. The input data DTA1=0.1in, H11=H22=0.01, DDH=0.002, NSTOP=5, and DDT=0.1in are chosen after a few trials. The H/P_y~Δ curve in this case has ascending and descending branches; the maximum horizontal force is $H = 0.01202P_y$. The frame fails due to the column buckling.

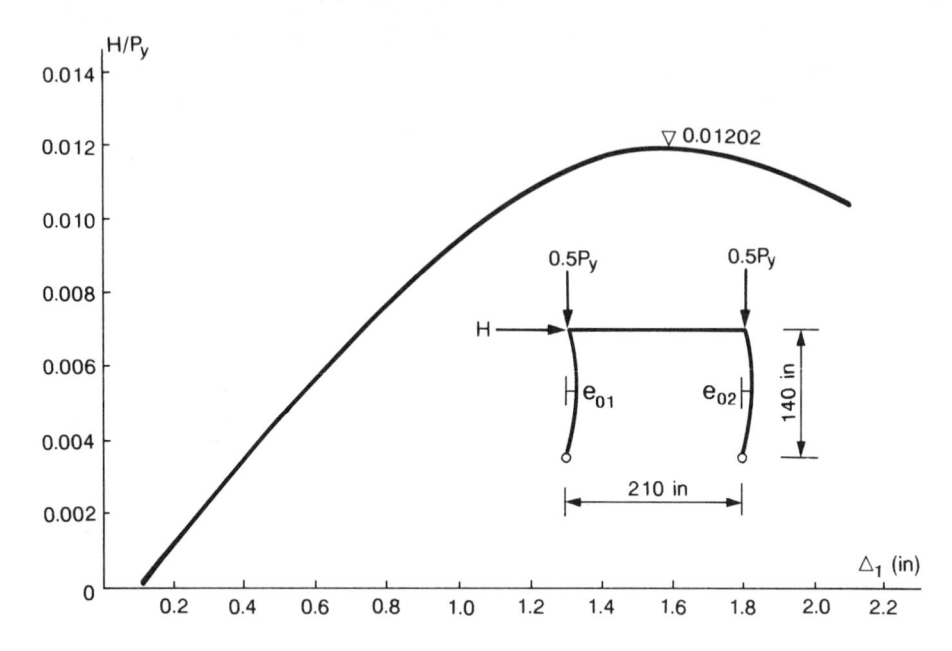

FIGURE 5.22. Behavior curves for problem 1 of FRAMH.

Figure 5.23 shows sample problem 3 in which the vertical load $P_t/P_y = 0.2$ acts on the beam ($q = 2*0.2P_y/L_b$). Accordingly, the input data are BAT=1.0 and PT=0.2. The input data DTA=0.1in, H11=H22=0.01, DDH=0.002, NSTOP=5, and DDT=0.1in are chosen after a few trials. The H/P_y-Δ curve in this case has only an ascending branch before the horizontal load H reaches the maximum value $0.02410P_y$. At the load $H/P_y = 0.02410$, the end moment on the top of the right column is $M_{cz} = 0.87988M_{pc}$ and the plastic moment of the right column at this loading stage is C1*M_{pc}=0.88165M_{pc} from the output file. Therefore, it can be seen that the plastic hinge at the top of right column has been formed for frame failure because $0.87988/0.88165=0.998\approx1$.

Sample problem 2 is the same as problem 3 except the increment of lateral displacement is chosen as DDT=0.5in. The calculation fails because DDT is too large and there will be a message for this. When DDT is reduced to 0.1, the calculation will be successful as shown in the sample problem 3.

The input and output for the sample problems are printed in Appendix C.

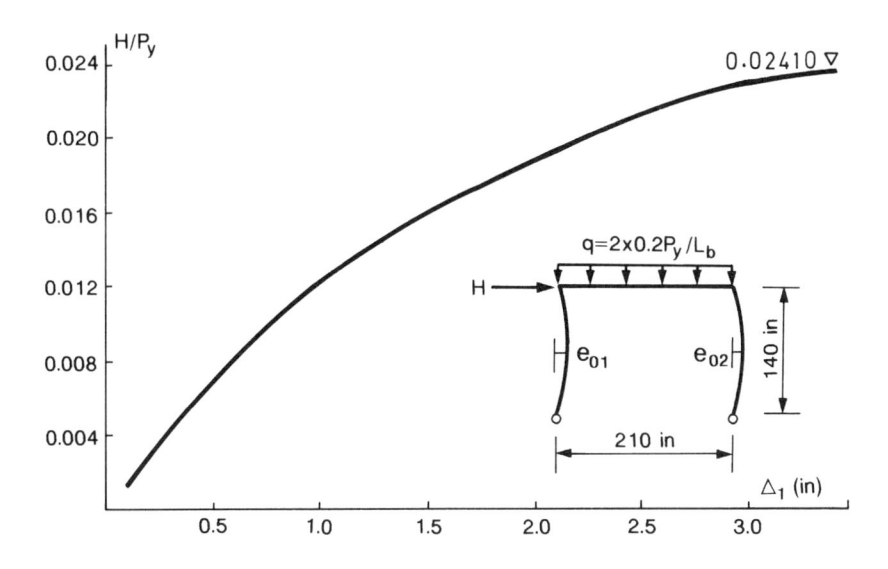

FIGURE 5.23. Behavior curves for problem 3 of FRAMH.

References

Chen, W. F. and Atsuta, T. (1977) *Theory of Beam-Columns*, Vol. 1, McGraw-Hill, New York.
Chen, W. F. and Zhou, S. P. (1987) Inelastic analysis of steel braced frames with flexible joints, *Int. J. Solids Struct.*, 23(5).
Cranston, W. B. (1967) A Computer Method for the Analysis of Restrained Columns, Technical Report, Cement and Concrete Association, London.
Zhou, S. P. and Chen, W. F. (1985) Design criteria for box columns under biaxial loading, *J. Struct. Eng.*, 111(12).
Zhou, S. P., Duan, L., and Chen, W. F. (1991) The P-Δ Effect on Portal Steel Frames, Proceedings of the Int. Conference on Steel and Aluminum Structures, Singapore, May 22–24.

APPENDIX A: Sample Problems of BCIN (File: "OBC")

1. Sample problem 1 for program BCIN

Input data from file d1

```
250.0 250.0 14.0 9.0 235.0 206000.0 0.001 5.0 0.3 0.005
1 1 100 10 4 11 6
```

Output data

```
              THE SECOND ORDER PLASTIC ZONE ANALYSIS OF   BEAM-COLUMNS
                              INPLANE BENDING
                               I SECTION
                          BENDING ABOUT WEAK AXIS
----------------------------------------------------------------------
              IN THIS PROGRAM WHEN N>10 THE AEE=0.001
     B-DDD-T-W-ALD              250.00    250.00    14.00     9.00    60.00
     FF1-FF2-AFA                  .10000    .60000   -.50000
     ADDY-AEE                    5.00000    .00500
     FY-EEE-RESID-EO             235.00  206000.00              .30      .00100
     MP-MJ-NF-NW-NSTOP-IJK-IJ1      1         1      100       10     4   11    6
     A-AIX-AIY                   .899800E+04  .105788E+09  .364718E+08
     ZX-ZY                       .936889E+06  .441996E+06
     PY-PMX-PMY                  .211453E+07  .220169E+09  .103869E+09
     ALDX-ALDY-CL                  35.23      60.00       3819.94
----------------------------------------------------------------------
                      END MOMENT F1-DISPLACEMENT CURVE

IN THIS CURVE, THE AXIAL FORCE F2 REMAIN CONSTANT & THE END MOMENTS
F1 & F3 CHANGE AS THE DISPLACEMENT INCREASE. THE Q1 IS ROTATION OF
THE END WITH F1. THE Y IS THE DEFLECTION OF SPECIAL POINT(IJ1).
----------------------------------------------------------------------

       N    NO1     F1         Q1           Y          T11       IT
----------------------------------------------------------------------
       1    87    .08437    .3701E-02     3.81061     .91443      6
    NNK= 1      DDY/2=  2.50000
       2    72    .13834    .6244E-02     6.33681     .84441      6
       3   100    .17517    .8724E-02     8.87155     .79773      6
       4    58    .20022    .1106E-01    11.33063     .75884      6
       5    60    .21949    .1334E-01    13.78703     .72772      6
       6    38    .23429    .1557E-01    16.24267     .70438      6
       7    34    .24590    .1776E-01    18.69195     .68104      5
       8    34    .25538    .1993E-01    21.14584     .66548      5
       9    33    .26333    .2208E-01    23.61720     .65770      6
      10    22    .26974    .2422E-01    26.09222     .64214      6
      11    33    .27504    .2633E-01    28.56557     .63436      6
      12    33    .27958    .2844E-01    31.04388     .61880      6
      13    46    .28238    .3054E-01    33.57484     .49222      6
      14   129    .27570    .3229E-01    36.05425     .33452      6
      15    56    .25875    .3373E-01    38.59122     .28784      6
      16   140    .23556    .3487E-01    41.05124     .24116      6
    NNK= 2      DDY/2=  1.25000
      17    80    .22224    .3535E-01    42.27333     .21783      6
      18    79    .20835    .3578E-01    43.49027     .20227      6
    F1MAX       .28238
----------------------------------------------------------------------
```

PLASTIC-ZONE ANALYSIS OF BEAM-COLUMNS AND PORTAL FRAMES 245

```
    2. Sample problem 2 for program BCIN

    Input data from file d1

    250.0 250.0 14.0 9.0 235.0 206000.0 0.001 2.0 0.3 0.005
    2 1 100 10 4 11 6
                 THE SECOND ORDER PLASTIC ZONE ANALYSIS OF  BEAM-COLUMNS
                                INPLANE BENDING
                                I SECTION
                          BENDING ABOUT WEAK AXIS
    -----------------------------------------------------------------
                     IN THIS PROGRAM WHEN N>10 THE AEE=0.001
    B-DDD-T-W-ALD             250.00   250.00    14.00     9.00    60.00
    FF1-FF2-AFA                .28238   .10000  -.50000
    ADDY-AEE                  2.00000   .00500
    FY-EEE-RESID-EO            235.00 206000.00              .30     .00100
    MP-MJ-NF-NW-NSTOP-IJK-IJ1    2     1    100    10      4     11     6
    A-AIX-AIY                 .899800E+04  .105788E+09  .364718E+08
    ZX-ZY                     .936889E+06  .441996E+06
    PY-PMX-PMY                .211453E+07  .220169E+09  .103869E+09
    ALDX-ALDY-CL                35.23    60.00    3819.94
    -----------------------------------------------------------------
                       AXIAL FORCE F2-DISPLACEMENT CURVE

    IN THIS CURVE, THE END MOMENTS F1 & F3 REMAIN CONSTANT &  THE AXIAL
    FORCE F2 CHANGES AS THE DISPLACEMENT INCREASE.THE Q1 IS ROTATION OF
    THE END WITH F1. THE Y IS THE DEFLECTION OF SPECIAL POINT(IJ1).
    -----------------------------------------------------------------

       N    NO1      F2        Q1           Y          T11       IT
    -----------------------------------------------------------------
       1     0    .10000    .6554E-02     5.74474    1.00000     11
    NNK= 1     DDY/2=  1.00000
       2    133   .31587    .7431E-02     6.76961     .96888      4
       3    135   .40906    .8408E-02     7.78376     .90665      5
       4    108   .45410    .9362E-02     8.78380     .86775      6
       5     68   .48402    .1030E-01     9.78472     .82885      5
       6     45   .50512    .1123E-01    10.79001     .80551      5
       7     44   .52096    .1214E-01    11.79283     .78217      5
       8     34   .53334    .1304E-01    12.79283     .76661      5
       9     33   .54335    .1392E-01    13.79283     .75106      5
      10     35   .55157    .1480E-01    14.79291     .73550      5
      11     39   .55844    .1567E-01    15.79321     .71994      5
      12     37   .56425    .1654E-01    16.79327     .71216      6
      13     22   .56898    .1739E-01    17.77842     .70438      6
      14     33   .57342    .1824E-01    18.77842     .69660      6
      15     33   .57714    .1909E-01    19.77842     .68104      5
      16     33   .58039    .1994E-01    20.77842     .67326      5
      17     21   .58319    .2078E-01    21.77303     .67326      6
      18     22   .58555    .2160E-01    22.75450     .66548      6
      19     32   .58796    .2244E-01    23.75450     .65770      6
      20     33   .58995    .2327E-01    24.75450     .64992      6
      21     23   .59152    .2409E-01    25.73238     .64992      6
      22     33   .59330    .2492E-01    26.73238     .64214      6
      23     31   .59473    .2575E-01    27.73238     .63436      6
      24     20   .59582    .2655E-01    28.70914     .63436      6
      25     31   .59717    .2738E-01    29.70914     .62658      6
      26     33   .59822    .2821E-01    30.70914     .62658      6
      27     21   .59898    .2901E-01    31.69244     .59660      6
      28     52   .59988    .2983E-01    32.69244     .55219      6
      29     37   .59984    .3063E-01    33.69294     .49222      6
      30     31   .59933    .3141E-01    34.69294     .43225      6
      31     38   .59804    .3217E-01    35.70411     .36450      6
      32     32   .59561    .3288E-01    36.70411     .32674      6
      33     34   .59286    .3357E-01    37.70522     .31118      6
    F2MAX     .59988
    -----------------------------------------------------------------
```

APPENDIX B: Sample Problems of FRAMP (File: "OFP")

1. Sample prolem 1 for program FRAMP

Input data form file d2

```
36.0 29000.0 0.3 1.0
8.0 0.28 8.0 0.43 140.0 0.001
16.26 0.38 7.07 0.63 210.0
100 10 11 6
```

Output data for rame

```
                    SECOND ORDER PLASTIC ZONE ANALYSIS
             FOR PORTAL BRACED FRAME WITH PINNED SUPPORTS

BENDING ABOUT STRONG AXIS OF I SECTION OF COLUMNS & BEAM

FACTOR OF BEAM LOAD__BAT           1.00
FY-E-RESID-EO             36.0 29000.0     .3    .0010
BEAM-D-W-B-T-L                 16.260    .380    7.070       210.000
COLUMN-D-W-B-T-L                8.000    .280    8.000       140.000
NW-NF-IJC-IHB                     100      10       11          6

            PT_QC CURVE OF PORTAL BRACED FRAME

HALF TOTAL VERTICAL LOAD OF FRAME = PT
LOAD ON TOP OF COLUMN =             PC
LOAD ON BEAM =                      QB
END MOMENT ON TOP OF COLUMN =       MC
END ROTATION ON TOP OF COLUMN =     QC
MOMENT IN MIDSPAN OF BEAM =         MBC & MBB
```

K	PT	PC	QB	MC	QC	MBC	MBB	FF1	ADDY	AEE	NCO	NDES
1	.1000	.0000	.2000	.27784	.45736E-02	1.29652	.42189	.10000	.02000	.00100	5	0
2	.2000	.0000	.4000	.55002	.91915E-02	2.59870	.84563	.10000	.02000	.00100	11	0
3	.2400	.0000	.4800	.74746	.13553E-01	3.03100	.98630	.10000	.02000	.00100	15	0

NO SOLUTION AT PT= .25000
SUGGESTION: CHOOSE SMALLER PT CALCULATE AGAIN IF NEED.

| 4 | .2450 | .0000 | .4900 | .79983 | .16430E-01 | 3.05736 | .99488 | .10000 | .02000 | .00100 | 17 | 0 |
| 5 | .2490 | .0000 | .4980 | .84947 | .31288E-01 | 3.07068 | .99921 | .10000 | .02000 | .00100 | 22 | 0 |

Output data for beam in the frame

(Only output for PT=0.25, 0.249 list here.)

THE FIRST ORDER INELASTIC ANALYSIS OF BEAM IN PORTAL BRACED FRAME
INPLANE BENDING
I SECTION
BENDING ABOUT STRONG AXIS

```
DB-BB-WB-TB-BL      16.26    7.07      .38       .63   210.00
PT-QB                .2500   .5000
FY-EEE-RESID         36.0  29000.0     .30
NF-NW-IHB            10    100     6
PY-PMC-PMB          .319651E+03  .106594E+04  .327573E+04
```

MB_QB CURVE OF BEAM IN PORTAL BRACED FRAME

```
END MOMENT OF BEAM =            MB
END ROTATION OF BEAM =          QB
DEFLECTION AT MIDSPAN =         Y
MINIMUM ELASTIC AREA RATIO =    AE
```

N	NO1	MB	QB	Y	AE
1	9	1.00000	.1045E-01	.7920E+00	.2887E+00
2	11	.99000	.1062E-01	.8053E+00	.2770E+00
3	10	.98000	.1080E-01	.8198E+00	.2692E+00
4	11	.97000	.1100E-01	.8360E+00	.2536E+00
5	11	.96000	.1121E-01	.8540E+00	.2419E+00
6	12	.95000	.1145E-01	.8744E+00	.2263E+00
7	11	.94000	.1173E-01	.8981E+00	.2146E+00
8	11	.93000	.1204E-01	.9258E+00	.1990E+00
9	11	.92000	.1242E-01	.9594E+00	.1873E+00
10	12	.91000	.1289E-01	.1002E+01	.1678E+00
11	12	.90000	.1350E-01	.1058E+01	.1522E+00
12	13	.89000	.1438E-01	.1141E+01	.1288E+00
13	14	.88000	.1588E-01	.1284E+01	.1014E+00
14	15	.87000	.1965E-01	.1650E+01	.6633E-01
15	16	.86500	.2823E-01	.2489E+01	.3512E-01
16	17	.86375	.3778E-01	.3424E+01	.2341E-01
17	17	.86312	.5932E-01	.5535E+01	.1561E-01
18	18	.86305	.6750E-01	.6336E+01	.1171E-01
19	18	.86301	.7178E-01	.6755E+01	.1171E-01
20	18	.86299	.7392E-01	.6965E+01	.1171E-01
21	18	.86298	.7499E-01	.7070E+01	.1171E-01

INPLANE BENDING
I SECTION
BENDING ABOUT STRONG AXIS

```
DB-BB-WB-TB-BL      16.26    7.07      .38       .63   210.00
PT-QB                .2490   .4980
FY-EEE-RESID         36.0  29000.0     .30
NF-NW-IHB            10    100     6
PY-PMC-PMB          .319651E+03  .106594E+04  .327573E+04
```

MB_QB CURVE OF BEAM IN PORTAL BRACED FRAME

```
END MOMENT OF BEAM =            MB
END ROTATION OF BEAM =          QB
DEFLECTION AT MIDSPAN =         Y
MINIMUM ELASTIC AREA RATIO =    AE
```

N	NO1	MB	QB	Y	AE
1	10	1.00000	.1025E-01	.7741E+00	.3049E+00
2	10	.98000	.1056E-01	.7985E+00	.2809E+00
3	11	.97000	.1073E-01	.8123E+00	.2692E+00
4	11	.96000	.1091E-01	.8275E+00	.2614E+00
5	11	.95000	.1112E-01	.8444E+00	.2497E+00
6	11	.94000	.1134E-01	.8634E+00	.2341E+00
7	11	.93000	.1160E-01	.8851E+00	.2224E+00
8	12	.92000	.1189E-01	.9103E+00	.2068E+00
9	11	.91000	.1223E-01	.9403E+00	.1951E+00
10	11	.90000	.1264E-01	.9773E+00	.1795E+00
11	12	.89000	.1316E-01	.1025E+01	.1600E+00
12	12	.88000	.1387E-01	.1091E+01	.1405E+00
13	13	.87000	.1495E-01	.1193E+01	.1171E+00
14	13	.86000	.1700E-01	.1390E+01	.8974E-01
15	16	.85000	.2566E-01	.2235E+01	.4292E-01
16	17	.84750	.5222E-01	.4837E+01	.1561E-01
17	18	.84734	.6312E-01	.5904E+01	.1171E-01
18	18	.84727	.7167E-01	.6742E+01	.1171E-01
19	18	.84725	.7380E-01	.6951E+01	.1171E-01
20	18	.84724	.7487E-01	.7056E+01	.1171E-01

248 ADVANCED ANALYSIS OF STEEL FRAMES

Output data for column in the frame

(Only output for PT=0.25, 0.249 list here.)

THE SECOND ORDER PLASTIC ZONE ANALYSIS OF COLUMN IN PORTAL BRACED FRAME
INPLANE BENDING
I SECTION
BENDING ABOUT STRONG AXIS

```
DC-BC-WC-TC-CL        8.00    8.00    .28     .43   140.00
FF1-PT                .10000         .25000
ADDY-AEE              .02000         .00100
FY-EEE-RESID-EO        36.00      29000.00      .30     .00100
NF-NW-IJC-IJ1         10  100    11    4
AC-CI                 .887920E+01  .107164E+03
PY-PMC                .319651E+03  .106594E+04
```

MC_QC CURVE OF COLUMN IN PORTAL BRACED FRAME

```
END MOMENT ON TOP OF COLUMN   =  MC
END ROTATION ON TOP OF COLUMN =  QC
DEFLECTION AT POINT IJ1       =  Y
MINIMUM ELASTIC AREA RATIO    =  AE
```

N	NO1	MC	QC	Y	AE	IT
1	0	.10000	.1826E-02	.4805E-01	.1000E+01	11
2	0	.14763	.2616E-02	.6803E-01	.1000E+01	11
3	0	.19530	.3405E-02	.8803E-01	.1000E+01	11
4	0	.24297	.4195E-02	.1080E+00	.1000E+01	11
5	0	.29064	.4985E-02	.1280E+00	.1000E+01	11
6	0	.33831	.5775E-02	.1480E+00	.1000E+01	11
7	0	.38598	.6565E-02	.1680E+00	.1000E+01	11
8	21	.43365	.7351E-02	.1879E+00	.9934E+00	1
9	32	.48132	.8145E-02	.2079E+00	.9554E+00	1
10	23	.52866	.8948E-02	.2279E+00	.9093E+00	1
11	44	.57571	.9773E-02	.2481E+00	.8569E+00	1
12	45	.62170	.1062E-01	.2683E+00	.7954E+00	1
13	44	.66602	.1151E-01	.2886E+00	.7164E+00	1
14	45	.70702	.1248E-01	.3086E+00	.6126E+00	1
15	46	.74384	.1365E-01	.3287E+00	.5766E+00	1
16	46	.77437	.1504E-01	.3489E+00	.5450E+00	1
17	44	.79891	.1667E-01	.3688E+00	.5135E+00	1
18	42	.81814	.1864E-01	.3885E+00	.4780E+00	1
19	44	.83239	.2118E-01	.4089E+00	.4235E+00	1
20	43	.84128	.2428E-01	.4286E+00	.3615E+00	1
21	58	.84653	.2818E-01	.4485E+00	.2935E+00	1
22	48	.84918	.3263E-01	.4681E+00	.2346E+00	1

THE SECOND ORDER PLASTIC ZONE ANALYSIS OF COLUMN IN PORTAL BRACED FRAME
INPLANE BENDING
I SECTION
BENDING ABOUT STRONG AXIS

```
DC-BC-WC-TC-CL        8.00    8.00    .28     .43   140.00
FF1-PT                .10000         .24900
ADDY-AEE              .02000         .00100
FY-EEE-RESID-EO        36.00      29000.00      .30     .00100
NF-NW-IJC-IJ1         10  100    11    4
AC-CI                 .887920E+01  .107164E+03
PY-PMC                .319651E+03  .106594E+04
```

MC_QC CURVE OF COLUMN IN PORTAL BRACED FRAME

```
END MOMENT ON TOP OF COLUMN   =  MC
END ROTATION ON TOP OF COLUMN =  QC
DEFLECTION AT POINT IJ1       =  Y
MINIMUM ELASTIC AREA RATIO    =  AE
```

N	NO1	MC	QC	Y	AE	IT
1	0	.10000	.1825E-02	.4802E-01	.1000E+01	11
2	0	.14769	.2615E-02	.6802E-01	.1000E+01	11
3	0	.19537	.3406E-02	.8802E-01	.1000E+01	11
4	0	.24305	.4196E-02	.1080E+00	.1000E+01	11
5	0	.29073	.4986E-02	.1280E+00	.1000E+01	11
6	0	.33841	.5776E-02	.1480E+00	.1000E+01	11
7	0	.38609	.6565E-02	.1680E+00	.1000E+01	11
8	21	.43377	.7351E-02	.1879E+00	.9938E+00	1

PLASTIC-ZONE ANALYSIS OF BEAM-COLUMNS AND PORTAL FRAMES

```
 8   21   .43365   .7351E-02   .1879E+00   .9934E+00   1
 9   32   .48132   .8145E-02   .2079E+00   .9554E+00   1
10   23   .52866   .8948E-02   .2279E+00   .9093E+00   1
11   44   .57571   .9773E-02   .2481E+00   .8569E+00   1
12   45   .62170   .1062E-01   .2683E+00   .7954E+00   1
13   44   .66602   .1151E-01   .2886E+00   .7164E+00   1
14   45   .70702   .1248E-01   .3086E+00   .6126E+00   1
15   46   .74384   .1365E-01   .3287E+00   .5766E+00   1
16   46   .77437   .1504E-01   .3489E+00   .5450E+00   1
17   44   .79891   .1667E-01   .3688E+00   .5135E+00   1
18   42   .81814   .1864E-01   .3885E+00   .4780E+00   1
19   44   .83239   .2118E-01   .4089E+00   .4235E+00   1
20   43   .84128   .2428E-01   .4286E+00   .3615E+00   1
21   58   .84653   .2818E-01   .4485E+00   .2935E+00   1
22   48   .84918   .3263E-01   .4681E+00   .2346E+00   1
```

```
        THE SECOND ORDER PLASTIC ZONE ANALYSIS OF COLUMN IN PORTAL BRACED FRAME
                              INPLANE BENDING
                                I SECTION
                          BENDING ABOUT STRONG AXIS
```

```
DC-BC-WC-TC-CL       8.00      8.00     .28      .43   140.00
FF1-PT               .10000    .24900
ADDY-AEE             .02000    .00100
FY-EEE-RESID-EO      36.00     29000.00          .30      .00100
NF-NW-IJC-IJ1        10   100     11     4
AC-CI                .887920E+01  .107164E+03
PY-PMC               .319651E+03  .106594E+04
```

```
              MC_QC CURVE OF COLUMN IN PORTAL BRACED FRAME

END MOMENT ON TOP OF COLUMN    =  MC
END ROTATION ON TOP OF COLUMN  =  QC
DEFLECTION AT POINT IJ1        =  Y
MINIMUM ELASTIC AREA RATIO     =  AE
```

```
 N   NO1    MC       QC          Y           AE         IT
 1    0   .10000   .1825E-02   .4802E-01   .1000E+01   11
 2    0   .14769   .2615E-02   .6802E-01   .1000E+01   11
 3    0   .19537   .3406E-02   .8802E-01   .1000E+01   11
 4    0   .24305   .4196E-02   .1080E+00   .1000E+01   11
 5    0   .29073   .4986E-02   .1280E+00   .1000E+01   11
 6    0   .33841   .5776E-02   .1480E+00   .1000E+01   11
 7    0   .38609   .6565E-02   .1680E+00   .1000E+01   11
 8   21   .43377   .7351E-02   .1879E+00   .9938E+00    1
 9   32   .48146   .8145E-02   .2079E+00   .9561E+00    1
10   23   .52882   .8948E-02   .2279E+00   .9102E+00    1
11   44   .57589   .9773E-02   .2481E+00   .8579E+00    1
12   45   .62192   .1062E-01   .2683E+00   .7965E+00    1
13   44   .66629   .1150E-01   .2885E+00   .7179E+00    1
14   45   .70737   .1248E-01   .3086E+00   .6126E+00    1
15   47   .74430   .1364E-01   .3287E+00   .5788E+00    1
16   47   .77497   .1503E-01   .3489E+00   .5450E+00    1
17   43   .79963   .1666E-01   .3688E+00   .5158E+00    1
18   43   .81894   .1862E-01   .3885E+00   .4778E+00    1
19   44   .83331   .2115E-01   .4089E+00   .4230E+00    1
20   43   .84229   .2425E-01   .4286E+00   .3605E+00    1
21   58   .84758   .2815E-01   .4485E+00   .2922E+00    1
22   48   .85025   .3258E-01   .4680E+00   .2330E+00    1
```

250 ADVANCED ANALYSIS OF STEEL FRAMES

```
2. Sample problem 2 for program FRAMP

Input data from file d2

36.0 29000.0 0.3 0.33333333
8.0 0.28 8.0 0.43 140.0 0.001
16.26 0.38 7.07 0.63 210.0
100 10 11 6

Output data for frame
-----------------------------------------------------------------
              SECOND ORDER PLASTIC ZONE ANALYSIS
         FOR PORTAL BRACED FRAME WITH PINNED SUPPORTS
    BENDING ABOUT STRONG AXIS OF I SECTION OF COLUMNS & BEAM
-----------------------------------------------------------------
FACTOR OF BEAM LOAD__BAT      .33
FY-E-RESID-EO           36.0 29000.0     .3     .0010
BEAM-D-W-B-T-L               16.260    .380    7.070    210.000
COLUMN-D-W-B-T-L              8.000    .280    8.000    140.000
NW-NF-IJC-IHB                  100       10      11        6
-----------------------------------------------------------------
          PT_QC CURVE OF PORTAL BRACED FRAME
HALF TOTAL VERTICAL LOAD OF FRAME = PT
LOAD ON TOP OF COLUMN =             PC
LOAD ON BEAM =                      QB
END MOMENT ON TOP OF COLUMN =       MC
END ROTATION ON TOP OF COLUMN =     QC
MOMENT IN MIDSPAN OF BEAM =         MBC & MBB
-----------------------------------------------------------------
 K   PT      PC     QB      MC       QC          MBC      MBB      FF1     ADDY    AEE    NCO  NDES
 1 .1500  .1000  .1000  .13522  .23087E-02    .65196  .21215    .10000  .02000  .00100   2    0
 2 .3000  .2000  .2000  .26595  .46441E-02   1.30841  .42576    .10000  .02000  .00100   5    0
 3 .4600  .3000  .3000  .38308  .69669E-02   1.97846  .64380    .10000  .02000  .00100   8    0
 4 .6000  .4000  .4000  .40342  .10225E-01   2.74530  .89333    .10000  .02000  .00100  11    0
 5 .6500  .4333  .4333  .37273  .14942E-01   3.03838  .98870    .10000  .01000  .00100  29    0
NO SOLUTION AT PT=  .65500
SUGGESTION: CHOOSE SMALLER PT CALCULATE AGAIN IF NEED.
 6 .6520  .4347  .4347  .37038  .16322E-01   3.05123  .99288    .10000  .01000  .00100  32    1
NO SOLUTION AT PT=  .65300
SUGGESTION: CHOOSE SMALLER PT CALCULATE AGAIN IF NEED.
-----------------------------------------------------------------

Output data for beam in the frame

(Only output for PT=0.652, 0.653 list here.)

  THE FIRST ORDER INELASTIC ANALYSIS OF BEAM IN PORTAL BRACED FRAME
                    INPLANE BENDING
                     I SECTION
               BENDING ABOUT STRONG AXIS
-----------------------------------------------------------------
DB-BB-WB-TB-BL        16.26      7.07         .38       .63    210.00
PT-QB                 .6520     .4347
FY-EEE-RESID          36.0    29000.0         .30
NF-NW-IHB              10       100            6
PY-PMC-PMB           .319651E+03 .106594E+04 .327573E+04
-----------------------------------------------------------------
           MB_QB CURVE OF BEAM IN PORTAL BRACED FRAME
```

PLASTIC-ZONE ANALYSIS OF BEAM-COLUMNS AND PORTAL FRAMES 251

Output data for beam in the frame

(Only output for PT=0.652, 0.653 list here.)

```
        THE FIRST ORDER INELASTIC ANALYSIS OF BEAM IN PORTAL BRACED FRAME
                            INPLANE BENDING
                              I SECTION
                         BENDING ABOUT STRONG AXIS
------------------------------------------------------------------------
DB-BB-WB-TB-BL      16.26    7.07     .38      .63   210.00
PT-QB               .6520   .4347
FY-EEE-RESID        36.0  29000.0     .30
NF-NW-IHB           10   100    6
PY-PMC-PMB          .319651E+03  .106594E+04  .327573E+04
------------------------------------------------------------------------
                    MB_QB CURVE OF BEAM IN PORTAL BRACED FRAME

      END MOMENT OF BEAM     =         MB
      END ROTATION OF BEAM   =         QB
      DEFLECTION AT MIDSPAN  =         Y
      MINIMUM ELASTIC AREA RATIO =     AE
------------------------------------------------------------------------

      N    NO1      MB          QB           Y            AE
------------------------------------------------------------------------
      1     4    1.00000    .7506E-02    .5734E+00    .8666E+00
      2     5     .95000    .7819E-02    .5904E+00    .8387E+00
      3     5     .90000    .8135E-02    .6078E+00    .8017E+00
      4     6     .85000    .8458E-02    .6258E+00    .7619E+00
      5     6     .80000    .8789E-02    .6444E+00    .7150E+00
      6     8     .75000    .9130E-02    .6640E+00    .6678E+00
      7     8     .70000    .9486E-02    .6849E+00    .6108E+00
      8     8     .65000    .9863E-02    .7076E+00    .5486E+00
      9     9     .63000    .1002E-01    .7174E+00    .5181E+00
     10     9     .61000    .1019E-01    .7278E+00    .4881E+00
     11     9     .59000    .1036E-01    .7388E+00    .4591E+00
     12     9     .57000    .1054E-01    .7505E+00    .4359E+00
     13     9     .55000    .1073E-01    .7634E+00    .4014E+00
     14     9     .53000    .1094E-01    .7779E+00    .3611E+00
     15    10     .51000    .1118E-01    .7958E+00    .3193E+00
     16    11     .49000    .1148E-01    .8189E+00    .2887E+00
     17    11     .48000    .1165E-01    .8321E+00    .2809E+00
     18    11     .47000    .1182E-01    .8465E+00    .2692E+00
     19    10     .46000    .1202E-01    .8624E+00    .2614E+00
     20    11     .45000    .1223E-01    .8800E+00    .2458E+00
     21    11     .44000    .1247E-01    .8999E+00    .2341E+00
     22    11     .43000    .1273E-01    .9226E+00    .2224E+00
     23    11     .42000    .1304E-01    .9491E+00    .2068E+00
     24    12     .41000    .1340E-01    .9808E+00    .1912E+00
     25    11     .40000    .1383E-01    .1020E+01    .1795E+00
     26    12     .39000    .1440E-01    .1071E+01    .1600E+00
     27    13     .38000    .1517E-01    .1143E+01    .1366E+00
     28    14     .37000    .1637E-01    .1256E+01    .1132E+00
     29    14     .36000    .1876E-01    .1486E+01    .8194E-01
     30    15     .35000    .3356E-01    .2931E+01    .2731E-01
     31    17     .34875    .6952E-01    .6455E+01    .1171E-01
     32    17     .34871    .7380E-01    .6874E+01    .1171E-01
     33    17     .34869    .7594E-01    .7084E+01    .1171E-01
     34    17     .34869    .7647E-01    .7136E+01    .7804E-02
------------------------------------------------------------------------
        THE FIRST ORDER INELASTIC ANALYSIS OF BEAM IN PORTAL BRACED FRAME
                            INPLANE BENDING
                              I SECTION
                         BENDING ABOUT STRONG AXIS
------------------------------------------------------------------------
DB-BB-WB-TB-BL      16.26    7.07     .38      .63   210.00
PT-QB               .6530   .4353
FY-EEE-RESID        36.0  29000.0     .30
NF-NW-IHB           10   100    6
PY-PMC-PMB          .319651E+03  .106594E+04  .327573E+04
------------------------------------------------------------------------
                    MB_QB CURVE OF BEAM IN PORTAL BRACED FRAME

      END MOMENT OF BEAM     =         MB
      END ROTATION OF BEAM   =         QB
      DEFLECTION AT MIDSPAN  =         Y
      MINIMUM ELASTIC AREA RATIO =     AE
------------------------------------------------------------------------

      N    NO1      MB          QB           Y            AE
------------------------------------------------------------------------
      1     4    1.00000    .7528E-02    .5749E+00    .8640E+00
      2     5     .95000    .7841E-02    .5920E+00    .8355E+00
      3     5     .90000    .8158E-02    .6094E+00    .7980E+00
      4     6     .85000                 .6274E+00
              .80000                     .   61E+00
```

APPENDIX C: Sample Problems of FRAMH (File: "OFH")

1. Sample problem 1 for program FRAMH

Input data from file d3

```
36.0 29000.0 0.3 0.0 0.5
8.0 0.28 8.0 0.43 140.0
16.26 0.38 7.07 0.63 210.0
100 10 11 11
```

Output data

```
             SECOND ORDER PLASTIC ZONE ANALYSIS
     FOR PORTAL UNBRACED FRAME WITH H & PINNED SUPPORTS

BENDING ABOUT STRONG AXIS OF I SECTION OF COLUMNS & BEAM

VERTICAL LOAD PT-BAT            .5000
FY-E-RESID-E01-E02     36.0 29000.0   .00     .3       .0010
BEAM-D-W-B-T-L         16.260   .380    7.070    .630   210.000
COLUMN-D-W-B-T-L        8.000   .280    8.000    .430   140.000
NW-NF-IJC-IJB-NSTOP    100     10      11       11       5
H11-H22-DDH             .1000E-01 .1000E-01  .2000E-02
```

H1 H2   = SHEAR FORCES AT ENDS OF LEFT & RIGHT COLUMNS
H   = HORIZONTAL LOAD AT TOP OF LEFT COLUMN
P1 P2   = AXIAL FORCES OF LEFT & RIGHT COLUMNS
MC1 MC2   = MOMENTS AT TOP OF LEFT & RIGHT COLUMNS
MB   = MAXIMUM MOMENT IN THE BEAM
C1 C2   = FULL PLASTIC MOMENTS OF LEFT & RIGHT COLUMNS INCLUDING EFFECT OF AXIAL FORCES
AE1 AE2   = MINIMUM ELASTIC AREA RATIO OF LEFT & RIGHT COLUMNS
DTA1 DTA2   = DISPLACEMENTS AT TOP OF LEFT & RIGHT COLUMNS
QC1 QC2   = ROTATIONS AT TOP OF LEFT & RIGHT COLUMNS
QF1 QF2   = ROTATIONS AT BOTTOM OF LEFT & RIGHT COLUMNS
IC   = TIMES OF ITERATION IN SUB. BROYD

| N | H1 | H2 | H | P1 | P2 | MC1 | MC2 | MB | C1 | C2 | AE1 | AE2 |
	DTA1	DTA2		QC1	QC2	QF1	QF2		IC			
1	.00009	.00009	.00017	.49941	.50059	-.01864	-.01859	.00607	.57961	.57829	.1000E+01	.1000E+01
	.10000E+00	.99986E-01		.36913E-04	.36632E-04	.12399E-02	.12404E-02		4			
2	.00062	.00061	.00123	.49823	.50177	-.05594	-.05577	.01820	.58093	.57697	.1000E+01	.1000E+01
	.20000E+00	.19990E+00		.11086E-03	.10982E-03	.22909E-02	.22919E-02		4			
3	.00116	.00114	.00229	.49704	.50296	-.09326	-.09293	.03035	.58225	.57565	.1000E+01	.1000E+01
	.30000E+00	.29982E+00		.18486E-03	.18295E-03	.33417E-02	.33435E-02		4			
4	.00169	.00166	.00335	.49586	.50414	-.13058	-.13009	.04249	.58357	.57433	.1000E+01	.1000E+01
	.40000E+00	.39974E+00		.25892E-03	.25602E-03	.43925E-02	.43953E-02		4			

PLASTIC-ZONE ANALYSIS OF BEAM-COLUMNS AND PORTAL FRAMES

```
 5   .00223    .00218    .00441    .49468    .50532   -.16792   -.16724    .05464            .58489    .57301    .1000E+01  .1000E+01
     .50000E+00 .49965E+00 .33305E-03          .32904E-03 .54431E-02           .54471E-02        4                 .9930E+00  .9834E+00
 6   .00278    .00270    .00549    .49349    .50651   -.20555   -.20463    .06689            .58622    .57168    .9930E+00  .9834E+00
     .60000E+00 .59957E+00 .40788E-03          .40242E-03 .64918E-02           .64970E-02        4                 .9620E+00  .9498E+00
 7   .00332    .00321    .00652    .49232    .50768   -.24257   -.24118    .07893            .58752    .57037    .9620E+00  .9498E+00
     .70000E+00 .69949E+00 .48193E-03          .47346E-03 .75396E-02           .75453E-02        4                 .9289E+00  .9142E+00
 8   .00383    .00369    .00752    .49118    .50882   -.27882   -.27670    .09073            .58879    .56910    .9289E+00  .9142E+00
     .80000E+00 .79942E+00 .55497E-03          .54244E-03 .85828E-02           .85884E-02        4                 .8938E+00  .8764E+00
 9   .00433    .00413    .00846    .49008    .50992   -.31404   -.31083    .10219            .59002    .56787    .8938E+00  .8764E+00
     .90000E+00 .89935E+00 .62671E-03          .60769E-03 .96204E-02           .96272E-02        4                 .8564E+00  .8356E+00
10   .00479    .00453    .00933    .48902    .51098   -.34785   -.34350    .11319            .59120    .56669    .8564E+00  .8356E+00
     .10000E+01 .99928E+00 .69577E-03          .66997E-03 .10653E-01           .10660E-01        4                 .8157E+00  .7910E+00
11   .00522    .00490    .01012    .48802    .51198   -.38030   -.37440    .12375            .59232    .56556    .8157E+00  .7910E+00
     .11000E+01 .10992E+01 .76293E-03          .72795E-03 .11679E-01           .11685E-01        4                 .7706E+00  .7406E+00
12   .00562    .00521    .01082    .48707    .51293   -.41102   -.40316    .13375            .59337    .56451    .7706E+00  .7406E+00
     .12000E+01 .11992E+01 .82745E-03          .78089E-03 .12696E-01           .12701E-01        4                 .7185E+00  .6800E+00
13   .00595    .00545    .01141    .48621    .51379   -.43954   -.42903    .14303            .59433    .56354    .7185E+00  .6800E+00
     .13000E+01 .12991E+01 .88903E-03          .83099E-03 .13702E-01           .13700E-01        4                 .6523E+00  .6129E+00
14   .00622    .00560    .01182    .48546    .51454   -.46510   -.45084    .15135            .59517    .56270    .6523E+00  .6129E+00
     .14000E+01 .13991E+01 .94684E-03          .86223E-03 .14689E-01           .14676E-01        4                 .6058E+00  .5901E+00
15   .00638    .00563    .01201    .48485    .51515   -.48606   -.46798    .15817            .59585    .56202    .6058E+00  .5901E+00
     .15000E+01 .14991E+01 .99523E-03          .88844E-03 .15645E-01           .15614E-01        4                 .5833E+00  .5698E+00
16   .00645    .00558    .01202    .48437    .51563   -.50303   -.48142    .16369            .59638    .56148    .5833E+00  .5698E+00
     .16000E+01 .15991E+01 .10362E-02          .90743E-03 .16570E-01           .16526E-01        4                 .5653E+00  .5518E+00
17   .00643    .00546    .01190    .48398    .51602   -.51677   -.49231    .16816            .59682    .56105    .5653E+00  .5518E+00
     .17000E+01 .16991E+01 .10687E-02          .92620E-03 .17473E-01           .17419E-01        4                 .5473E+00  .5360E+00
18   .00636    .00530    .01166    .48365    .51635   -.52823   -.50110    .17189            .59718    .56069    .5473E+00  .5360E+00
     .18000E+01 .17992E+01 .10955E-02          .93565E-03 .18359E-01           .18300E-01        4                 .5338E+00  .5225E+00
19   .00624    .00510    .01134    .48339    .51661   -.53760   -.50833    .17494            .59747    .56039    .5338E+00  .5225E+00
     .19000E+01 .18992E+01 .11199E-02          .94694E-03 .19232E-01           .19168E-01        4                 .5203E+00  .5113E+00
20   .00609    .00487    .01096    .48317    .51683   -.54542   -.51421    .17748            .59771    .56015    .5203E+00  .5113E+00
     .20000E+01 .19992E+01 .11336E-02          .95032E-03 .20094E-01           .20029E-01        4                 .5090E+00  .5023E+00
21   .00590    .00462    .01052    .48299    .51701   -.55202   -.51923    .17963            .59792    .55994    .5090E+00  .5023E+00
     .21000E+01 .20993E+01 .11334E-02          .96099E-03 .20946E-01           .20878E-01        4

HMAX=    .01202
```

2. Sample problem 2 for program FRAMH

Input data from file d3

```
36.0 29000.0 0.3 1.0 0.2
8.0 0.28 8.0 0.43 140.0
16.26 0.38 7.07 0.63 210.0
100 10 11 11
```

Output data

```
              SECOND ORDER PLASTIC ZONE ANALYSIS
          FOR PORTAL UNBRACED FRAME WITH H & PINNED SUPPORTS
BENDING ABOUT STRONG AXIS OF I SECTION OF COLUMNS    BEAM

VERTICAL LOAD PT-BAT         .2000      1.00
VERTICAL LOAD PT-BAT         .2000      1.00
FY-E-RESID-EO1-EO2       36.0 29000.0       .3    .0010
BEAM-D-W-B-T-L              16.260         .380    7.070   .0010   210.000
COLUMN-D-W-B-T-L             8.000         .280    8.000   .630    140.000
NW-NF-IJC-IJB-NSTOP         100    10     11      .430
H11-H22-DDH                 .1000E-01  .1000E-01  .2000E-02   5

H1 H2     = SHEAR FORCES AT ENDS OF LEFT & RIGHT COLUMNS
H         = HORIZONTAL LOAD AT TOP OF LEFT COLUMN
P1 P2     = AXIAL FORCES OF LEFT & RIGHT COLUMNS
MC1 MC2   = MOMENTS AT TOP OF LEFT & RIGHT COLUMNS
MB        = MAXIMUM MOMENT IN THE BEAM
C1 C2     = FULL PLASTIC MOMENTS OF LEFT & RIGHT COLUMNS INCLUDING EFFECT OF AXIAL FORCES
AE1 AE2   = MINIMUM ELASTIC AREA RATIO OF LEFT & RIGHT COLUMNS
DATA1 DTA2 = DISPLACEMENTS AT TOP OF LEFT & RIGHT COLUMNS
QC1 QC2   = ROTATIONS AT TOP OF LEFT & RIGHT COLUMNS
QF1 QF2   = ROTATIONS AT BOTTOM OF LEFT & RIGHT COLUMNS
IC        = TIMES OF ITERATION IN SUB. BROTD

 N     H1       H2         H        P1       P2       MC1       MC2       MB       C1       C2     AE1       AE2
      DTA1     DTA2                QC1      QC2      QF1       QF2                 IC
 1  -.01265   .01381    .00116   .19904   .20096   .52515   -.58559   .84389   .90581   .90399  .9598E+00  .8988E+00
    .10000E+00  .97812E-01  .92223E-02  -.90956E-02  -.35607E-02  .57501E-02   4
 2  -.00879   .01706    .00827   .19335   .20665   .33443   -.75339   .84762   .91113   .89847  .1000E+01  .6081E+00
    .60000E+00  .59730E+00  .96541E-01  -.88914E-02   .16114E-02  .10797E-01   4

BEAM STOP NO1-T1   13   .00000E+00
SUGGESTION: CHANGE THE DDT OR DDH AND TRY AGAIN!
```

3. Sample problem 3 for program FRAMH

Input data from file d3

```
36.0 29000.0 0.3 1.0 0.2
8.0 0.28 8.0 0.43 140.0
16.26 0.38 7.07 0.63 210.0
100 10 11 11
```

Output data

```
                 SECOND ORDER PLASTIC ZONE ANALYSIS
           FOR PORTAL UNBRACED FRAME WITH H & PINNED SUPPORTS

BENDING ABOUT STRONG AXIS OF I SECTION OF COLUMNS & BEAM

VERTICAL LOAD PT-BAT           .2000     1.00
FY-E-RESID-EO1-EO2         36.0 29000.0      .3    .0010    .0010
BEAM-D-W-B-T-L              16.260    .380        7.070     .630    210.000
COLUMN-D-W-B-T-L             8.000    .280        8.000     .430    140.000
NW-NF-IJC-IJB-NSTOP           100       10          11      11      5
H11-H22-DDH                 .1000E-01  .1000E-01  .2000E-02

H1 H2    = SHEAR FORCES AT ENDS OF LEFT & RIGHT COLUMNS
H        = HORIZONTAL LOAD AT TOP OF LEFT COLUMN
P1 P2    = AXIAL FORCES OF LEFT & RIGHT COLUMNS
MC1 MC2  = MOMENTS AT TOP OF LEFT & RIGHT COLUMNS
MB       = MAXIMUM MOMENT IN THE BEAM
C1 C2    = FULL PLASTIC MOMENTS OF LEFT & RIGHT COLUMNS INCLUDING EFFECT OF AXIAL FORCES
AE1 AE2  = MINIMUM ELASTIC AREA RATIO OF LEFT & RIGHT COLUMNS
DATA1 DTA2 = DISPLACEMENTS AT TOP OF LEFT & RIGHT COLUMNS
QC1 QC2  = ROTATIONS AT TOP OF LEFT & RIGHT COLUMNS
QF1 QF2  = ROTATIONS AT BOTTOM OF LEFT & RIGHT COLUMNS
IC       = TIMES OF ITERATION IN SUB. BROYD
```

N	H1	H2	H	P1	P2	MC1	MC2	MB	C1	C2	AE1	AE2
	DTA1	DTA2		QC1	QC2	QF1	QF2		IC			
1	-.01265	.01381	.00116	.19904	.20096	.52515	-.58559	.84389	.90581	.90399	.9598E+00	.8988E+00
	.10000E+00	.97812E-01	.92223E-02			-.35607E-02	.57501E-02		4			
2	-.01187	.01454	.00267	.19784	.20216	.48654	-.62234	.84419	.90694	.90284	.9912E+00	.8558E+00
	.20000E+00	.19770E+00	.92980E-02			-.25219E-02	.67776E-02		4			
3	-.01110	.01524	.00415	.19667	.20333	.44813	-.65806	.84463	.90805	.90171	.1000E+01	.8074E+00
	.30000E+00	.29758E+00	.93814E-02			-.14853E-02	.77998E-02		4			
4	-.01032	.01591	.00558	.19552	.20448	.40998	-.69212	.84529	.90912	.90060	.1000E+01	.7511E+00
	.40000E+00	.39748E+00	.94679E-02			-.45070E-03	.88120E-02		4			
5	-.00956	.01651	.00695	.19441	.20559	.37208	-.72385	.84630	.91014	.89952	.1000E+01	.6784E+00
	.50000E+00	.49738E+00	.95589E-02			.58147E-03	.98095E-02		4			
6	-.00880	.01701	.00821	.19339	.20661	.33471	-.75103	.84796	.91109	.89851	.1000E+01	.6103E+00
	.60000E+00	.59731E+00	.96586E-02			.16091E-02	.10775E-01		4			

256 ADVANCED ANALYSIS OF STEEL FRAMES

7	-.00806	.01741	.00935	.19243	.20757	.29779	-.77422	.85019	.91196	.89757	.1000E+01	.5901E+00
	.70000E+00	.69724E+00	.97660E-02		-.88121E-02	.26328E-02		.11713E-01	4			
8	-.00732	.01772	.01041	.19154	.20846	.26128	-.79398	.85291	.91278	.89669	.1000E+01	.5698E+00
	.80000E+00	.79719E+00	.98803E-02		-.88139E-02	.36529E-02		.12627E-01	4			
9	-.00659	.01795	.01136	.19072	.20928	.22524	-.80991	.85619	.91353	.89587	.1000E+01	.5518E+00
	.90000E+00	.89716E+00	.10002E-01		-.88310E-02	.46690E-02		.13517E-01	4			
10	-.00587	.01811	.01224	.18994	.21006	.18954	-.82309	.85985	.91423	.89510	.1000E+01	.5358E+00
	.10000E+01	.99713E+00	.10130E-01		-.88622E-02	.56821E-02		.14388E-01	4			
11	-.00516	.01822	.01306	.18920	.21080	.15414	-.83420	.86380	.91490	.89436	.1000E+01	.5174E+00
	.11000E+01	.10971E+01	.10263E-01		-.88986E-02	.66927E-02		.15244E-01	4			
12	-.00445	.01828	.01383	.18849	.21151	.11899	-.84352	.86800	.91553	.89365	.1000E+01	.4969E+00
	.12000E+01	.11971E+01	.10400E-01		-.89438E-02	.77010E-02		.16085E-01	4			
13	-.00375	.01831	.01456	.18782	.21218	.08412	-.85104	.87245	.91613	.89297	.1000E+01	.4715E+00
	.13000E+01	.12971E+01	.10542E-01		-.89984E-02	.87068E-02		.16912E-01	4			
14	-.00305	.01830	.01525	.18717	.21283	.04948	-.85730	.87707	.91671	.89231	.1000E+01	.4487E+00
	.14000E+01	.13971E+01	.10688E-01		-.90593E-02	.97107E-02		.17728E-01	4			
15	-.00236	.01826	.01590	.18655	.21345	.01509	-.86238	.88184	.91727	.89168	.1000E+01	.4249E+00
	.15000E+01	.14971E+01	.10838E-01		-.91266E-02	.10712E-01		.18533E-01	4			
16	-.00167	.01820	.01653	.18594	.21406	-.01910	-.86652	.88673	.91781	.89106	.1000E+01	.4009E+00
	.16000E+01	.15971E+01	.10991E-01		-.91994E-02	.11712E-01		.19330E-01	4			
17	-.00099	.01812	.01713	.18535	.21465	-.05301	-.86978	.89172	.91833	.89046	.1000E+01	.3788E+00
	.17000E+01	.16971E+01	.11149E-01		-.92703E-02	.12710E-01		.20118E-01	4			
18	-.00031	.01801	.01770	.18477	.21523	-.08668	-.87229	.89679	.91883	.88987	.1000E+01	.3571E+00
	.18000E+01	.17971E+01	.11311E-01		-.93589E-02	.13705E-01		.20899E-01	4			
19	.00036	.01790	.01826	.18421	.21579	-.12011	-.87434	.90189	.91933	.88929	.1000E+01	.3379E+00
	.19000E+01	.18972E+01	.11477E-01		-.94511E-02	.14699E-01		.21675E-01	4			
20	.00103	.01778	.01880	.18366	.21634	-.15325	-.87595	.90702	.91981	.88873	.1000E+01	.3185E+00
	.20000E+01	.19972E+01	.11648E-01		-.95451E-02	.15690E-01		.22447E-01	4			
21	.00168	.01765	.01933	.18312	.21688	-.18604	-.87719	.91215	.92028	.88817	.1000E+01	.2994E+00
	.21000E+01	.20972E+01	.11824E-01		-.96449E-02	.16678E-01		.23216E-01	4			
22	.00233	.01751	.01984	.18259	.21741	-.21846	-.87819	.91727	.92074	.88762	.1000E+01	.2801E+00
	.22000E+01	.21972E+01	.12007E-01		-.97504E-02	.17662E-01		.23980E-01	4			
23	.00297	.01736	.02033	.18207	.21793	-.25045	-.87896	.92235	.92119	.88708	.1000E+01	.2609E+00
	.23000E+01	.22972E+01	.12197E-01		-.98568E-02	.18644E-01		.24742E-01	4			
24	.00360	.01721	.02081	.18156	.21844	-.28193	-.87936	.92740	.92163	.88656	.1000E+01	.2448E+00
	.24000E+01	.23973E+01	.12395E-01		-.99833E-02	.19620E-01		.25500E-01	4			
25	.00422	.01705	.02127	.18106	.21894	-.31291	-.87979	.93237	.92206	.88604	.1000E+01	.2291E+00
	.25000E+01	.24973E+01	.12602E-01		-.10184E-01	.20593E-01		.26262E-01	7			

26	.00482	.01689	.02171	.18058	.21942	−.34316	−.88003	.93726	.92248	.88554	.1000E+01	.2136E+00
	.26000E+01	.25973E+01	.12820E-01		−.10345E-01	.21559E-01		.27019E-01	8			
27	.00539	.01673	.02212	.18011	.21989	−.37214	−.88010	.94196	.92287	.88505	.1000E+01	.2050E+00
	.27000E+01	.26973E+01	.13060E-01		−.10403E-01	.22515E-01		.27770E-01	4			
28	.00592	.01656	.02248	.17968	.22032	−.39940	−.88019	.94703	.92325	.88460	.1000E+01	.1936E+00
	.28000E+01	.27974E+01	.13327E-01		−.10584E-01	.23457E-01		.28526E-01	5			
29	.00640	.01640	.02280	.17928	.22072	−.42446	−.88024	.95192	.92359	.88418	.1000E+01	.1833E+00
	.29000E+01	.28974E+01	.13631E-01		−.10794E-01	.24380E-01		.29282E-01	4			
30	.00684	.01623	.02307	.17891	.22109	−.44807	−.88032	.95652	.92391	.88379	.1000E+01	.1726E+00
	.30000E+01	.29974E+01	.13959E-01		−.11022E-01	.25291E-01		.30039E-01	4			
31	.00725	.01607	.02332	.17855	.22145	−.47052	−.88027	.96091	.92421	.88341	.1000E+01	.1633E+00
	.31000E+01	.30975E+01	.14306E-01		−.11357E-01	.26192E-01		.30798E-01	6			
32	.00764	.01590	.02355	.17821	.22179	−.49193	−.88031	.96508	.92450	.88305	.9968E+00	.1550E+00
	.32000E+01	.31975E+01	.14673E-01		−.11532E-01	.27082E-01		.31552E-01	6			
33	.00799	.01573	.02373	.17790	.22210	−.51157	−.88015	.96894	.92476	.88273	.9876E+00	.1490E+00
	.33000E+01	.32975E+01	.15065E-01		−.11835E-01	.27960E-01		.32208E-01	5			
34	.00831	.01557	.02387	.17761	.22239	−.52991	−.88010	.97253	.92500	.88242	.9739E+00	.1412E+00
	.34000E+01	.33975E+01	.15478E-01		−.12125E-01	.28827E-01		.33066E-01	4			
35	.00859	.01540	.02399	.17734	.22266	−.54680	−.88002	.97584	.92523	.88214	.9590E+00	.1367E+00
	.35000E+01	.34976E+01	.15914E-01		−.12484E-01	.29682E-01		.33823E-01	5			
36	.00883	.01523	.02406	.17710	.22290	−.56197	−.87994	.97881	.92543	.88188	.9452E+00	.1304E+00
	.36000E+01	.35976E+01	.16377E-01		−.12682E-01	.30522E-01		.34576E-01	4			
37	.00904	.01507	.02410	.17689	.22311	−.57572	−.87988	.98150	.92561	.88165	.9322E+00	.1255E+00
	.37000E+01	.36976E+01	.16862E-01		−.13150E-01	.31351E-01		.35338E-01	4			

```
IT= -1
INTERRUPT DUE TO PLASTIC HINGE FORMED AT THE TOP OF RIGHT COLUMN
HMAX=  .02410
```

6: Plastic-Zone Analysis of Frames

Murray J. Clarke, *School of Civil and Mining Engineering, The University of Sydney, New South Wales, Australia*

6.1 Introduction

In the previous chapter, the plastic-zone analysis of beam-columns and portal frames was discussed, and the associated computer programs were presented. The method of analysis was based on numerical integration of the differential equation of equilibrium of a beam-column using Newmark's integration procedure (Newmark, 1943).

In this chapter, a refined plastic-zone analysis based on the finite element method is described. The finite element method is well established in structural mechanics as a general and versatile analytical technique that is readily extended from linear elastic analysis to the rigorous inclusion of geometric and material nonlinear effects. The plastic-zone finite element analysis presented in this chapter is applicable to the in-plane analysis of framed structures of arbitrary geometry where the behavior of the members can be described adequately by beam-column theory. The nonlinear analysis can be used to solve in-plane stability problems of columns, beam-columns, arches, and frames.

A finite element computer program, NIFA (*N*onlinear *I*nelastic *F*rame *A*nalysis), based on the theory described in this chapter has been written by the author in the FORTRAN 77 language. However, due to the size and complexity of the program, and also because the proper use of the program requires considerable experience and expertise, a program listing and user's manual is not included with this book. However, the theoretical basis of the program is described fully in this chapter, and several examples illustrating the capability and accuracy of the program are also provided.

In the development of computer programs for the second-order inelastic analysis of frames, there are two basic approaches to material inelasticity which have been used (Wright and Gaylord, 1968 and SSRC, 1988). In the first approach, which is termed *concentrated plasticity*, *plastic-hinge*, or *elastic-plastic analysis*, it is assumed that plastic hinges form at discrete points in the elements that comprise the model, with the structure remaining elastic between the plastic hinges. In the second approach to plasticity, the spread of inelasticity through the cross-sections and along the lengths of members is modeled. This type of analysis was termed *compatibility analysis* by Wright and Gaylord (1968) but has since been given other terms by various researchers, including *distributed plasticity analysis*, *plastic-zone analysis*, *spread-of-plasticity analysis*, and *elasto-plastic analysis*. In this chapter, this type of analysis in which the spread of plasticity is considered is termed *plastic-zone analysis*. A refined plastic-zone analysis is a useful research tool that can be used to calibrate and check the accuracy of the simplified methods of inelastic analysis, such as those based on the plastic hinge concept.

There are two main approaches which have been used by researchers to model the partial plastification of members in a second-order inelastic analysis, although in both cases the member needs to be subdivided into several elements along its length to model the behavior accurately. Cross-sectional behavior can be described by moment-thrust-curvature (M-P-Φ) relations (Wright and Gaylord, 1968) derived for each cross-section in the analysis. The M-P-Φ relations may be approximated by closed-form expressions. Alternatively, the cross-sections can be subdivided into elemental areas and the state of strain, stress, and yield stress monitored explicitly for each of the elemental areas during the analysis. One of the first analyses of the latter type appears to be that of Alvarez and Birnstiel (1969); later analyses include those of Swanger and Emkin (1979) who developed a three-dimensional analysis, White (1985), Kitipornchai et al. (1988), Chan (1988), and Bild and Trahair (1989).

6.2 Historical Sketch

An extensive research program on inelastic analysis in limit states design using a concentrated plasticity formulation has been undertaken at Cornell University under the direction of McGuire (1988). McGuire and his co-workers have developed analytical tools embedded in a user-friendly graphical interface capable of analyzing two- and three-dimensional frames composed of compact sections (Ziemian et al., 1990). An updated Lagrangian formulation is used to account for the geometric nonlinearity. The formation of plastic hinges at the ends of elements is governed by an interaction equation which is a function of the axial force and the major and minor axis bending moments. The gradual softening along the length of a beam-column due to residual stresses is approximated by an equivalent tangent modulus. The concentrated plastic hinge theory in force-displacement space described by Orbison et al. (1982) was based on a consistent extension of the concepts of plasticity theory in stress-strain space and resulted in the formulation of a plastic reduction matrix that accounts for the presence of plastic hinges. The generalized concentrated plasticity theory adopted by Orbison et al. (1982) was based on earlier work by Nigam (1970), and Porter and Powell (1971). In a recent paper by Deierlein et al. (1991), the concentrated plasticity model based on a single yield surface was extended to a two-surface model to represent the gradual plastification of members from first yield to the fully plastic cross-sectional strength. The second-order inelastic analysis has also recently been extended to include the effects of semi-rigid connections on frame behavior (Hsieh and Deierlein, 1991).

Methods of analysis for which the simple plastic hinge concept has been modified to include implicitly the effects of residual stresses, gradual yielding, and also possibly member geometrical imperfections by the proper calibration of behavioral models, are termed modified plastic hinge analyses or refined plastic-hinge analyses. Apart from the recently described Cornell research, refined plastic-hinge beam-column elements have also been presented by Powell and Chen (1986) and King et al. (1992).

Over the years, several methods of analysis have been presented in which the effects of partial plastification of members are considered through the cross-sectional moment-thrust-curvature (M-P-Φ) behavior. These methods of analysis are more refined than methods based on plastic hinges. A slight inconvenience of the method is that the effects of residual stresses on M-P-Φ behavior differ for tensile and compressive axial force. Separate M-P-Φ relations are therefore required for tensile and compressive loading if residual stresses are to be considered.

One of the earliest studies of inelastic frame behavior including the spread of plasticity was by Chu and Pabarcius (1964). The M-P-Φ relations, derived for a W8 × 31 section and expressed in nondimensional form, were used in conjunction with a numerical integration technique to determine the load-deflection behavior and ultimate loads of portal frames. Procedures for the inelastic analysis of portal frames with hinged bases considering spread of inelastic zones and the effects of axial force were also presented by Moses (1964).

Wright and Gaylord (1968) presented a rigorous compatibility analysis of regular multistory frames under nonproportional loads to take the P-Δ effect into account. The analysis permitted consideration of extended regions of plastic deformation by using M-P-Φ relations and dividing each member of the frame into several equal segments. The finite difference method was used to determine the effective stiffness coefficients of the partially plastic members for incorporation in the frame stiffness matrix of an equivalent linear elastic frame. A system of fictitious elastic supports was introduced to aid in obtaining the unloading portion of the load-deformation response.

Second-order elastic-plastic frame analyses that utilize M-P-Φ data to model the partial plastification of the cross-section generally provide an improved prediction of behavior over the discrete plastic hinge methods. However, M-P-Φ data, even if expressed in a nondimensional form, are theoretically unique for a particular cross-sectional geometry and stress-strain characteristic, and separate relations are required for tension and compression if the effects of residual stresses wish to be considered. A further subtlety is that the inherent path dependence of an inelastic analysis, and the concept of elastic unloading, cannot be captured rigorously in M-P-Φ relations.

To overcome these aforementioned disadvantages, various researchers have developed nonlinear analyses in which the cross-sections are subdivided as part of the nonlinear analysis, and only a knowledge of the fundamental stress-strain relation is required as input to the analysis. Such analyses may be regarded as the most refined possible, although also the most computationally intensive, for investigating the full-range behavior of steel frames within the confines of beam-column theory. Analyses of this type are variously referred to in the literature as distributed plasticity analysis, plastic-zone analysis, spread-of-plasticity analysis, and elasto-plastic analysis. These analyses are characterized by explicit modeling of the spread of yielding within the members of a frame. Other factors that affect strength and stability, such as residual stresses and geometrical imperfections, are also modeled explicitly. Compared to the plastic-hinge approaches, plastic-zone analyses are more generally applicable to cross-sectional shapes of arbitrary geometry, and the effects of specialized stress-strain behavior and distributions of residual stresses, such as those occurring in cold-formed tubes, are more readily accommodated. Elastic unloading is also readily included in analyses of the plastic-zone type, since the member cross-sections are subdivided into a grid of elemental areas and the state of stress and strain is tracked explicitly at each point incrementally throughout the analysis. A refined plastic-zone analysis is a useful research tool which can be used to calibrate and check the accuracy of simplified methods of inelastic analysis, such as the modified plastic-hinge methods. The finite element nonlinear analysis described in this chapter utilizes a plastic-zone (distributed plasticity) formulation.

One of the earliest presentations of a refined plasticity analysis of steel frames was by Alvarez and Birnstiel (1969), who included the spread of inelastic zones within a member, and strain reversal. Their analysis was restricted to planar rectangular frames composed of I-sections bent about their major axis. The fundamental material behavior was assumed to be elastic-perfectly plastic. A distributed plasticity analysis for three-dimensional structures was presented by Swanger and Emkin (1979), wherein it was termed the fiber element model. A more recent distributed inelasticity analysis based on the finite element method and suitable for analyzing three-dimensional structures was presented by Kitipornchai et al. (1988), Chan and Kitipornchai (1988), and Chan (1988). In their work, particular emphasis was placed on the analysis of tubular beam-columns and struts. The material behavior was assumed to be elastic-perfectly plastic and the effects of strain unloading were considered for accurate prediction of the post-buckling behavior.

6.3 Finite Element Formulation

6.3.1 General

In the finite element nonlinear analysis described henceforth, a curved line element (arch element) is used. The curved element geometrical description and displacement fields are a simplification of those employed by Teng and Rotter (1989) for the finite element analysis of axisymmetric thin shells. As a consequence of the curved element geometry and the distributed plasticity formulation, the element stiffness matrices and nodal residual force vectors are integrated numerically along the element length using three-point Gaussian quadrature.

The assumptions employed in the analysis are generally those of conventional beam-column and shallow arch theory, and are summarized as follows:

(1) the beam thickness is small compared to its length and radius of curvature;
(2) the normal strain perpendicular to the beam axis is negligible, thus the cross-section preserves its shape during deformation;

(3) normals to the beam axis remain plane and normal during the deformation, so that the strain at any point in the beam can be expressed in terms of the membrane strain and curvature (Bernoulli-Euler hypothesis);
(4) the strains are small.

In the *Lagrangian formulations* of the continuum mechanics equations of motion, all the static and kinematic variables are referred to a *reference configuration*. The following description of the plastic-zone analysis is based on the *total Lagrangian formulation*, which means that the reference configuration corresponds to the original, undeformed configuration. Different kinematic formulations of the plastic-zone analysis, including the *updated Lagrangian* and *co-rotational* formulations (Bathe, 1982; Mattiasson and Samuelsson, 1984; and Mattiasson et al., 1986) are described fully in Clarke (1992).

The stress and strain measures utilized in Lagrangian geometric nonlinearity are the *second Piola-Kirchhoff* stress tensor and the *Green-Lagrange* strain tensor (Bathe, 1982 and Fung, 1965). These stress and strain tensors are energy-conjugate and possess the property of objectivity which means that their components are invariant under rigid body motion of the material. Also, for small strains, the components of the second Piola-Kirchhoff stress tensor and the Green-Lagrange strain tensor have a physical significance close to that of the conventional engineering stress and strain measures. Elasto-plastic constitutive models developed using engineering stress and strain measures are therefore directly applicable to the total Lagrangian formulation, provided the strains are small. A detailed treatment of stress and strain tensors can be found in Bathe (1982) and Fung (1965).

The finite element formulation process involves two basic steps, which are:

(1) Derivation of a set of governing equations that describe the nonlinear problem. This step is not in fact unique to the finite element process, but is common to all problems of structural mechanics. Derivation of the governing equations involves the following procedures, which are described in detail in Sections 6.3.2 to 6.3.4:
  (a) selection of an appropriate theory to relate strains to displacements (Section 6.3.2);
  (b) adoption of an appropriate constitutive model to deduce stresses from strains (Section 6.3.3);
  (c) application of the principle of virtual displacements to derive the governing equilibrium equations (Section 6.3.4).
(2) Solution of these nonlinear equations by the finite element process involving discretization of the structure and adoption of nodal displacements as the fundamental unknowns. Solution of the governing equations by the finite element method involves the following procedures:
  (a) selection of an appropriate finite element and set of nodal displacements;
  (b) definition of the displacement functions expressing general element displacements in terms of nodal displacements;

(c) expression of the governing equilibrium equations in terms of nodal displacements;
(d) solution of the resulting nonlinear equations for the nodal displacements.

Steps (a), (b), and (c) above are described in Section 6.3.5. The basis of step (d), involving derivation of the tangent stiffness matrix, is given in Section 6.3.6.

6.3.2 Strain-Displacement Relations

For in-plane beam-column and arch problems, the only strain component of relevance is the longitudinal strain, denoted here by ε_s, which may be expressed

$$\varepsilon_s = \left(\bar{\varepsilon}_s - \bar{\varepsilon}_{s0}\right) + y\left(\kappa_s - \kappa_{s0}\right) \tag{6.1}$$

or, alternatively in matrix and vector notation as

$$\varepsilon_s = [1\ y]\left(\begin{Bmatrix}\bar{\varepsilon}_s \\ \kappa_s\end{Bmatrix} - \begin{Bmatrix}\bar{\varepsilon}_{s0} \\ \kappa_{s0}\end{Bmatrix}\right) \tag{6.2}$$

$$= [Y]\left(\{\varepsilon\} - \{\varepsilon_0\}\right) \tag{6.3}$$

in which y is the distance above the reference line in the cross-section (see Figure 6.1), $\bar{\varepsilon}_s$ and κ_s are the membrane strain and curvature, respectively, of the reference line that are caused by deformations, and $\bar{\varepsilon}_{s0}$ and κ_{s0} are the generalized *initial strains*. The *generalized* strain vectors for strains related to deformations, and strains related to prescribed initial strains, are denoted $\{\varepsilon\}$ and $\{\varepsilon_0\}$, respectively.

The Green-Lagrange strain measure of the reference line is defined by

$$\bar{\varepsilon}_s = \left(\frac{d\bar{u}}{ds} + \bar{v}\frac{d\phi}{ds}\right) + \frac{1}{2}\left(\frac{d\bar{u}}{ds} + \bar{v}\frac{d\phi}{ds}\right)^2 + \frac{1}{2}\left(\frac{d\bar{v}}{ds} - \bar{u}\frac{d\phi}{ds}\right)^2 \tag{6.4}$$

and the curvature of the reference line is given for moderate rotations by

$$\kappa_s = -\frac{d}{ds}\left(\frac{d\bar{v}}{ds} - \bar{u}\frac{d\phi}{ds}\right) \tag{6.5}$$

$$= -\frac{d^2\bar{v}}{ds^2} + \frac{d\bar{u}}{ds}\frac{d\phi}{ds} + \bar{u}\frac{d^2\phi}{ds^2} \tag{6.6}$$

in which $\bar{u}$, $\bar{v}$, and ϕ are as defined in Figure 6.1, and all derivatives are evaluated with respect to the initial configuration.

The strain-displacement relations defined by Eqs. (6.4) to (6.6) above include the effect of the initial element curvature, and are a simplification of those employed by Teng and Rotter (1989) for the analysis of axisymmetric thin shells in a total Lagrangian framework. The bold term in Eq. (6.4) is insignificant for small rotations

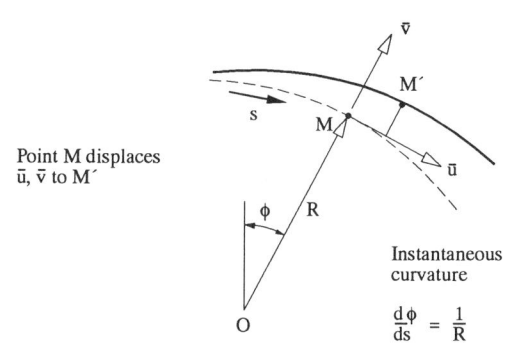

FIGURE 6.1. Definition of terms used in strain-displacement relations for a curved element.

and is therefore often neglected by researchers. In the present formulation, however, the complete Green-Lagrange strain expression has been retained in Eq. (6.4), since the resulting strain measure can then represent a rigid body rotation without the spurious straining that would occur if the bold term in Eq. (6.4) was omitted.

To express the above strain-displacement relations in vector and matrix notation, the generalized strain vector $\{\varepsilon\}$ can be expressed as the sum of linear and nonlinear components as

$$\{\varepsilon\} = \{e\} + \{\eta\} \tag{6.7}$$

in which

$$\{\varepsilon\} = \begin{Bmatrix} \bar{\varepsilon}_s \\ \kappa_s \end{Bmatrix} \tag{6.8}$$

is defined in Eq. (6.2),

$$\{e\} = \begin{Bmatrix} \theta_{us} \\ -d\theta_{vs}/ds \end{Bmatrix} \tag{6.9}$$

$$\{\eta\} = \frac{1}{2} \begin{bmatrix} \theta_{us} & \theta_{vs} \\ 0 & 0 \end{bmatrix} \begin{Bmatrix} \theta_{us} \\ \theta_{vs} \end{Bmatrix} \tag{6.10}$$

$$= \frac{1}{2}[A]\{\theta\} \tag{6.11}$$

and the displacement gradients θ_{us} and θ_{vs} are defined by

$$\theta_{us} = \frac{d\bar{u}}{ds} + \bar{v}\frac{d\phi}{ds} \tag{6.12}$$

$$\theta_{vs} = \frac{d\bar{v}}{ds} - \bar{u}\frac{d\phi}{ds} \tag{6.13}$$

Taking a variation of Eq. (6.7) with respect to displacements results in

$$d\{\varepsilon\} = d\{e\} + [A]d\{\theta\} \qquad (6.14)$$

since it can be shown readily by expanding out terms that

$$d[A]\{\theta\} = [A]d\{\theta\} \qquad (6.15)$$

6.3.3 Stress-Strain Relations

A finite element nonlinear analysis is usually performed incrementally, and interest is therefore focused on the relationship between stress increments and strain increments. An *incremental* or *flow* theory of plasticity (Mendelson, 1968) has been adopted in the plastic-zone analysis described here. In the present work, only longitudinal strains (ε_s) are considered and so at any point in the cross-section, yielding is assumed to occur as a result of longitudinal stress (σ_s) only. In addition, the material stress-strain curve is assumed to be identical in tension and compression, and the yield surface after initial yielding is defined by the isotropic hardening rule (Mendelson, 1968 and Armen, 1979). The assumption of isotropic hardening is usually satisfactory provided the loading regime is static and does not lead to substantial elastic unloading.

Following the onset of yielding, an increment in longitudinal strain ($d\varepsilon_s$) can be considered to be composed of an elastic component ($d\varepsilon_{se}$) and a plastic component ($d\varepsilon_{sp}$),

$$d\varepsilon_s = d\varepsilon_{se} + d\varepsilon_{sp} \qquad (6.16)$$

By analogy with the terminology employed in plasticity theory for multi-axial stress states, the nonnegative equivalent plastic strain increment is defined by

$$d\bar{\varepsilon}_p = \sqrt{d\varepsilon_{sp}^2} \qquad (6.17)$$

and the nonnegative effective stress is defined by

$$\bar{\sigma} = \sqrt{\sigma_s^2} \qquad (6.18)$$

The initial and subsequent yield surfaces can be expressed in the form

$$F(\sigma_s, \kappa) = \bar{\sigma} - \sigma_Y = 0 \qquad (6.19)$$

in which κ is a 'hardening parameter', and σ_Y is the instantaneous yield stress. For isotropic hardening, κ can be interpreted as the total equivalent plastic strain $\bar{\varepsilon}_p = \int d\bar{\varepsilon}_p$. The instantaneous yield stress σ_Y is given for isotropic hardening by

$$\sigma_Y = \sigma_{Y0} + \int d\sigma_Y = \sigma_{Y0} + \int_0^{\bar{\varepsilon}_p} H' d\bar{\varepsilon}_p \qquad (6.20)$$

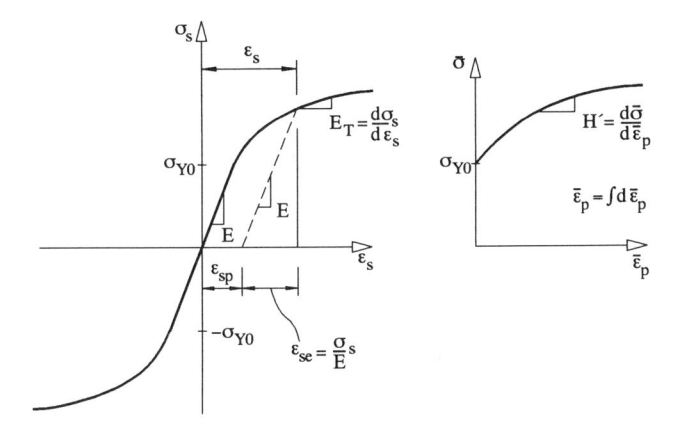

FIGURE 6.2. Uniaxial stress-strain relation and effective stress-equivalent plastic strain relation.

in which σ_{Y0} is the initial yield stress and H' is the slope of the effective stress-equivalent plastic strain curve, as shown in Figure 6.2. The parameter H' can be computed from

$$H' = \frac{EE_T}{E - E_T} \qquad (6.21)$$

in which E and E_T are the elastic and tangent moduli of the material, respectively.

Prior to yielding, or during elastic unloading from the yield surface, the strain increments $d\varepsilon_s$ are entirely elastic and the stress increments $d\sigma_s$ are computed simply as

$$d\sigma_s = E d\varepsilon_s \qquad (6.22)$$

in which E is the elastic modulus (Young's modulus) of the material. If the point is currently undergoing plastic straining, the incremental stress-strain relation can be expressed

$$d\sigma_s = E_T d\varepsilon_s \qquad (6.23)$$

in which E_T is the tangent modulus of the uniaxial stress-strain curve, computed as a function of the equivalent plastic strain.

Stress Resultants

For the in-plane analysis of arches and beam-columns, a vector of stress resultants $\{\sigma\}$ can be defined as

$$\{\sigma\} = \begin{Bmatrix} N \\ M \end{Bmatrix} = \int_A [Y]^T \sigma_s \, dA \qquad (6.24)$$

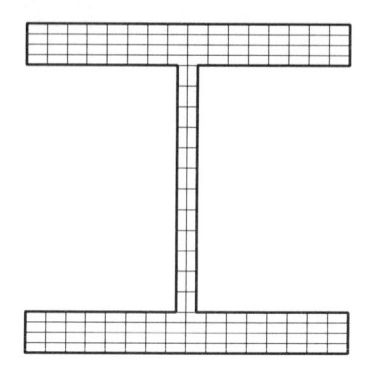

FIGURE 6.3. Cross-section discretization in the nonlinear analysis.

in which $[Y] = [1\ y]$ and A denotes cross-sectional area (assumed constant throughout the analysis). The stress resultants of axial force (N) and bending moment (M) are therefore defined by

$$N = \int_A \sigma_s\, dA \tag{6.25}$$

$$M = \int_A y\sigma_s\, dA \tag{6.26}$$

If the cross-section is partially yielded then the stress distribution is nonlinear and the integrals of Eqs. (6.25) and (6.26) are most easily calculated numerically. Further details on the cross-sectional integration procedure are given in Section 6.4.

Taking a variation of Eq. (6.24) furnishes

$$d\{\sigma\} = \int_A [Y]^T E_T [Y]\, dA\ d(\{\varepsilon\} - \{\varepsilon_0\}) \tag{6.27}$$

$$= [D_T] d(\{\varepsilon\} - \{\varepsilon_0\}) \tag{6.28}$$

in which $[D_T]$ is the tangent modulus matrix, given explicitly by

$$[D_T] = \begin{bmatrix} \int_A E_T\, dA & \int_A E_T y\, dA \\ \int_A E_T y\, dA & \int_A E_T y^2\, dA \end{bmatrix} \tag{6.29}$$

In Eq. (6.29), y is the coordinate in the cross-section above the element reference line. If the material behavior is elastic and the reference line is chosen as the centroidal axis, the tangent modulus matrix simplifies to the elastic constitutive matrix

$$[D] = \begin{bmatrix} EA & 0 \\ 0 & EI \end{bmatrix} \tag{6.30}$$

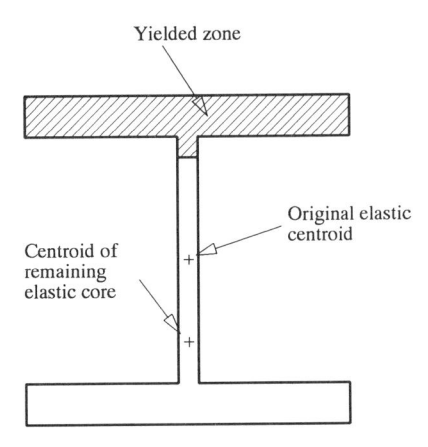

FIGURE 6.4. Shift in elastic centroid for a partially yielded section.

in which A and I are the cross-sectional area and second moment of area about the centroidal axis, respectively, of the considered cross-section. If the material behavior is inelastic and y in Eq. (6.29) is measured from the original elastic centroid (the geometric centroid), the off-diagonal terms in Eq. (6.29) become nonzero on account of the shift in position of the effective elastic centroid for a section which is partially yielded, as shown in Figure 6.4. Using the same cross-sectional subdivision that was used for computing the stress resultants [Eqs. (6.25) and (6.26)], the integrals in $[D_T]$ are also calculated numerically (see Section 6.4).

6.3.4 The Principle of Virtual Displacements

For a total Lagrangian formulation of geometrically nonlinear analysis with *elastic* material behavior, the equilibrium equations may be formulated equivalently either with the *principle of virtual displacements*, or the *principle of minimum total potential* Π (i.e., $\delta\Pi = 0$). Similarly, the incremental equilibrium equations can be formulated by taking a variation of the virtual work equation, or by the requirement that the second variation of the total potential vanishes (i.e., $\delta^2\Pi = 0$). However, for *inelastic* material behavior, the equilibrium equations are most expediently formulated using the principle of virtual displacements, and so this method has been adopted here.

In the total Lagrangian formulation, equilibrium is expressed using the principle of virtual displacements (Bathe, 1982) as

$$\int_V \sigma_s \delta\varepsilon_s \, dV = \pi \tag{6.31}$$

in which σ_s is the longitudinal stress, ε_s is the longitudinal strain, δ denotes 'variation in', and V denotes the original volume of the element in the undeformed configuration.

Since the variation in the strain component $\delta\varepsilon_s$ is equivalent to the use of virtual strains (Bathe, 1982), the left-hand side of Eq. (6.31) constitutes the *internal virtual work* performed when the element is subjected to a virtual displacement field δu_i. The term π represents the corresponding *external virtual work* performed in moving the external load system, consisting of body forces per unit mass q_i and surface tractions p_i (both defined in the original configuration), through the virtual displacement field δu_i, and is given by

$$\pi = \int_V \rho q_i \delta u_i \, dV + \int_S p_i \delta u_i \, dS \tag{6.32}$$

in which ρ is the mass density, and V and S are volume and surface area, respectively, all with reference to the original, undeformed configuration.

Using Eqs. (6.3) and (6.24), the virtual work equation [Eq. (6.31)] can be expressed in terms of *generalized* strains and stresses as

$$\int_s \delta\{\varepsilon\}^T \{\sigma\} \, ds = \int_s \delta\{u\}^T \{p\} \, ds \tag{6.33}$$

in which $\{u\}$ is the vector of displacements of any point on the reference axis, $\{p\}$ is the vector of distributed forces at any point on the reference axis, and s denotes arc-length along the reference axis. The next step in the finite element formulation is to discretize Eq. (6.33) so that the resulting equation is expressed in terms of *nodal* displacements.

6.3.5 Discretization of the Virtual Work Equation

Element Geometric Description

The element described here is termed isoparametric (Bathe, 1982; Zienkiewicz, 1977; and Cook, 1981) which means that the following two conditions are satisfied:

(1) nodal coordinates and nodal degrees of freedom are in one-to-one correspondence;
(2) the same shape functions are used to interpolate the coordinates of a point within the element from the nodal coordinates, and the displacements of a point within the element from the nodal displacements.

The curved eight-degree of freedom element used for the plastic-zone analysis is illustrated in Figure 6.5. The element geometry is defined in terms of a reference line. The element nodal coordinates are defined in a global Cartesian axis system by (X,Y) coordinates. The slope of the element is described by the angle ϕ and the rates of change of the X and Y coordinates with respect to arc-length s, given by dX/ds and dY/ds, respectively. The gradients dX/ds and dY/ds are not independent of ϕ, but are related by

PLASTIC-ZONE ANALYSIS OF FRAMES   **271**

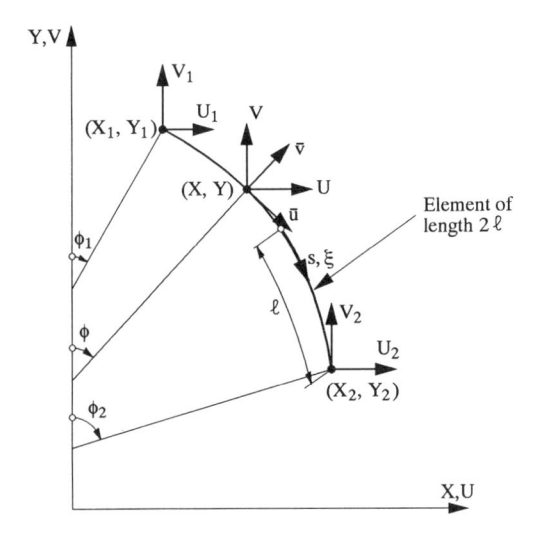

FIGURE 6.5. Geometry and local and global displacement fields of the element.

$$\frac{dX}{ds} = \cos\phi \tag{6.34}$$

$$\frac{dY}{ds} = -\sin\phi \tag{6.35}$$

The geometrical curvature of the element and its rate of change with respect to arc-length s in the reference configuration are described by $d\phi/ds$ and $d^2\phi/ds^2$ respectively. The sign convention for ϕ results in positive values of curvature $d\phi/ds$ for the element oriented as shown in Figure 6.5. This element description, although complex, enables the initial structure geometry to be defined precisely, and also has additional subtle advantages for the updated Lagrangian variant of the nonlinear analysis (Clarke, 1992).

The geometry within the element is interpolated from the nodal parameters using cubic Hermitian polynomial interpolating functions as

$$X(\xi) = \sum_{i=1}^{2} N_{0i}(\xi) X_i + N_{1i}(\xi)\left(\frac{dX}{ds}\right)_i \tag{6.36}$$

$$Y(\xi) = \sum_{i=1}^{2} N_{0i}(\xi) Y_i + N_{1i}(\xi)\left(\frac{dY}{ds}\right)_i \tag{6.37}$$

$$\frac{d\phi}{ds}(\xi) = \sum_{i=1}^{2} N_{0i}(\xi)\left(\frac{d\phi}{ds}\right)_i + N_{1i}(\xi)\left(\frac{d^2\phi}{ds^2}\right)_i \tag{6.38}$$

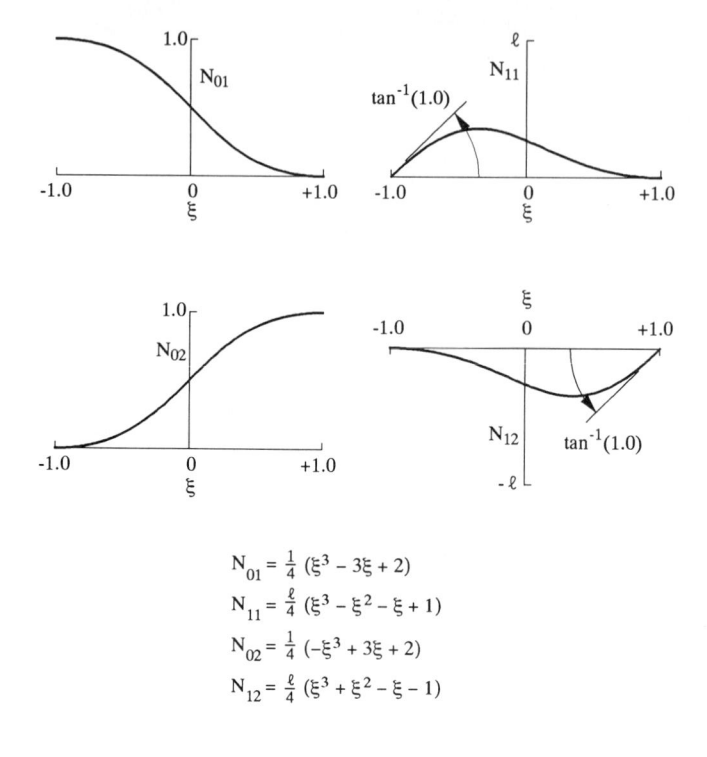

$$N_{01} = \tfrac{1}{4}(\xi^3 - 3\xi + 2)$$
$$N_{11} = \tfrac{l}{4}(\xi^3 - \xi^2 - \xi + 1)$$
$$N_{02} = \tfrac{1}{4}(-\xi^3 + 3\xi + 2)$$
$$N_{12} = \tfrac{l}{4}(\xi^3 + \xi^2 - \xi - 1)$$

FIGURE 6.6. Hermitian cubic polynomial functions.

in which N_{01}, N_{11}, N_{02}, N_{12} are the element shape functions, illustrated in Figure 6.6 and defined by

$$N_{01} = \frac{1}{4}\left(\xi^3 - 3\xi + 2\right) \tag{6.39}$$

$$N_{11} = \frac{l}{4}\left(\xi^3 - \xi^2 - \xi + 1\right) \tag{6.40}$$

$$N_{02} = \frac{1}{4}\left(-\xi^3 + 3\xi + 2\right) \tag{6.41}$$

$$N_{12} = \frac{l}{4}\left(\xi^3 + \xi^2 - \xi - 1\right) \tag{6.42}$$

In Eqs. (6.39) to (6.42), l is the half-length of the element in the initial configuration, and ξ is a dimensionless curvilinear coordinate ranging in value from -1 at node 1 of the element ($s = -l$) to $+1$ at node 2 of the element ($s = +l$), so that

$$s = l\xi \tag{6.43}$$

$$ds = l d\xi \tag{6.44}$$

A subtlety in this element geometric description is that the definition provided by Eqs. (6.36) to (6.38) is slightly nonunique in that the element curvature $d\phi/ds$ [Eq. (6.38) is defined as one of the parametric variables independently of the Cartesian coordinates X and Y [Eqs. (6.36) and (6.37)]. The alternative derivation of curvature from the parametric definitions of the X and Y coordinates may provide a less accurate measure of curvature within the element than the independent definition of Eq. (6.38), resulting in a less precise strain-displacement matrix (Teng and Rotter, 1989).

The element reference line local displacements in the curvilinear coordinate system consist of the local tangential displacement $\bar{u}$ and the local transverse displacement $\bar{v}$ (see Figure 6.5). For a curved element, simple polynomial displacement fields defined in terms of the local displacements may produce spurious straining when the element undergoes rigid body translation (Cook, 1981), although convergence towards strain-free rigid body motion occurs as the angle subtended by each element decreases to zero. To overcome this problem of spurious straining, the element nodal displacements and strain-displacement matrix are expressed in global coordinates. Such a definition also allows compatibility and equilibrium requirements at complex junctions of elements to be achieved simply.

The vector of element nodal displacements in global coordinates is defined as

$$\{a\} = \{\{a\}_1 \{a\}_2\}^T \tag{6.45}$$

with

$$\{a\}_i = \left\{ U_i \quad \left(\frac{dU}{ds}\right)_i \quad V_i \quad \left(\frac{dV}{ds}\right)_i \right\}^T \tag{6.46}$$

The element displacement field is written in terms of the shape functions defined in Eqs. (6.39) to (6.42) as

$$U(\xi) = \sum_{i=1}^{2} N_{0i}(\xi) U_i + N_{1i}(\xi) \left(\frac{dU}{ds}\right)_i \tag{6.47}$$

$$V(\xi) = \sum_{i=1}^{2} N_{0i}(\xi) V_i + N_{1i}(\xi) \left(\frac{dV}{ds}\right)_i \tag{6.48}$$

which can be expressed succinctly in matrix notation as

$$\{u\} = [N]\{a\} \tag{6.49}$$

in which $\{u\} = \{U\ V\}^T$ and $[N]$ is a matrix of shape functions given by

$$[N] = [[N]_1 \quad [N]_2] \tag{6.50}$$

with

$$[N]_i = \begin{bmatrix} N_{0i} & N_{1i} & 0 & 0 \\ 0 & 0 & N_{0i} & N_{1i} \end{bmatrix} \qquad (6.51)$$

It can therefore be seen from Eqs. (6.47) and (6.48) that both the global displacement components are interpolated by cubic polynomials. At any point in the element, the transformation between the local and global displacement components is

$$\begin{Bmatrix} \bar{u} \\ \bar{v} \end{Bmatrix} = \begin{bmatrix} \cos\phi & -\sin\phi \\ \sin\phi & \cos\phi \end{bmatrix} \begin{Bmatrix} U \\ V \end{Bmatrix} \qquad (6.52)$$

$$\{\bar{u}\} = [T]\{u\} \qquad (6.53)$$

The interpolation scheme of Eqs. (6.47) and (6.48) is therefore equivalent to interpolating the local element longitudinal and transverse displacements, $\bar{u}$ and $\bar{v}$, by cubic polynomials; this differs from the more usual practice of assuming linear and cubic interpolation for the longitudinal and transverse displacements, respectively.

Strain-Displacement Matrices

It was stated previously [Eq. (6.7)] that the vector of generalized strains could be expressed as the sum of linear and nonlinear components,

$$\{\varepsilon\} = \{e\} + \{\eta\} \qquad (6.54)$$

The linear strain components $\{e\}$ are related to the element nodal displacements $\{a\}$ through the infinitesimal strain-displacement matrix $[B_0]$ as

$$\{e\} = [B_0]\{a\} \qquad (6.55)$$

The infinitesimal strain-displacement matrix $[B_0]$ (2×8) is a matrix of shape functions defined by

$$[B_0] = \begin{bmatrix} [B_0]_1 & [B_0]_2 \end{bmatrix} \qquad (6.56)$$

in which

$$[B_0]_i = \begin{bmatrix} \dfrac{N'_{0i}}{l}\cos\phi & \dfrac{N'_{1i}}{l}\cos\phi & -\dfrac{N'_{0i}}{l}\sin\phi & -\dfrac{N'_{1i}}{l}\sin\phi \\ -\dfrac{N'_{0i}}{l}\dfrac{d\phi}{ds}\cos\phi & -\dfrac{N'_{1i}}{l}\dfrac{d\phi}{ds}\cos\phi & \dfrac{N'_{0i}}{l}\dfrac{d\phi}{ds}\sin\phi & \dfrac{N'_{1i}}{l}\dfrac{d\phi}{ds}\sin\phi \\ -\dfrac{N''_{0i}}{l^2}\sin\phi & -\dfrac{N''_{1i}}{l^2}\sin\phi & -\dfrac{N''_{0i}}{l^2}\cos\phi & -\dfrac{N''_{1i}}{l^2}\cos\phi \end{bmatrix} \quad (6.57)$$

In the above equation, ( )' denotes differentiation with respect to the dimensionless variable ξ.

To express the nonlinear strain components $\{\eta\}$ in terms of nodal displacements, the vector of displacement gradients $\{\theta\}$ defined in Eq. (6.11) can be written in terms of the nodal displacements $\{a\}$ as

$$\{\theta\} = [G]\{a\} \quad (6.58)$$

in which $[G]$ (2×8) is a matrix of shape functions defined by

$$[G] = \begin{bmatrix} [G]_1 & [G]_2 \end{bmatrix} \quad (6.59)$$

with

$$[G]_i = \begin{bmatrix} \dfrac{N'_{0i}}{l}\cos\phi & \dfrac{N'_{1i}}{l}\cos\phi & -\dfrac{N'_{0i}}{l}\sin\phi & -\dfrac{N'_{1i}}{l}\sin\phi \\ \dfrac{N'_{0i}}{l}\sin\phi & \dfrac{N'_{1i}}{l}\sin\phi & \dfrac{N'_{0i}}{l}\cos\phi & \dfrac{N'_{1i}}{l}\cos\phi \end{bmatrix} \quad (6.60)$$

Combining Eqs. (6.7), (6.11), (6.55), and (6.58), the generalized strains resulting from element nodal displacements $\{a\}$ and measured relative to the reference configuration can therefore be expressed as

$$\{\varepsilon\} = \left[[B_0] + \dfrac{1}{2}[B_L]\right]\{a\} \quad (6.61)$$

in which

$$[B_L] = [A][G] \quad (6.62)$$

and is a linear function of the nodal displacements. Noting that $[B_0]$ and $[G]$ are independent of the nodal displacements, and making use of Eq. (6.14), the variation of Eq. (6.61) may be determined as

$$\delta\{\varepsilon\} = \left[[B_0] + [B_L]\right]\delta\{a\} \quad (6.63)$$

$$= [B]\delta\{a\} \quad (6.64)$$

in which [B] is termed the complete strain matrix referred to the reference configuration and includes nonlinear terms.

Equilibrium Equations

The aim here is to express the virtual work equation [Eq. (6.33)] in discretized form. The variation of the element displacement field defined by Eq. (6.49) can be written

$$\delta\{u\} = [N]\delta\{a\} \tag{6.65}$$

in which [N] is the matrix of shape functions defined by Eqs. (6.39) to (6.42). Substituting Eqs. (6.64) and (6.65) into Eq. (6.33) leads to the following equation for the element

$$\delta\{a\}^T \int_s [B]^T \{\sigma\} ds = \delta\{a\}^T \{R\} \tag{6.66}$$

in which $\{R\}$ is the vector of equivalent nodal loads due to the distributed loads $\{p\}$, and also includes concentrated nodal loads. Noting that the virtual nodal displacements $\delta\{a\}$ are arbitrary, Eq. (6.66) can be expressed in the slightly modified form of

$$\{\psi\} = \int_s [B]^T \{\sigma\} ds - \{R\} \tag{6.67}$$

in which $\{\psi\}$ can be interpreted as the nodal residual force vector for the element, with element equilibrium expressed by the condition $\{\psi\} = \{0\}$ (Zienkiewicz, 1977).

The equilibrium equation [Eq. (6.67)] is nonlinear in the element nodal displacements and so cannot be solved directly. By far the most common technique for solving Eq. (6.67), especially when the material behavior is inelastic, is to derive an incremental equation of equilibrium and use an incremental-iterative scheme to solve it. The complete nonlinear load-deflection response of the structure can thereby be obtained. The derivation of the incremental equilibrium equations and tangent stiffness matrix is described in Section 6.3.6 following.

6.3.6 Solving the Nonlinear Equilibrium Equations

The basis of solution algorithms for the nonlinear equations [Eq. (6.67)] is the Newton-Raphson method (Zienkiewicz, 1977) involving a series of solutions to linear incremental equilibrium equations. The procedure utilizes a tangent stiffness matrix to determine the nodal displacement vector for use in the subsequent iteration. It should be emphasised here that the fundamental aim is to obtain the set of nodal displacements $\{a\}$ for which $\{\psi(\{a\})\} = \{0\}$ [Eq. (6.67)], and that the Newton-Raphson iteration technique in conjunction with a tangent stiffness matrix is only one such means of achieving this. The actual tangent stiffness matrix used in the iteration process need not be theoretically exact, and need not be updated for every iteration.

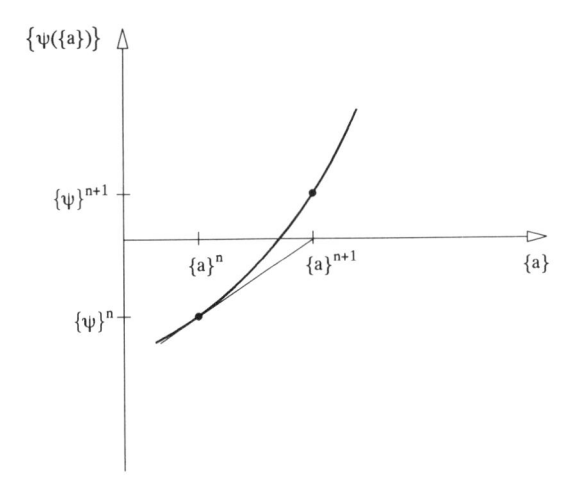

FIGURE 6.7. Solving the nonlinear equilibrium equations.

Approximate tangent stiffness matrices are often used, and the specific form of the tangent affects the rate of convergence but not the final solution. In the modified Newton-Raphson method, the tangent stiffness matrix is formed at the commencement of the load step and is then held constant throughout the equilibrium iterations.

There are several approaches that can be used to derive the incremental equilibrium equations and consistent tangent stiffness matrix. The approach presented here follows Zienkiewicz (1977). An alternative approach based on incremental decomposition of the strain and stress tensors has been presented by Bathe and his coworkers (1975, 1979, 1982).

Suppose an initial estimate $\{a\}^n$ of the nodal displacements is known for which the structure is not in equilibrium. The equilibrium function $\{\psi\}^n = \{\psi(\{a\}^n)\}$ is then nonzero, as shown in Figure 6.7, in which the coordinate axes are denoted symbolically by $\{a\}$ and $\{\psi(\{a\})\}$. For an increment in nodal displacements of $\{\Delta a\}^{n+1}$ the function $\{\psi\}$ is given by the Taylor's series expansion of $\{\psi\}$ about $\{a\}^n$, ignoring third and succeeding terms as

$$\left\{\psi\left(\{a\}^n + \{\Delta a\}^{n+1}\right)\right\} = \left\{\psi\left(\{a\}^n\right)\right\} + \frac{\partial\{\psi\}}{\partial\{a\}}\bigg|_{\{a\}=\{a\}^n} \{\Delta a\}^{n+1} + \ldots \tag{6.68}$$

From the requirement $\{\psi(\{a\}^n + \{\Delta a\}^{n+1})\} = \{\psi(\{a\}^{n+1})\} = \{0\}$, the linearized approximation to the equilibrium equations is

$$\left\{\psi\left(\{a\}^n\right)\right\} + [K_T]\{\Delta a\}^{n+1} = \{0\} \tag{6.69}$$

in which

$$[K_T] = \frac{\partial\{\psi\}}{\partial\{a\}}\bigg|_{\{a\}=\{a\}^n} \tag{6.70}$$

is the tangent stiffness matrix. From Eq. (6.69), the estimate of $\{a\}^{n+1}$ therefore becomes

$$\{a\}^{n+1} = \{a\}^n - [K_T]^{-1}\{\psi\}^n \tag{6.71}$$

To evaluate $[K_T]$, a variation of Eq. (6.67) with respect to the nodal displacements $\{a\}$ is taken to furnish

$$d\{\psi\} = \int_s [B]^T d\{\sigma\} ds + \int_s d[B]^T \{\sigma\} ds - d\{R\} \tag{6.72}$$

If the loading is assumed to be conservative (i.e., not dependent on the deformation) then $d\{R\} = \{0\}$. A particular case of nonconservative loading is that of the reaction forces exerted on a structure by continuous or discrete elastic restraints (springs). For completeness, the case of $d\{R\} \neq \{0\}$ is assumed in the following derivation of the tangent stiffness matrix.

Using Eqs. (6.62) to (6.64) and expanding terms in $[A]$ and $\{\sigma\}$ it can be shown that the second integrand in Eq. (6.72) can be expressed in the form

$$d[B]^T\{\sigma\} = d\big([G]^T[A]^T\big)\{\sigma\} = [G]^T d[A]^T\{\sigma\} = [G]^T[S][G]d\{a\} \tag{6.73}$$

in which $[S]$ is a 2×2 matrix of stress resultants given by

$$[S] = \begin{bmatrix} N & 0 \\ 0 & N \end{bmatrix} \tag{6.74}$$

Substituting Eqs. (6.28), (6.64), and (6.73) into Eq. (6.72) gives

$$d\{\psi\} = [K_T]d\{a\} \tag{6.75}$$

in which $[K_T]$ is the element tangent stiffness matrix

$$[K_T] = [\overline{K}] + [K_\sigma] + [K_R] \tag{6.76}$$

with

$$[\overline{K}] = [K_0] + [K_L] = \int_s [B]^T[D_T][B]ds \tag{6.77}$$

$$[K_0] = \int_s [B_0]^T[D_T][B_0]ds \tag{6.78}$$

$$[K_L] = \int_s [B_0]^T[D_T][B_L] + [B_L]^T[D_T][B_0] + [B_L]^T[D_T][B_L] \, ds \qquad (6.79)$$

$$[K_\sigma] = \int_s [G]^T[S][G] \, ds \qquad (6.80)$$

$$[K_R] = -\frac{\partial\{R\}}{\partial\{a\}} \qquad (6.81)$$

In Eq. (6.77), $[K_0]$ is the small displacement stiffness matrix, and $[K_L]$ is due to the large displacement and is hence termed the initial displacement matrix, or large displacement matrix. The matrix $[K_\sigma]$ is the geometric stiffness matrix and accounts for the effects of internal stresses on the instantaneous stiffness of the structure. The matrix $[K_R]$ is the contribution of the nonconservative loading to the tangent stiffness matrix, as might arise from the imposition of elastic spring restraints, for example.

The derivation of incremental equilibrium equations and a tangent stiffness matrix constitute the basis of commonly used schemes for solving the nonlinear equilibrium equations. Early methods of solving the incremental equilibrium equations used simple Euler integration (Zienkiewicz, 1977). However, this technique requires that load increments be kept small so that the solution does not deviate too far from the satisfaction of equilibrium. Larger load steps can be accommodated if the purely incremental method is combined with conventional or modified Newton-Raphson iterations (Zienkiewicz, 1977), as described above, to enable the total equilibrium equations to be satisfied to within a specified tolerance.

Conventional Newton-type iterative strategies hold the load parameter constant while iterating to convergence and, due to the near-singular nature of the tangent stiffness matrix in the neighborhood of a load limit point, are therefore only suited to obtaining the pre-ultimate equilibrium path. However, successful attainment of the load-deflection response of a structure, including traversal of load and displacement limit points, can be achieved if more advanced incremental-iterative solution strategies are employed. These advanced incremental-iterative solution techniques are also based on the Newton-Raphson procedure, but share the common feature that iterations are performed on the load parameter as well as the nodal displacements (i.e., the load parameter can vary during the equilibrium iteration phase). Examples of these advanced incremental-iterative solution techniques include the displacement control algorithm presented by Powell and Simons (1981), the 'arc-length' constraint method described by Crisfield (1981), and the 'minimum residual displacement' method of Chan (1988). A summary of incremental-iterative strategies, and an investigation into their relative effectiveness has been presented by Clarke and Hancock (1990). In the same paper, Clarke and Hancock review and discuss strategies for determining the size of the initial load increment (i.e., at the commencement of a new load step), and strategies for determining when load limit points have been traversed, thus necessitating a change in sign of the initial load increment.

280 ADVANCED ANALYSIS OF STEEL FRAMES

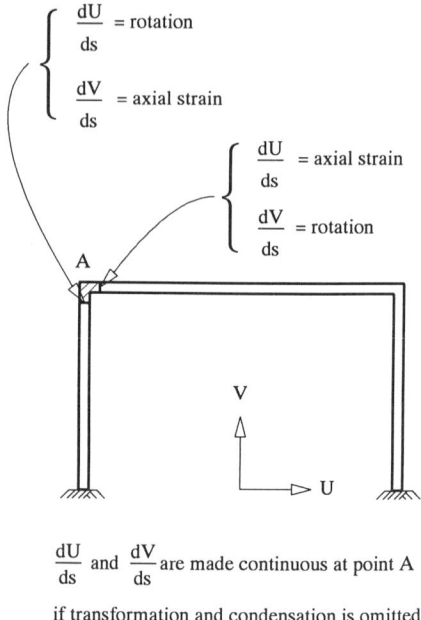

$\dfrac{dU}{ds}$ = rotation

$\dfrac{dV}{ds}$ = axial strain

$\dfrac{dU}{ds}$ = axial strain

$\dfrac{dV}{ds}$ = rotation

$\dfrac{dU}{ds}$ and $\dfrac{dV}{ds}$ are made continuous at point A
if transformation and condensation is omitted

FIGURE 6.8. Compatibility at a discontinuity of element slope.

The strain-displacement matrices $[B]$ and $[G]$ used in the formation of the element tangent stiffness matrix $[K_T]$ are expressed in global coordinates. The conventional transformation of $[K_T]$ from local to global coordinates is therefore *not* required before assembly at the structure level. However, an alternative transformation and condensation procedure *is* required in order to enforce the appropriate connectivity of nodal variables at junctions of elements, as described in Section 6.3.7 following.

6.3.7 Transformation, Condensation, and Recovery of Nodal Variables

Transformation of Nodal Variables

The four nodal variables used in the derivation of the element tangent stiffness matrix $[K_T]$ and the nodal residual force vector $\{\psi\}$ were U_i, $(dU/ds)_i$, V_i, and $(dV/ds)_i$ (defined in a global axis system). These four variables are satisfactory for structures consisting of a single uniform straight or smoothly curving member, such as beams and arches. However, excessive continuity at element junctions occurs if these four nodal variables are retained for use in problems where a discontinuity in element slope (see Figure 6.8) or cross-section occurs at a node (Teng and Rotter, 1989). The only variables which should be connected between all members framing into a common node are the two translations U_i and V_i and the in-plane rotation θ_i. This is achieved by firstly transforming the global displacement derivatives $(dU/ds)_i$ and $(dV/ds)_i$ as

$$\begin{Bmatrix} \varepsilon_i \\ \theta_i \end{Bmatrix} = \begin{bmatrix} \cos\phi_i & -\sin\phi_i \\ \sin\phi_i & \cos\phi_i \end{bmatrix} \begin{Bmatrix} (dU/ds)_i \\ (dV/ds)_i \end{Bmatrix} \qquad (6.82)$$

The incremental equilibrium equations given by Eq. (6.69) are therefore transformed as

$$[\hat{K}_T]\{\Delta\hat{a}\} = -\{\hat{\psi}\} \qquad (6.83)$$

in which the transformed tangent stiffness matrix, transformed nodal load vector, and transformed nodal degree of freedom vector are given, respectively, by

$$[\hat{K}_T] = [\hat{T}]^T [K_T][\hat{T}] \qquad (6.84)$$

$$\{\hat{\psi}\} = [\hat{T}]^T \{\psi\} \qquad (6.85)$$

$$\{\hat{a}\} = \{\{\hat{a}\}_1 \quad \{\hat{a}\}_2\}^T \qquad (6.86)$$

$$\{\hat{a}\}_i = \{U_i \quad \varepsilon_i \quad V_i \quad \theta_i\}^T \qquad (6.87)$$

$$[\hat{T}] = \begin{bmatrix} [\hat{T}]_1 & [0] \\ [0] & [\hat{T}]_2 \end{bmatrix} \qquad (6.88)$$

$$[\hat{T}]_i = \begin{bmatrix} 1 & 0 & 0 & 0 \\ 0 & \cos\phi_i & 0 & \sin\phi_i \\ 0 & 0 & 1 & 0 \\ 0 & -\sin\phi_i & 0 & \cos\phi_i \end{bmatrix} \qquad (6.89)$$

Condensation of Nodal Variables

The transformed variables ε_i should not be connected between different elements meeting at a common node and so must be condensed out of the element stiffness matrix before assembly of the global system. Suppose the transformed degree of freedom vector $\{\hat{a}\}$ is partitioned as $\{\hat{a}\} = \{\{a_r\} \{a_e\}\}^T$, in which $\{a_r\}$ are the retained degrees of freedom U_i, V_i, θ_i, and $\{a_e\}$ are the eliminated degrees of freedom, ε_i. The element stiffness equations can then be written in partitioned form as

$$\begin{bmatrix} [K_{rr}] & [K_{re}] \\ [K_{re}]^T & [K_{ee}] \end{bmatrix} \begin{Bmatrix} \{\Delta a_r\} \\ \{\Delta a_e\} \end{Bmatrix} = \begin{Bmatrix} -\{\psi_r\} \\ -\{\psi_e\} \end{Bmatrix} \qquad (6.90)$$

The lower partition of Eq. (6.90) may be solved for $\{\Delta a_e\}$, giving

$$\{\Delta a_e\} = [K_{ee}]^{-1}\left(-\{\psi_e\} - [K_{re}]^T \{\Delta a_r\}\right) \quad (6.91)$$

Substituting Eq. (6.91) into the upper partition of Eq. (6.90) yields

$$[K_{Tc}]\{\Delta a_r\} = -\{\psi_c\} \quad (6.92)$$

in which $[K_{Tc}]$ and $\{\psi_c\}$ are the condensed element tangent stiffness matrix and condensed vector of residual forces, given, respectively, by

$$[K_{Tc}] = [K_{rr}] - [K_{re}][K_{ee}]^{-1}[K_{re}]^T \quad (6.93)$$

$$\{\psi_c\} = \{\psi_r\} - [K_{re}][K_{ee}]^{-1}\{\psi_e\} \quad (6.94)$$

The eliminated degrees of freedom $\{a_e\}$ are allowed to 'float' in the analysis and so small discontinuities may occur at the nodes of a single uniform straight or smoothly curved member.

Recovery of Nodal Variables

Evaluation of the element tangent stiffness matrix and total equilibrium equations depends on the nonlinear strain matrix $[B_L]$, which in turn requires the displacements at any point in the element to be known. These internal displacements are expressed in terms of the nodal displacements using Eqs. (6.47) and (6.48). Recovery of the complete set of nodal displacements $\{a\}$ for each element is therefore necessary for evaluation of $[B_L]$.

The increments of eliminated variables $\{\Delta a_e\}$ are recovered through Eq. (6.91), and the transformed nodal variables $(dU/ds)_i$ and $(dV/ds)_i$ are evaluated using the inverse of Eq. (6.82).

Internal Rotational Releases

The condensation procedure described above, in which the ε_i are the only eliminated variables, is applicable to rigidly connected frames for which the rotations θ_i of a node are equal for all interconnecting members. In some frames, however, the joints between members may be more appropriately modeled as perfectly pinned (internal hinges) rather than perfectly rigid.

Connectivity of nodal rotations at end i of an element can be avoided by condensing the rotational degree of freedom θ_i from the element before assembly. Thus θ_i is permitted to 'float' and no moment is transferred to the other elements framing into the node. The condensation and recovery procedure is performed exactly as described above in Eqs. (6.90) to (6.94) except that the released degree of freedom θ_i is transferred from the vector of retained degrees of freedom $\{a_r\}$ to the vector of eliminated degrees of freedom $\{a_e\}$. The condensed tangent stiffness matrix remains 6×6, but the row and column corresponding to θ_i are filled with zeros.

(a) I-section

(b) Square hollow section

(c) Circular hollow section

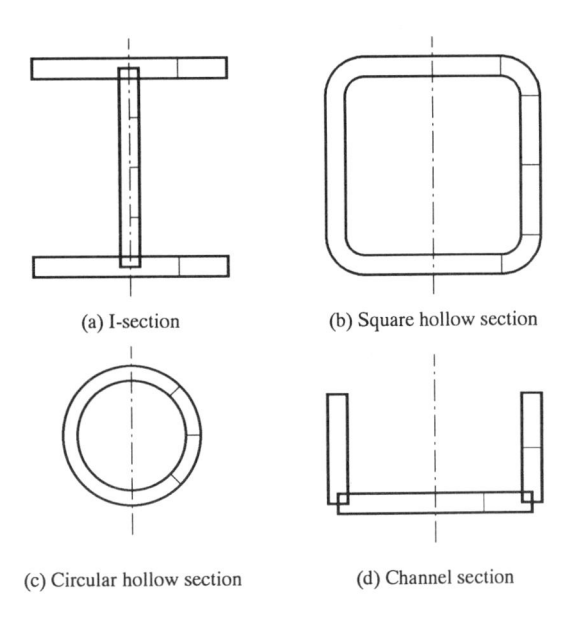

(d) Channel section

FIGURE 6.9.   Cross-section subdivision into strips.

The condensation procedure employed for internal rotational releases can also be used for releasing the translational degrees of freedom of the element.

6.4 Cross-Sectional Analysis

6.4.1 General

The plastic-zone (distributed plasticity) formulation is characterized by the subdivision of the member cross-sections into elemental areas, or into grids of 'monitoring points' or 'fibers'. As a consequence of the path-dependent nature of plasticity, the current stress, current yield stress, and the current level of equivalent plastic strain must be stored at each monitoring point and updated incrementally throughout the analysis. In this way, the spread of yielding throughout the cross-section can be captured, and the member stress resultants and effective cross-sectional properties can be determined by numerical integration over the cross-section.

The grid of monitoring points is obtained by firstly dividing a cross-section into straight or uniformly curved strips, in the manner illustrated in Figure 6.9 The use of curved strips enables the accurate geometrical specification of circular hollow sections as well as precise modeling of the rounded corners in square hollow sections. As a consequence of the assumed cross-sectional symmetry, and the fact that the analysis is in-plane, only half of the cross-sectional geometry needs to be stored.

The layout of monitoring points for a typical strip is illustrated in Figure 6.10. The monitoring point layout in each strip required for an accurate analysis depends on the

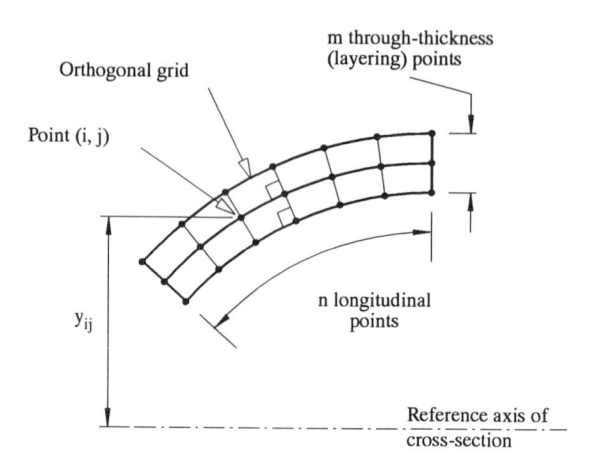

FIGURE 6.10. Subdivision of a cross-section strip into a grid of "monitoring points" or "fibers".

position and orientation of the strip relative to the axis of bending of the cross-section, the stress gradient through the section, and whether residual stresses are included in the analysis or not (see Section 6.6).

6.4.2 Numerical Integration Procedure

The stress resultants and instantaneous cross-sectional properties are obtained by numerical integration over the area of the cross-section. In the present plastic-zone analysis, the numerical integration is performed by discretizing the cross-section into an *orthogonal* grid of monitoring points as shown in Figure 6.3, and applying Simpson's rule in two dimensions. An integration scheme based on a two-dimensional application of Simpson's rule can depict yielding of the extreme fiber of the cross-section, and is therefore believed to be preferable to Gaussian quadrature. There is also no reason to suspect that Gaussian quadrature is superior to Simpson's rule when applied to a discontinuous function, such as the stress distribution in a partially yielded section. It is also interesting to note that Simpson's rule results in exact calculation of the second moment of area of a rectangular region oriented with its sides parallel and perpendicular to the axis of bending, and is therefore superior to the simpler trapezoidal rule for which the integration is not exact. For more complex structural shapes for which an orthogonal grid of monitoring points can be used, such as square hollow sections with rounded corners, and circular hollow sections, the area and second moment of area are also computed exactly using Simpson's rule.

The integral of a general two-dimensional function $f(x,y)$ over an area which can be subdivided into an orthogonal grid of $m \times n$ monitoring points (Figure 6.10) can be approximated using Simpson's rule as

$$\int_A f(x,y)\,dA \approx \sum_{i=1}^{m}\sum_{j=1}^{n} f\left(x_{ij}, y_{ij}\right) w_{ij} \tag{6.95}$$

In the above equation, m is the number of through-thickness monitoring points ($m \geq 3$, m odd), n is the number of longitudinal monitoring points ($n \geq 3$, n odd), (x_{ij}, y_{ij}) are the coordinates of monitoring point (i, j), and w_{ij} is the (i, j)th element of the two-dimensional Simpson's rule 'weight' array $[W]$, defined by

$$[W] = [H_s][\overline{W}][H_t] \quad (6.96)$$

in which

$$[H_s] = \begin{bmatrix} \Delta s_1 & & & \\ & \Delta s_2 & & \\ & & \ddots & \\ & & & \Delta s_m \end{bmatrix} \quad (6.97)$$

$$[H_t] = \begin{bmatrix} \Delta t_1 & & & \\ & \Delta t_2 & & \\ & & \ddots & \\ & & & \Delta t_n \end{bmatrix} \quad (6.98)$$

$$[\overline{W}] = \frac{1}{9} \begin{bmatrix} 1 & 4 & 2 & 4 & 2 & \cdots & 4 & 1 \\ 4 & 16 & 8 & 16 & 8 & \cdots & 16 & 4 \\ 2 & 8 & 4 & 8 & 4 & \cdots & 8 & 2 \\ 4 & 16 & 8 & 16 & 8 & \cdots & 16 & 4 \\ 2 & 8 & 4 & 8 & 4 & \cdots & 8 & 2 \\ \vdots & \vdots & \vdots & \vdots & \vdots & \ddots & \vdots & \vdots \\ 4 & 16 & 8 & 16 & 8 & \cdots & 16 & 4 \\ 1 & 4 & 2 & 4 & 2 & \cdots & 4 & 1 \end{bmatrix} \quad (6.99)$$

In Eqs. (6.97) and (6.98), Δs_i and Δt_j are the spacings of the longitudinal monitoring points in row i and the through-thickness monitoring points in column j, respectively, of the strip discretization (see Figure 6.10).

6.4.3 Numerical Integration of Stress Resultants

Using Simpson's rule in two dimensions, the axial force and bending moment acting on a single strip can be expressed, respectively, as

$$N_{strip} = \sum_{i=1}^{m} \sum_{j=1}^{n} \sigma_{ij} w_{ij} \quad (6.100)$$

$$M_{strip} = \sum_{i=1}^{m}\sum_{j=1}^{n} \sigma_{ij} y_{ij} w_{ij} \tag{6.101}$$

in which m, n, and w_{ij} are as described above, σ_{ij} is the longitudinal stress acting at monitoring point (i, j), and y_{ij} is the coordinate of monitoring point (i, j) above the reference axis of the cross-section (see Figure 6.10). The stress resultants acting on the entire cross-section are simply obtained by summing the contributions of all strips,

$$N = \sum_{strips} N_{strip} \tag{6.102}$$

$$M = \sum_{strips} M_{strip} \tag{6.103}$$

6.4.4 Numerical Integration of the Tangent Modulus Matrix

The elastic-plastic (instantaneous) cross-section properties in the tangent modulus matrix [Eq. (6.29)] are also computed using Simpson's rule. The individual terms in the tangent modulus matrix are computed using Simpson's rule as

$$\int_A E_T \, dA = \sum_{strips}\sum_{i=1}^{m}\sum_{j=1}^{n} E_{Tij} w_{ij} \tag{6.104}$$

$$\int_A E_T y \, dA = \sum_{strips}\sum_{i=1}^{m}\sum_{j=1}^{n} E_{Tij} y_{ij} w_{ij} \tag{6.105}$$

$$\int_A E_T y^2 \, dA = \sum_{strips}\sum_{i=1}^{m}\sum_{j=1}^{n} E_{Tij} y_{ij}^2 w_{ij} \tag{6.106}$$

6.5 Some Aspects of the Computer Implementation

6.5.1 General

The theory described in this chapter has been programmed in FORTRAN 77 as the finite element nonlinear analysis program NIFA. Due to the size and complexity of program NIFA, a user's manual and program listing is not provided with this book. However, some of the specific implementation details and capabilities of the program are outlined in this section. In addition to the main analysis program, a pre- and post-processor has been developed to automate the data generation process, to allow graphical verification of the input data, and to view and interpret the results of a nonlinear analysis through the plotting of deflected shapes and the distributions of strains, stress resultants, and yielded areas throughout the frame.

6.5.2 Node Numbering and Numerical Integration

As is typical of all frame analysis and finite element analysis programs, the structure geometry is subdivided into nodes and elements. To reduce the bandwidth of the stiffness matrix, the finite element nonlinear analysis incorporates a node renumbering algorithm due to Collins (1973), and Durocher and Gasper (1979). This renumbering procedure does not guarantee an optimum reordering, but usually results in a system which is at least close to optimal. The node renumbering procedure becomes an important feature of the analysis when an automatic meshing algorithm is used to generate the structure geometry.

As a consequence of the curved element geometry, the incorporation of general nonlinear material behavior into the analysis, and the discontinuous nature of material yielding, numerical rather than analytical integration is used to evaluate the tangent stiffness matrix and to assess total equilibrium. Gaussian quadrature is used so as to obtain maximum efficiency from a minimum number of stations. In general, Gaussian integration with n sampling points will integrate a polynomial of degree less than or equal to $(2n - 1)$ exactly. A three-point Gauss rule, which can therefore integrate a polynomial of order five exactly, was adopted for all the examples reported in Sections 6.7 to 6.10 following.

Computation of the small displacement stiffness matrix $[K_0]$ [Eq. (6.78)] and the geometric stiffness matrix $[K_\sigma]$ [Eq. (6.80)] requires the integration of polynomials of order four which can be achieved exactly using three-point Gaussian quadrature. The large displacement stiffness matrix $[K_L]$ [Eq. (6.79)] requires the integration of polynomials of up to order eight which cannot be achieved exactly using a three-point Gauss rule. Overall therefore, a three-point rule results in a slightly under-integrated element stiffness matrix, although in practice the differences in results obtained by using a higher-order rule compared to a three-point rule are negligible.

6.5.3 Nonlinear Formulations and Incremental-Iterative Strategies

Although only the total Lagrangian formulation is described in this chapter, program NIFA actually includes three formulations of geometric nonlinearity: the total Lagrangian, updated Lagrangian, and co-rotational formulations (Bathe, 1982; Mattiasson et al., 1984, 1986; Clarke, 1992). The total Lagrangian formulation is satisfactory for typical problems in steel structures involving small to moderate displacements/rotations, while for large or gross displacement/rotation problems either the updated Lagrangian or co-rotational formulations are necessary to model the response adequately.

A particular feature of the finite element nonlinear analysis is its nonlinear solution procedures. Program NIFA incorporates several advanced incremental-iterative solution techniques (Clarke and Hancock, 1990), as well as rigorous strategies for material inelasticity based on an assumption of incremental reversibility (Nyssen, 1981) and a 'double loop' strategy (Bushnell, 1977). By virtue of these numerical techniques, the attainment of the complete equilibrium path, including negotiation of load and displacement limit points, can be achieved for most problems in a fairly straightforward and automated manner.

As recommended by Bergan and Clough (1972), a convergence criterion based on the structure displacements is employed in program NIFA. Of the three convergence criteria defined by Bergan and Clough (1972), the modified Euclidean norm has been used for all the examples presented in this chapter. The modified Euclidean norm $\|\varepsilon\|$ is defined by

$$\|\varepsilon\| = \sqrt{\frac{1}{N} \sum_{k=1}^{N} \left|\frac{\Delta a_k}{(a_k)_{ref}}\right|^2} \tag{6.107}$$

in which N is the total number of degrees of freedom, Δa_k is the change in the displacement component k during the current iteration cycle, and $(a_k)_{ref}$ is the largest displacement component of the corresponding type. Convergence is attained when

$$\|\varepsilon\| < \zeta \tag{6.108}$$

where ζ is the convergence tolerance, usually in the range 10^{-3} to 10^{-5} depending on the desired accuracy and the characteristics of the particular problem.

6.5.4 Material Inelasticity

The cross-sectional subdivision procedure and the numerical integration techniques have been described in Section 6.4. In the geometrical break-up of the cross-section into straight or uniformly curved strips, each strip is described by the coordinates of its two end nodes, the curvature of the strip, the number of longitudinal and through-thickness (layering) monitoring stations or fibers, and the stress-strain behavior. Longitudinal residual stresses in each strip are specified by separate membrane, bending and layering components (Key and Hancock, 1993 and Key, 1988), all of which may vary linearly along the strip.

The stress-strain curve applicable to each strip may be specified as either:

(1) multi-linear stress-strain curve
(2) Ramberg-Osgood stress-strain curve.

The Ramberg-Osgood curves (Ramberg and Osgood, 1943) were included because they may be used to represent the rounded stress-strain behavior typical of highly worked cold-formed steel, such as the corner zones in square hollow sections (Key and Hancock, 1993). The rounded stress-strain behavior of other materials such as aluminum and stainless-steel may also be modeled accurately with the Ramberg-Osgood curves. The Ramberg-Osgood equation expresses strain ε as a function of stress σ as

$$\varepsilon = \frac{\sigma}{E} + k\left(\frac{\sigma}{E}\right)^n \tag{6.109}$$

or alternatively

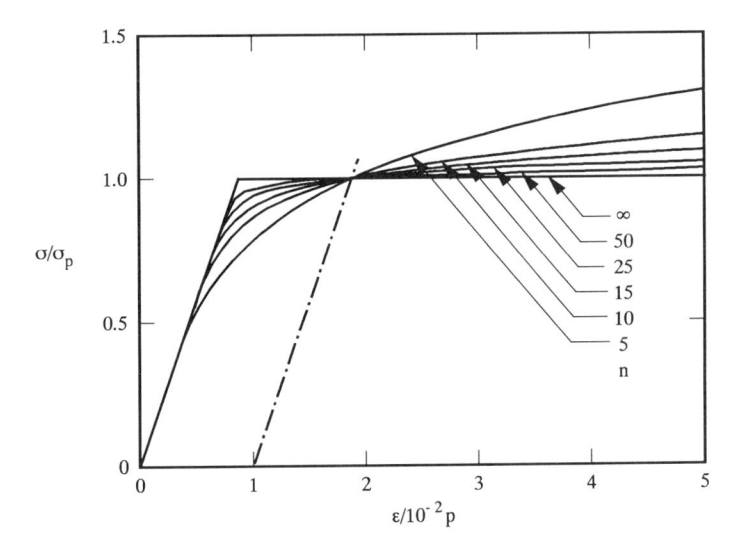

FIGURE 6.11. Ramberg-Osgood stress-strain curves for $p = 0.2$.

$$\varepsilon = \frac{\sigma}{E} + \frac{p}{100}\left(\frac{\sigma}{\sigma_p}\right)^n \tag{6.110}$$

in which k, p, σ_p, and n are constants. The constant n dictates the sharpness of the 'knee' of the stress-strain curve, and σ_p denotes the stress at which the plastic component of the total strain is p percent. A value of $p = 0.2$, corresponding to the 0.2% proof stress $\sigma_{0.2}$, is commonly used. Typical Ramberg-Osgood curves for $p = 0.2$ and various values of n are illustrated in Figure 6.11.

6.5.5 Computer Requirements

A refined plastic-zone analysis, such as program NIFA, requires considerably more computing power for its effective operation than do plastic hinge analysis programs. In plastic-hinge analyses, each frame member is usually subdivided into one, or at most a few, elements. Inelasticity is captured through the plastic-hinge concept using a stress resultant-based yield surface. Conversely, in plastic-zone analysis programs, subdivision of each member into several elements is necessary to accurately model frame behavior. Additionally, the use of a stress-based yield criterion and the associated subdivision of the cross-section into a grid of fibers, and the need to store the stress, plastic strain, and yield stress at each fiber, places large demands on computer memory and storage.

While program NIFA can be used on a PC, the most effective use of the program, including its graphical pre- and post-processor, is obtained from using a UNIX workstation running the X Window System. For the PC platform, the minimum suggested hardware requirements are an 80386 CPU and 4 Mb of memory, although significantly larger problems can be tackled if the memory is doubled to 8 Mb and the more powerful 80486 CPU is used. It is believed that these hardware demands are not excessive and fall within the realms of 'standard' PC configurations currently (1993) available. For the UNIX workstation platform employing the X Window System, it is suggested that a practical minimum of 16 Mb of memory is required.

6.6 Investigation of Analysis Parameters

6.6.1 Introductory Comments

As an introduction to illustrating the capabilities of the plastic-zone analysis, it was thought prudent to perform some preliminary analyses to investigate the sensitivity of the predicted behavior to variations in some of the nonlinear analysis modeling parameters. The main parameters considered in the studies were the number of elements used to model a member, the order of Gaussian integration used to compute the element stiffness matrices and nodal residual force vectors, and the degree of refinement of the cross-section subdivision. The particular problems used for the parametric investigations are an I-section pin-ended column, and a rectangular section beam with fully constrained ends subjected to a central concentrated load.

6.6.2 I-Section Column

A pin-ended I-section column of the type analyzed by Little (1982) was employed to investigate the number of elements and monitoring points required to model column behavior accurately. The column was characterized by the normalized slenderness and imperfection parameters $\lambda = 1$ and $\Delta_0^* = 0.14$, in which

$$\lambda = \sqrt{\frac{P_Y}{P_E}} = \frac{1}{\pi}\left(\frac{L}{r}\right)\sqrt{\frac{\sigma_Y}{E}} \tag{6.111}$$

$$\Delta_0^* = \left(\frac{\Delta_0}{L}\right)\left(\frac{L}{r}\right) \tag{6.112}$$

and σ_Y is the yield stress in simple tension or compression, $P_Y = A\sigma_Y$ denotes the squash load, P_E is the elastic critical (Euler) buckling load, E is the elastic modulus, r is the radius of gyration of the cross-section, and Δ_0 is the out-of-straightness at midspan of the column of overall length L. The initial geometrical imperfection varied as a half-sine curve along the length of the member and was therefore in the fundamental buckling mode. The material behavior was assumed elastic-perfectly plastic with no residual stresses. The cross-sectional geometry used in the study, which is representative of the shape of a typical universal column, is shown in Figure 6.12.

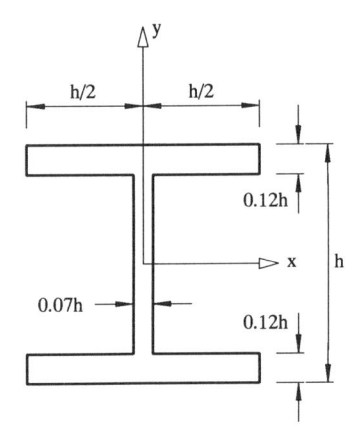

FIGURE 6.12. Cross-sectional geometry employed in analysis of I-section column.

The effect on the ultimate load and load-axial shortening response of the number of elements used to model the column is illustrated in Figure 6.13a. In these analyses, nine monitoring points over the depth of the web and three monitoring points over the depth of each flange were used, and three-point Gaussian quadrature was used for integration along each element. In some further studies, five-point Gaussian quadrature was employed and the results were virtually indistinguishable from those of a three-point rule. The effects of increasing the density of cross-section monitoring points is shown in Figure 6.13b. In these analyses, four elements over half the column and a three-point Gauss rule were employed. The results shown in Figure 6.13 can be summarized as follows:

(1) The differences in ultimate loads for the cases of one and two elements per half column, and two and four elements per half column, were 0.30 and 0.27%, respectively. Significant variations in the post-buckling load-shortening response were observed for the cases of one, two, and four elements per half column, as shown in Figure 6.13a. The post-ultimate response became steeper as the number of elements, and consequently the accuracy of the buckled shape, increased.

(2) For this example of overall column buckling, virtually identical results were obtained for the two cases of (see Figure 6.13b):
   (a) three monitoring points through the depth of the flanges, nine monitoring points through the depth of the web;
   (b) seven monitoring points through the depth of the flanges, 17 monitoring points through the depth of the web.

For a cross-section of a different shape and subjected to a more substantial stress gradient, the layout of monitoring points will be more critical in obtaining an accurate load-displacement response, as discussed hereafter.

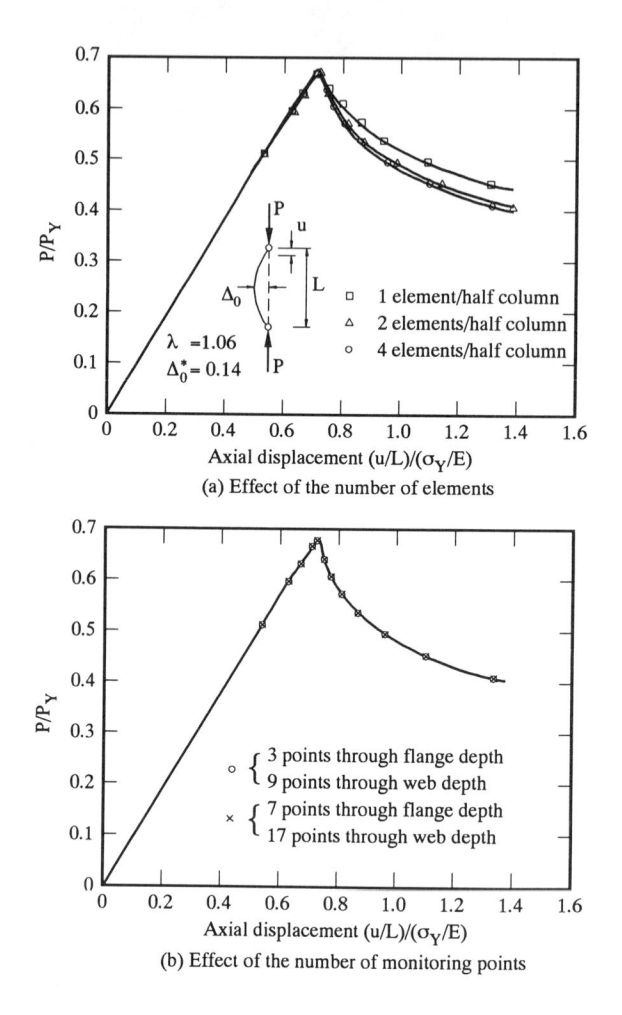

FIGURE 6.13. Investigation of analysis parameters for I-section column. (a) Effect of the number of elements. (b) Effect of the distribution of monitoring points.

6.6.3 Beam with Fully Constrained Ends

The case of a rectangular section beam of length L with fully constrained ends and a central concentrated load P has been analyzed theoretically by Haythornthwaite (1957) in which the deformed configuration was considered and a rigid-plastic material was assumed. The classical limit (plastic collapse) value for this problem is $P_p = 8M_p/L$, in which M_p is the plastic moment capacity. The example is challenging as the stress state at the hinges changes from one of nearly pure flexure to one of nearly pure tension as the beam deflects, stiffens, and becomes essentially a two bar truss. This problem therefore facilitates a verification of the elastic unloading in the analysis.

Investigations into the effects of the number of elements, the order of Gaussian

PLASTIC-ZONE ANALYSIS OF FRAMES 293

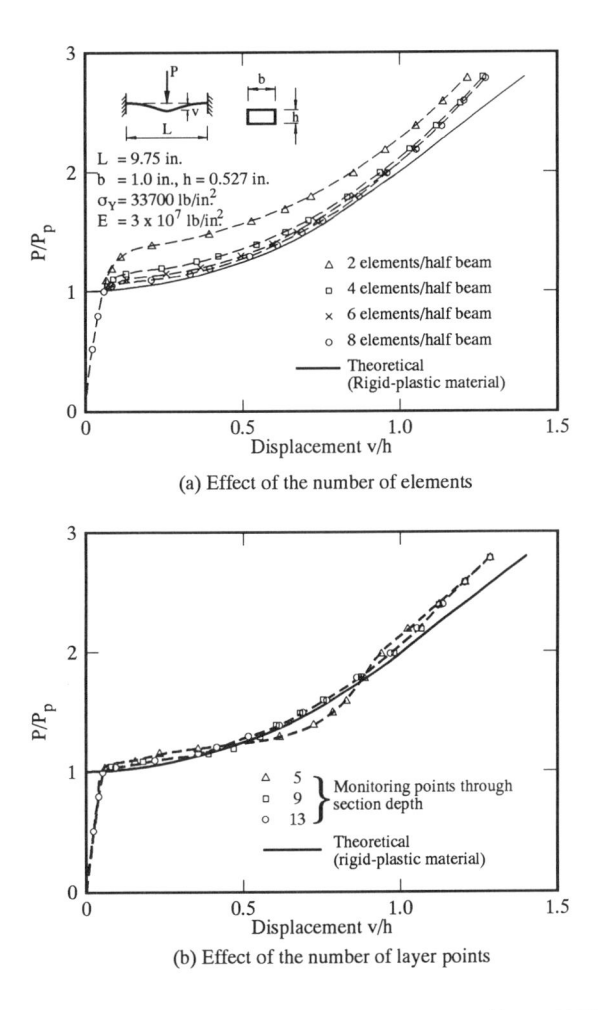

FIGURE 6.14. Investigation of analysis parameters for fully constrained beam. (a) Effect of the number of elements. (b) Effect of the distribution of monitoring points.

integration, and the number of cross-section monitoring points on the response of the beam have been performed. Selective results are shown in Figure 6.14, and the findings can be summarized as follows:

(1) Eight elements for half the beam were necessary to obtain an accurate representation of the load versus central displacement response (see Figure 6.14a). Seventeen monitoring points through the depth of the section were employed for the analyses plotted in Figure 6.14a.

(2) There was virtually no difference in the solutions for three-point and five-point Gaussian integration when eight elements were used to model half the beam and 17 monitoring points through the depth of the cross-section were employed (these results are not shown in Figure 6.14).

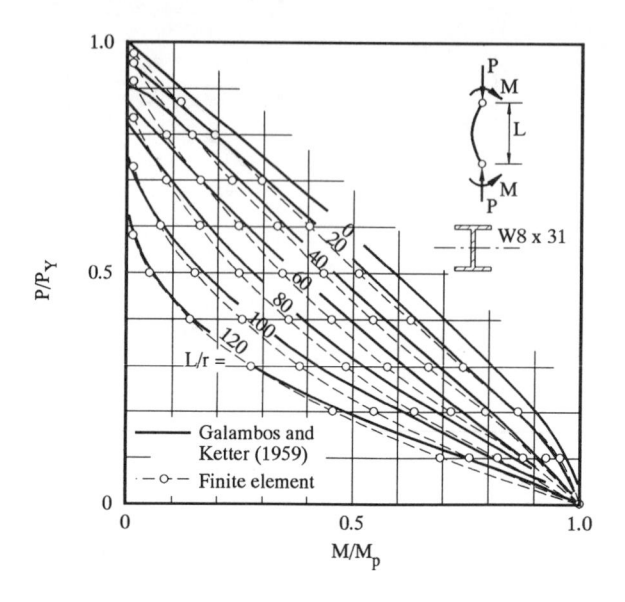

FIGURE 6.15. Ultimate strength interaction curves for straight beam-columns with end-moment ratio $\beta = -1.0$.

(3) The importance of an adequate number of monitoring points is illustrated in Figure 6.14b, in which eight elements for half the beam were employed. At least 13 points through the section depth were needed to model accurately the gradual plastification and elastic unloading that occurs. In contrast, for the previous I-section column example, three monitoring points through the depth of the flange were satisfactory (see Figure 6.13b). It is therefore apparent that both the nature of the through-depth stress distribution and the shape factor of the cross-section affect the monitoring point layout required to achieve a smooth and accurate transition from elastic to plastic behavior.

6.7 Analysis of Beam-Columns

Numerical studies of the strength of I-section beam-columns have been reported by Galambos and Ketter (1959), and Ketter (1961). In these studies, the beam-columns were assumed to be initially straight and to have the distribution of residual stresses shown in Figure 7.2 of Chapter 7. The strength interaction curves were obtained by Galambos and Ketter (1959) using a numerical integration procedure with nine equally spaced stations along the length of the member, in conjunction with moment-curvature relations derived for a W8 × 31 section. A Young's modulus $E = 30,000$ ksi and a yield stress $\sigma_Y = 33$ ksi were assumed in the studies. The material behavior was idealized as elastic-perfectly plastic. A more complete description of beam-column problems studied in the US is provided in Section 7.3 of Chapter 7.

Apart from the member geometrical slenderness L/r, another major parameter

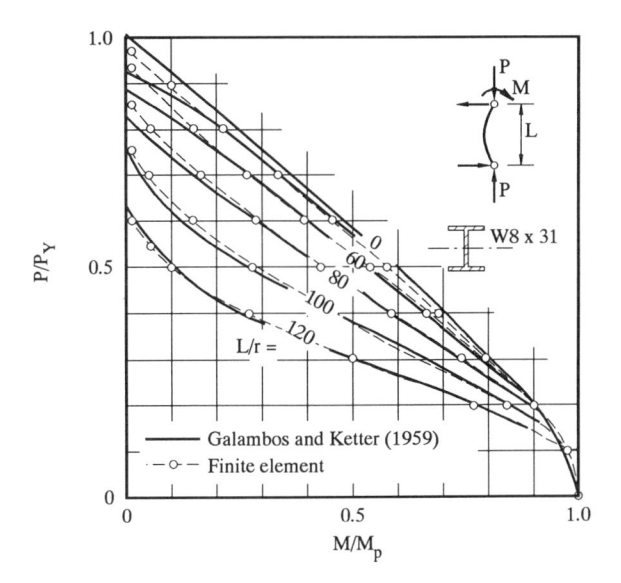

FIGURE 6.16. Ultimate strength interaction curves for straight beam-columns with end-moment ratio $\beta = 0.0$.

influencing the strength of beam-columns is the end-moment ratio β, which is defined here as the ratio of the smaller to the larger applied end moments, taken positive when the member is bent in reverse curvature. This sign convention for β is different from that adopted by Ketter (1961).

For comparison with the interaction curves of Galambos and Ketter (1959), the $\beta = -1.0$ and $\beta = 0.0$ cases have been analyzed with the plastic-zone finite element analysis. In the analyses, the beam-columns were modeled with ten equal-length elements. The distribution of monitoring points over the cross-section comprised five points through the thickness of the flanges, five points across the half-width of the flanges, 13 points over the web depth, and three points through the web thickness. The maximum strength interaction curves obtained with the plastic-zone analysis are compared with those of Galambos and Ketter in Figures 6.15 and 6.16 for the $\beta = -1.0$ and $\beta = 0.0$ cases, respectively. Due to the numerical difficulties associated with computing the strength of a perfectly straight column with residual stresses and a concentric axial load, the finite element solution points plotted adjacent the P/P_Y axis in Figures 6.15 and 6.16 were computed by first applying the end moments at the level $M/M_p = 0.01$ and then increasing the axial load P to ultimate. All the other finite element solutions indicated in Figures 6.15 and 6.16 were obtained by firstly applying a prescribed level of axial load P/P_Y and then incrementing the end moments until ultimate. This latter loading sequence was consistent with the procedure adopted by Galambos and Ketter. It is interesting to note in Figures 6.15 and 6.16 that the finite element solution points adjacent to the P/P_Y axis, which were computed with constant end moments and increasing axial force, appear to be consistent with the remainder of the finite element solutions, which were computed with the reverse

296 ADVANCED ANALYSIS OF STEEL FRAMES

loading sequence. It is therefore believed that the results shown in Figures 6.15 and 6.16 were not affected significantly by the proportionality of the applied loading.

It can be seen in Figures 6.15 and 6.16 that reasonably good agreement between the two methods of analysis is obtained, with the finite element strength curves tending to be slightly below the interaction curves of Galambos and Ketter (1959) at the lower levels of axial load, and slightly above at the higher levels of axial load.

6.8 European Calibration Frames
6.8.1 General

One of the earliest attempts to develop a defined and systematic approach to the verification of computer programs for second-order inelastic analysis was presented by Vogel (1985). As a result of discussions with Technical Committee 8 (Structural Stability) of the European Convention for Constructional Steelwork (ECCS), Vogel defined three different sway frames which are representative of common types of building frames and which can be used for the calibration of second-order inelastic analysis. These 'calibration frames' comprise a rectangular portal frame with fixed column bases, a pitched-roof portal frame with pinned column bases, and a six-story two-bay unbraced frame. Vogel analyzed the frames using both plastic-hinge and plastic-zone methods. These three frames have also been studied recently by Ziemian (1992) using plastic-hinge and plastic-zone analysis programs developed at Cornell University, thus providing an independent comparison with Vogel's solutions. A complete description of the European calibration frames is given in Section 7.4 of Chapter 7.

All of the members used in the frames are of I-section and oriented with their webs in the plane of the frame (major axis bending). It is assumed that all members are fully braced out of plane. A trilinear elastic-plastic-strain hardening relationship, as shown in Figure 7.24b of Chapter 7, is assumed for the steel. Residual stresses are assumed to be uniform along the member length and to be distributed over the cross-section as shown in Figure 7.4 of Chapter 7 (ECCS, 1983). Out-of-plumb frame imperfections were incorporated in the analyses according to the guidelines of the ECCS Publication No. 33 (ECCS, 1983; see Section 7.4.2, Chapter 7).

For all the analyses described henceforth, the size of the load increments prior to yielding was limited to about 10% of the ultimate load, with the steps becoming progressively smaller as the degree of nonlinearity increased. It is therefore believed that the load increments used for the following problems were sufficiently small to accurately track the elastic-plastic behavior of the frames throughout the loading regime, and that the results are therefore independent of load increment size.

6.8.2 Rectangular Portal Frame with Fixed Bases

The geometry, member sizes, material properties, geometrical imperfections, and applied loading for the fixed-base rectangular portal frame analyzed by Vogel (1985) and subsequently by Ziemian (1992) are shown in Figure 7.21 of Chapter 7. The

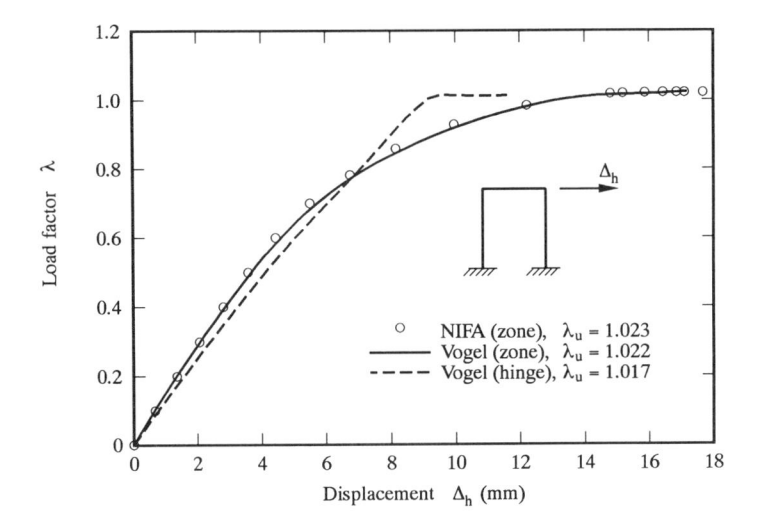

FIGURE 6.17. Load-deflection response of rectangular portal frame.

loads are increased proportionally until the ultimate strength of the frame is attained. For the present plastic-zone analysis, a refined mesh of 50 elements per column and 20 elements for the beam was used. Although not necessary to accurately determine the ultimate load and load-deformation behavior of the frame, a mesh of this refinement was chosen as it comprises the same member-to-element subdivision that was used by Ziemian (1992), as well as enabling the bending moment distribution and spread of yielding throughout the structure to be determined to a high degree of accuracy. In the cross-section subdivision, nine fibers across the flange half-width, five fibers through the flange thickness, 17 fibers through the web depth, and three fibers through the web thickness were used.

Using the modeling parameters described above, the ultimate strength limit state of the frame occurred at an applied load ratio of $\lambda_u = 1.02$, which is in excellent agreement with the plastic-zone analysis results of Vogel (1985; $\lambda_u = 1.02$) and Ziemian (1992; $\lambda_u = 1.00$). The *plastic-hinge* solutions of Vogel ($\lambda_u = 1.02$) and Ziemian ($\lambda_u = 1.05$) are also in good agreement with the plastic-zone solutions. As explained by Ziemian (1992), the difference in the ultimate load of the two hinge analyses may be attributed to the fact that different generalized yield surfaces in force space are used in the two analyses.

To investigate the significance of residual stresses and strain hardening, an additional plastic-zone analysis ignoring these effects was performed. The ultimate strength was computed to be $\lambda_u = 1.04$, which is very close to Ziemian's hinge analysis result of $\lambda_u = 1.05$.

The load-deflection response of the frame is shown in Figure 6.17, in which the responses of the Vogel zone and hinge analyses are also given for comparison. A significant observation from Figure 6.17 is that the response predicted by the zone analyses, which model the gradual spread of yielding through the volume of the members, is significantly more rounded at the higher levels of applied load than the response predicted by the plastic-hinge analysis.

The axial force and bending moment distributions at the strength limit state computed using the plastic-zone analysis (program NIFA) are given in Figure 6.18.* The values of bending moment quoted in Figure 6.18, and in the bending moment diagrams pertaining to subsequent examples in this chapter, represent nodal values which have been obtained by extrapolation from Gauss point values and then, where appropriate, nodal averaging. The extent of yielding through the structure can be gauged from Figure 6.19, in which the percentage of yielded cross-sectional areas is plotted.** The corresponding percentages computed by Vogel's zone analysis are also given in Figure 6.19 for comparison.

Comments on Sensitivity of Analysis Results to Modeling Assumptions

To gain an appreciation of the sensitivity of the finite element analysis results to the number of elements used, plastic-zone analyses have been performed for the following two additional meshes: (1) ten elements per column and five elements for the beam, and (2) 20 elements per column and ten elements for the beam. The ultimate load factor for both these analyses, which assume proportional loading, remained unchanged at $\lambda_u = 1.02$.

The effect of different loading sequences on the strength and behavior of the frame has also been examined. To investigate nonproportional loading, an analysis was performed in which the frame was loaded initially by the vertical loads only, to a load factor of $\lambda_{vu} = 1.02$, and then loaded by the horizontal load until the ultimate strength was attained. The ultimate load factor on the horizontal loading was computed to be $\lambda_{hu} = 1.06$. This is slightly above the value of $\lambda_u = 1.02$ computed assuming proportional loading, but it should be borne in mind that the magnitude of the lateral load is only 0.625% of the total vertical load and that the frame is therefore quite sensitive to lateral loads. The reverse loading sequence (horizontal load to a load factor $\lambda_{hu} = 1.02$, followed by vertical load) has also been investigated, from which the ultimate load factor on the vertical loads was computed to be $\lambda_{vu} = 1.02$. On the basis of these results, it can be concluded that the sequence of load application does not affect the strength of this portal frame to any significant extent.

---

* In the present plastic-zone finite element analysis, the stress resultants of axial force and bending moment are initially calculated at the Gauss points within the elements. Nodal values of the stress resultants are obtained by fitting a polynomial function through the Gauss point values and extrapolating to the nodes. As a consequence of this procedure, small discontinuities in the bending moment diagram may occur at the boundary between two adjacent elements, particularly in regions of substantial yielding; these discontinuities become smaller as the finite element mesh is refined. The bending moment diagram can be 'smoothed' by the appropriate averaging of the nodal values.

** The percentages of yielded cross-sectional areas quoted in Figure 6.19, and in other such diagrams in later examples in this chapter, represent values computed at the Gauss point nearest the positions indicated in the figure (e.g., in Figure 6.19, the percentages quoted are those at the Gauss points nearest the ends of the columns).

(a) Axial force (kN)	(b) Bending moment (kNm)

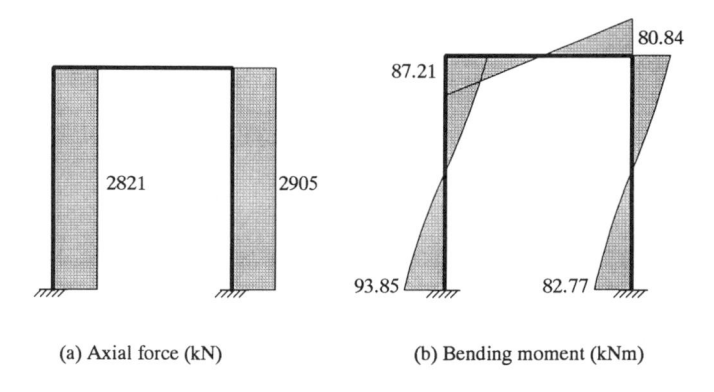

FIGURE 6.18. Axial force and bending moment distributions at ultimate ($\lambda_u = 1.023$) for rectangular portal frame.

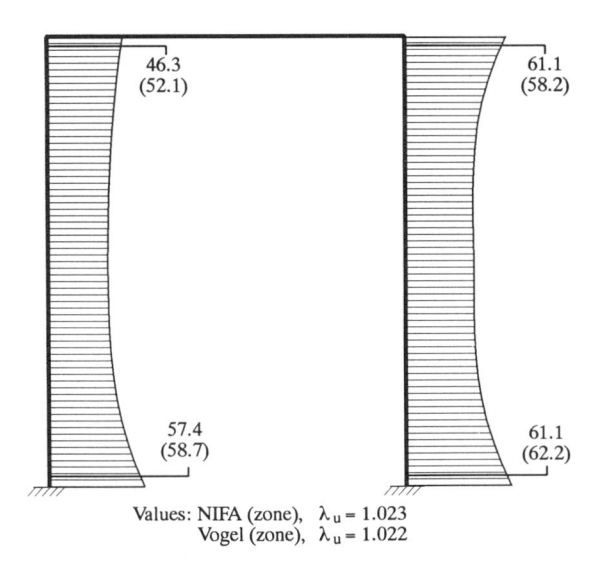

Values: NIFA (zone), $\lambda_u = 1.023$
Vogel (zone), $\lambda_u = 1.022$

FIGURE 6.19. Percentages of section areas yielded at ultimate for rectangular portal frame.

6.8.3 Pitched-Roof Portal Frame

The configuration of the pinned-base pitched-roof portal frame proposed as a calibration problem is shown in Figure 7.22 of Chapter 7. For the present plastic-zone analysis, 20 elements per column and 40 elements per beam were used. The cross-sectional subdivision used was the same as that described in Section 6.8.2 for the rectangular portal frame. The uniformly distributed load was modeled by work equivalent nodal actions acting on the 41 nodes across each beam.

The ultimate load factor computed by the present zone analysis was $\lambda_u = 1.06$, which is in excellent agreement with the zone analysis results of Vogel (1985) and Ziemian (1992) (both $\lambda_u = 1.07$). These zone analysis ultimate strengths are about

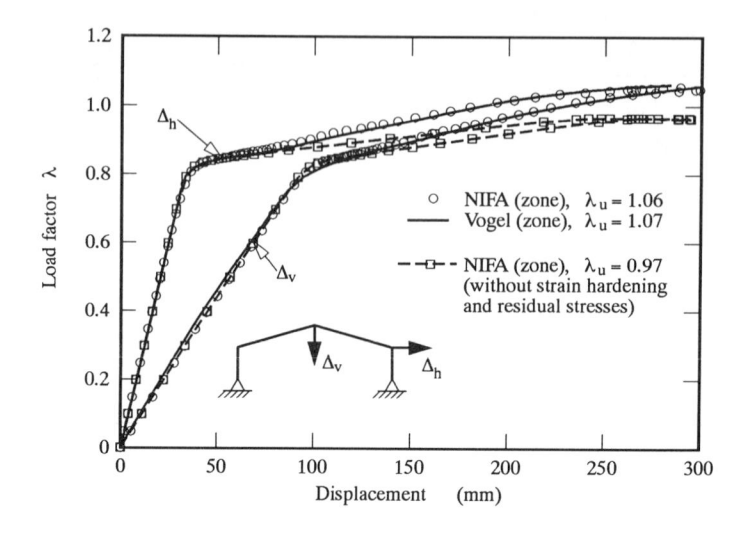

FIGURE 6.20. Load-deflection response of pitched-roof portal frame.

10% higher than the strengths predicted by the *hinge* analyses of Vogel and Ziemian ($\lambda_u = 0.96$ and 0.97, respectively). The appreciable difference between the zone and hinge ultimate strengths is due primarily to the fact that this portal frame is dominated by flexure and is little affected by instability, resulting in straining well into the strain hardening range. Strain hardening was included in the zone analyses, but neglected in the hinge analyses. For interest, a plastic-zone analysis of the frame was performed assuming elastic-perfectly plastic material behavior (i.e., neglecting strain hardening) and without residual stresses. Under this assumption, the frame ultimate strength corresponded to $\lambda_u = 0.97$ which is in agreement with the hinge analyses.

The load-deflection response of the frame, both including and neglecting strain hardening, is shown in Figure 6.20. The axial force and moment distributions at the ultimate load are given in Figure 6.21, and the extent of yielding throughout the frame at ultimate is depicted in Figure 6.22. Generally, the distribution of yielding calculated by the present zone analysis appears to match the results of Vogel's zone analysis quite well.

The sensitivity of the ultimate strength of the frame to the number of elements used has also been investigated for the case where strain hardening and residual stresses were included. When ten elements per column and 20 elements per beam were employed, the ultimate strength remained at $\lambda_u = 1.06$. When the number of elements per beam and column were halved again, the ultimate load dropped marginally to $\lambda_u = 1.05$.

6.8.4 Six-Story Two-Bay Frame

The multistory frame shown in Figure 7.23 of Chapter 7 may be regarded as a reasonably typical multistory frame of intermediate slenderness, having an elastic critical buckling load factor of 8.79 for the loads as defined. The loads on the frame

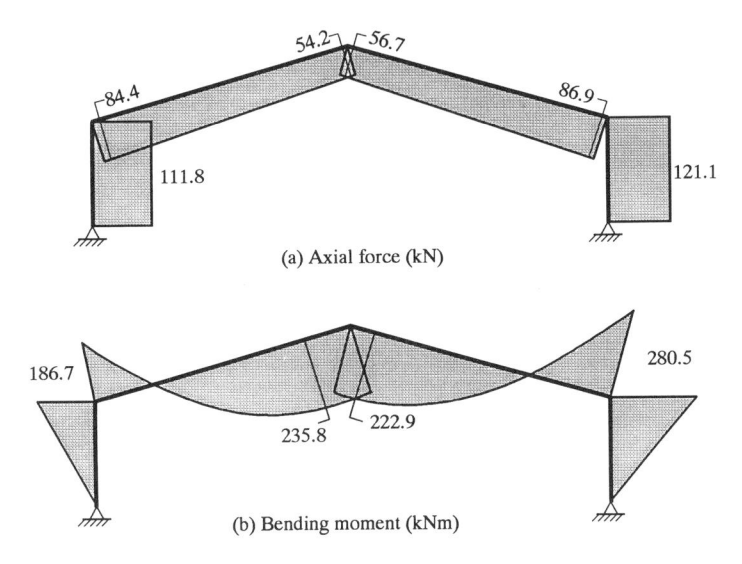

FIGURE 6.21. Axial force and bending moment distributions at ultimate ($\lambda_u = 1.06$) for pitched-roof portal frame.

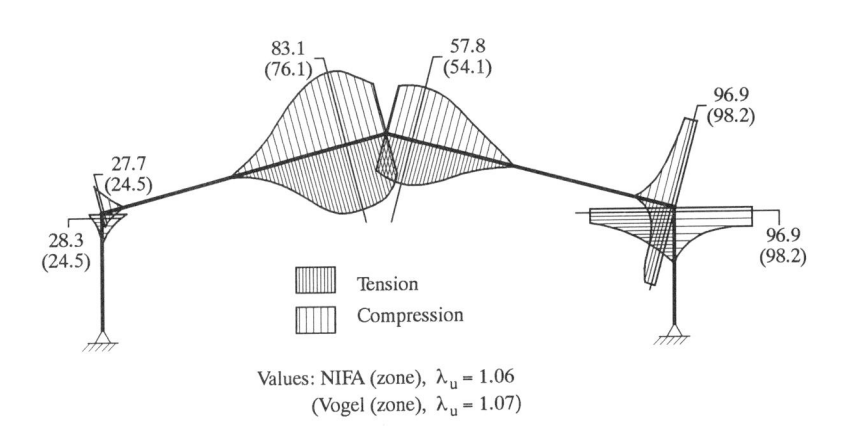

FIGURE 6.22. Percentages of section areas yielded at ultimate for pitched-roof portal frame.

are applied proportionally, and the uniformly distributed loads were modeled by work-equivalent nodal actions.

For the present plastic-zone analysis, each column and beam in the frame was subdivided into ten equal-length elements. This mesh is somewhat coarser than that used by Ziemian (1992), who used 20 elements for each beam and column, and is also different to the mesh used by Clarke et al. (1993) in a previous analysis of this frame. The cross-sectional subdivision used was the same as that described in Section 6.8.2 for the rectangular portal frame.

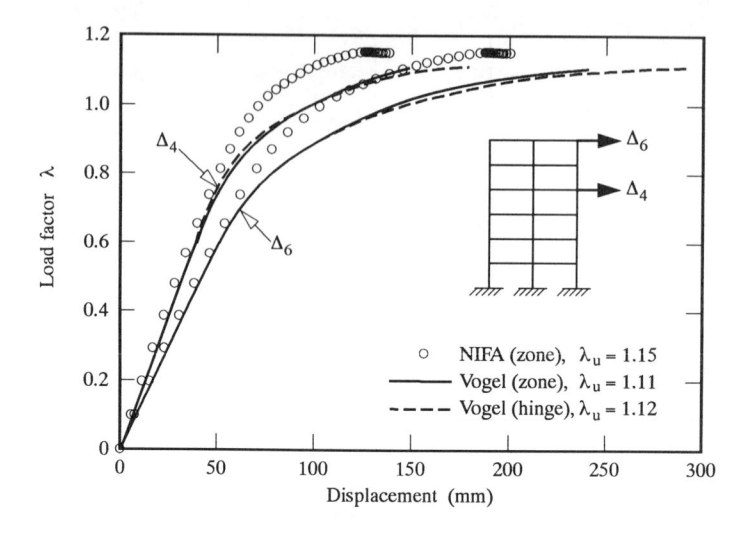

FIGURE 6.23. Load-deflection response of six-story frame.

In the previous analysis of Clarke et al. (1993), each column was subdivided into six elements, with the ratio of the element lengths along the member being 1:25:25:25:25:1, and each beam was subdivided into eight elements with the ratio of the element lengths being 1:25:25:1:1:25:25:1. This finite element mesh therefore employed short-length elements at the tops and bottoms of the columns, and at the ends and centers of the beams, with the aim of capturing the formation of 'plastic hinges' at these locations. Using this mesh, the ultimate load factor for the frame was computed to be $\lambda_u = 1.17$. When strain hardening and residual stresses were neglected, the frame strength decreased to $\lambda_u = 1.13$.

Using the present finite element mesh of ten elements per member, the ultimate load factor computed by the plastic-zone analysis including strain hardening was $\lambda_u = 1.15$. This result is between the ultimate strengths of Vogel (1985; $\lambda_u = 1.11$) and Ziemian (1992; $\lambda_u = 1.18$). Upon ignoring strain hardening and residual stresses, the ultimate load factor increased marginally to $\lambda_u = 1.16$.

The load-deflection behavior of the frame is shown in Figure 6.23, in which the results of Vogel (1985) and the present analysis have been plotted for comparison. Slight differences in the load-deflection behavior exist for the two zone analyses incorporating residual stresses and strain hardening.

The axial forces in the columns of the frame computed using the plastic-zone analysis are given in Figure 6.24. The extent of yielding throughout the frame at the ultimate load is indicated in Figure 6.25.

In a further investigation of the sensitivity of the frame strength to the degree of refinement of the mesh, two further analyses were performed: one using five equal-length elements per member, and another using 20 equal-length elements per member. For the coarser mesh, the ultimate strength rose slightly from $\lambda_u = 1.15$ to $\lambda_u = 1.17$. This latter figure is in agreement with the result of Clarke et al. (1993), in

```
        ┌─────┐ ┌─────┐ ┌─────┐
        │ 103 │ │ 229 │ │ 108 │
        │     │ │     │ │     │
        │ 256 │ │ 580 │ │ 283 │
        │     │ │     │ │     │
        │ 408 │ │ 923 │ │ 469 │
        │     │ │     │ │     │
        │ 547 │ │1267 │ │ 666 │
        │     │ │     │ │     │
        │ 673 │ │1616 │ │ 873 │
        │     │ │     │ │     │
        │ 776 │ │1996 │ │1071 │
```

Axial force (kN)

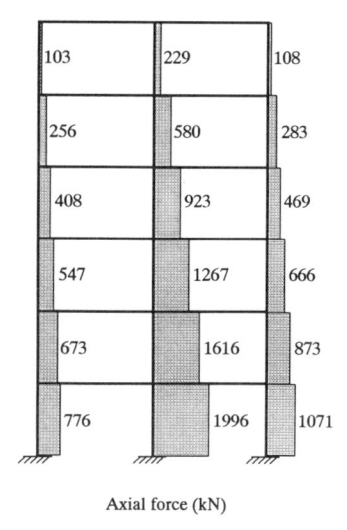

FIGURE 6.24. Axial force distribution at ultimate ($\lambda_u = 1.15$) for six-story frame.

which four equal-length elements per member were used over the majority of the member lengths, with additional short elements being used at the ends of the columns, and at the ends and centers of the beams. For the finer mesh using 20 elements per member, the computed ultimate load was unchanged at $\lambda_u = 1.15$.

6.9 A North American Calibration Frame

The two-story frame shown in Figure 6.26 has been proposed by Ziemian (1992) as a calibration frame. The frame is subjected to vertical loads only. The frame has been studied previously by Iffland and Birnstiel (1982) in an American Institute of Steel Construction report on frame stability. Ziemian conducted both plastic-hinge and plastic-zone analyses of the frame, which was assumed to be geometrically perfect.

For the present zone analyses, each column was subdivided into 20 elements, beams B1 and B3 were each subdivided into 20 elements, and beams B2 and B4 were each subdivided into 40 elements. This finite element mesh is somewhat coarser than that used by Ziemian (1992), who used 60 elements per beam and 50 elements per column. The cross-sectional subdivision used in the present zone analysis was the same as that described in Section 6.8.2 for the rectangular portal frame.

Load-displacement responses for the frame as computed by the hinge and zone analyses of Ziemian (1992) and the present zone analysis are shown in Figure 6.27. The ultimate load factor was computed to be $\lambda_u = 0.985$ which is slightly below the corresponding results of Ziemian's zone and hinge analyses ($\lambda_u = 1.01$ and 1.00, respectively). The percentages of section-area yielded at the respective ultimate loads of the plastic-zone analyses are compared in Figure 6.28.

To investigate the sensitivity of the frame strength to the refinement of the mesh, an additional analysis was conducted using only half the number of elements (i.e., ten

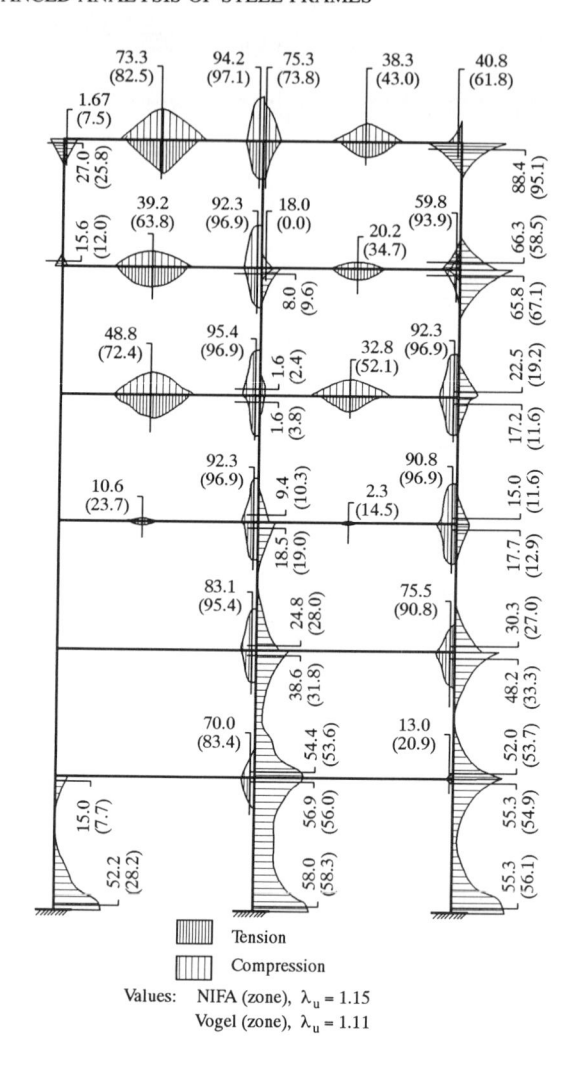

FIGURE 6.25. Percentages of section areas yielded at ultimate for six-story frame.

elements per member for the columns and beams B1 and B3, 20 elements for each of beams B2 and B4). The ultimate load factor remained virtually unchanged at $\lambda_u = 0.984$.

6.10 Australian Calibration Frames

6.10.1 Rigid-Jointed Truss

In recent papers (Clarke et al., 1992 and 1993), the plastic-zone analysis program described in this chapter was used to undertake advanced analyses of steel building frames composed of Australian sections and subjected to realistic design loads. The

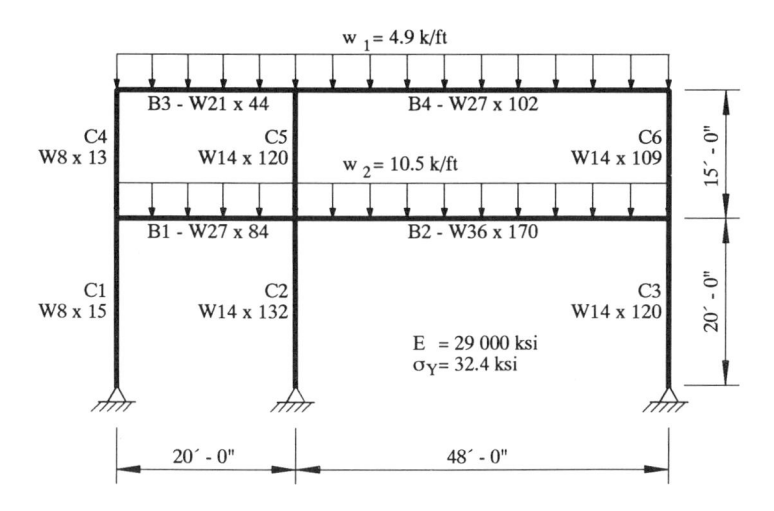

FIGURE 6.26. North American calibration frame.

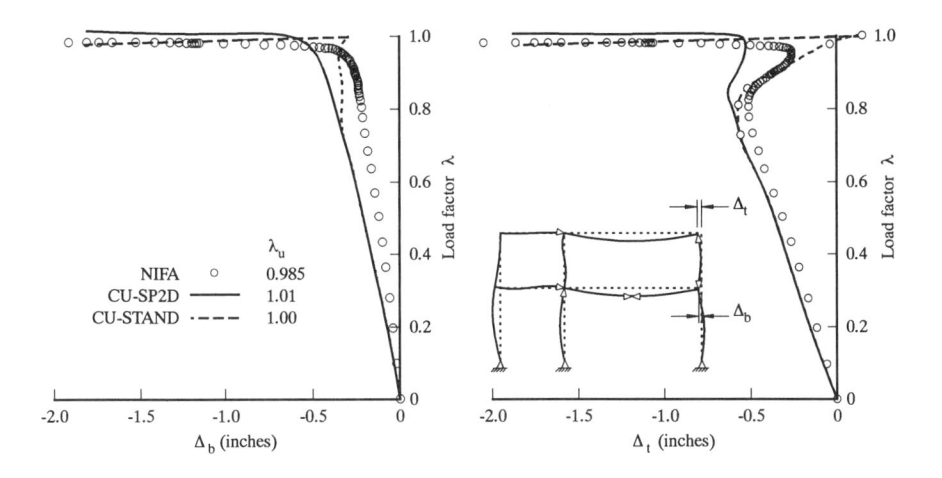

FIGURE 6.27. Load-deflection response of North American calibration frame.

	λ_u
NIFA	0.985
CU-SP2D	1.01
CU-STAND	1.00

effects of residual stresses, strain hardening, and both story out-of-plumbness and member out-of-straightness geometrical imperfections, were considered in the advanced analyses.

One of the frames studied by Clarke et al. (1992, 1993) was a rigid-jointed roof truss, shown in Figure 6.29, with transverse loads applied between as well as at the panel points of the top chord members. This frame therefore acts as a braced flexural frame under primary bending actions, and constitutes an atypical application of advanced analysis compared to the more usual unbraced portal frames or rectangular multistory frames subjected to gravity and lateral loads. An additional distinguishing

306 ADVANCED ANALYSIS OF STEEL FRAMES

FIGURE 6.28. Percentages of section areas yielded for North American calibration frame.

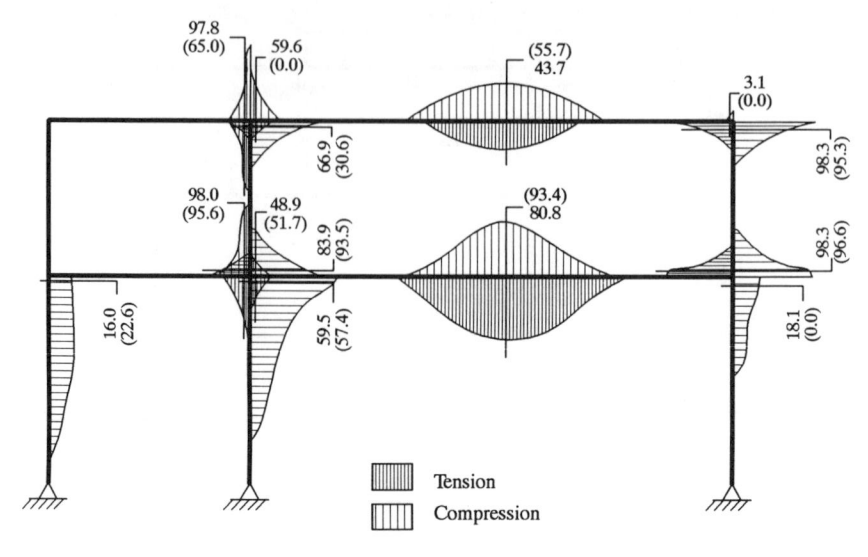

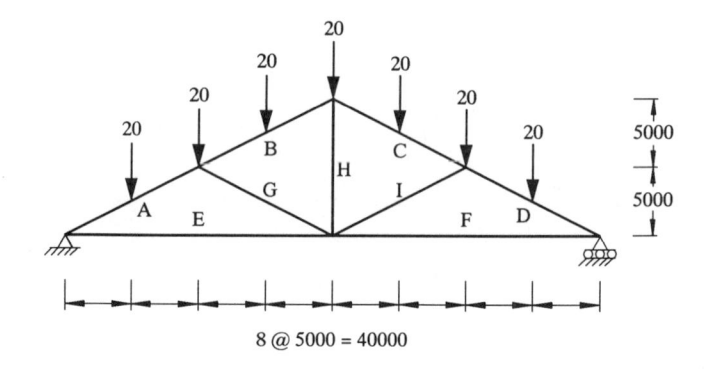

Values: NIFA (zone), $\lambda_u = 0.985$
(CU-SP2D (zone), $\lambda_u = 1.01$)

FIGURE 6.29. Rigid-jointed truss.

Member	Location	Section	A (mm²)	I_y (mm⁴)	S_y (mm³)	F_y (MPa)
A, D	Top chord	250UC72.9	9290	38.7×10^6	462×10^3	250
B, C	Top chord	200UC52.2	6640	17.7×10^6	264×10^3	250
E, F	Bottom chord	100UC14.8	1890	1.14×10^6	35.3×10^3	260
G, I	Web	150UC23.4	2980	4.03×10^6	80.9×10^3	260
H	Web	100UC14.8	1890	1.14×10^6	35.3×10^3	260

Note: (E) 200 000 MPa; (I_y) second moment of area about minor axis; (S_y) plastic section modulus about minor axis; and (F_y) yield stress.

$(\delta_0/L)_{member} = 1/1000$ $\Delta_0/H = 1/500$

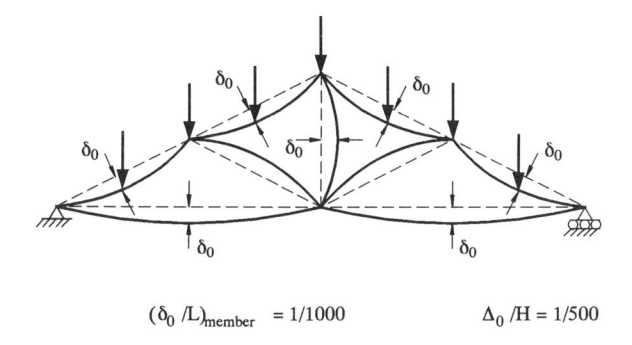

FIGURE 6.30. Geometrical imperfections for rigid-jointed truss.

feature of the truss frame of Figure 6.29 is that the I-section members are oriented with their flanges in the vertical plane and so are subjected to primary bending actions about the minor axis. I-section members bent about the minor rather than the major axis are more sensitive to residual stresses. This truss frame was initially proportioned using conventional elastic analysis/design procedures. The elastic buckling load factor for the truss is approximately 3.6, and first-order plastic collapse occurs through beam mechanisms in the upper top chord members at a load factor of 2.64.

To investigate the strength reduction caused by imperfections, plastic zone analyses of the rigid-jointed truss have been performed for the two cases of: (1) perfect frame geometry and neglecting residual stresses; and (2) imperfect frame geometry and including residual stresses. When included, the residual stresses followed the model of Figure 7.2 in Chapter 7. Strain hardening according to the stress-strain curve of Figure 7.24b (Chapter 7) was included in both analyses, although, as shown in Clarke et al. (1993), its influence on the frame strength is negligible. The geometrical imperfections comprised an assumed out-of-straightness for each member between panel points of 0.001 of the member length; the assumed orientation of the imperfections is shown in Figure 6.30. Twenty elements were used for each member of the truss between panel points. For the cross-sectional subdivision, the number of monitoring points used across the flange width, through the flange thickness, across the web half-width, and through the web thickness were 17, 3, 9, and 3, respectively.

When material and geometrical imperfections were not included, the ultimate strength of the truss corresponded to a load factor $\lambda_u = 2.04$. Upon inclusion of residual stresses and geometrical imperfections, the load factor at collapse decreased by 5.4% to $\lambda_u = 1.93$. The load-deflection behavior of the truss for the two analysis cases is shown in Figure 6.31. For the case where imperfections were included, the axial force and bending moment distributions at the ultimate load are shown in Figure 6.32. The extent of yielding throughout the imperfect frame at its ultimate strength is depicted in Figure 6.33. It is evident that the yielding is fairly localized around the ends and centers of each top chord segment between panel points.

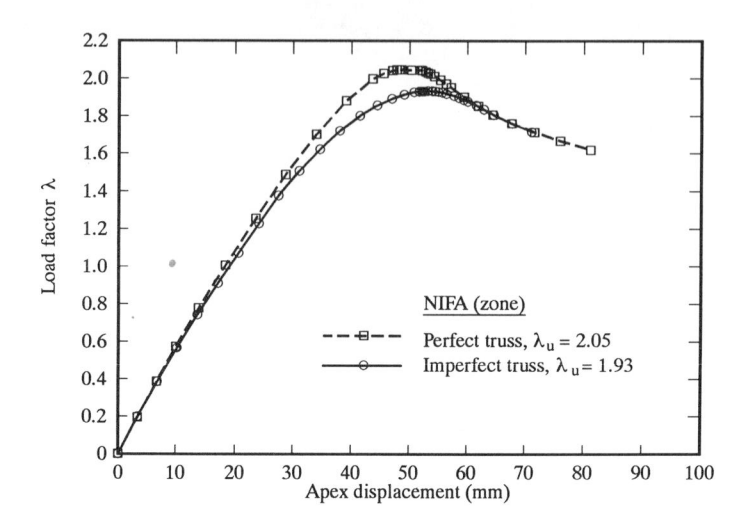

FIGURE 6.31. Load-deflection response of rigid-jointed truss.

6.10.2 Stressed-Arch Frame

The stressed-arch ('Strarch') frame of the type shown in Figure 6.34 is an unusual application of second-order inelastic (advanced) analysis, but one which has a practical application and for which test results are available (Clarke and Hancock, 1991). The frame shown in Figure 6.34 represents a subassemblage of a complete stressed-arch frame. The distinguishing feature of stressed-arch frames is their erection without the use of cranes or scaffolding, but by a post-tensioning stressing procedure, resulting in curving of the top chord into its final configuration. For some highly curved structures, the top chord may be plastically deformed during the erection procedure.

The geometry of the stressed-arch frame shown in Figure 6.34 is identical to the frames tested experimentally at the University of Sydney (Clarke and Hancock, 1991). However, for the theoretical results presented here, nominal material properties based on an elastic-perfectly plastic stress-strain curve and zero residual stresses, and pinned connections of the web members to the top and bottom chords, were assumed rather than the measured properties. A precise simulation of the experimental behavior of the frames has been performed previously (Clarke and Hancock, 1991) using measured data for the material stress-strain curves, residual stress distributions, measured cross-sectional dimensions, and joint flexibilities.

From the viewpoint of demonstrating plastic-zone analysis, the stressed-arch calibration frame incorporates distinct nonproportional loading and the interaction of axial force and bending in square tubes.

For the finite element analyses of the stressed-arch frame, each segment of the square hollow section top chord between its connections to the web members was subdivided into ten elements of equal length. The top chord outstands and rigid end bearings (Figure 6.34) were each modeled with a single element. The circular hollow

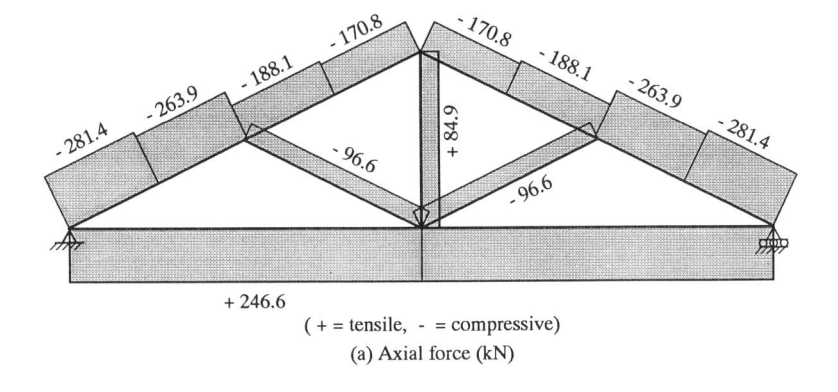

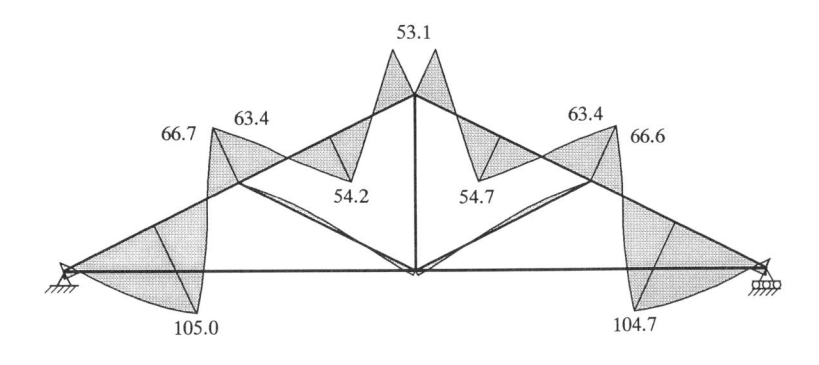

FIGURE 6.32. Axial force and bending moment distributions at ultimate ($\lambda_u = 1.93$) for imperfect rigid-jointed truss.

section web members were each subdivided into four elements, and the bottom chord was subdivided into two elements. Pinned connections were assumed between the web members and the top and bottom chords, and as a result, the material nonlinear behavior was confined to the top chord. To model the spread of yielding through the top chord, the fine grid of elemental areas shown in Figure 6.35a was used. However, as the analysis was in-plane and there were no residual stresses included, no refinement of the analysis was gained by employing the fine grid across the width of the top and bottom flanges. The dimensions and properties of the sections employed in the analysis, including the area (A), second moment of area (I), and yield stress (F_y) are given in Figure 6.35b. The elastic modulus (E) was assumed to be 200 000 MPa for all members.

To simulate the behavior of prototype stressed-arch frames, the loading on the top chord was nonproportional and was applied in two stages as follows:

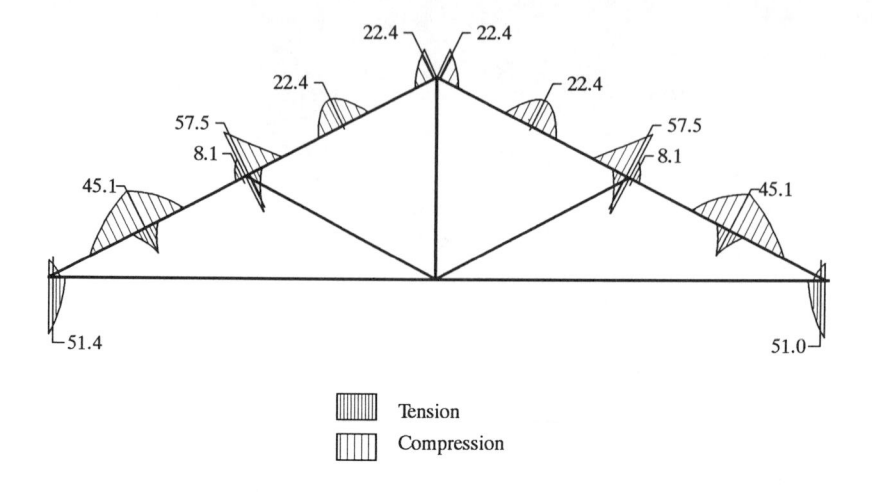

FIGURE 6.33. Percentages of section areas yielded at ultimate ($\lambda_u = 1.93$) for imperfect rigid-jointed truss.

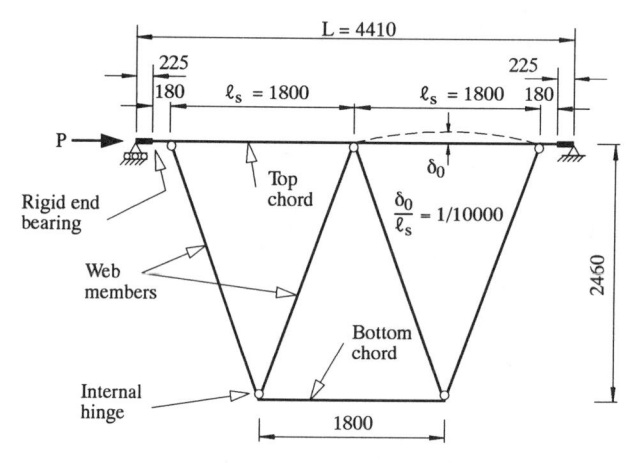

FIGURE 6.34. Stressed-arch (Strarch) calibration frame.

All dimensions in mm
Top chord: 102x102x4.9 SHS or 76x76x4.9 SHS
Bottom chord: 76x76x3.2 SHS   Web members: 48.3x3.2 CHS

(1) The top chord was curved by initially straining ('shrinking') the bottom chord. The initial midspan lateral deflection of the top chord introduced by the initial straining process is denoted v_{10} (refer to Figure 6.36 and Table 6.1).
(2) With the applied bottom chord initial strain held constant, the top chord was loaded by the axial load P until the maximum load of the frame was attained.

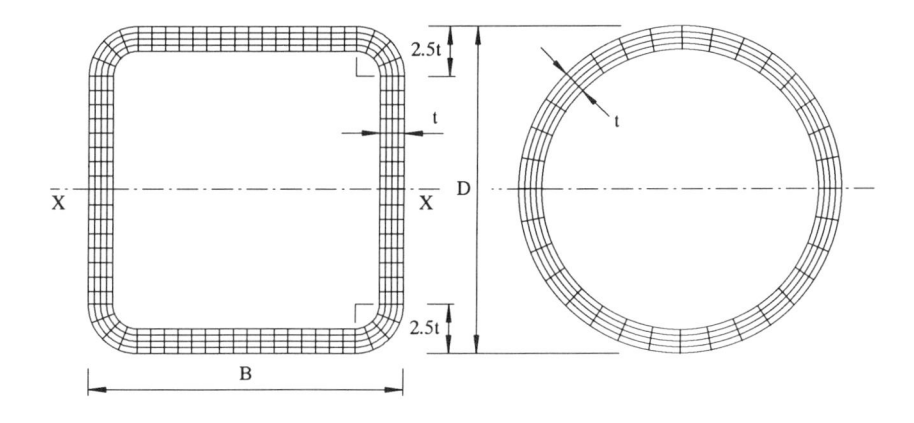

FIGURE 6.35. Cross-sectional subdivision and properties used in stressed-arch calibration frame. (a) Cross-sectional subdivision used for top chord and web members.

Designation	D (mm)	B (mm)	t (mm)	A (mm^2)	I_x (mm^4)	F_y (MPa)
102 × 102 × 4.9 SHS	101.6	101.6	4.9	1810	2.75 × 10^6	350
76 × 76 × 4.9 SHS	76.2	76.2	4.9	1320	1.07 × 10^6	350
76 × 76 × 3.2 SHS	76.2	76.2	3.2	899	0.782 × 10^6	350
48.3 × 3.2 CHS	48.3	—	3.2	453	0.116 × 10^6	200

(b) Dimensions, properties and yield stresses of cross-sections.

It is interesting to note that for the perfectly symmetric problem ($\delta_0 = 0$ in Figure 6.34), the failure mode involves excessive deformations in both top chord segments and is symmetric. However, in practice, excessive deformations will occur in one span only and result in an asymmetric failure mode at a lower load than the symmetric mode. To trigger the asymmetric mode of failure in the numerical analyses, a geometrical imperfection in the shape of a half-sine curve and of magnitude $\delta_0 = l_s/10000$ was superimposed on the right-hand top chord segment (Figure 6.34).

For a 102 × 102 × 4.9 SHS top chord, and the two cases of $v_{10} = 55.4$ mm and $v_{10} = 123.4$ mm, the load-deflection behavior for both the symmetric and asymmetric modes is shown in Figure 6.36. It can be seen in this figure that the asymmetric failure mode results in large deformations of the imperfect top chord segment (v_3), while the perfect segment (v_2) 'switches back' and reverses its sense of deformation just prior to the ultimate load. This 'switch back' behavior was also observed experimentally (Clarke and Hancock, 1991). Such complex behavior also constitutes a reasonable test of the numerical capabilities of the advanced analysis to traverse load and displacement limit points on the response path.

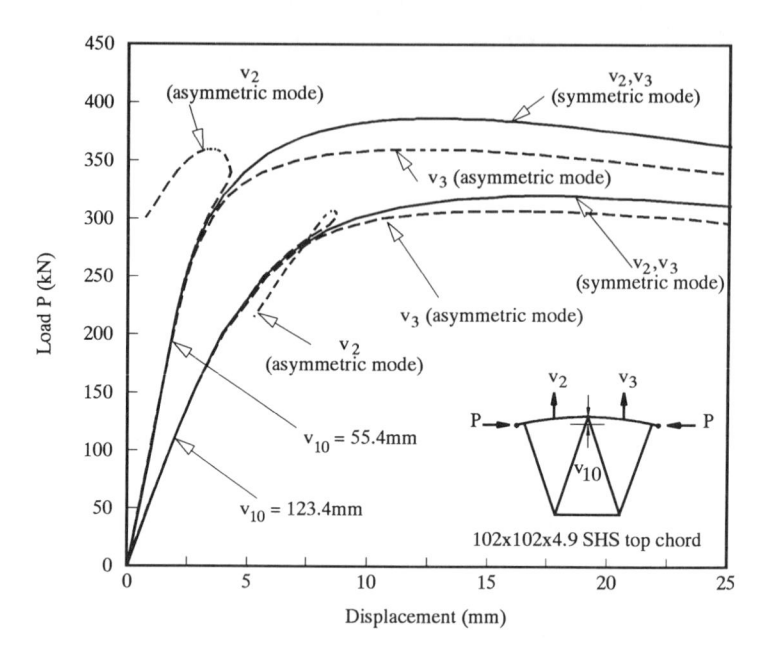

FIGURE 6.36. Load-deflection response of stressed-arch frame.

Several analyses with different initial midspan deflections v_{10} have been performed for the two top chord sizes of $76 \times 76 \times 4.9$ SHS and $102 \times 102 \times 4.9$ SHS. The ultimate loads P_u for both the symmetric and asymmetric modes of failure are given in Table 6.1. The results in Table 6.1 have been plotted in Figure 6.37, in which the degradation in top chord strength with increasing initial lateral deflection can be observed. The test results from Clarke and Hancock (1991) have also been plotted in Figure 6.37 for comparison with the theoretical solutions.

6.11 Application to Engineering Practice
6.11.1 Current Practice

It is well known than the strength and stability of a structural system and its members are interdependent. However, conventional methods of steel design based on the use of elastic analysis combined with member strength interaction equations do not take this interdependence into account rigorously, but consider the strength and stability of individual members separately from the stability of the frame as a whole. Although the majority of typical steel structures can be, and have been in the past, designed perfectly adequately using conventional methods of analysis and member design, the use of advanced analysis will result in structures which are more *rationally proportioned* according to the limit states design philosophy, and which hopefully use less material and are therefore more economical.

In recent years, several methods of advanced analysis based on the refined plastic hinge concept, which consider the effects of gradual yielding, residual stresses, and

Table 6.1 Ultimate Strength Results for Stressed-Arch Calibration Frame

76 × 76 × 4.9 SHS Top Chord			102 × 102 × 4.9 SHS Top Chord		
Initial displacement v_{10} (mm)	Symmetric mode P_u (kN)	Asymmetric mode P_u (kN)	Initial displacement v_{10} (mm)	Symmetric mode P_u (kN)	Asymmetric mode P_u (kN)
13.2	353	347	13.1	542	530
26.6	296	295	26.3	472	449
35.6	269	261	35.2	438	415
44.7	246	237	44.2	411	384
58.5	219	206	55.4	387	360
81.8	191	179	75.3	361	338
99.7	181	170	98.1	341	324
118.1	173	162	123.4	320	307

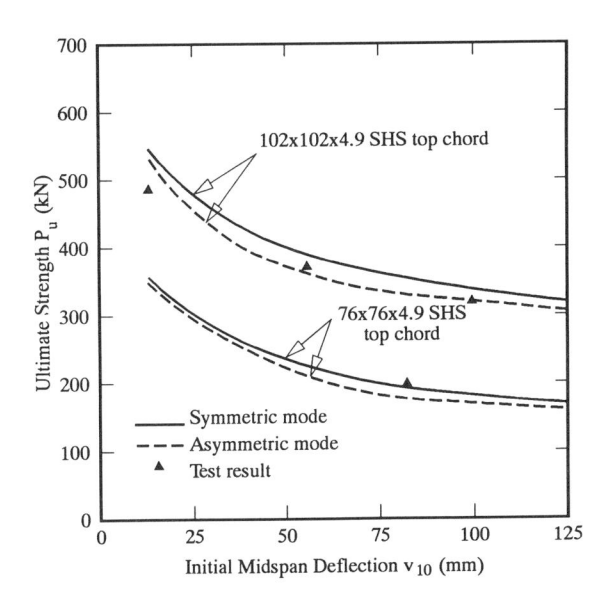

FIGURE 6.37. Variation of strength with initial midspan deflection for stressed-arch frame.

geometrical imperfections in some phenomenonological or behavioral way, have been developed (White et al., 1993). One of the major aims in developing these refined hinge methods was that they be of practical use to the structural engineering profession. A major advantage of refined plastic-hinge analysis over plastic-zone analysis is the considerably lower computational effort required; for typical unbraced multistory rectangular frames, refined hinge analysis results based on a one element per member discretization of the structure can achieve results of comparable accuracy to plastic-zone analysis (and which are certainly adequate for engineering design purposes) (White et al., 1993).

At the present time, the computational intensity of plastic-zone analysis precludes its wide use in routine design. The role of plastic-zone analysis in structural analysis and design is therefore seen to be complementary to refined plastic-hinge analysis. The behavioral models incorporated in refined hinge analyses are often only applicable to hot-rolled I-section members bent about the major axis. Plastic-zone analyses, however, are completely general in the sense that the cross-sectional geometry, residual stress distributions, and material stress-strain characteristics can be modeled explicitly. Structures composed of rectangular and circular tubular members, which are often cold-formed rather than hot-rolled, can therefore be analyzed accurately with plastic-zone analysis, while the application of refined hinge analyses to sections of this type may be of uncertain accuracy. Plastic-zone analysis may also have a role in the design of a special, unique, or exceptional structure, in the investigation of an existing structure whose capacity is in doubt, or in the investigation of a structural failure. Plastic-zone analysis also has a clear role in research, especially in the development of benchmark problems and solutions, the calibration of the refined plastic-hinge analyses, and the development of simplified design rules.

The stressed-arch framing system (Clarke, 1992; Clarke and Hancock, 1991; and Hancock et al., 1988), which was described briefly in Section 6.10.2 in relation to a calibration problem, is an example of a unique structural form which in fact *relies* on the use of plastic-zone analysis to develop rational and economical design procedures, particularly for the top chord. The irrationality of the conventional elastic analysis/design methodology for stressed-arch frames, which was demonstrated in Clarke (1992), is primarily a result of the fact that the mechanism of load resistance of the top chord under superimposed load differs markedly from its behavior during erection. Furthermore, inelastic analysis based on the plastic hinge concept is believed to be inappropriate for the analysis of stressed-arch frames, primarily because of the inability of a hinge analysis method to adequately model the extensive yielding of the top chord which occurs during the erection procedure and under service loading. It is therefore apparent that plastic-zone analysis is the only analysis method which can be used to facilitate the rational and economical design of the top chord of stressed-arch frames.

The recommended approach to the design of the top chord of stressed-arch frames described in Clarke (1992) was therefore based on the advanced analysis/design provisions of Appendix D of the Australian Standard AS4100–1990 *Steel Structures* (1990). These provisions state that advanced analysis may only be performed on two-dimensional frames comprising members of compact section with lateral restraints that prevent flexural-torsional buckling. An advanced analysis is required to take into account all the relevant factors that influence the behavior of this class of frame, including material properties, residual stresses, geometrical imperfections, second-order (instability) effects, connection behavior, erection procedures, and interaction with the foundations. To the author's knowledge, the AS4100–1990 is the only steel design standard or specification worldwide which explicitly addresses advanced analysis in this way, and, even more importantly, waives the requirement of separate checks of member in-plane strength when the action effects have been determined using advanced analysis.

The basis of the recommended design procedure for the top chord of stressed-arch frames (Clarke, 1992), involved the application of the plastic-zone analysis to perform advanced analyses of frame *subassemblages*. The parametric study quantified the top chord strength as a function of the initial curvature, the geometrical slenderness, and the yield stress. The resulting top chord strength charts can then be used for member design in lieu of the conventional design rules of specifications and standards. The top chord strength charts are suitable for use in conjunction with a first-order elastic analysis of the erected frame to determine the distribution of axial force in the top chord.

6.11.2 Future Research

One area of plastic-zone analysis research in which further work is required is concerned with the appropriate modeling of imperfections and how to most logically incorporate a resistance (capacity) factor into the overall design process. The AS4100–1990 specifies erection tolerance limits for story out-of-plumbness and overall frame nonverticality, and fabrication tolerance limits for member out-of-straightness. However, no guidance is given on the distribution and orientation of the imperfections to be used; one approach would be to assume that all imperfections are of magnitude as given by code limits, and are orientated so as to produce the lowest frame strength (Clarke et al., 1992, 1993). These assumptions for the geometrical imperfections, combined with an assumption of no strain hardening, represent maximum design limits for a worst case scenario. The application of a resistance factor for the total frame uncertainties which is the same as is used in member design ($\phi = 0.9$ in AS4100–1990) produces further conservatism.

For the practical application of advanced analysis, however, it would be more appropriate to adopt mean, rather than extreme, values for the imperfections (e.g., length/1500 for column out-of-straightness rather than length/1000) and to consider a random distribution of average imperfections rather than full detrimental orientation. Equally well, it would be appropriate to include the beneficial effects of strain hardening. Under these more appropriate assumptions, it would appear to be reasonable to apply a capacity factor of $\phi = 0.9$ to the strength of the whole frame to reflect the uncertainties in such aspects as the imperfections, the material behavior, the joints, and the section properties.

Alternatively, inclusion in the advanced analysis of extreme-value assumptions for the imperfection magnitudes and orientations (all detrimental), and the material behavior (neglection of strain hardening), seems to justify the use of a higher capacity factor, perhaps $\phi = 0.95$, to reflect the reduced set of applicable frame uncertainties, particularly for redundant systems. If indeed a higher capacity factor ϕ is warranted under these circumstances, then the required value would need to be determined by calibration so as to give advanced analysis/design a competitive edge over conventional design methods. Further research is required to investigate the theoretical question of mean versus extreme values, and appropriate capacity factors, within the context of advanced analysis.

6.12 Summary and Concluding Remarks

This chapter has described the plastic-zone analysis of frames. The plastic-zone analysis was formulated using a classical finite element approach, and the capabilities and accuracy of the analysis was demonstrated through its application to several beam-columns and frames. Within the scope of classical in-plane beam-column theory (i.e., ignoring local buckling and out-of-plane behavior) plastic-zone analysis may be regarded as the most refined possible, but also the most computationally intensive.

Compared to refined plastic-hinge analyses, which nonetheless qualify as 'advanced analyses' for the majority of practical unbraced multistory rigid frames, plastic-zone analyses have the advantage of *generality* in the sense that the actual cross-sectional geometry, stress-strain characteristics, and residual stresses of the members is modeled explicitly in the analysis. A typical structure such as stressed-arch frames can therefore be analyzed accurately with plastic-zone analysis, while the performance of the refined hinge analyses for this type of structural system is of uncertain accuracy. Stressed-arch frames are a unique example of the value of advanced analysis, and plastic-zone analysis in particular; the top chord of stressed-arch frames *cannot* be designed in a rational and economic manner using any method of structural analysis for which compliance with conventional member design rules is mandatory.

In recognition of the potential of advanced analysis, the method is documented in Appendix D of the Australian Standard AS4100–1990 *Steel Structures*. The AS4100–1990 is believed to be unique among steel design standards and specifications worldwide in that separate checks of member in-plane capacity are not mandatory if advanced analysis is used.

In addition to the nonlinear member behavior, accurate prediction of the overall frame response requires modeling of the strength and behavior of the connections (primarily the nonlinear moment-rotation behavior). Most of the standard structural connections used in steel frames behave as semi-rigid or flexible connections (see Chapter 3), yet are often tacitly assumed to be fully rigid or perfectly flexible from a structural analysis viewpoint. In all the analysis examples presented in this chapter, idealized models of connection response (fully rigid or perfectly flexible) were assumed. Although the advanced analysis provisions of the AS4100–1990 do not require separate verification of member in-plane strength, the Standard's requirements for connection capacity *do* have to be checked independently.

Guidance on the requirements and use of inelastic analysis, including recommended geometrical imperfections, residual stresses, and stress-strain curves to include in plastic-zone theory, were documented by the European Convention for Constructional Steelwork in Publication No. 33 (ECCS, 1983). Eurocode 3 (1990) also provides guidance on methods of global analysis, including 'elastic-perfectly plastic' (plastic-hinge) methods, and 'elasto-plastic' (plastic-zone) methods. Extensive provisions relating to the modeling of imperfections in the global analysis are also given in Eurocode 3. However, unlike the advanced analysis provisions of AS4100–1990, Eurocode 3 appears to enforce separate cross-section and member resistance checks, although it does contain provisions for the determination of column strength using direct analysis of the member with an initial bow imperfection.

The use of inelastic analysis is not specifically addressed in the current AISC-LRFD Specification (1986). There has, however, been much recent effort by researchers in the US and elsewhere to establish guidelines for the use of practical methods of second-order inelastic analysis in conjunction with limit states standards and specifications. Most of the research work has been performed in conjunction with Task Group 29, 'Second-Order Inelastic Analysis for Frame Design' of the Structural Stability Research Council (SSRC) under the chairmanship of Professor D. W. White. The recent SSRC publication "Plastic Hinge Based Methods for Advanced Analysis and Design of Steel Frames: An Assessment of the State-of-the-Art" (1993) documents much of the research effort of Task Group 29. Particular issues addressed in the SSRC monograph are the demonstration and clarification of the capabilities and limitations of contemporary plastic-hinge based methods for advanced analysis and design of steel frames, and the background to and implications of current and potential design provisions relating to advanced analysis in codes and specifications from North America, Europe, and Australia.

References

AISC (1986) *Load and Resistance Factor Design Specification for Structural Steel Buildings*, 1st ed., American Institute of Steel Construction, Chicago, IL.

Alvarez, R. J. and Birnstiel, C. (1969) Inelastic analysis of multistory multibay frames, *J. Struct. Div.*, ASCE, 95(ST11), 2477–2503.

Armen, H. (1979) Assumptions, models, and computational methods for plasticity, *Computers Struct.*, 10(1/2), 161–174.

AS4100–1990, *Steel Structures*, Standards Australia, Sydney, 1990.

Bathe, K. J. (1982) *Finite Element Procedures in Engineering Analysis*, Prentice-Hall, Englewood Cliffs, NJ.

Bathe, K. J. and Bolourchi, S. (1979) Large displacement analysis of three-dimensional beam structures, *Int. J. Numerical Methods Eng.*, 14(7), 961–986.

Bathe, K. J., Ramm, E., and Wilson, E. L. (1975) Finite element formulations for large deformation dynamic analysis, *Int. J. Numerical Methods Eng.*, 9(2), 353–386.

Bergan, P. G. and Clough, R. W. (1972) Convergence criteria for iterative processes, *AIAA J.*, 10(8), 1107–1108.

Bild, S. and Trahair, N. S. (1989) In-plane strengths of steel columns and beam-columns, *J. Constr. Steel Res.*, 13(1), 1–22.

Bushnell, D. (1977) A strategy for the solution of problems involving large deflections, plasticity and creep, *Int. J. Numerical Methods Eng.*, 11(4), 683–708.

Chan, S. L. (1988) Geometric and material non-linear analysis of beam-columns and frames using the minimum residual displacement method, *Int. J. Num. Methods Eng.*, 26(12), 2657–2669.

Chan, S. L. and Kitipornchai, S. (1988) Inelastic post-buckling behavior of tubular struts, *J. Struct. Eng. ASCE*, 114(5), 1091–1105.

Chu, K. H. and Pabarcius, A. (1964) Elastic and inelastic buckling of portal frames, *J. Eng. Mech. Div. ASCE*, 90(EM5), 221–249.

Clarke, M. J. (1992) The Behaviour of Stressed-Arch Frames, Ph.D. thesis, School of Civil and Mining Engineering, University of Sydney, Australia.

Clarke, M. J., Bridge, R. Q., Hancock, G. J., and Trahair, N. S. (1992) Advanced analysis of steel building frames, *J. Constr. Steel Res.*, 23(1–3), 1–29.

Clarke, M. J., Bridge, R. Q., Hancock, G. J., and Trahair, N. S. (1993) Australian trends in the plastic analysis and design of steel building frames, in *Plastic Hinge Based Methods for Advanced Analysis and Design of Steel Frames: An Assessment of the State-of-the-Art*, White, D. W. and Chen, W. F., Eds., Structural Stability Research Council, Lehigh University, Bethlehem, PA, 65–93.

Clarke, M. J., Bridge, R. Q., Hancock, G. J., and Trahair, N. S. (1993) Benchmarking and verification of second-order elastic and inelastic frame analysis programs, in *Plastic Hinge Based Methods for Advanced Analysis and Design of Steel Frames: An Assessment of the State-of-the-Art*, White, D. W. and Chen, W. F., Eds., Structural Stability Research Council, Lehigh University, Bethlehem, PA, 245–274.

Clarke, M. J. and Hancock, G. J. (1990) A study of incremental-iterative strategies for non-linear analyses, *Int. J. Numerical Methods Eng.*, 29, 1365–1391.

Clarke, M. J. and Hancock, G. J. (1991) Finite-element nonlinear analysis of stressed-arch frames, *J. Struct. Eng. ASCE*, 117(10), 2819–2837.

Collins, R. J. (1973) Bandwidth reduction by automatic renumbering, *Int. J. Numerical Methods Eng.*, 6(3), 345–356.

Cook, R. D. (1981) *Concepts and Applications of Finite Element Analysis*, 2nd ed., John Wiley & Sons, New York.

Crisfield, M. A. (1981) A fast incremental/iterative solution procedure that handles 'snap-through', *Computers Struct.*, 13(1–3), 55–62.

Deierlein, G. G., Zhao, Y., and McGuire, W. (1991) A Two-Surface Concentrated Plasticity Model for Analysis of 3D Framed Structures, Proceedings Annual Technical Session, Structural Stability Research Council, Chicago, IL, 423–432.

Durocher, L. L. and Gasper, A. (1979) A versatile two-dimensional mesh generator with automatic bandwidth reduction, *Computers Struct.*, 10(4), 561–575.

ECCS (1983) Ultimate Limit State Calculation of Sway Frames with Rigid Joints, Technical Working Group 8.2-Systems, Publication No. 33, European Convention for Constructional Steelwork, Brussels, November, 1983.

EUROCODE 3, Design of Steel Structures, Part 1-General Rules and Rules for Buildings, Edited draft, Issue 5, Commission of the European Communities, Brussels, November, 1990.

Fung, Y. C. (1965) *Foundations of Solid Mechanics*, Prentice-Hall, Englewood Cliffs, NJ.

Galambos, T. V. and Ketter, R. L. (1959) Columns under combined bending and thrust, *J. Engineering Mechanics Div. ASCE*, 85(EM2), 1–30.

Hancock, G. J., Key, P. W., and Olsen, C. J. (1988) Structural behaviour of a stressed arch structural system, Recent Research and Developments in Cold-Formed Steel Design and Construction, Ninth International Specialty Conference on Cold-Formed Steel Structures, University of Missouri-Rolla, St. Louis, MO, 273–294.

Haythornthwaite, R. M. (1957) Beams with full end fixity, *Engineering*, 183, 110–112.

Hsieh, S. H. and Deierlein, G. G. (1991) Nonlinear analysis of three-dimensional frames with semi-rigid connections, *Computers Struct.*, 41(5), 995–1009.

Iffland, J. S. B. and Birnstiel, C. (1982) Stability Design Procedures for Building Frameworks, Research Report, AISC Project No. 21.62, American Institute of Steel Construction, Chicago, IL.

Ketter, R. L. (1961) Further studies of the strength of beam-columns, *J. Structural Div. ASCE*, 87(ST6), 135–152.

Key, P. W. (1988) The Behaviour of Cold-Formed Square Hollow Section Columns, Ph.D. thesis, School of Civil and Mining Engineering, University of Sydney, Australia.

Key, P. W. and Hancock, G. J. (1993) A theoretical investigation of the column behaviour of cold-formed square hollow sections, *Thin-Walled Struct.*, 16(1-4), 31–64.

King, W. S., White, D. W. and Chen, W. F. (1992) Second-order inelastic analysis methods for steel-frame design, *J. Struct. Eng. ASCE*, 118(2), 408–428.

Kitipornchai, S., Al-Bermani, F. G. A., and Chan, S. L. (1988) Geometric and material nonlinear analysis of structures comprising rectangular hollow sections, *Eng. Struct.*, 10, 13–23.

Little, G. H. (1982) Complete collapse analysis of steel columns, *Int. J. Mechanical Sci.*, 24(5), 279–298.

Mattiasson, K., Bengtsson, A., and Samuelsson, A. (1986) On the Accuracy and Efficiency of Numerical Algorithms for Geometrically Nonlinear Structural Analysis, *Finite Element Methods for Nonlinear Problems*, Bergan, P. G., Bathe, K. J., and Wunderlich, W., Eds., Springer-Verlag, Berlin, 3–23.

Mattiasson, K. and Samuelsson, A. (1984) Total and Updated Lagrangian Forms of the Co-rotational Finite Element Formulation in Geometrically and Materially Nonlinear Analysis, *Proceedings, International Conference on Numerical Methods for Nonlinear Problems*, Vol. 2, Taylor, et al., Eds., Pineridge Press, Swansea, 134–151.

McGuire, W. (1988) Research and Practice in Computer-Aided Structural Engineering, Arthur J. Boase Lecture, Department of Civil, Environmental and Architectural Engineering, University of Colorado, Boulder, CO.

Mendelson, A. (1968) *Plasticity: Theory and Application*, Macmillan, New York.

Moses, F. (1964) Inelastic frame buckling, *J. Struct. Div. ASCE*, 90(ST6), 105–121.

Newmark, N. M. (1943) Numerical procedure for computing deflections, moments and buckling loads, *Trans. ASCE*, 108, 1161.

Nigam, N. C. (1970) Yielding in framed structures under dynamic loads, *J. Eng. Mechanics Div. ASCE*, 96(EM5), 687–709.

Nyssen, C. (1981) An efficient and accurate iterative method, allowing large incremental steps, to solve elasto-plastic problems, *Computers Struct.*, 13(1–3), 63–71.

Orbison, J. G., McGuire, W., and Abel, J. F. (1982) Yield surface applications in non-linear steel frame analysis, *Computer Methods Appl. Mechanics Eng.*, 33, 557–573.

Porter, F. L. and Powell, G. H. (1971) Static and Dynamic Analysis of Inelastic Frame Structures, Report No. EERC 71-2, Earthquake Engineering Research Center, University of California, Berkeley, CA.

Powell, G. H. and Chen, P. F. (1986) 3D Beam-column element with generalized plastic hinges, *J. Eng. Mech. ASCE*, 112(7), 627–641.

Powell, G. and Simons, J. (1981) Improved iteration strategy for nonlinear structures, *Int. J. Numerical Methods Eng.*, 17(10), 1455–1467.

Ramberg, W. and Osgood, W. R. (1943) Description of stress-strain curves by three parameters, *Tech. Note, NACA*, No. 902a.

SSRC (1988) *Guide to Stability Design Criteria for Metal Structures*, 4th ed., Galambos, T.V., Ed., Structural Stability Research Council, John Wiley & Sons, New York.

Swanger, M. H. and Emkin, L. Z. (1979) A Fiber Element Model for Nonlinear Frame Analysis, Proceedings of the Seventh Congress on Electronic Computation, ASCE, St. Louis, MO, 510–536.

Teng, J. G. and Rotter, J. M. (1989) Elastic-plastic large deflection analysis of axisymmetric shells, *Computers Struct.*, 31(2), 211–233.

Vogel, U. (1985) Calibrating frames, *Stahlbau*, 54(10), 295–301.

White, D. W. (1985) Material and Geometric Nonlinear Analysis of Local Planar Behavior in Steel Frames Using Interactive Computer Graphics, M.S. thesis, School of Civil and Environmental Engineering, Cornell University, Ithaca, NY.

White, D. W. and Chen, W. F., Eds. (1993) *Plastic Hinge Based Methods for Advanced Analysis and Design of Steel Frames: An Assessment of the State-of-the-Art*, Structural Stability Research Council, Lehigh University, Bethlehem, PA.

White, D. W., Liew, J. Y. R., and Chen, W. F. (1993) Toward advanced analysis in LRFD, in *Plastic Hinge Based Methods for Advanced Analysis and Design of Steel Frames: An Assessment of the State-of-the-Art*, White, D. W. and Chen, W. F., Eds., Structural Stability Research Council, Lehigh University, Bethlehem, PA, 95–173.

Wright, E. W. and Gaylord, E. H. (1968) Analysis of unbraced multistory steel rigid frames, *J. Struct. Div.*, ASCE, 94(ST5), 1143–1163.

Ziemian, R. D. (1992) A Verification Study for Methods of Second-Order Inelastic Analysis, Proceedings, Annual Technical Session, Structural Stability Research Council, Pittsburgh, PA, 315–326.

Ziemian, R. D., White, D. W., Deierlein, G. G., and McGuire, W. (1990) One Approach to Inelastic Analysis and Design, Proceedings 1990 National Steel Construction Conference, American Institute of Steel Construction, Chicago, IL, 19.1–19.19.

Zienkiewicz, O. C. (1977) *The Finite Element Method*, 3rd ed., McGraw-Hill, New York.

7: Benchmark Problems and Solutions

Shouji Toma, *Department of Civil Engineering, Hokkai-Gakuen University, Sapporo, Japan*

W. F. Chen, *School of Civil Engineering, Purdue University, West Lafayette, Indiana*

7.1 Introduction

A trend of design codes around the world moves from the allowable stress design to the limit states design. The limit states design involves nonlinear analysis: the material nonlinearity is referred to as "inelastic effect", while the geometrical nonlinearity is called "second-order effect". In the allowable stress design, an elastic analysis is sufficient, while an inelastic analysis is necessary for the limit states design because the design allows the material to be stressed beyond the elastic limit into the inelastic range.

The second-order inelastic analysis enables designers to assess directly the ultimate strength and behavior of structures based on limit states design. In this analysis, both the strength and the behavior such as load-deformation, resultant member forces, strain distribution, or spread of plasticity are traced at the ultimate and serviceability states of structures. Future design codes are expected to permit the direct use of second-order inelastic analysis without member capacity checks.

All analysis requires an experimental verification in order to confirm its validity. Test results are the only "facts" that give the actual behavior and strength of structures. If the second-order inelastic analysis can adequately predict the actual behavior and strength of the structure, it can be a powerful tool for providing these "facts" in place of testing which demands a substantial cost.

Some sophisticated second-order inelastic analysis programs (plastic-zone theory) have been developed and are currently available for predicting accurately the strength and behavior of frame structures. However, these programs are not intended for use in daily engineering practice. Attempts have been focused in recent years on developing simplified, yet accurate, second-order inelastic analysis for practical use. A general problem encountered by developers is how to confirm the validity for their programs. Simplified methods must be verified by calibrating with benchmark problems. To this end, a literature survey has been made of the research in North America, Europe, and Japan, and some representative frame structures are selected as benchmark problems.

A selection of these benchmark frames is based on the folowing criteria:

(1) papers for the frames are well documented and instructed,
(2) the frames have been widely cited in the past,
(3) tests cover the critical ranges of behavior such as post-peak, maximum loads, yield pattern of material, etc.,
(4) the frames range in type from simple to complex,
(5) connections are rigid, and
(6) no local buckling is involved.

In Europe, the European Convention for Constructional Steelworks (ECCS) proposed calibration frames for second-order inelastic analysis (Vogel, 1985). Three types of frames, a portal frame, a gable frame, and a six-bay frame, have been adopted in Europe (Toma, 1992a). The Structural Stability Research Council (SSRC) makes no such recommendations, but Task Group 29 (Second-Order Analysis for Frame Design) and Task Group 28 (Computer Applications) emphasized the necessity to select calibration frames at the 1990 Annual Meeting (White, 1990 and Basu, 1990). In Australia, Bridge et. al. (1991) also suggested a need for establishing a suitable set of benchmark problems to validate computer programs. Frames in Japan were not readily available for English-speaking countries, and these will be introduced.

In this chapter, the requirements and criteria in selecting the benchmark frames are first described. The selected frames satisfy these requirements. The benchmark frames in North America, Europe, and Japan are then presented to accommodate researchers with the type of information needed for the calibration of the second-order inelastic analysis. Some editorial changes are made to be more informative for calibration frames.

7.2 Requirements for Benchmark Problems

7.2.1 Physical Attributes and Behavioral Phenomena of Frames

A typical frame design may involve the proportioning of many different structural components such as: (1) beams; (2) columns; (3) beam-columns; (4) connections — fully restrained or partially restrained; (5) struts and bracing members; (6) structural walls; (7) floor slabs; and (8) secondary systems such as cladding and partitions. In practice, many of these types of components would not be included in the analysis of an overall structural system. For simplicity, we shall focus our discussion on the basic concept and issues associated with the direct use of second-order inelastic analysis and concentrate on the first three types of structural components: beams, columns, and beam-column members.

There are a number of behavioral phenomena and physical attributes that will influence the strength and stability of these three types of components. These are summarized as follows (White, 1991):

(1) Geometric nonlinearity
- P-Δ effect — effect of axial force acting through displacements associated with member chord rotation
- P-δ effect — effect of axial force acting through displacements associated with member curvature
- Wagner effect — effect of bending moments and axial forces acting through displacements associated with member twisting
- Curvature shortening — effect of curvature on longitudinal displacements at member ends
- Sway shortening — effect of large chord rotation on longitudinal displacements at member ends
- Lateral-torsional buckling
- Local buckling and local distortion
- Interaction of local and member instability

(2) Material nonlinearity
- Yielding — concentrated or spreaded yielding
- Strain-hardening
- Elastic unloading
- Multi-dimensional plasticity effect
- Influence of loading sequence on path-dependent plasticity
- Cyclic plasticity effect — Bauschinger effect, cyclic hardening, elastic shakedown, etc.
- Strain-aging

(3) Physical attributes
- Initial geometric imperfections — out-of-plumbness, out-of-straightness, cross-section distortion, connection eccentricities, etc.
- Initial residual stresses
- Positive member end restraint
- Negative member restraint
- Cross-section symmetry/nonsymmetry
- Prismatic/nonprismatic member profile
- Location and stiffness of bracing
- Composite interconnection with floor slabs

It is not practical at the present time to consider all of these aspects directly in an analysis for design. Among these, some basic aspects are taken into consideration in the analysis of calibration frames. The criteria for selecting calibration frames are discussed in the following.

7.2.2 Criteria for Selecting Calibration Frames

Among the frames analyzed by different analytical approaches, the plastic-zone analysis may be considered to give an exact solution (see Figure 7.12), while the experimental tests give actual results. The following conditions are met in the selection of calibration frames:

(1) Members are compact sections (no local buckling).
(2) In-plane behavior is only considered (no torsional buckling).

7.2.3 Required Information

The following information must be provided for calibration frames:

(1) Input information
- Analytical procedures
  Material nonlinearity (first-order or second-order), iterative procedures, local buckling effect, torsional buckling effect, shear effect on yielding, type of elements, number of elements, etc.
- Configuration of the frame
  Skeleton of the frame, braced or unbraced, sway or nonsway, etc.
- Sizes of the members
  Cross-sectional dimensions, member length, bending with respect to strong axis or weak axis, etc.
- Stress-strain relations of the material
  Yield strength, elastic and strain hardening modulus, stress-strain curve, yield surface, etc.
- Material imperfections
  Residual stress pattern, etc.
- Geometrical imperfections
  Out-of-plumbness, out-of-straightness, etc.
- Loading conditions
  Gravity and horizontal loads, concentrated and distributed loads, monotonic or cyclic loading, loading sequence (proportional increase or constant loading), etc.
- Joint conditions
  Rigid, semi-rigid, or flexible, stiffness of joints, moment-rotation curves, etc.

(2) Output information
- Load-deformation relations
  Load-deflection curves, load-displacement curves, load-rotation curves, ultimate loads, etc.
- Distribution of member forces
  Axial forces, bending moments, shear forces, etc.
- Strain distribution
  Strain distribution at sections
- Spread of plasticity
  Spread of plastic zones along members
- Computing time
  Computing time, type of computer, etc.

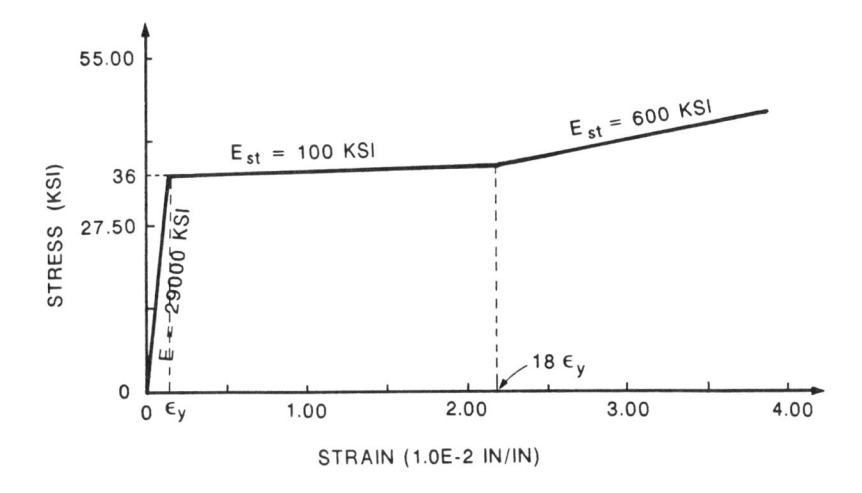

FIGURE 7.1. Stress-strain relation. (White, 1985)

7.2.4 Some Comments on the Analytical Assumptions

Constitutive Relations of Material

The stress-strain relation of the material is most commonly assumed as elastic-perfectly plastic in the analysis. A tri-linear type of curve such as shown in Figure 7.1 is also often used in many cases. These assumptions are appropriate for mild steel such as A36 steel. For higher strength steels, the effect of strain hardening has to be considered.

Residual Stresses (Material Imperfection)

Hot-rolled members usually contain significant residual stresses because of the nonuniform heating and cooling processes during manufacturing. A larger mass at the junction of the flange and web results in slower cooling than other parts of the section. This temperature difference in cooling causes the residual stresses locked inside the member.

The magnitude and distribution of residual stresses in wide flange shapes depend on the type of cross section and manufacturing processes (SSRC, 1988). It is not possible to assign one pattern of distribution for all wide flange cross sections. In the US, the residual stress pattern shown in Figure 7.2 is widely used for analysis of frames containing rolled wide flange sections, in which the residual stress is constant in the web. However, when the depth of a wide flange section is large, the residual stress in the web is not necessarily constant, rather it varies more or less parabolically as shown by actual measured data in Figure 7.3. Another possible residual stress pattern in the web is the one simplified by a linear variation as used in European calibration frames (Vogel, 1985 and Toma, 1992a). This is shown in Figure 7.4. One of these three residual stress patterns for the web, i.e., constant, parabolic, or linear distribution, may be chosen for calibration frames. Distribution of the residual stresses in the flanges is generally expressed by either a linear function (Figures 7.2,

326 ADVANCED ANALYSIS OF STEEL FRAMES

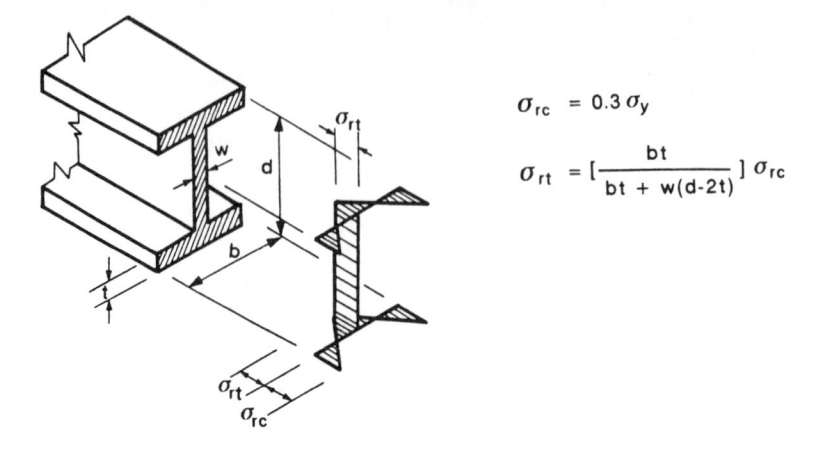

$$\sigma_{rc} = 0.3\,\sigma_y$$

$$\sigma_{rt} = \left[\frac{bt}{bt + w(d-2t)}\right]\sigma_{rc}$$

FIGURE 7.2. Residual stress pattern in wide flange (US).

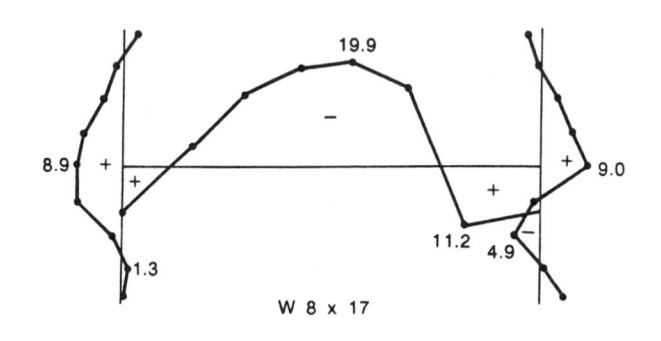

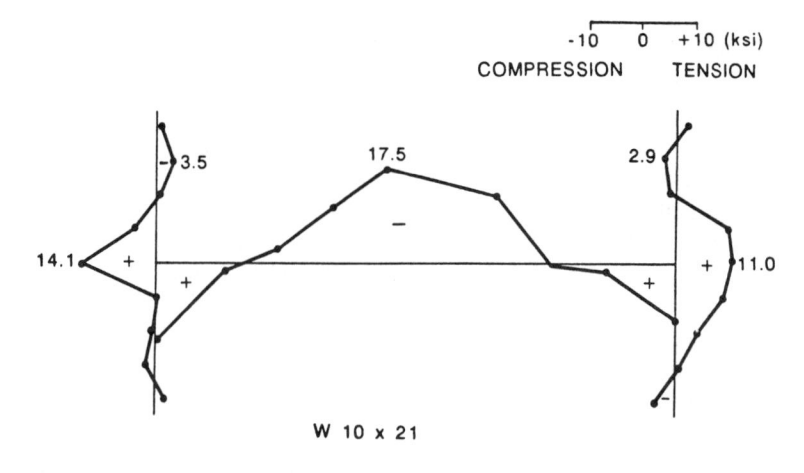

W 8 × 17

-10　　0　　+10 (ksi)
COMPRESSION　　TENSION

W 10 × 21

FIGURE 7.3. Measured residual stresses. (Courtesy of T. Kanchanalai, Civil Engineering/Structures Research Laboratory, University of Texas, Austin.)

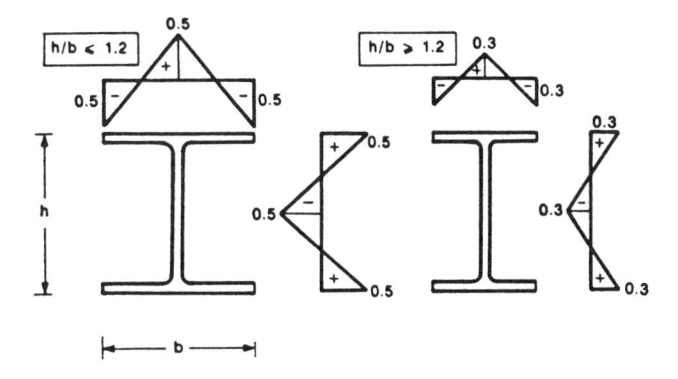

FIGURE 7.4. Residual stress pattern for rolled I-sections (σ_y) (Europe).

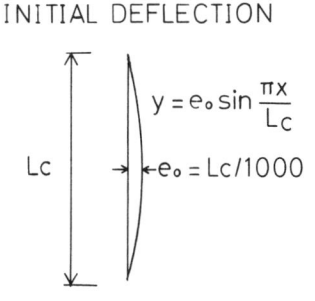

FIGURE 7.5. Initial deflection.

7.3, and 7.4) or a parabolic function. The residual stresses must be self-balanced in the cross-section: the total axial force due to residual stresses is zero.

Geometrical Imperfections

There are two kinds of geometrical imperfections in a frame analysis: out-of-straightness of a member and out-of-plumbness of a structure. In sway frames, the out-of-straightness is not significant, and usually ignored. However, when it is necessary for braced frames or beam-columns, 0.1% of member length (L/1000) is usually adopted as shown in Figure 7.5.

The effect of out-of-plumbness is significant when the gravity load is large. Since the value of $\Delta = h/500$ (h is the story height) is the AISC erection tolerance for columns (SSRC, 1988), it is recommended for the use in the calibration frames. Beaulieu (1977) proposed a statistically based formula, $\Delta = 0.006h/n^{0.455}$, where n is the number of columns (SSRC, 1988).

Joint Conditions

Joint flexibility is one of the key factors in frame design. Strictly speaking, all connections behave nonlinearly when subjected to bending moment. There are no perfectly pinned or fully rigid connections in real frames. The joint conditions are important in the second-order inelastic analysis which considers all nonlinearities in the frame design.

Typical moment-rotation behaviors of connections are shown in Figure 7.6. Curve 2 corresponds to a fully restrained connection (Type FR), while Curves 1 and 2' are partially restrained connections (Type PR). AISC-LRFD Specifications (AISC, 1986) define these two types of connections, but give little guidance on design. This is attributed to the lack of knowledge about how the flexibility of connections affects frame behaviors. In this chapter, joints in the calibration frames are considered rigid.

Many types of analytical models have been proposed to express the moment-rotation behavior of connections. A number of recent models are reviewed by King (1990). Kishi et. al. (1986a, b) have developed a computerized date-based system on the connections, which facilitates the frame analysis to include the nonlinearity of connections (see Chapter 4).

7.3 North American Calibration Frames
7.3.1 Introduction

In the past, a large number of frames have been analyzed by the second-order inelastic approach in North America. A representative sample of frames which have been utilized in research studies is outlined in this section. The detailed configurations and loadings of these frames are summarized by Toma et al. (1991).

Yarimci (1966) analyzed two large frames; a 3-story one-bay frame and a two-bay frame. He also performed full-scale tests to verify his analytical procedures. The load-deflection relations measured in the tests have been used often by various researchers for checking computer programs. Kanchanalai (1977) made an extensive study on portal frames and leaned frames with columns by plastic-zone analysis and contributed largely to the development of the beam-column equations in the current load and resistance factor design (LRFD) specifications (AISC, 1986). He also tested one-story two-bay frames to verify his analyses.

Lu et. al. (1975) analyzed multi-story frames to study frame stability and design of columns based on the slope-deflection analysis. The member stiffness is modified to take into account the effect of axial force and yielding. El-Zanaty et. al. (1980) developed a computer program for nonlinear frame analysis using the plastic-zone approach and made extensive behavioral studies for a variety of frames from simple cantilever beams to multi-story frames. Dhillon et.al. (1990) recently performed analytical work for 3-story and multi-story frames with semi-rigid connections.

McGuire and his co-workers at Cornell University have made a series of analytical studies on steel frames (Orbison, 1982; White, 1985; and Deierlein, 1990). The aim of their research has been to implement the second-order inelastic analysis for practical application. Orbison (1982) analyzed Yarimci's 3-story frames and used Yarimci's results for verification. White (1985) used El-Zanaty's portal frame to

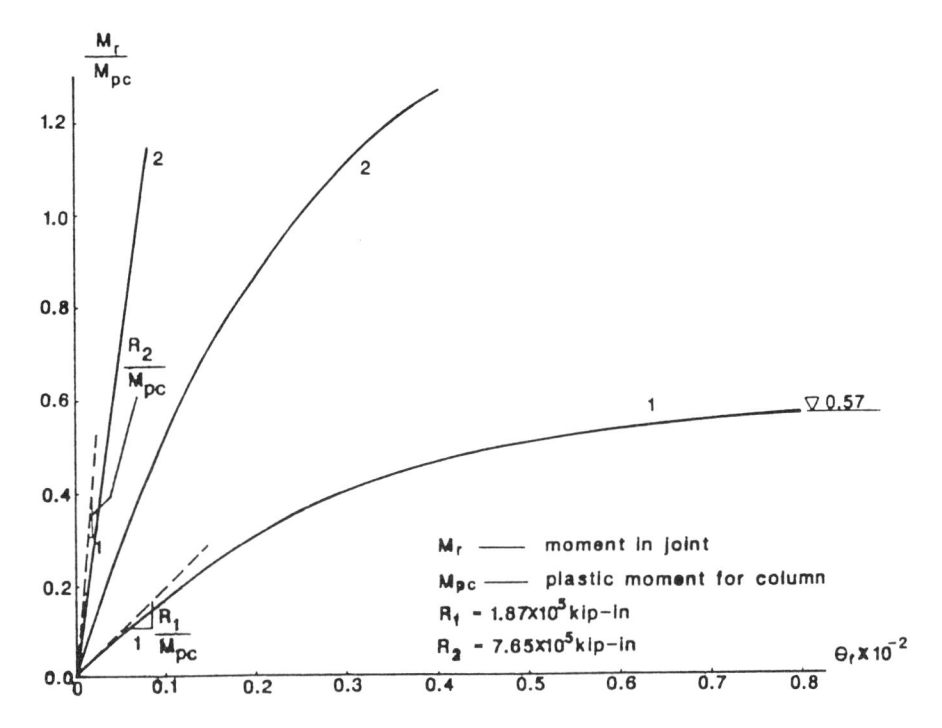

FIGURE 7.6. Moment-rotation curves of joints. (Courtesy of S. P. Zhou and W. F. Chen, Purdue University, West Lafayette, IN.)

verify his plastic-zone analysis. Recently, Cornell researchers have reported a study on the design application of second-order inelastic analysis, using a modified plastic hinge approach for a two-story two-bay frame (Deierlein, 1990).

Chen and his co-workers at Purdue University have focused on the development of simplified second-order inelastic analysis for frames with flexible joints. Goto et. al. (1987) used two-story one-bay and two-story four-bay frames to verify their second-order elastic analysis for flexibly jointed frames, which were analyzed previously by Moncarz et. al. (1981) and Lindsey et. al. (1985), respectively. Al-Mashary (1989) analyzed the rigid frames defined previously by El-Zanaty, Yarimci, and Orbison. King (1990) obtained extensive analytical results for these frames with flexible joints using a modified plastic hinge approach. For more refined analysis of flexibly jointed frames, Zhou et. al. (1986) analyzed braced portal frames using Newmark's numerical approach.

Among these frames, suitable representative frames which meet the requirements for calibration frames will be described in the following sections. These calibration frames were widely cited by researchers in the past.

7.3.2 Beam-Columns

Galambos (1957) studied the interaction curves of beam-columns with W8×31 section using Newmark's numerical method which follows a concept of the plastic-zone theory. Kanchanalai (1977) also studied the behavior and strength of portal

(a) Strong axis bending

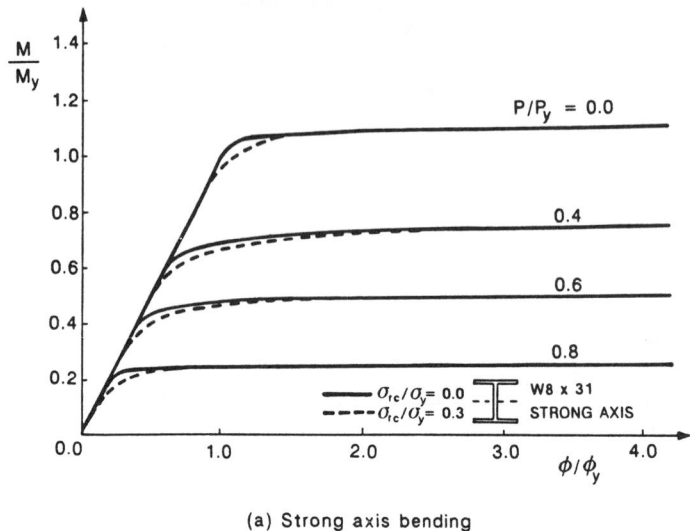

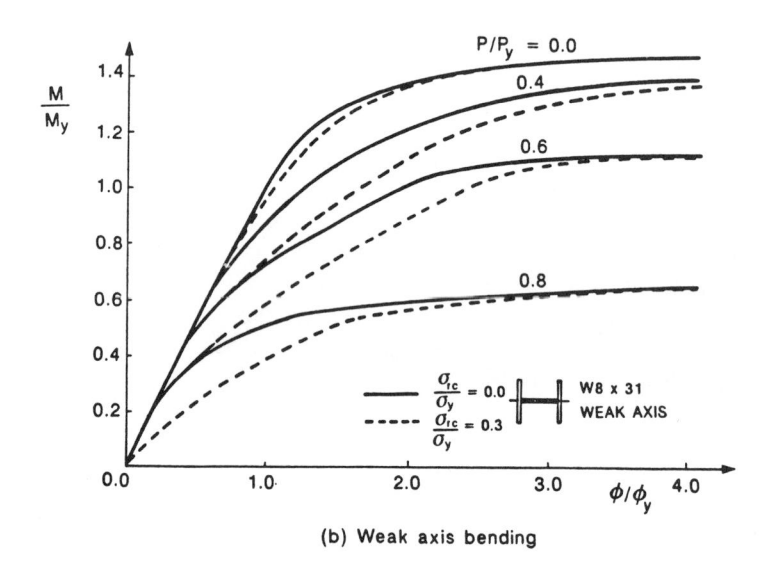

(b) Weak axis bending

FIGURE 7.7. M-P-ϕ relationship of section W8×31. (Courtesy of T. Kanchanalai, Civil Engineering/Structures Research Laboratory, University of Texas, Austin.) (E) = 29,000 ksi (σ_y) = 36 ksi.

frames using the same method and member. The moment-curvature relations of the W8×31 section with respect to the strong and weak axes are shown in Figure 7.7 with the use of the residual stress pattern shown in Figure 7.2.

Analytical results of the interaction curves are shown in Figures 7.8 and 7.9 for loading cases with equal end-moments and one end-moment, respectively. In these figures, the effect of residual stresses is compared for different slenderness ratios.

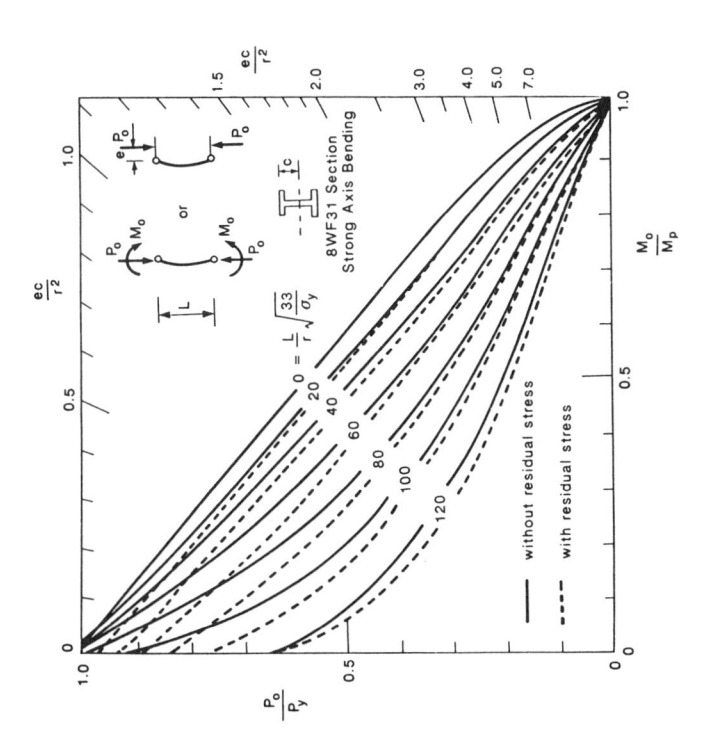

FIGURE 7.8. Interaction curves of section W8×31, equal end-moments. (Courtesy of T. V. Galambos and R. L. Ketter, Fritz Engineering Laboratory, Lehigh University, Bethlehem, PA.)

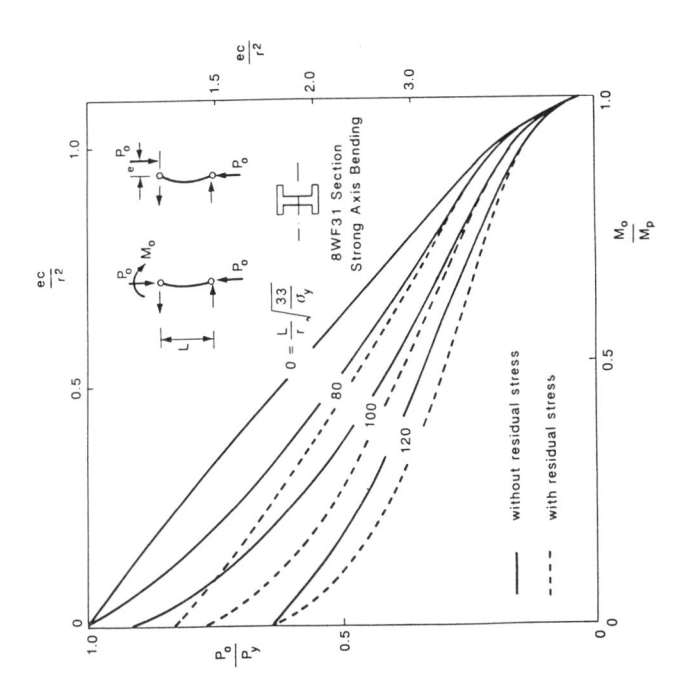

FIGURE 7.9. Interaction curves of section W8×31, one end-moment. (Courtesy of T. V. Galambos and R. L. Ketter, Fritz Engineering Laboratory, Lehigh University, Bethlehem, PA.)

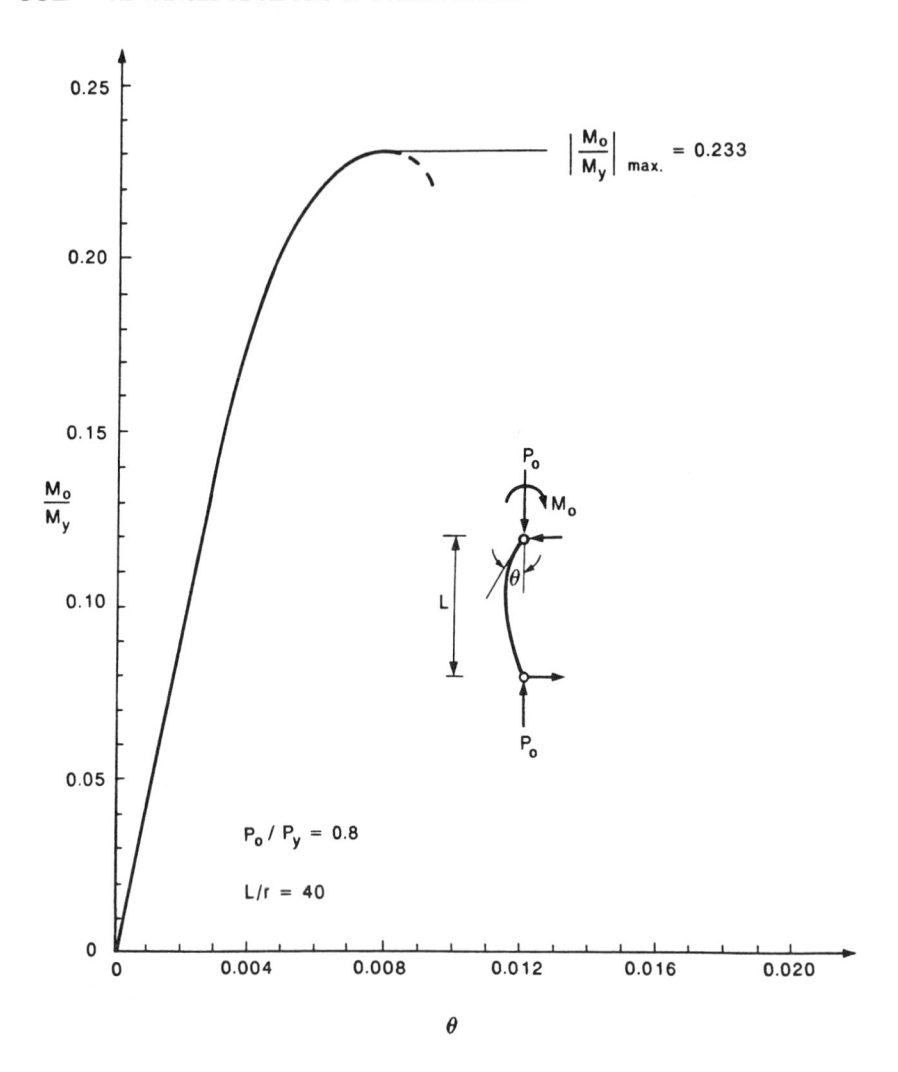

FIGURE 7.10. M-θ relationship of section W8×31, one end-moment. (Courtesy of T. V. Galambos and R. L. Ketter, Fritz Engineering Laboratory, Lehigh University, Bethlehem, PA.)

A typical moment-end rotation relation is plotted in Figure 7.10 for one end-moment case. Geometrical imperfections are not considered. The stress-strain curve is assumed to be elastic-perfectly plastic with the 33 ksi (227 MPa) yield stress. For steels different from 33 ksi (227 Mpa), the slenderness ratio may be modified by the factor $\sqrt{33/\sigma_y}$. The eccentricity values (ec/r^2) are also given in the figures for convenience.

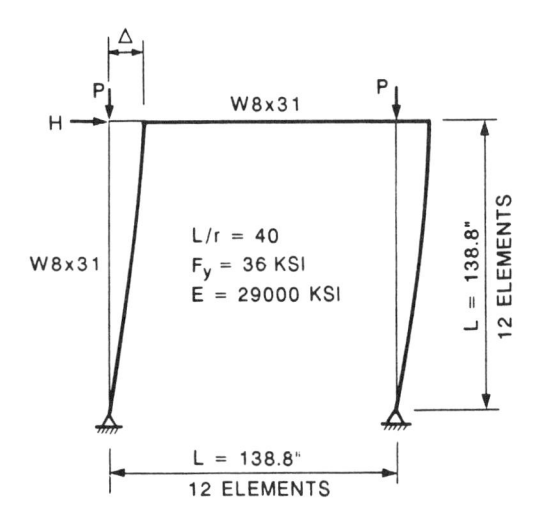

FIGURE 7.11. Unbraced portal frame. (Courtesy of M. H. El-Zanaty et al., University of Alberta, Edmonton, Canada.)

7.3.3 Portal Frames

The frame shown in Figure 7.11 was first analyzed by El-Zanaty (1980) followed by White (1985), Al-Mashary (1989), and King (1990) among others. The frame is composed of the same wide flange section as Galambos' beam-column, that is W8×31.

White (1985) applied the plastic-zone theory in which the frame is divided into finite elements, and the section of a member is further subdivided into fiber elements as illustrated in Figure 7.12. In this case, each member is divided into 12-bar elements with the sub-division of the section shown in Figure 7.13. The residual stress pattern used is also shown. A discretization of the section makes it possible to obtain inelastic sectional stiffness including the effect of residual stress. No initial deflection is taken into account here. White used the stress-strain relation shown in Figure 7.1. The frame is analyzed in the plane with respect to the strong axis of the members. Loads are applied in such a way that the gravity loads P are first applied, then held constant, while the horizontal load H is increased monotonically.

The analytical results are shown in Figures 7.14 to 7.16 for the gravity loads $P/P_y = 0.6$, 0.4 and 0.2, respectively. In these figures, comparisons are made for two different analytical methods: SP2D is based on the plastic-zone theory or the second-order inelastic analysis, and SLD3D is based on the elastic-plastic hinge theory or the concentrated plasticity analysis. "The finite joint" has a rigid link element representing the member depth at the joint (joint panel zone), and the relative depth to the length of the members is considered. It is noted from the figures that the stiffness is significantly increased when the rigid links at joints (rigid joint panel zones) are assumed.

334 ADVANCED ANALYSIS OF STEEL FRAMES

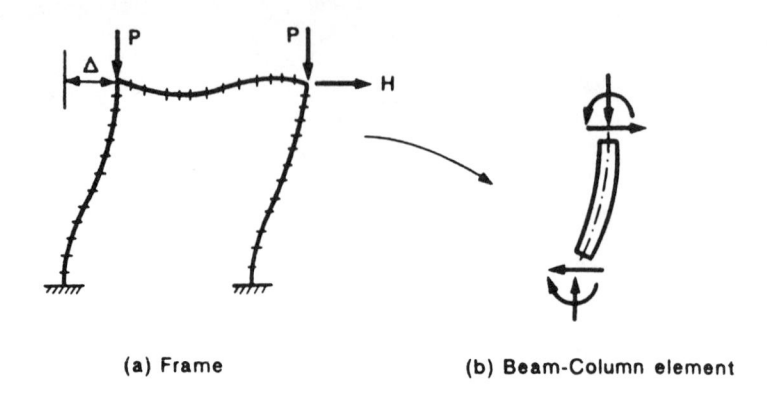

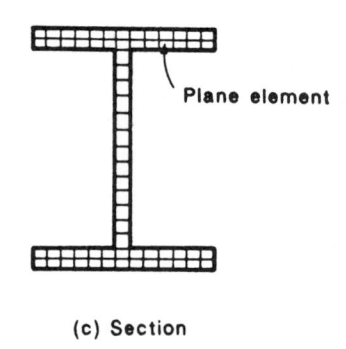

(a) Frame

(b) Beam-Column element

(c) Section

FIGURE 7.12. Plastic-zone theory for frame analysis.

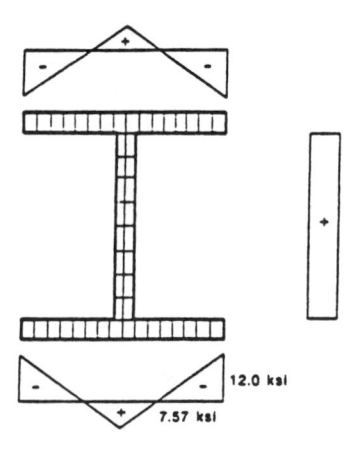

12.0 ksi
7.57 ksi

FIGURE 7.13. Subdivision of section and residual stress distribution. (White, 1985)

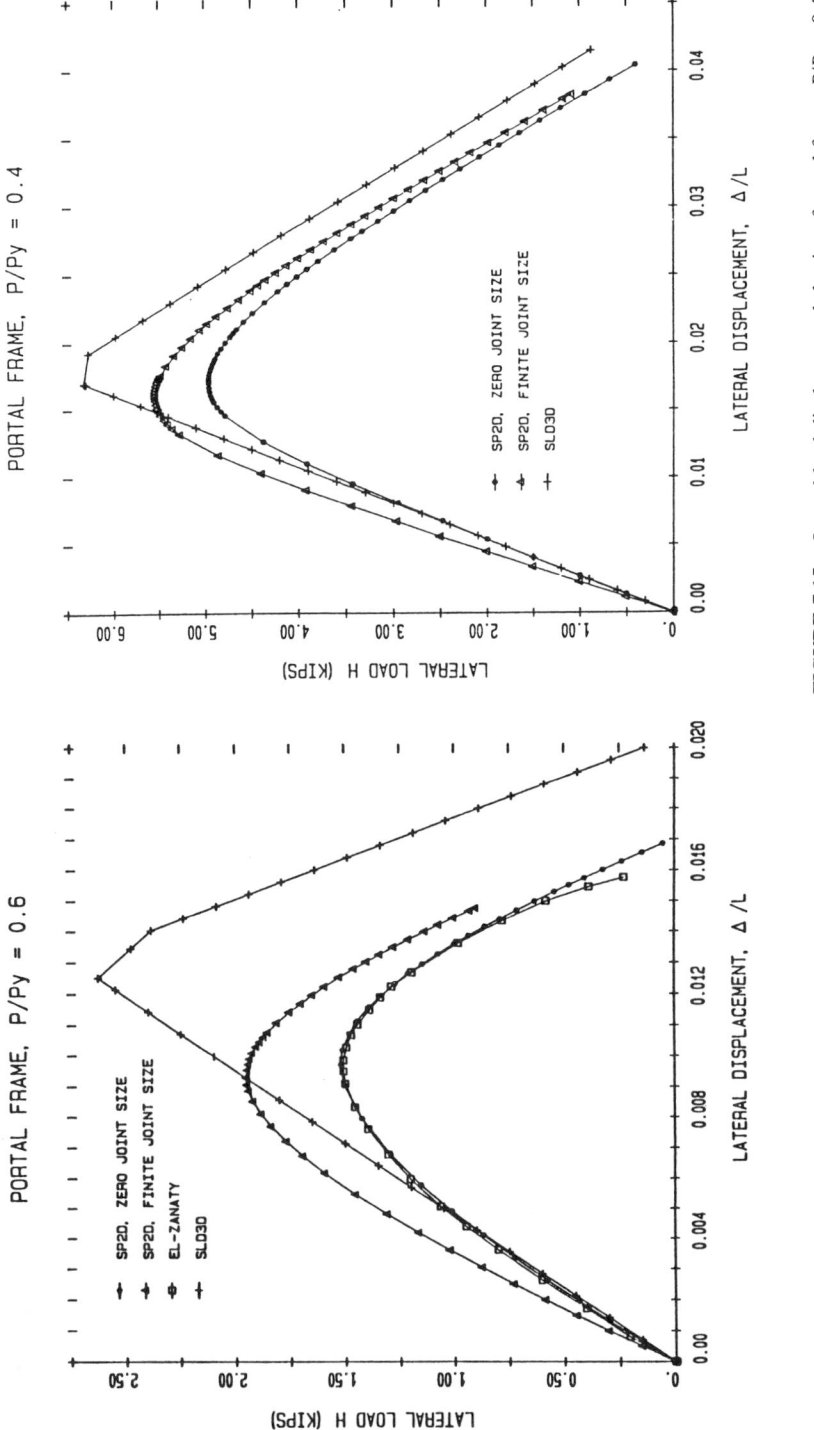

FIGURE 7.14. Lateral load-displacement behavior of portal frame, $P/P_y = 0.6$. (White, 1985)

FIGURE 7.15. Lateral load-displacement behavior of portal frame, $P/P_y = 0.4$. (White, 1985)

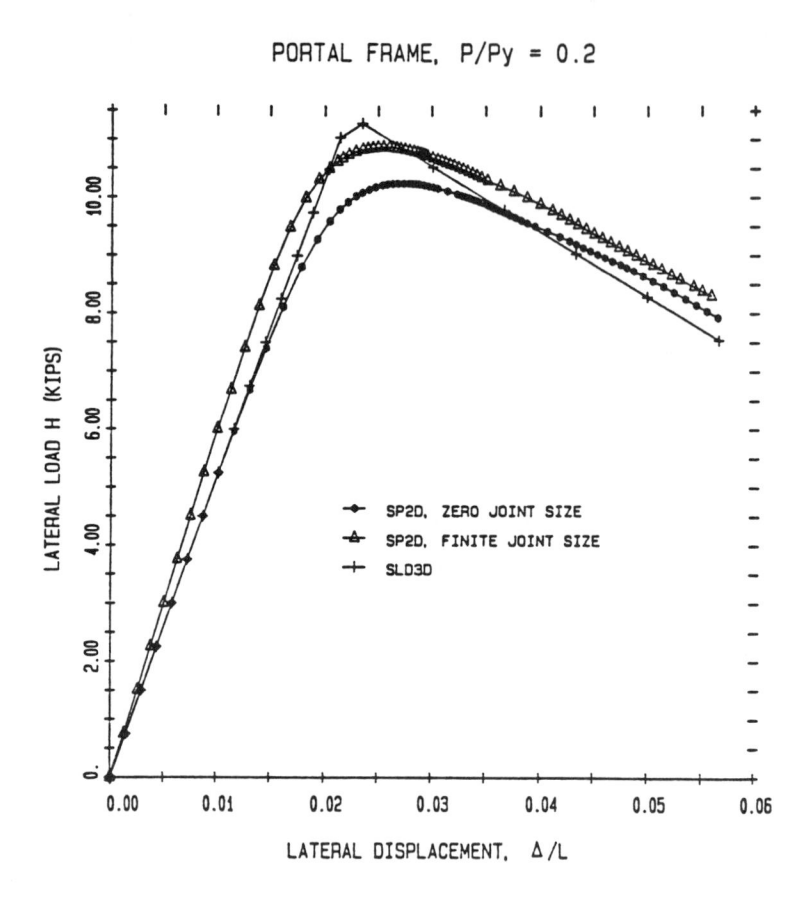

FIGURE 7.16. Lateral load-displacement behavior of portal frame, $P/P_y = 0.2$. (White, 1985)

7.3.4 Interaction Curves for Portal Frames

Kanchanalai (1977) performed an extensive analysis on portal frames and established specific procedures for second-order elastic analysis. His studies formed the base of the current AISC-LRFD design specifications (AISC, 1986) for frame design. He obtained the moment-curvature curves of the beam-column shown in Figure 7.7. Using the moment-curvature curves for beam-columns, the deflection is computed from the equilibrium conditions at the joints between columns and beam, while the beam is assumed to remain elastic. The beam is assumed to be long enough such that the change in axial force in the columns due to lateral load can be neglected. Elastic-perfectly plastic stress-strain relation for A36 steel and the residual stress pattern shown in Figure 7.2 are used. Initial geometrical imperfections are not considered.

Figures 7.17 and 7.18 show the horizontal force and sway displacement relations for typical portal frames. The figures also compare the deflections with and without the influence of residual stresses. The interaction curves between vertical (gravity) and horizontal (wind) loads for different L/r ratios are shown in Figures 7.19 and

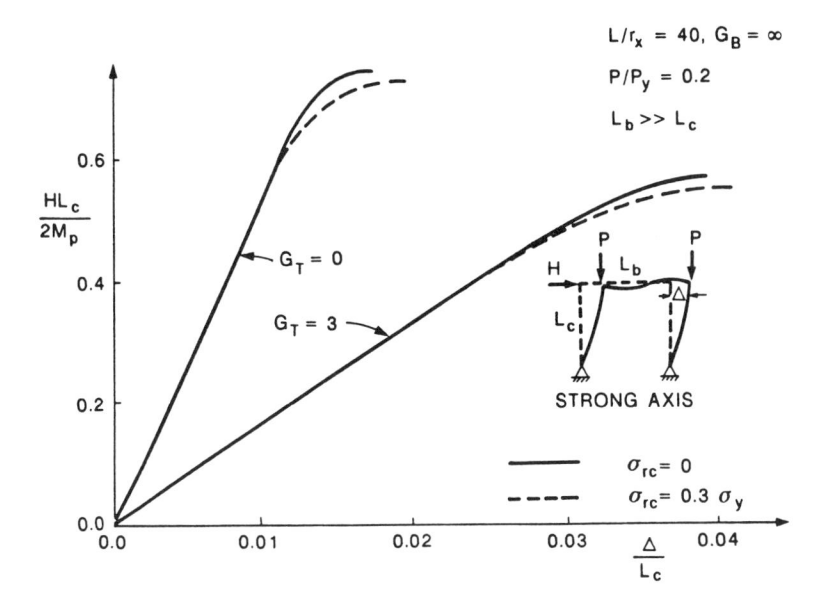

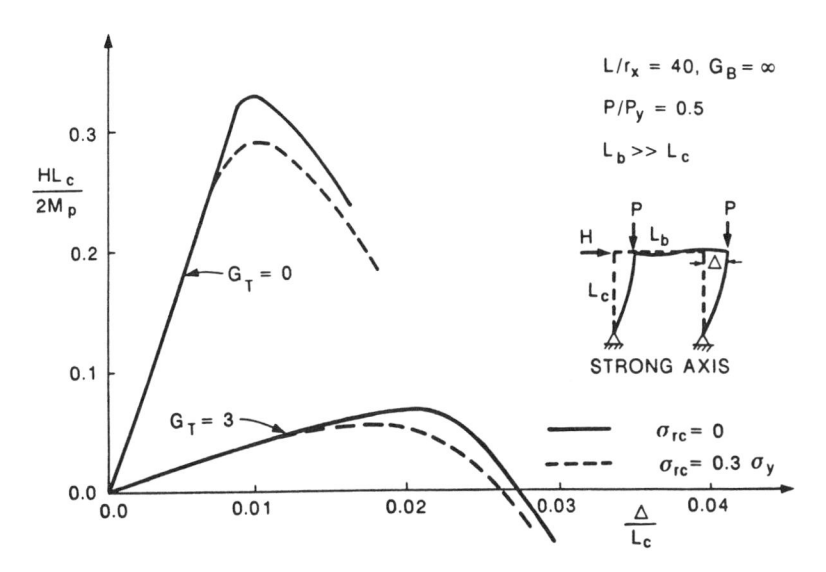

FIGURE 7.17. Lateral load-displacement behavior of portal frame, strong axis bending. (Courtesy of T. Kanchanalai, Civil Engineering/Structures Research Laboratory, University of Texas, Austin.)

7.20. Kanchanalai also studied leaned column frames. He further conducted experimental work to verify his analysis (1977). The frames analyzed by Kanchanalai have been used as a benchmark to assess the design formulas of AISC-LRFD by White et. al. (1991).

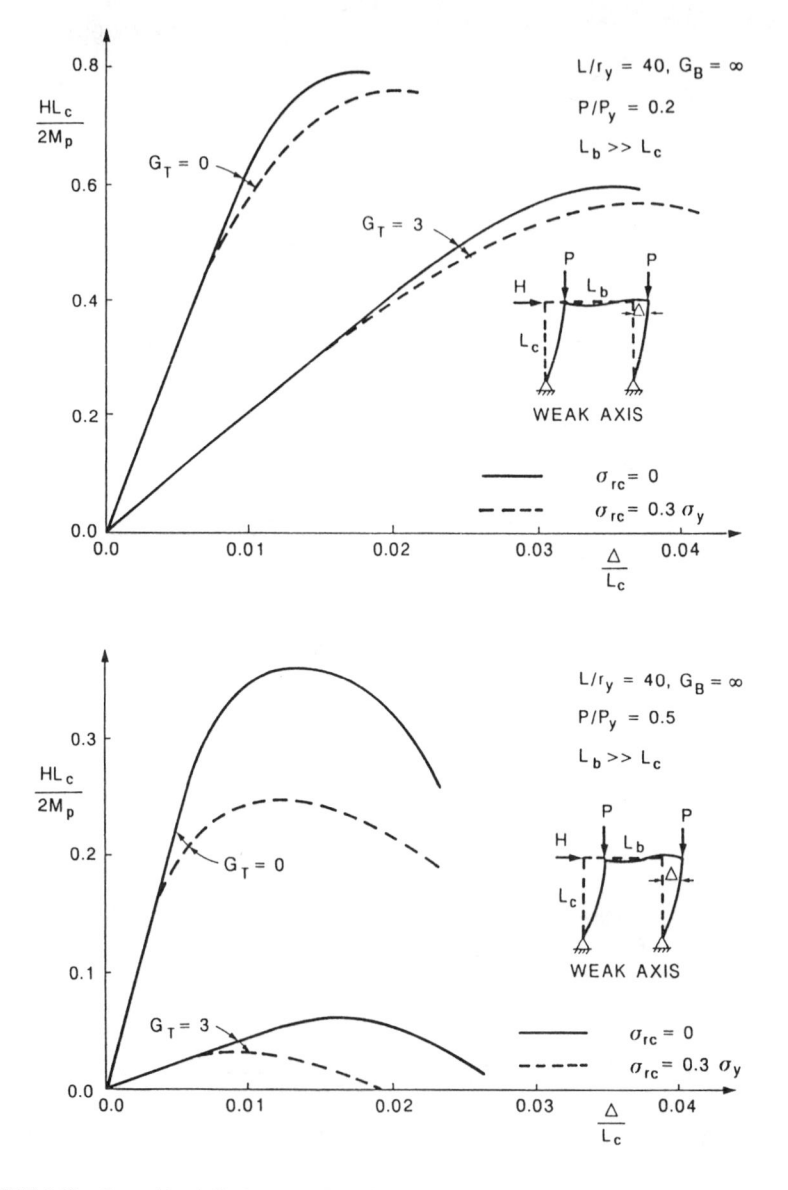

FIGURE 7.18. Lateral load-displacement behavior of portal frame, weak axis bending. (Courtesy of T. Kanchanalai, Civil Engineering/Structures Research Laboratory, University of Texas, Austin.)

7.4 European Calibration Frames
7.4.1 Introduction

The ECCS selected three types of frames for the calibration of second-order inelastic analysis (Vogel et. al., 1984). These are:

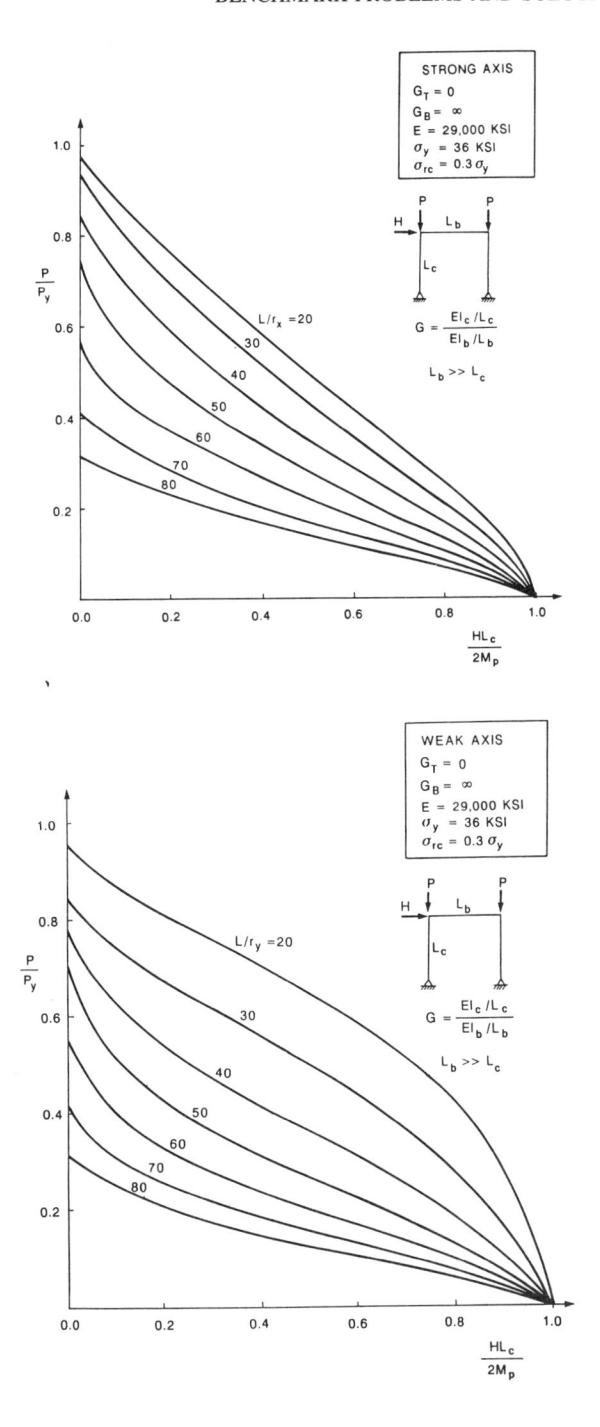

FIGURE 7.19. Interaction curves for portal frames, $G_T = 0$. (Courtesy of T. Kanchanalai, Civil Engineering/Structures Research Laboratory, University of Texas, Austin.)

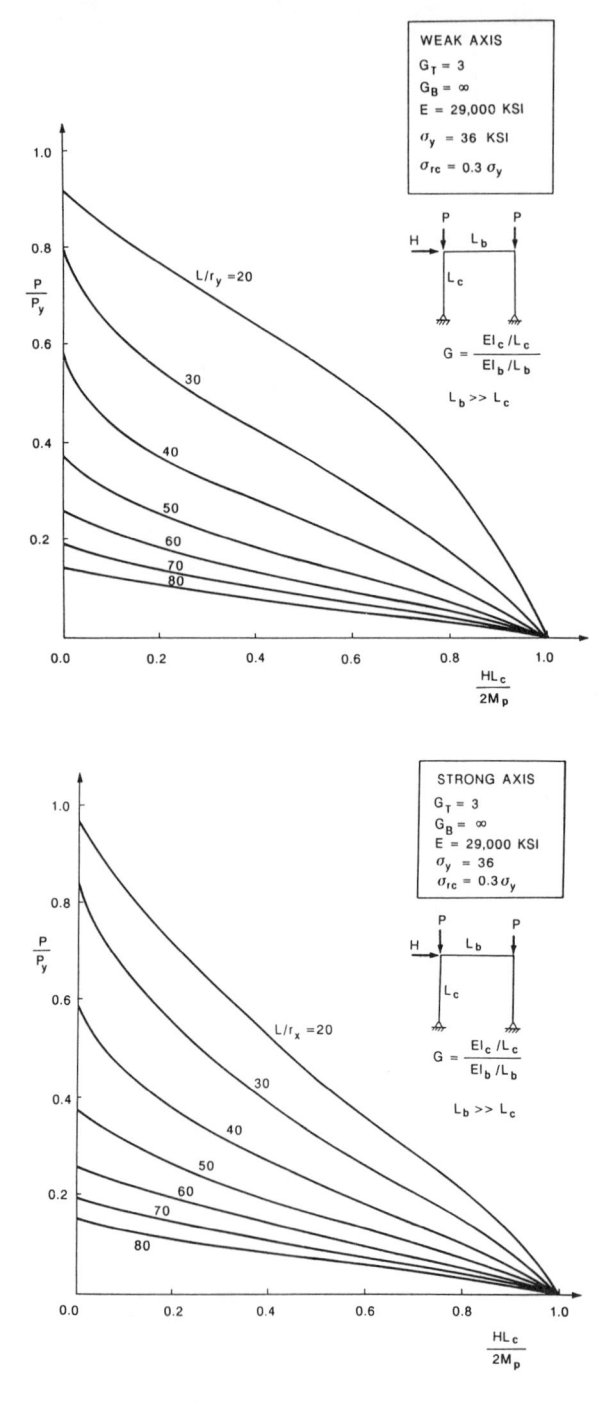

FIGURE 7.20. Interaction curves for portal frames, $G_T = 3$. (Courtesy of T. Kanchanalai, Civil Engineering/Structures Research Laboratory, University of Texas, Austin.)

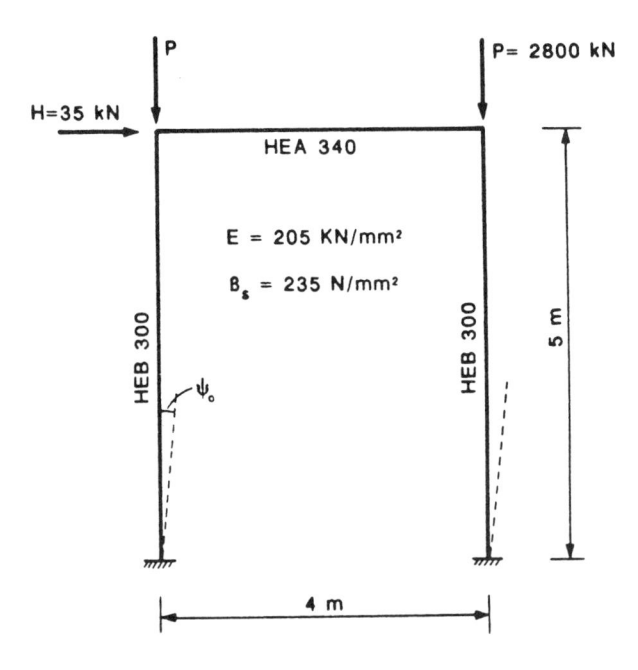

FIGURE 7.21. Calibration frame: (1) portal frame. (From Vogel, U. *Stahlbau*, 10, 1, 1985. With permission.)

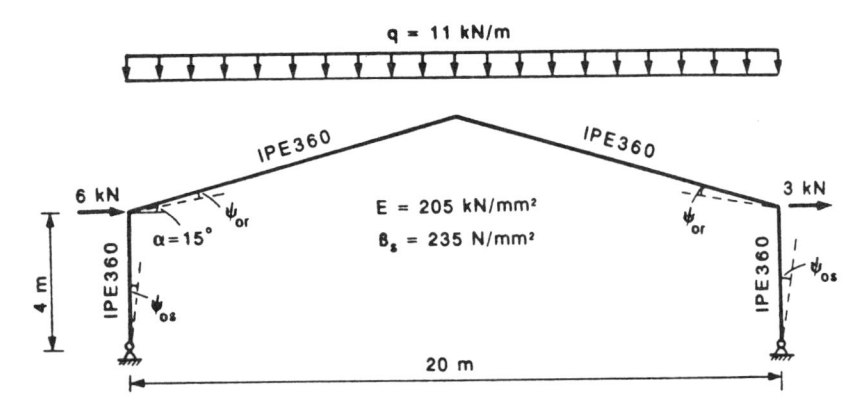

FIGURE 7.22. Calibration frame: (2) gable frame. (From Vogel, U. *Stahlbau*, 10, 1, 1985. With permission.)

(1) A portal frame
(2) A gable frame
(3) A six-story two-bay frame

Configurations of the frames are shown in Figures 7.21 to 7.23, respectively. Members of the frames are all hot-rolled shapes, and sectional dimensions of each member are given in Table 7.1. The portal frame in Figure 7.21 is supported with

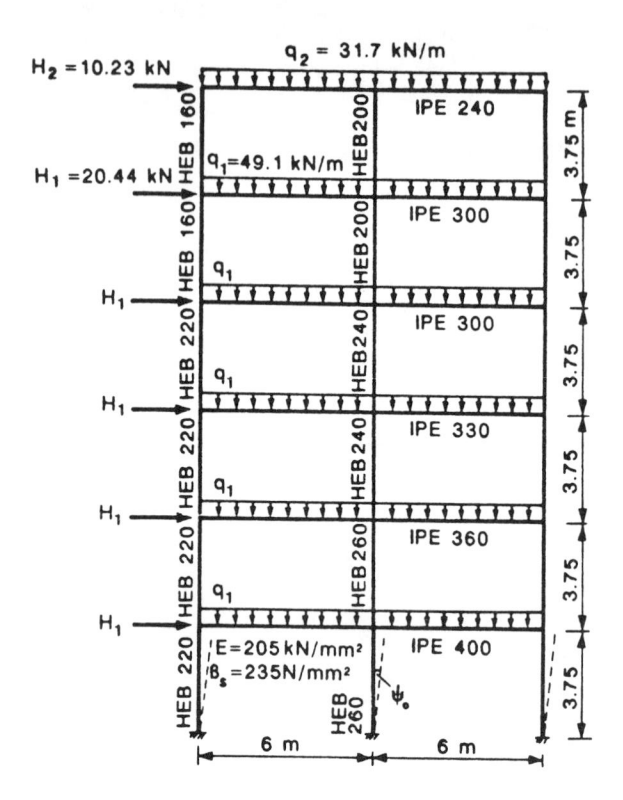

FIGURE 7.23. Calibration frame: (3) six-story two-bay frame. (From Vogel, U. *Stahlbau*, 10, 1, 1985. With permission.)

rigid joints at its base ends and subjected to vertical and horizontal loads at the top of the frame. The horizontal load produces a side-sway and brings a significant secondary effect on the members. The loads shown in Figure 7.21 are the case when the load factor $\gamma = 1.0$. The factored loads will be multiplied by the load factor γ: both vertical and horizontal loads are proportionally increased by the load factor γ.

The second calibration frame is a gable frame in Figure 7.22. The loads are applied both vertically and horizontally, of which magnitudes are shown in Figure 7.22 for the load factor $\gamma = 1.0$. The vertical load is given as a load acting on the projected area of the roof. The horizontal load is assumed such that the wind force at the windward side has a double intensity of the leeward side at which negative (suction) wind force acts.

The third frame is a six-bay two-bay frame shown in Figure 7.23. Loadings are horizontal wind and vertical gravity forces as the previous frames. The wind load at top of the frame is a half of the other intermediate levels because the projected area for the top level is half. Also, the distributed gravity load at top of the frame is smaller than other floors. The sections for columns and beams are selected such that members at lower stories have a larger resistance.

Table 7.1 Sectional Dimensions of Members (Europe)

Members	h (mm)	b (mm)	a (mm)	e (mm)	r (mm)	A (cm²)	I_x (cm⁴)	I_y (cm⁴)	Z_{px} (cm³)	Z_{py} (cm³)	σ_{res}/σ_y
IPE240	240	120	6.2	9.8	15	39.1	3892	284	367	73.9	0.3
IPE300	300	150	7.1	10.7	15	53.8	8356	604	628	125	0.3
IPE330	330	160	7.5	11.5	18	62.6	11770	788	804	154	0.3
IPE360	360	170	8.0	12.7	18	72.7	16270	1043	1019	191	0.3
IPE400	400	180	8.6	13.5	21	84.5	23130	1318	1307	229	0.3
HEB160	160	160	8.0	13.0	15	54.3	2492	889	354	170	0.5
HEB200	200	200	9.0	15.0	18	78.1	5696	2003	643	306	0.5
HEB220	220	220	9.5	16.0	18	91.0	8091	2843	827	394	0.5
HEB240	240	240	10.0	17.0	21	106.0	11260	3923	1053	498	0.5
HEB260	260	260	10.0	17.5	24	118.0	14920	5135	1283	602	0.5
HEB300	300	300	11.0	19.0	27	149.0	25170	8563	1869	870	0.5
HEB340	330	300	9.5	16.5	27	133.0	27690	7436	1850	756	0.5

Note: See Figure 7.4 for residual stress pattern.

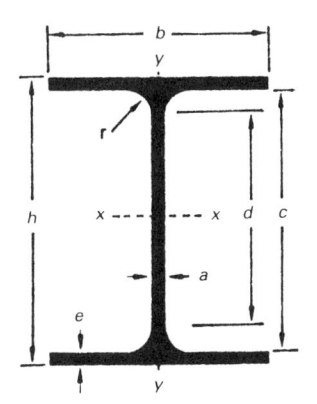

7.4.2 Analytical Assumptions

The second-order inelastic analysis based on the plastic-zone method is used to analyze the calibration frames in Europe (Vogel, 1985). In the plastic-zone theory for frame analysis, the members are divided into a number of beam-column elements. Furthermore, the sections of a member are cut into a number of plane elements (see Figure 7.12). Stiffness of the section is obtained by integrating the stiffness of those plane elements which remain in the elastic or in the strain-hardening range. In this way, the plastic-zone theory can trace the gradual spreading of plastic-zone and the load-deflection relations.

This analytical method includes both material and geometrical nonlinearities such as initial imperfections, secondary effects induced by deflections, etc. The assumptions used in the analysis are listed as follows (Vogel, 1985):

344 ADVANCED ANALYSIS OF STEEL FRAMES

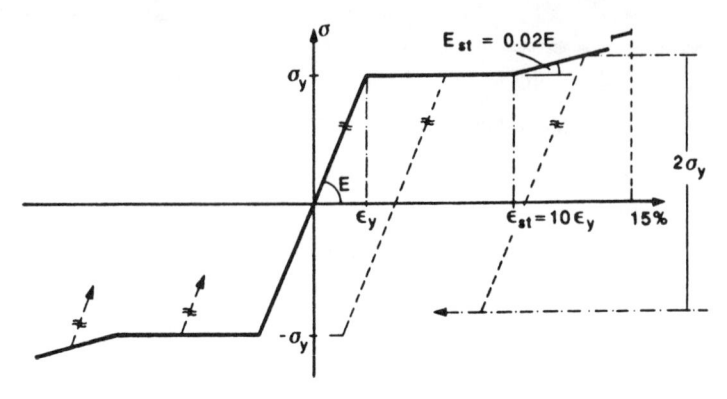

(a) Generalized stress-strain relation

(b) Stress-strain used in the analysis

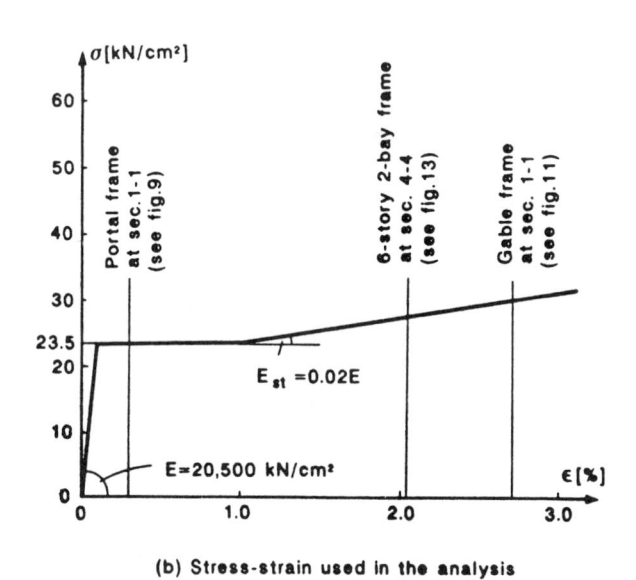

FIGURE 7.24. Stress-strain relationship. (From Vogel, U. *Stahlbau*, 10, 1, 1985. With permission.)

(1) Plane section remains plane after deformation.
(2) Out-of-plane instability (lateral torsional buckling) is prevented.
(3) Cross-sections do not buckle locally.
(4) Connections are rigid.
(5) Effect due to column leaning (P-Δ effect) is included.
(6) Effect due to member deflection (P-δ effect) is included.
(7) Tri-linear stress-strain relation is used but unloading is neglected (see Figure 7.24).

(8) Residual stresses are considered (see Figure 7.4).
(9) Effect of shear on yielding of material and deformation is neglected: stress is considered as uniaxial in plane elements.
(10) Residual stress is uniformly distributed over the entire length of a member.

The calibration frames are analyzed as a plane frame such that the bending moments act on the members with respect to their strong axis. Some discussion of these assumptions will be explained further in what follows.

Stress-Strain Relation

A tri-linear stress-strain relationship is adopted in the analysis as illustrated in Figure 7.24 (Vogel, 1989). Strain-hardening starts at the strain $\varepsilon_{st} = 10\varepsilon_y$, in which ε_y is the yield strain, and its stiffness E_{st} is taken as 2% of the elastic stiffness E. Stress-strain relation on the unloading branch is based on the kinematic hardening rule: the stiffness during unloading is parallel to the initial stiffness only in the range of $2\sigma_y$, in which σ_y is the yield stress. This unloading is, however, not considered in monotonic loading case. The stress-strain relationship of the material used in the calibration frame analyses in Europe is shown in Figure 7.24b. The maximum fiber strains reached at the ultimate loading are also shown in Figure 7.24b for each calibration frame. The corresponding sections where the maximum strains occur will be described later.

Residual Stresses

The ECCS adopts the residual stress pattern as shown in Figure 7.4. When the depth-width ratio h/b is less than 1.2, the maximum residual stress is $\sigma_{res} = 0.5\sigma_y$. When h/b is larger than 1.2, the heat mass rate of the junction portion becomes small; thus, the residual stress will be reduced to the maximum $\sigma_{res} = 0.3\sigma_y$. According to this classification, the residual stress adopted in the calibration analysis is summarized in Table 7.1.

Geometrical Imperfections

If the column inclines with respect to its vertical line, the vertical gravity load will induce a significant secondary moment to the column. The ECCS agreed to have a consistent out-of-plumb geometric imperfection in order to have comparable analytical results among researchers. This geometric initial imperfection is given by the following equation (Vogel, 1989):

$$\psi_0 = \frac{1}{300} r_1 r_2 \qquad (7.1)$$

where ψ_0 = initial angle inclined

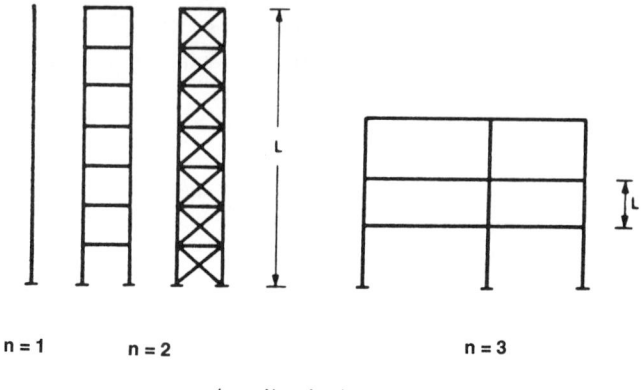

FIGURE 7.25. Definition of the height for geometric imperfection. (From Vogel, U. *Frame and Slab Structures*, Butterworths, London, 29. With permission.)

Table 7.2 Geometrical Imperfections (Europe)

Calibration frames	r_1	r_2	ψ_0	Deflections (mm)
Portal frame	1.0	$\frac{1}{2}\left(1+\frac{1}{2}\right)=\frac{3}{4}$	$\frac{1}{400}$	$\frac{5.0}{400}=12.5$
Gable frame				
Column	1.0	1.0	$\frac{1}{300}$	$\frac{4.0}{300}=13.3$
Roof	1.0	$\sqrt{\frac{5}{10/\cos 15°}}=0.695$	$\frac{1}{432}$	$\frac{10/\cos 15°}{432}=24.0$
6-Story 2-bay frame	1.0	$\frac{1}{2}\left(1+\frac{1}{3}\right)=\frac{2}{3}$	$\frac{1}{450}$	$\frac{22.0}{450}=50.0$

Note: (1) Angle of initial deflection $\psi_0 = \frac{1}{300} r_1 r_2$. (2) Deflections are at the top of the frame.

$$r_1 = \begin{cases} \sqrt{5/L} : & L > 5\ m \\ 1 & : L \leq 5\ m \end{cases} \tag{7.2}$$

$$r_2 = \frac{1}{2}\left(1+\frac{1}{n}\right) \tag{7.3}$$

L = height of the frame (m, see Figure 7.25)
n = number of columns in the plane of frame

The coefficient r_1 reflects the effect of height of the frame. The higher the frame, r_1 becomes smaller. Note that the height L should be taken as a story height of the frame when the frame has more than two bays ($n > 3$). When the frame has only one

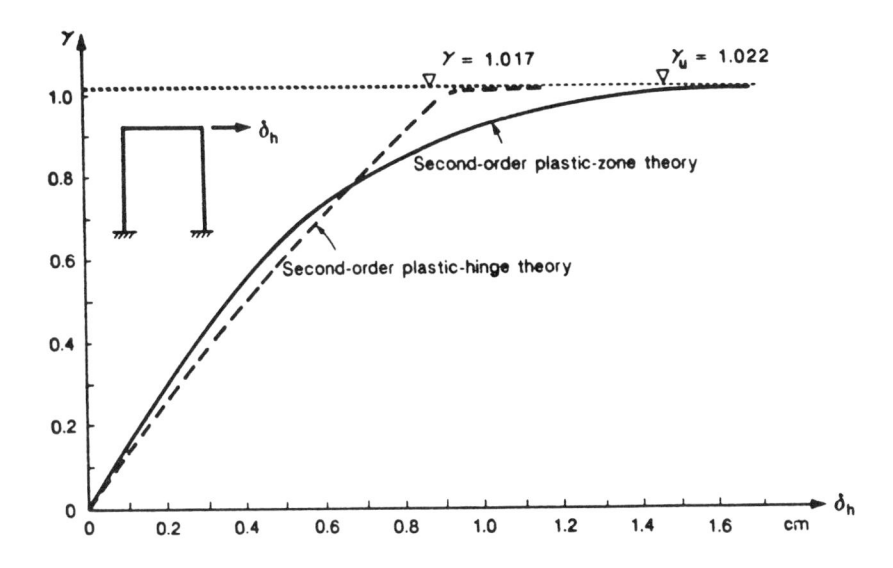

FIGURE 7.26. Load-deflection behavior by plastic-zone theory: (1) portal frame. (From Vogel, U. *Stahlbau*, 10, 1, 1985. With permission.)

bay ($n = 2$), L should be taken as the overall height of the frame. The coefficient r_2 takes into account the width-spread effect of the frame. A number of columns could probabilistically reduce the vertical geometric imperfection. The initial geometric imperfections used for the calibration frames are given in Table 7.2. These are obtained by Eq. (7.1).

The initial out-of-straightness of beam members is assumed to have a parabolic configuration and the maximum offset 0.1% of the member length at midspan.

Based on the assumptions described above, the second-order inelastic analysis (plastic-zone theory) was performed by the European group (Vogel, 1984 and 1985) and the results are shown in Figures 7.26 to 7.31 for the calibration frames in terms of the load deflection relation and the rate of spread of plastic-zones. Also, shown in the figures are the comparisons to the second-order plastic hinge method (Vogel, 1985).

7.4.3 A Portal Frame

The load-deflection curve for the portal frame is shown in Figure 7.26, in which the second-order inelastic analysis does not have a clear linear limit and the stiffness of the frame decreases gradually. This is because the P-Δ effect due to large vertical loads amplifies the bending moment in the members as the horizontal displacement Δ increases. The ultimate load-carrying capacity does not differ from the plastic hinge theory, but the stiffness is quite different, especially in the transitional range from elastic to plastic.

Figure 7.27 shows the rate of spread of yielded areas (plastic-zones) for the portal frame at the ultimate load ($\gamma_u = 1.03$). It can be seen from the figure that around 60% of section of the column is partially yielded in compression almost throughout the

348 ADVANCED ANALYSIS OF STEEL FRAMES

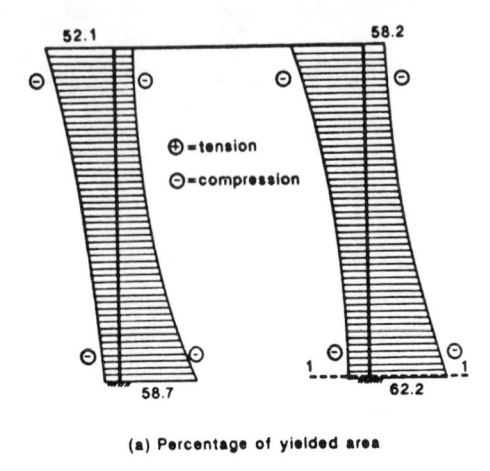

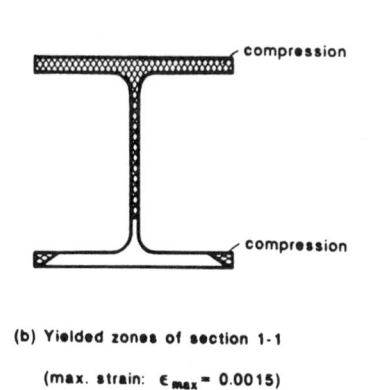

(a) Percentage of yielded area

(b) Yielded zones of section 1-1
(max. strain: ϵ_{max} = 0.0015)

FIGURE 7.27. Spread of plastic-zones: (1) portal frame. (From Vogel, U. *Stahlbau*, 10, 1, 1985. With permission.)

members. This is because, in this particular case, the vertical loads are dominant. Figure 7.27a shows a skeleton lined by the neutral axis (centroid) of the members. The width of the shaded area indicates the percentage of the yielded zone to the whole section, taking each side from the neutral axis. Note that in this case the members are bent with respect to their strong axis. The residual stress in the flanges enhances an early yielding of the material and produces yielded zones at the two flanges and part of the web as shown in Figure 7.27b.

7.4.4 A Gable Frame

Figure 7.28 is the load-deflection relations for the gable frame, in which the horizontal and vertical displacements at typical points are plotted. When the ultimate loads

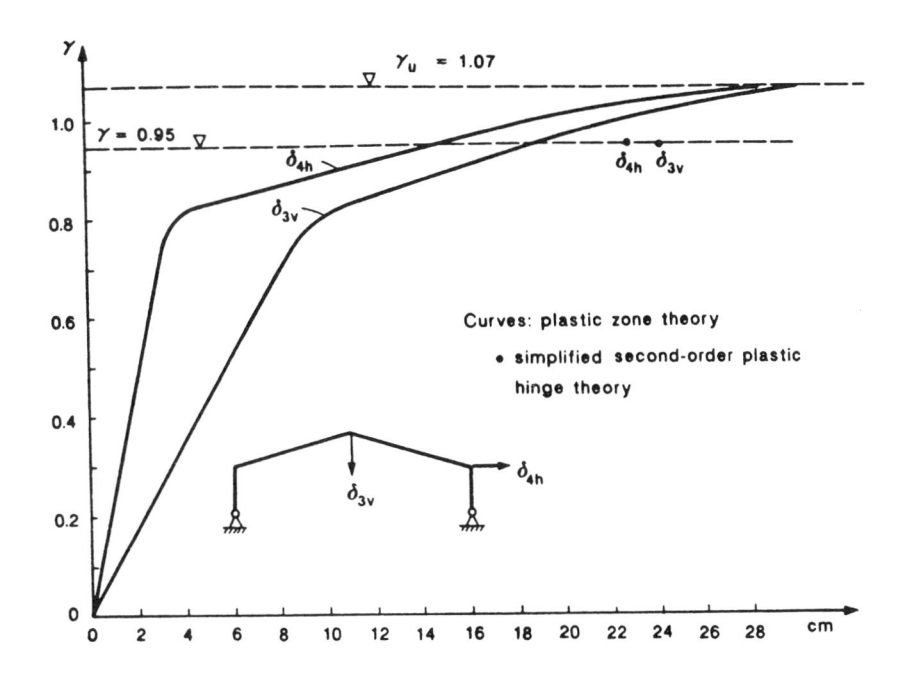

FIGURE 7.28. Load-deflection behavior by plastic-zone theory: (2) gable frame. (From Vogel, U. *Stahlbau*, 10, 1, 1985. With permission.)

are reached ($\gamma_u = 1.07$), the plastic-zone occurs around the top of the roof and the right knee joint, as shown in Figure 7.29a: especially at the knee joint, the nearly fully plastic hinge has been formed (Figure 7.29b).

7.4.5 A Six-Story Two-Bay Frame

In Figure 7.30, the load-sidesway relation for the six-story two-bay building is plotted. The figure shows that the elastic stiffness at a higher story is less than that of a lower story. This is because of the P-Δ effect induced by the initial geometrical imperfections. The comparison between the second-order inelastic analysis (plastic-zone theory) and the second-order plastic hinge theory shows that both analyses result in a similar load-deformation behavior. It is known that in general the highly redundant structures can be well predicted by a simple analysis such as plastic hinge theory. The rate of spread of yielded zones in Figure 7.31 clearly illustrates a collapse mode of the structure: especially, the columns at lower levels will collapse by an almost full plasticity. Seen in the portal frame (Figure 7.27) as well, more than 50% of the column section has yielded and does not have much remaining strength to sustain further loading.

350 ADVANCED ANALYSIS OF STEEL FRAMES

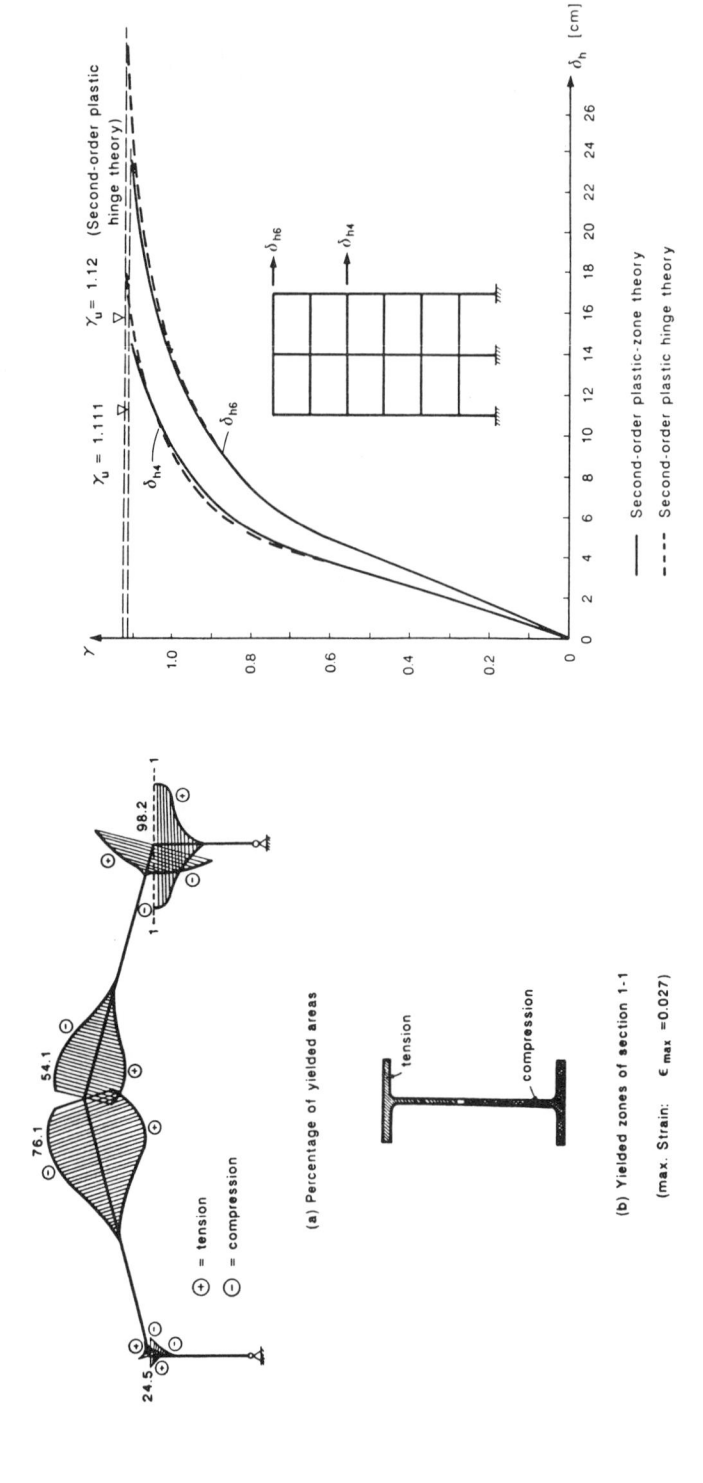

FIGURE 7.29. Spread of plastic-zones: (2) gable frame. (From Vogel, U. *Stahlbau*, 10, 1, 1985. With permission.)

FIGURE 7.30. Load-deflection behavior by plastic-zone theory: (3) six-story two-bay frame. (From Vogel, U. *Stahlbau*, 10, 1, 1985. With permission.)

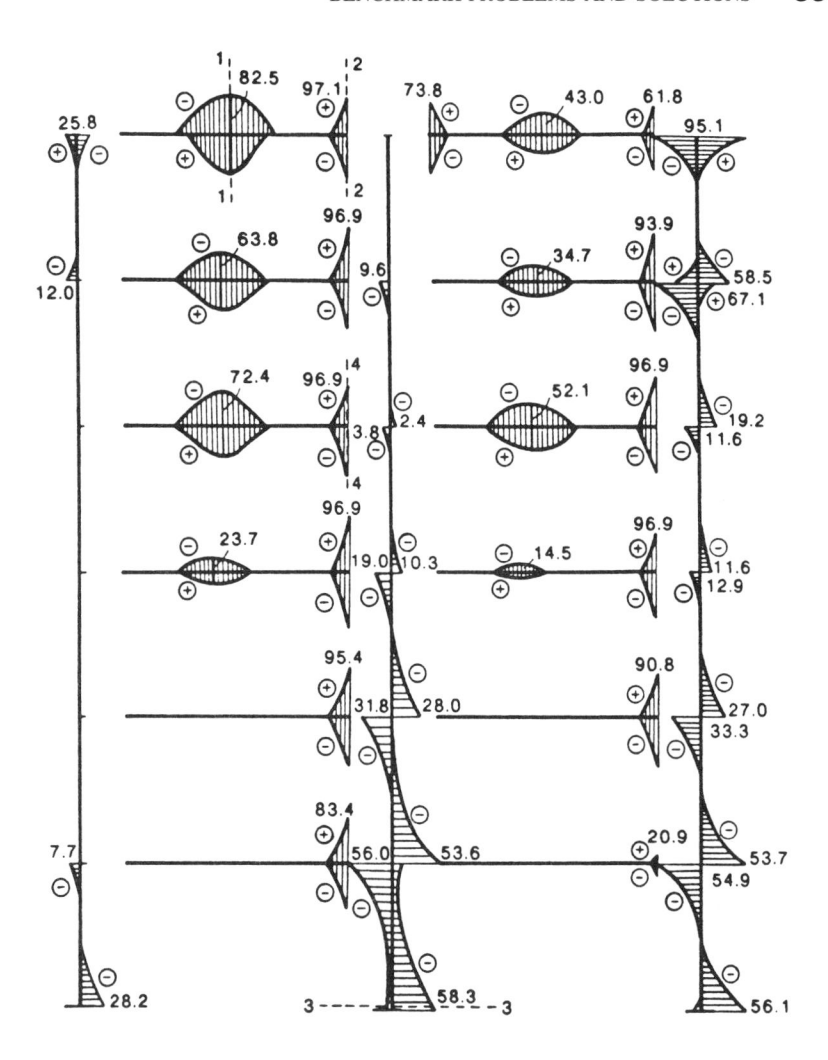

(a) Percentage of yielded areas

FIGURE 7.31. Spread of plastic-zones: (3) six-story two-bay frame. (From Vogel, U. *Stahlbau*, 10, 1, 1985. With permission.)

7.5 Japanese Calibration Frames

7.5.1 Introduction

Since Japan is in a severe earthquake zone, Japan developed its own design procedures. Researchers in Japan consider the energy absorption of a structure more important than its static ultimate strength when subjected to earthquake forces. Thus, a rigorous analysis does not really attract much interest to researchers, rather simple procedures to enable designers to assess cyclic behavior are of great interest. These

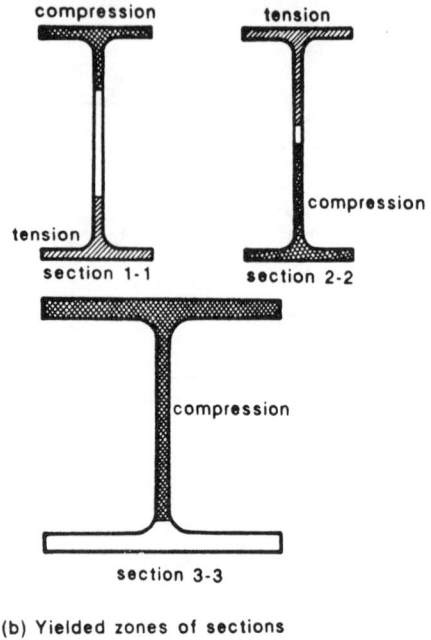

(b) Yielded zones of sections

(max. strain: $\epsilon_{max} = 0.0203$ at section 4-4)

FIGURE 7.31. *(continued)*

simple analytical procedures are generally compared with experimental results for their final verifications.

Researches in Japan were reviewed for the last 20 years, and the investigation has concluded that there are some excellent experimental researches in Japan that have studied the ultimate strength and behavior of steel frames. Most of the experimental works on this subject were made in the early 1970s, and few in recent years. In this section, some of these frames are selected as the proposed calibration frames for second-order inelastic analysis. The description will be made to provide the necessary information for the following tests: full size tests of a portal frame (Wakabayashi, 1972a) and two series of one-quarter scaled frames (Wakabayashi, 1972b). Analytical results given by the authors are also shown for comparison, but the analyses may not be considered as exact solutions because of some inadequate assumptions for second-order inelastic analysis. Analytical results in this section should be regarded as a reference. The full-size test for a gable frame (Abe et al., 1983) is also reported in the reference (Toma et al., 1992b).

7.5.2 Full-Size Test of Portal Frame

Test Specimens

Four full-size specimens of portal frame were tested for monotonic and cyclic loading. Configurations of the specimens are shown in Figure 7.32. Member sizes

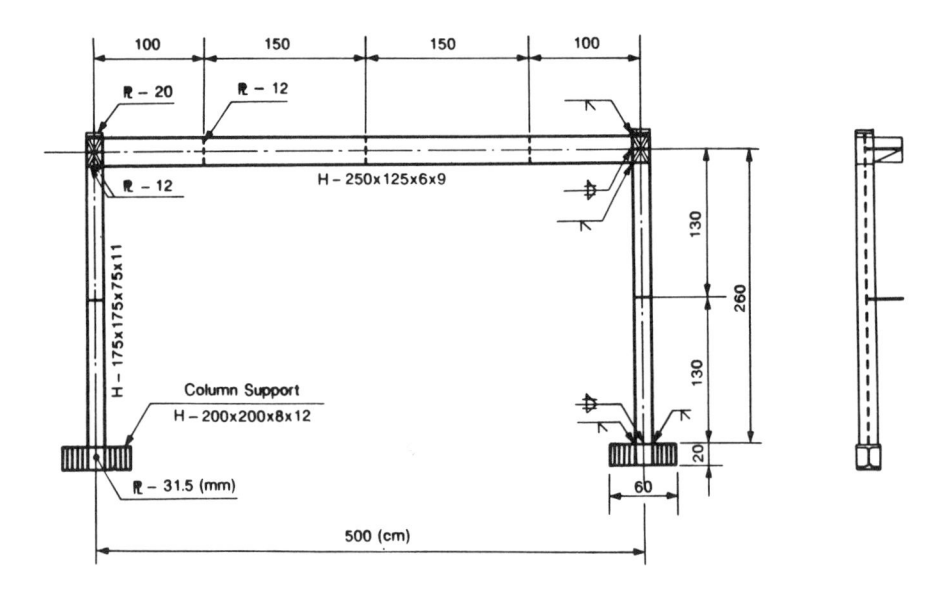

FIGURE 7.32. Full-size portal frame. (From Wakabayashi et al., *Trans. Arch. Inst. Jpn.*, 198, 7, 1972. With permission.)

and loading conditions are summarized in Table 7.3. Specimens FM0 and FM5 are loaded monotonically and FC0 and FC5 are cyclicly. The panel of the connections between column and beam are stiffened with horizontal and diagonal plates to prevent shear buckling so that the bending of the members controls the ultimate strength of the frame.

The specimens are supported in out-of-plane direction at four equal points of the beam and the mid-point of the columns by a hinge device that can rotate in the plane of the frame but is fixed in out-of-plane. Sectional properties of the members are measured as given in Table 7.4. Material properties were obtained by the standard coupon test for which two test pieces were taken out from the flanges of each H-shape member. The results are summarized in Table 7.5.

Loading Procedures

No vertical load is applied to specimens FM0 and FC0. For specimens FM5 and FC5, a constant vertical load is first applied on top of the columns, then the horizontal load at the top of the frame is increased gradually. Due to swaying of the frame by the horizontal load, the loading points of vertical loads move horizontally. In order that the loading points be kept on the center of the columns, the loading frame which supports vertical loading jacks is devised in such a way that it follows the horizontal movement of the frame. For the cyclic loading, the horizontal load is controlled by deflection: the deflection is added by 20 mm at each new cycle to the maximum deflection of the previous cycle.

354 ADVANCED ANALYSIS OF STEEL FRAMES

Table 7.3 Test Program of Full-Size Portal Frames

Specimen name	P (ton)	P/P_y	P/P_e	h (cm)	l (cm)	h/r	Column (mm)	Beam (mm)	$\dfrac{I_b \times h}{I_c \times l}$	Loading conditions
F M 0	0	0	0	260	500	34.7	H-175 × 175 × 7.5 × 11	H-250 × 125 × 6 × 9	0.768	Monotonic
F M 5	70	0.489	0.12	260	500	34.7	H-175 × 175 × 7.5 × 11	H-250 × 125 × 6 × 9	0.742	Monotonic
F C 0	0	0	0	260	500	34.5	H-175 × 175 × 7.5 × 11	H-250 × 125 × 6 × 9	0.737	Repeated
F C 5	70	0.516	0.12	260	500	34.5	H-175 × 175 × 7.5 × 11	H-250 × 125 × 6 × 9	0.745	Repeated

Note: (P) Column load, (P_e) elastic buckling of loaded frame, (P_y) yield load of column, (h) column height, (l) beam length, (r) radius of gyration of a column, (I) sectional moment of inertia, and $I_b \times h/(I_c \times l)$ beam-to-column stiffness ratio.
From Wakabayashi, M. et al. (1972a) Elastic-plastic behaviors of full size steel frame, *Trans. Arch. Inst. Jpn.*, 198 (August), 7–17 (in Japanese). With permission.

Table 7.4 Actual Section Properties of Full-Size Portal Frames

Specimen name	Column				Beam			
	A (cm²)	I (cm⁴)	Z (cm³)	Z_p (cm³)	A (cm²)	I (cm⁴)	Z (cm³)	Z_p (cm³)
F M 0	48.8	2740	314	351	37.9	4050	325	367
F M 5	50.6	2840	323	363	37.3	4050	322	363
F C 0	50.8	2880	328	366	38.3	4080	326	370
F C 5	50.2	2840	324	362	37.6	4070	325	367

Note: (A) Cross-sectional area, (I) sectional moment of inertia, (Z) section modulus, and (Z_p) plastic section modulus.
From Wakabayashi, M. et al. (1972a) Elastic-plastic behaviors of full size steel frame, *Trans. Arch. Inst. Jpn.*, 198 (August), 7–17 (in Japanese). With permission.

Test Results

Monotonic Loading

Load-deflection curves and load-rotation curves for monotonic loading are shown in Figures 7.33 and 7.34, respectively. The measured results are also given in Table 7.6. An ordinate is normalized by the horizontal collapse load H_{pc} at which the frame reaches the mechanism load. The horizontal collapse load considers the effect of axial force on the plastic moment and is given in Table 7.6. An abscissa is normalized by the story height h. The occurrence of local buckling (L.B.) and lateral buckling (Lat. B.) is indicated in these figures.

Because the specimens contain imperfections, a considerable amount of bending moment is initially induced to the members when the specimens are set on the test bed. This causes early yielding and the stiffness decreases at an early stage of horizontal loading. While Specimen FM0, after reaching the mechanism, shows a gradual increase in horizontal load due to strain hardening, Specimen FM5 shows a degradation of stiffness as the lateral deflection increases. The local buckling appears not to affect much on the horizontal load-carrying capacity.

Cyclic Loading

The load-deflection and load-rotation curves for cyclic loading are shown in Figures 7.35 and 7.36, respectively. Degradation of the stiffness is also observed at an early stage of the first cycle of loading because of the imperfections of these specimens. While specimen FC0 shows a stable loop, the behaviors of specimen FC5 are quite different from its monotonic loading counterpart, i.e., the maximum strength at each cycle increases.

Theoretical Analysis

The test results are compared with the theories in Figures 7.33 to 7.36 by the plastic-zone method and the plastic hinge method. It is assumed in these analyses that both

Table 7.5 Material Properties of Full-Size Portal Frames

Specimen name	Column					Beam				
	σ_y (t/cm²)	σ_u (t/cm²)	ε_u (%)	$\varepsilon_{st}/\varepsilon_y$	E_{st}/E	σ_y (t/cm²)	σ_u (t/cm²)	ε_u (%)	$\varepsilon_{st}/\varepsilon_y$	E_{st}/E
F M 0	2.70	4.42	29.3	14.0	0.016	2.70	4.23	26.5	15.7	0.013
F M 5	2.78	4.44	32.2	13.7	0.014	2.88	4.35	30.5	14.9	0.013
F C 0	2.68	4.28	34.1	15.7	0.014	2.86	4.18	25.6	12.1	0.011
F C 5	2.70	4.31	34.4	15.2	0.010	2.56	3.95	24.6	13.6	0.011

Note: (σ_y) Yield stress, (σ_u) tensile strength, (ε_u) maximum elongation, (ε_y) strain at first yield, (ε_{st}) strain at start of strain hardening, (E) modulus of elasticity, and (E_{st}) strain hardening modulus.
From Wakabayashi, M. et al. (1972a) Elastic-plastic behaviors of full size steel frame, *Trans. Arch. Inst. Jpn.*, 198 (August), 7–17 (in Japanese). With permission.

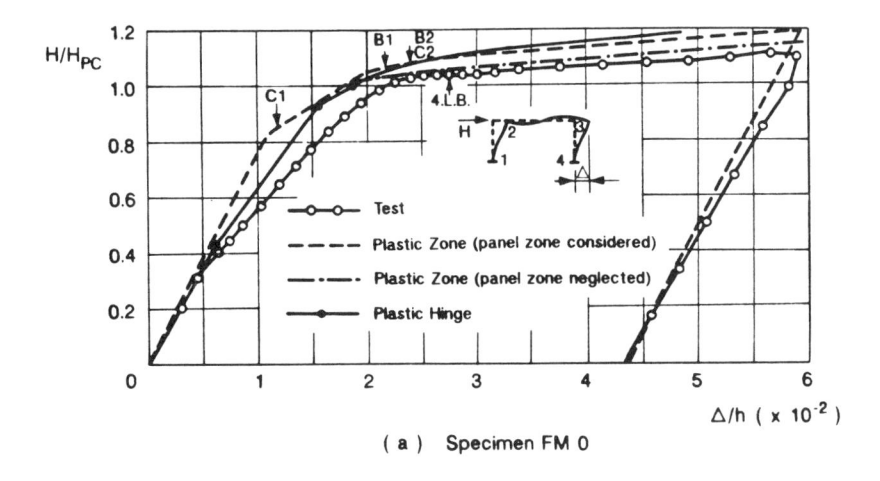

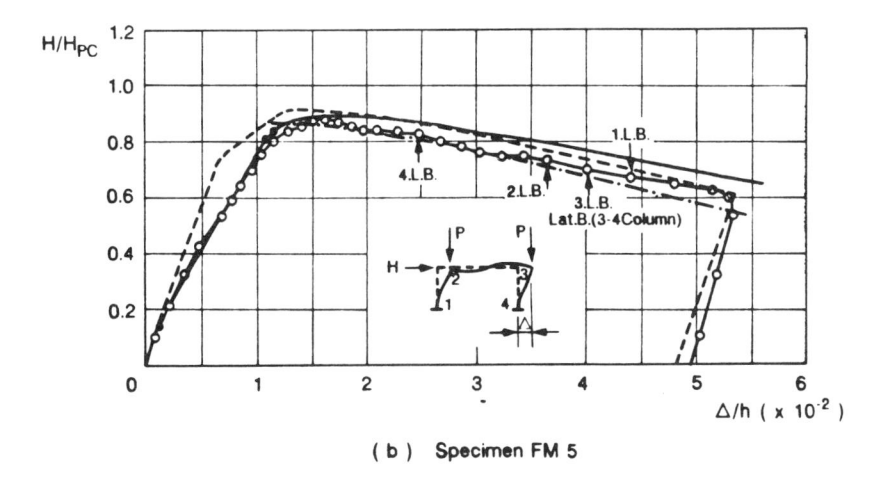

FIGURE 7.33. Horizontal force-displacement behaviors of full-size portal frame, monotonic loading. (From Wakabayashi et al., *Trans. Arch. Inst. Jpn.*, 198, 7, 1972. With permission.)

local buckling and lateral buckling do not occur and displacements are small (first-order). In the plastic-zone method, the portal frame is assumed to be composed of two L-shape frames by anti-symmetry. Furthermore, the L-shape frame is separated into two members, i.e., a beam and a column. Applying the equilibrium and compatibility conditions at the joint of the beam and the column, the load-deformation behavior of the frame is obtained. The members are divided into 25 segments. The rotation of the member is calculated by integrating the curvature along the length. The effects of shear force and stiffness of the joint panel are considered. In Figure 7.33a, the yielded segments are indicated: for example, C1 or B2 means that the yielding occurs in the

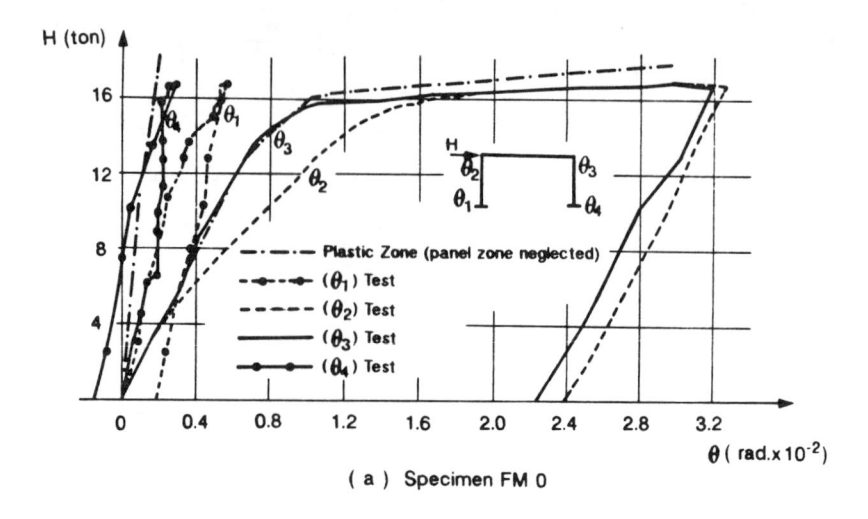

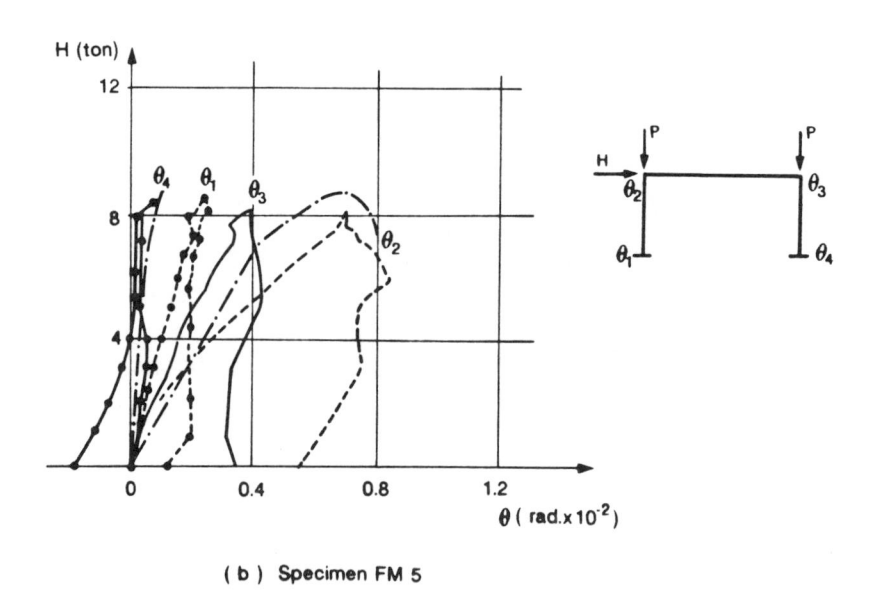

FIGURE 7.34. Horizontal force-rotation behaviors of full-size portal frame, monotonic loading. (From Wakabayashi et al., *Trans. Arch. Inst. Jpn.*, 198, 7, 1972. With permission.)

first segment of the column or the second segment of the beam, respectively, from the member end.

The moment-curvature relationship is assumed to be a bi-linear curve as shown in Figure 7.37. In Figure 7.37a, an ordinate and abscissa are normalized by the plastic moment M_p and the plastic curvature M_p/EI, respectively. For members with axial force (Figure 7.37b), the coordinates are normalized by the plastic moment M_{pc} and the plastic curvature M_{pc}/EI considering the effect of axial force, and the elastic limit

Table 7.6 Test Results of Full-Size Portal Frames

Specimen name	Number of cycles	Test results				H_{pc} (ton)
		H_f (ton)	H_f/H_{pc}	Δ_f/h	Δ_f/Δ_{fy}	
F M 0		15.8	1.03	0.059	4.04	15.3
F M 5		8.5	.89	0.015	2.20	9.59
F C 0	0/2	12.4	.78	0.014	—	15.9
	1/2	26.9	1.69	0.030	—	
	2/2	31.5	1.99	0.038	1.47	
	3/2	34.4	2.17	0.047	1.65	
	4/2	35.2	2.22	0.055	1.84	15.9
	5/2	36.3	2.29	0.060	1.97	
	6/2	37.0	2.33	0.067	2.18	
	7/2	37.9	2.39	0.073	2.27	
F C 5	0/2	6.7	0.75	0.011	—	8.97
	1/2	15.0	1.68	0.022	1.64	
	2/2	16.6	1.85	0.023	1.56	8.97
	3/2	17.4	1.94	0.023	1.49	
	4/2	20.5	2.29	0.019	1.24	
	5/2	20.8	2.32	0.027	1.49	

Note: (H) Maximum horizontal force, (H_f) experimental, (H_{pc}) rigid plastic, (Δ_f) maximum displacement (FM0, FC0), displacement at maximum force (FM5, FC5), and (Δ_{fy}) displacement at yield force.
From Wakabayashi, M. et al. (1972a) Elastic-plastic behaviors of full size steel frame, *Trans. Arch. Inst. Jpn.*, 198 (August), 7–17 (in Japanese). With permission.

is increased until it reaches the bending moment at the previous cycle. The plastic moments with and without the effect of axial force are designated as M_p and M_{pc}, respectively. The term S is the ratio of M_p to M_{pc}. The strain hardening factor is 0.01 for the member with no axial force (τ_0) and 0.02 with axial force (τ).

In the plastic hinge method, the effect of strain hardening of material and rigid stiffness of the joint panel are considered. Since the column bottom is not perfectly fixed in real structures, a fictitious member is inserted between the column supports to form a rectangle closed frame and include the effect of semi-rigid supports. The stiffness of the fictitious member is assumed to be 1.8 times that of the column. The analysis also includes the effect of the bending moment induced when the specimen is set on the test bed: the moment is assumed to be a half of the plastic moment M_p at the column bottom. This results in an early degradation of the stiffness.

7.5.3 One-Quarter Scaled Test of Portal Frames

Test Specimens

Two series of test are conducted in this study for a one-story frame (portal frame) and a two-story frame. Configurations of the frames are shown in Figure 7.38, and nominal dimensions of these frames and of members are given in Table 7.7. The specimens consist of rolled H-shapes. The connections are welded and stiffened to

360 ADVANCED ANALYSIS OF STEEL FRAMES

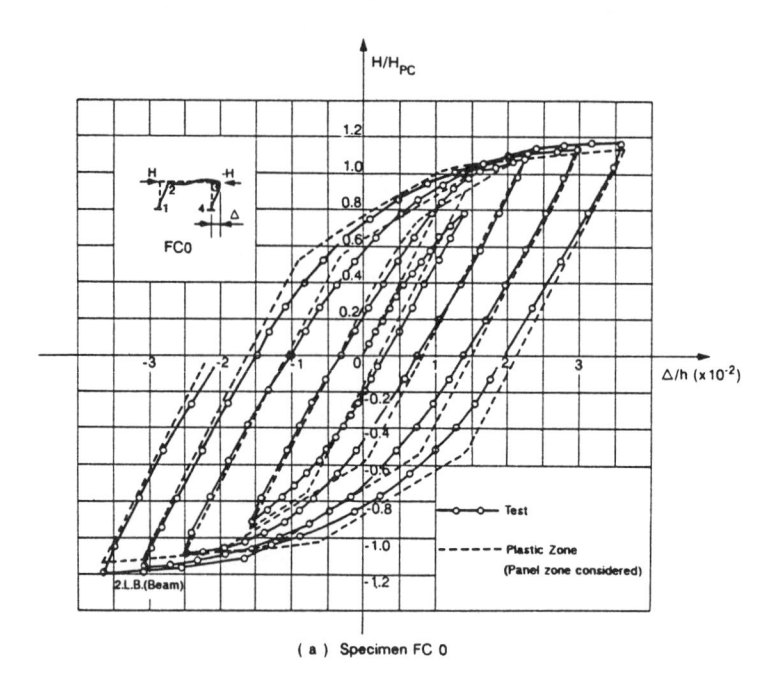

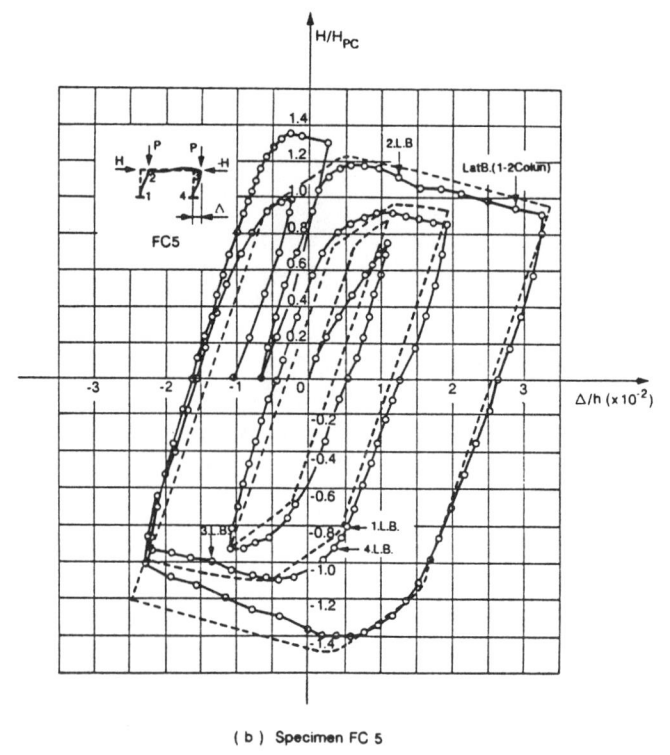

(a) Specimen FC 0

(b) Specimen FC 5

FIGURE 7.35. Horizontal force-displacement behaviors of full-size portal frame, cyclic loading. (From Wakabayashi et al., *Trans. Arch. Inst. Jpn.*, 198, 7, 1972. With permission.)

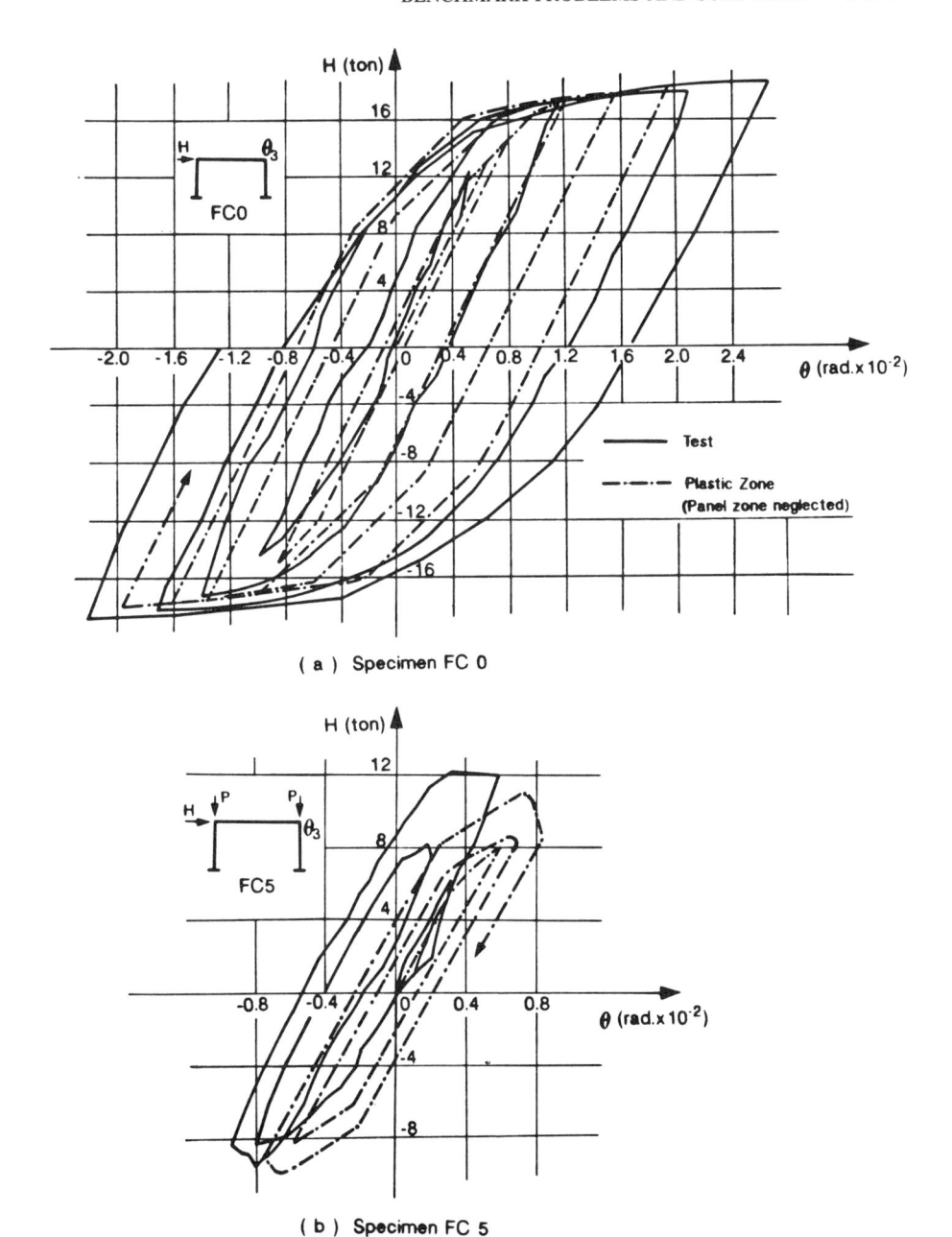

FIGURE 7.36. Horizontal force-rotation behaviors of full-size portal frame, cyclic loading. (From Wakabayashi et al., *Trans. Arch. Inst. Jpn.*, 198, 7, 1972. With permission.)

362 ADVANCED ANALYSIS OF STEEL FRAMES

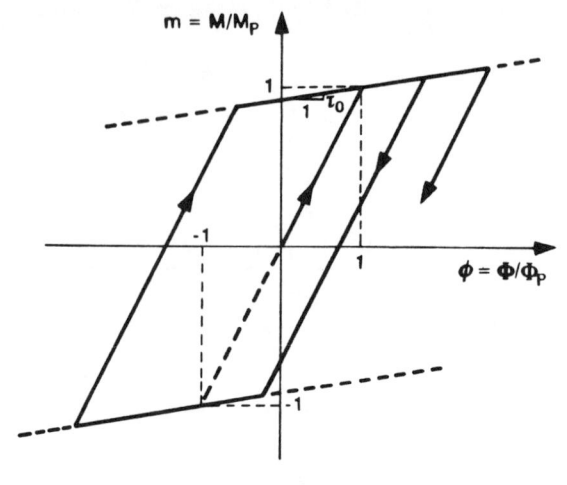

(a) With no axial force

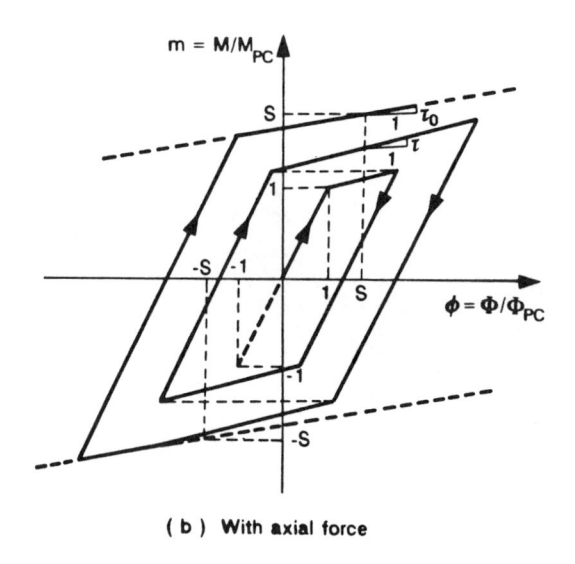

(b) With axial force

FIGURE 7.37. Assumed moment-curvature curves. (From Wakabayashi et al., *Trans. Arch. Inst. Jpn.*, 198, 7, 1972. With permission.)

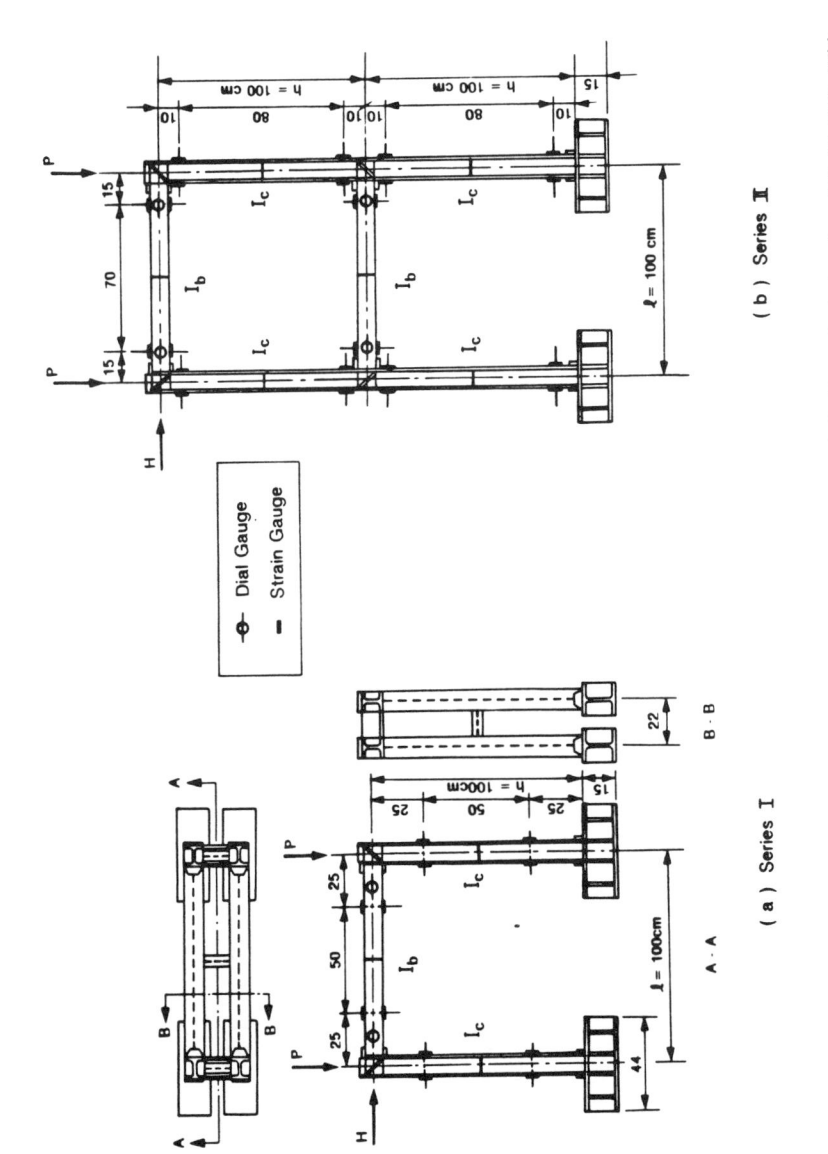

FIGURE 7.38. One-quarter scaled frames. (From Wakabayashi, M. and Matsui, C., *Trans. Arch. Inst. Jpn.*, 193, 17, 1972. With permission.)

Table 7.7 Test Program of One-Quarter Scaled Frames

Specimen number	P (ton)	P/P_y	P/P_e	h (cm)	l (cm)	h/r	Column (mm)	Beam (mm)	$\dfrac{I_b \times h}{I_c \times l}$	Heat treatment
I-1	20	0.31	0.04	100	100	24	H-100 × 100 × 6 × 8	H-100 × 50 × 4 × 6	0.43	none
I-2	20	0.31	0.04					H-100 × 50 × 4 × 6	0.43	annealed
I-3	20	0.29	0.03					H-100 × 100 × 6 × 8	1.00	none
I-4	20	0.30	0.03					H-100 × 100 × 6 × 8	1.00	annealed
II-1	10	0.17	0.04	100	100	24	H-100 × 100 × 6 × 8	H-100 × 50 × 4 × 6	0.43	none
II-2	20	0.34	0.07					H-100 × 50 × 4 × 6	0.43	
II-3	10	0.17	0.02					H-100 × 100 × 6 × 8	1.00	
II-4	20	0.33	0.05					H-100 × 100 × 6 × 8	1.00	

Note: (P) Column load, (P_e) elastic buckling load of frame, (P_y) yield load of column, (h) column height, (l) beam length, (r) radius of gyration of a column, (I) sectional moment of inertia, and $I_b \times h/(I_c \times l)$ beam-to-column stiffness ratio.

From Wakabayashi, M. and Matsui, C. (1972b) Experimental study on elasto plastic stability of steel frames. II. Portal frames composed of H-shape members, *Trans. Arch. Inst. Jpn.*, 193 (March), 17–27 (in Japanese). With permission.

prevent local buckling in the joint panels. Measured sectional properties of the members are given in Table 7.8. Material properties, which are obtained by the standard coupon test, are listed in Table 7.9.

These tests are intended to study the effects of the: (1) number of stories; (2) axial force in columns; (3) stiffness ratio of column to beam; and (4) residual stress. The stiffness ratio of the beam to the column is smaller in specimens I-1 and I-2 than that of specimens I-3 and I-4 as seen in Table 7.7. In the two-story frames, the axial force ratio and the stiffness ratio vary in order to study their effect. To prevent the out-of-plane buckling, two of the same specimens are set in parallel and connected at the joints and the mid-length of the members. In other words, twin specimens are tested simultaneously.

Loading Procedures

The vertical load given in Table 7.7 is first applied at top of four columns by a fixed testing machine. (Parallel twin specimens are loaded simultaneously.) Then, the horizontal load at the top of the frame is increased gradually. When the frame sways by the horizontal loading, the base beam that supports the test specimen and the horizontal loading jack follows a horizontal movement so that the vertical loading points can be kept on the center of the columns. The loads are measured by the load cells which are installed between the hydraulic jacks and the specimen.

Test Results of One-Story Frames

Test results of the one-story frame are shown in Figure 7.39 and summarized in Table 7.10. An ordinate is normalized by the horizontal collapse load H_{pc}. The collapse load H_{pc} is obtained such that the plastic moment is reached at the critical sections of the beam and the columns. The critical sections are taken at the ends of the covered plates (40-mm length) which are attached at both ends of the beam and the bottom end of the columns. The critical section of the column top is considered at the end of the joint panel.

Since the beam stiffness of specimens I-1 and I-2 is small, the plastic hinges are formed at the ends of the beam and the bottom of the columns. On the other hand, in specimens I-3 and I-4, which have larger stiffness of the beam, the plastic hinges are formed at the bottom and top of the columns, but not in the beam. Noticing the difference in stiffness of the beam, larger stiffness of the beam results in a stable load-deflection behavior after yielding. Local buckling or lateral buckling does not occur in any test until near the end of loading. The effect of annealing is not seen clearly in specimens I-2 and I-4. This implies that the residual stresses do not affect significantly the load-deflection behavior of these frames.

Test Results of Two-Story Frames

The load-deflection curves of the two-story frames are shown in Figure 7.40, and the collapse loads are given in Table 7.10. Comparisons of these tests show the effects of axial force and stiffness of the beam on the frame behavior. The larger the axial

Table 7.8 Actual Section Properties of One-Quarter Scaled Frames

Specimen number	h (cm)	l (cm)	Column					Beam					$\frac{I_b \times h}{I_c \times l}$
			A (cm²)	I (cm⁴)	Z (cm³)	Z_p (cm³)		A (cm²)	I (cm⁴)	Z (cm³)	Z_p (cm³)		
I-1	100.0	100.0	22.0	385	76.2	87.3		11.0	181	35.6	41.6		0.47
I-2	100.0	99.9	22.2	386	76.5	88.2		10.9	181	35.5	41.5		0.47
I-3	99.8	100.1	23.1	407	80.3	92.5		23.1	408	80.3	92.5		1.00
I-4	99.9	99.9	23.1	405	79.9	92.1		23.1	406	80.1	92.2		1.00
II-1	99.9	98.8	21.8	391	77.4	88.5		10.6	177	35.0	40.6		0.46
II-2	99.9	99.6	22.1	397	78.4	89.7		10.6	177	35.0	40.6		0.45
II-3	100.3	99.6	22.0	392	77.7	88.9		22.1	397	78.4	89.7		1.02
II-4	99.9	99.8	22.2	396	78.4	89.8		22.1	396	78.3	89.6		1.00

Note: (A) Cross-sectional area, (I) sectional moment of inertia, (Z) elastic section modulus, and (Z_p) plastic section modulus.

From Wakabayashi, M. and Matsui, C. (1972b) Experimental study on elasto plastic stability of steel frames. II. Portal frames composed of H-shape members, *Trans. Arch. Inst. Jpn.*, 193(March), 17–27 (in Japanese). With permission.

Table 7.9 Material Properties of One-Quarter Scaled Frames

Specimen number	Column					Beam				
	σ_y (t/cm²)	σ_u (t/cm²)	ε_u (%)	$\varepsilon_{st}/\varepsilon_y$	E_{st}/E	σ_y (t/cm²)	σ_u (t/cm²)	ε_u (%)	$\varepsilon_{st}/\varepsilon_y$	E_{st}/E
I-1	2.92	4.55	26.5	14.1	0.019	3.54	4.72	25.0	11.2	0.013
I-2	2.90	4.52	26.0	12.8	0.016	3.44	4.63	20.2	13.2	0.011
I-3	2.98	4.74	24.3	10.7	0.021	2.98	4.74	24.3	10.7	0.021
I-4	2.86	4.63	26.2	11.0	0.019	2.86	4.63	26.2	11.0	0.019
II-1	2.64	4.22	28.3	13.0	0.017	3.04	4.27	20.1	11.0	0.014
II-2	2.67	4.21	28.9	13.3	0.017	3.08	4.40	20.1	13.1	0.014
II-3	2.73	4.31	28.2	11.8	0.017	2.63	4.23	28.4	12.3	0.017
II-4	2.73	4.26	28.3	12.0	0.017	2.73	4.26	28.6	12.8	0.017

Note: (σ_y) Yield stress, (σ_u) tensile strength, (ε_u) maximum elongation, (ε_y) strain at first yield, (ε_{st}) strain at start of strain hardening, (E) modulus of elasticity, and (E_{st}) strain hardening modulus.
From Wakabayashi, M. and Matsui, C. (1972b) Experimental study on elasto plastic stability of steel frames. II. Portal frames composed of H-shape members, *Trans. Arch. Inst. Jpn.*, 193(March), 17–27 (in Japanese). With permission.

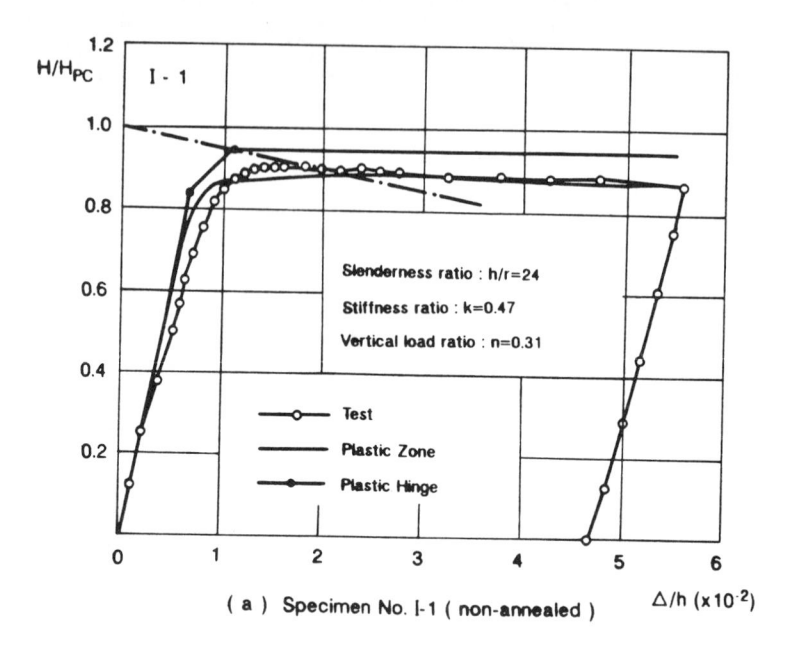

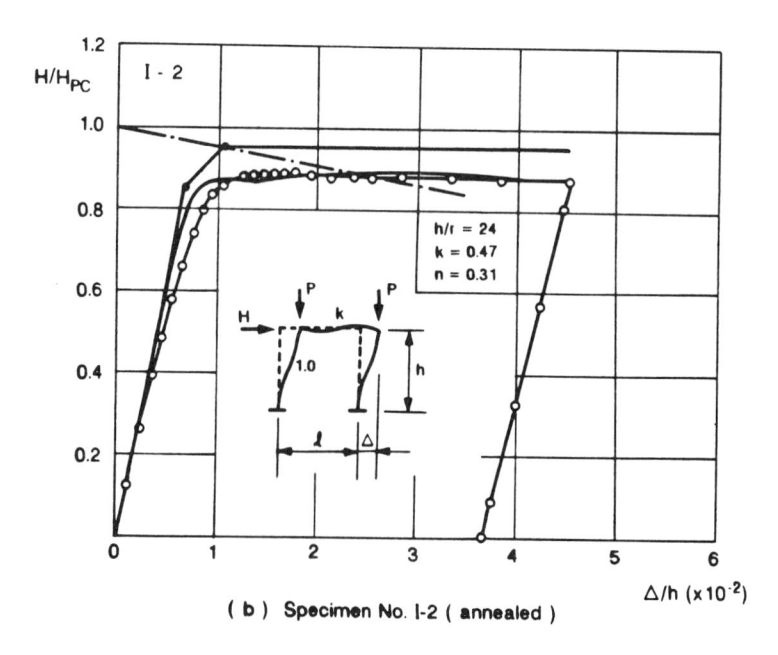

FIGURE 7.39. Horizontal force-displacement behaviors of one-quarter scaled frame (one story). (From Wakabayashi, M. and Matsui, C., *Trans. Arch. Inst. Jpn.*, 193, 17, 1972. With permission.)

(c) Specimen No. I-3 (non-annealed)

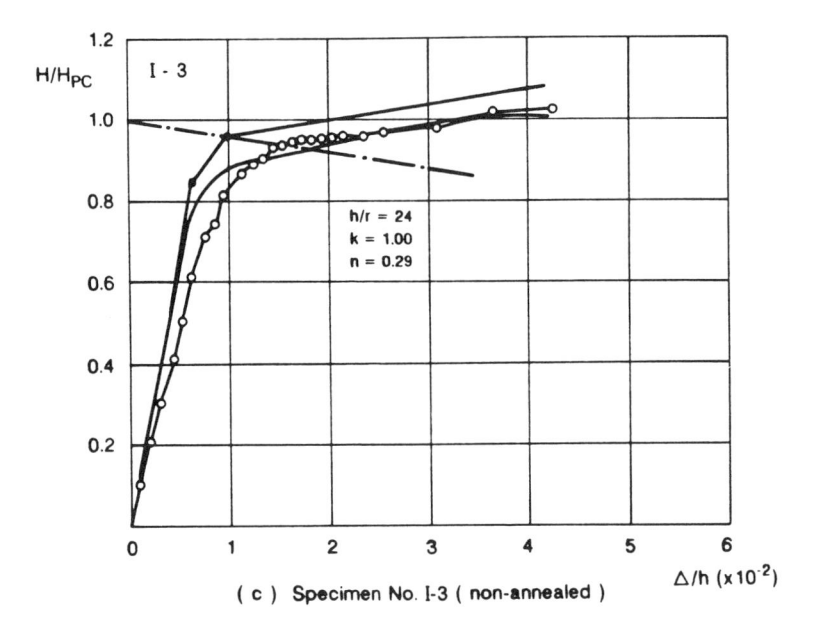

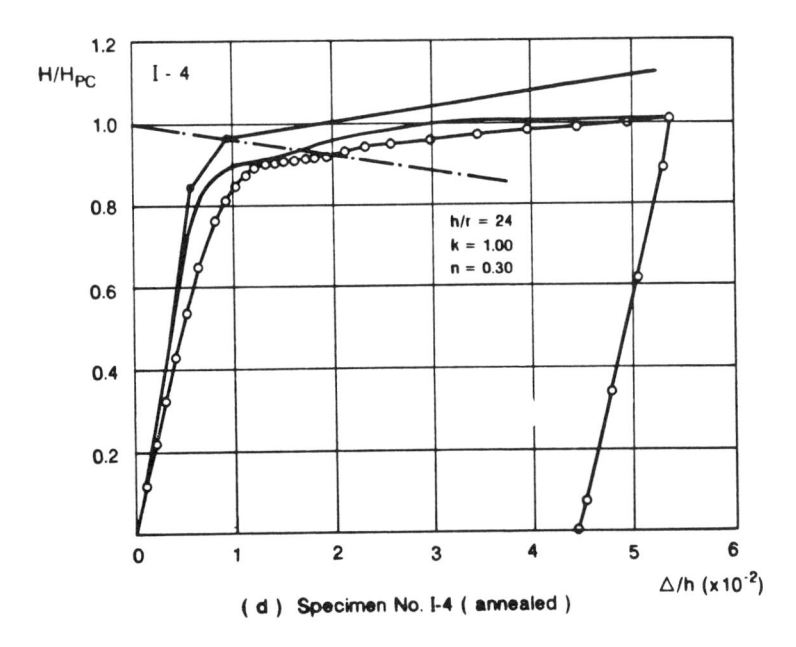

$h/r = 24$
$k = 1.00$
$n = 0.29$

(d) Specimen No. I-4 (annealed)

$h/r = 24$
$k = 1.00$
$n = 0.30$

FIGURE 7.39. *(continued)*

Table 7.10 Test Results of One-Quarter Scaled Frames

Specimen number	Test results					Rigid plastic H_{pc} (ton)
	H_f (ton)	H_f/H_{pc}	Δ_{f1}/h (×10^{-2})	Δ_{f2}/h (×10^{-2})	Δ_f/Δ_{fy}	
I-1	7.48	0.91	1.40		1.7	8.24
I-2	7.20	0.89	1.45		1.8	8.12
I-3	9.55	0.94	1.50		4.7	10.19
I-4	8.60	0.90	1.50		6.1	9.56
II-1	5.50	1.02	1.50	3.35	4.5	5.40
II-2	5.00	0.98	1.30	2.85	1.7	5.10
II-3	7.70	0.97	1.50	3.00	4.0	7.91
II-4	7.20	0.99	1.50	3.05	4.1	7.26

From Wakabayashi, M. et al. (1972a) Elastic-plastic behaviors of full size steel frame, *Trans. Arch. Inst. Jpn.*, 198(August), 7–17 (in Japanese). With permission.

force in columns and the smaller the stiffness of the beam, the more unstable the frames become.

Theoretical Analysis

The frames are analyzed by the plastic-zone method and by the plastic hinge method. The basic analytical concept is similar to the one-story frame in the previous section. In the plastic-zone method, one half of the frames is analyzed taking anti-symmetry into consideration. The moment-curvature relationship is approximated by a closed form expression. The members are divided into 20 segments. The joint panel is assumed to be rigid (50-mm length). The effect of shear force on the deformation is considered.

In the plastic hinge method, the moment-curvature relationship before the formation of the failure mechanism is assumed to be elastic-perfectly plastic. After the formation of the failure mechanism, the strain hardening starts at each plastic hinge, one at a time.

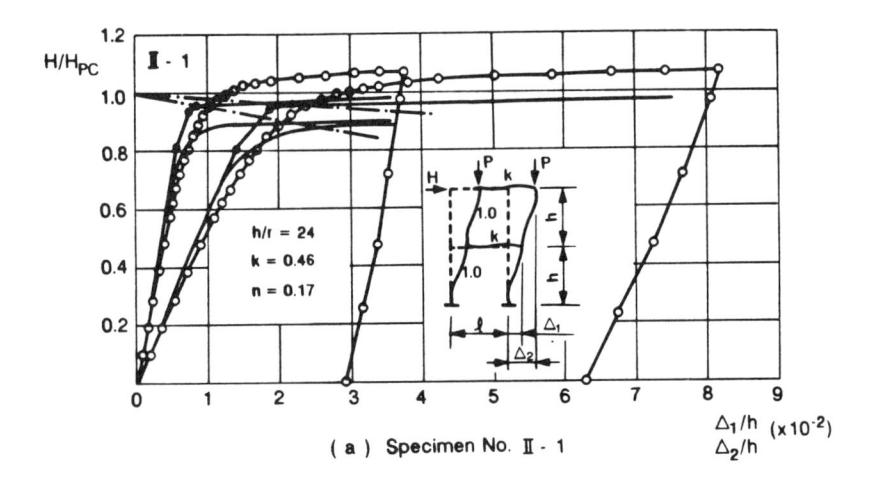

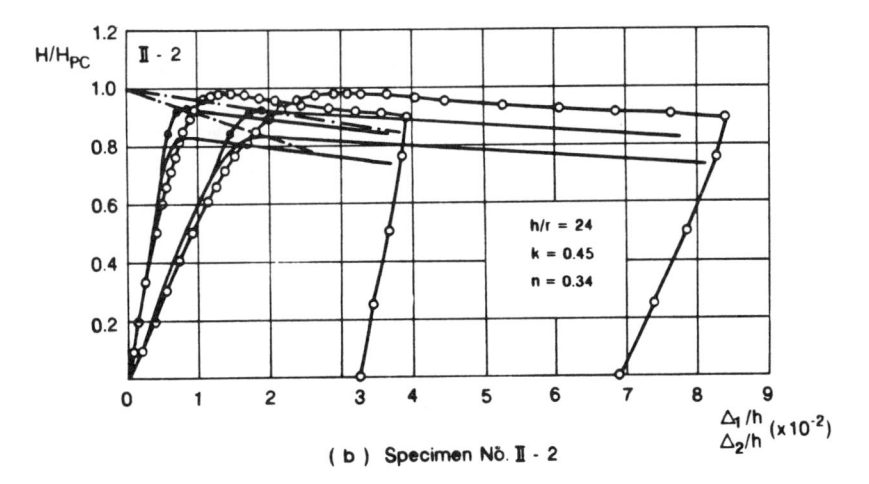

FIGURE 7.40. Horizontal force-displacement behaviors of one-quarter scaled frame (two story). (From Wakabayashi, M. and Matsui, C., *Trans. Arch. Inst. Jpn.*, 193, 17, 1972. With permission.)

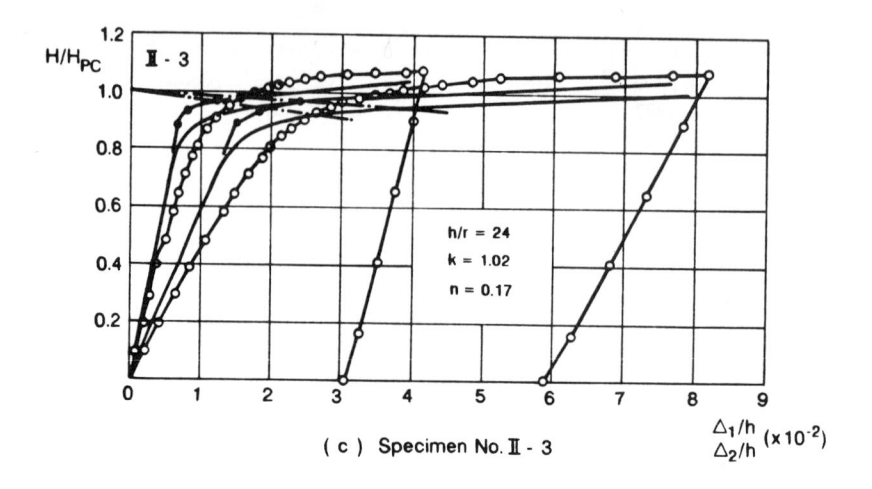

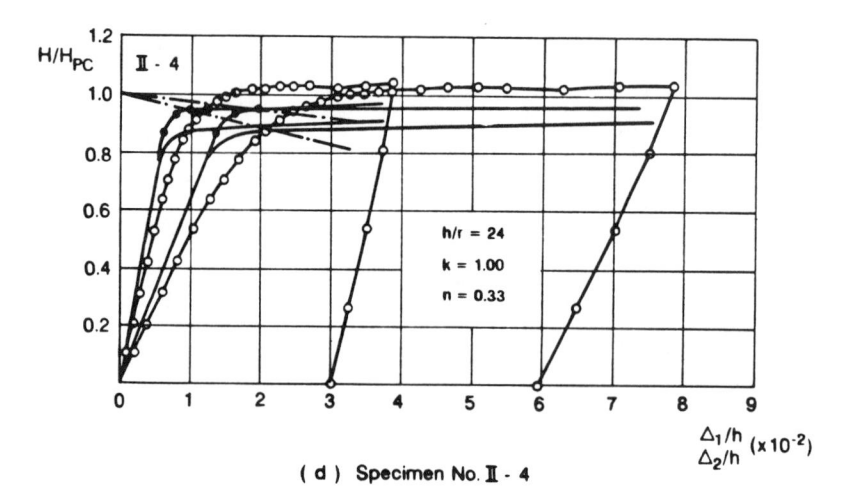

FIGURE 7.40. *(continued)*

References

Abe, K., Tagawa, K., and Sakai, M. (1983) A study of elasto-plastic strength of gable frame composed by H-shapes, *Trans. Arch. Inst. Jpn.*, 325(March), 25–31 (in Japanese).

AISC (1986) Load and Resistance Factor Design for Structural Steel Buildings, American Institute of Steel Construction, Chicago, IL.

Al-Mashary, F. A. (1989) Simplified Non-linear Analysis for Steel Frames, Ph.D. dissertation, School of Civil Engineering, Purdue University, West Lafayette, IN, December, 193 pp.

Basu, P. K. (1990) Draft Guidelines for Evaluating Existing Stability Analysis Software, SSRC 1990 Annual Meeting TG28 Chairman's Report, April, Structural Stability Research Council.

Beaulieu, D. and Adams, P. F. (1977) The Destabilizing Forces Caused by Gravity Loads Acting on Initially Out-of-Plumb Members in Structures, Structural Eng. Report, No. 59, Department of Civil Engineering, University of Alberta, Edmonton, Canada, February.

Bridge, R. Q., Clarke, M. J., Hancock, G. J., and Trahair, N. S. (1991) Verification of Approximate Methods of Structural Analysis, ASCE Structural Congress 1991 Compact Papers, Indianapolis, April 29 to May 1, pp. 753–756.

Deierlein, G. G. and McGuire, W. (1990) The Use of 2nd-order Inelastic Analysis to Evaluate Frame Stability for Design, Proc. of 1990 Annual Tech. Session and Meet., Structural Stability Research Council.

Dhillon, B. S. and Abdel-Majad, S. (1990) Interactive analysis and design of flexibly connected frames, *Comp. Struct.*, 36(2), 189–202.

El-Zanaty, M. H., Murray, D. W., and Bjorhovde, R. (1980) Inelastic Behavior of Multistory Steel Frames, Structural Engineering Report, No. 83, The University of Alberta, Edmonton, Canada, April, 248 pp.

Galambos, T. V. and Ketter, R. L. (1957) Further Studies of Columns Under Combined Bending and Thrust, Fritz Engineering Laboratory Report, No. 205A. 19, Lehigh University, Bethlehem, PA, 82 pp.

Goto, Y. and Chen, W. F. (1987) Second-order elastic analysis for frame design, *J. Struct. Eng.*, 113(7), 1501–19.

Kanchanalai, T. (1977) The Design and Behavior of Beam-Columns in Unbraced Steel Frames, AISI Project No. 189, Report No. 2, Civil Engineering/Structures Research Laboratory, University of Texas at Austin, 300 pp.

King, W. (1990) Simplified Second-Order Inelastic Analysis for Semi-Rigid Frame Design, Ph.D. dissertation, School of Civil Engineering, Purdue University, West Lafayatte, IN, 200 pp.

Kishi, N. and Chen, W. F. (1986a) Steel Connection Data Bank Program, Structural Engineering Report, No. CE-STR-86–18, School of Civil Engineering Purdue University, West Lafayette, IN, May, 163 pp.

Kishi, N. and Chen, W. F. (1986b) Data Base of Steel Beam-Column Connections, Vol. I and II, Structural Engineering Report, No. CE-STR-86–26, Purdue University, West Lafayette, IN, August, 653 pp.

Lindsey, S. D., Ioannides, S. A., and Goverdhan, A. V. (1985) The effect of connection flexibility on steel members and frame stability, in Connection Flexibility and Steel Frames, Chen, W. F., Ed., Proc. of a Session Sponsored by Structural Division of ASCE, Detroit, October, pp. 6–12.

Lu, L. W., Ozer, E., Daniels, J. H., Okten, O. S., and Morino, S. (1975) Frame Stability and Design of Columns in Unbraced Multi-Story Steel Frames, Fritz Engineering Lab., Report No. 375.2, Lehigh University, Bethlehem, PA, 134 pp.

Orbison, J. G. (1982) Nonlinear Static Analysis of Three-Dimensional Steel Frames, Report No. 82–6, Department of Structural Engineering, Cornell University, Ithaca, NY, March, 243 pp.

SSRC (1988) Guide to Stability Design Criteria for Metal Structures, 4th ed., Galambos, T. V., Ed., John Wiley & Sons, New York, 786.

Toma, S., Chen, W. F., and White, D. W. (1991) Calibration Frames for Second-Order Inelastic Analysis in North America, Structural Engineering Report, No. CE-STR-91–23, School of Civil Engineering, Purdue University, West Lafayette, IN, 70 pp.

Toma, S. and Chen, W. F. (1992a) European calibration frames for second-order inelastic analysis, *Eng. Struct.*, 14(1), 7–14.

Toma, S. and Chen, W. F. (1992b) Calibration Frames for Second-Order Inelastic Analysis in Japan, Structural Engineering Report, No. CE-STR-92–16, School of Civil Engineering, Purdue University, West Lafayette, IN, 34 pp.

Vogel, U. et. al. (1984) Ultimate Limit State Calculation of Sway Frames with Rigid Joints, ECCS-CECM-EKS-Publication No. 33, 1st ed., Rotterdam.

Vogel, U. (1985) Calibration frames, *Stahlbau*, 10, 1–7.

Vogel, U. (1989) The stability of framed structures, in *Frame and Slab Structures*, Armer, G. S. T. and Moore, D. B., Eds., Butterworths, London, 29–56.

Wakabayashi, M., Matsui, C., Minami, K., and Mitani, I. (1972a) Elastic-plastic behaviors of full size steel frame, *Trans. Arch. Inst. Jpn.*, 198(August), 7–17 (in Japanese).

Wakabayashi, M. and Matsui, C. (1972b) Experimental study on elasto-plastic stability of steel frames. II. Portal frames composed of H-shape members, *Trans. Arch. Inst. Jpn.*, 193(March), 17–27 (in Japanese).

White, D. W. (1985) Material and Geometric Nonlinear Analysis for Local Planar Behavior in Steel Frames, Master thesis, Cornell University, Ithaca, NY, August, 281 pp.

White, D. W. (1990) Chairman's Report to SSRC Executive Committee, TG29: Second-Order Inelastic Analysis for Frame Design, April, Structural Stability Research Council.

White, D. W., Liew, J. Y. R., and Chen, W. F. (1991) Second-Order Inelastic Analysis for Frame Design: A Report to SSRC Task Group 29 on Recent Research and the Perceived Sate-of-the-Art, Structural Engineering Report, No. CE-STR-91–12, School of Civil Engineering, Purdue University, West Lafayette, IN, 116 pp.

Yarimci, E. (1966) Incremental Inelastic Analysis of Framed Structures and Some Experimental Verifications, Ph.D. dissertation, Department of Civil Engineering, Lehigh University, Bethlehem, PA, 157 pp.

Zhou, S. P. and Chen W. F. (1986) Inelastic Analysis of Steel Braced Frames with Flexible Joints, Structural Engineering Report, No. CE-STR-86–6, School of Civil Engineering, Purdue University, West Lafayette, IN.

Index

Advanced analysis
  benchmarks for verification of, 40
  computer programs classified as, 41
  design concept, 5
  effect of residual stresses considered in, 305
    member capacity checks for, 142
  method, 26
  plastic-hinge based methods for, 317
  verification of refined-plastic hinge approach for, 33
Advanced analysis, trends toward, 1–45
  benchmarking verification, 40–42
  design formats, 2–6
    advanced analysis format, 4–6
    allowable stress design, 3
    load and resistance factor design, 4
    plastic design, 3–4
  elastic methods of analysis, 6–16
    AISC-LRFD beam-columns interaction equations, 6–8
    effective length factor, 8–12
    second-order elastic analysis for frame design, 12–16
  organization, 6
  second-order inelastic analysis, 26–40
    elastic-plastic hinge method, 27–28
    notional load plastic hinge method, 28–31
    plastic-zone method, 26–27
    refined-plastic hinge analysis, 31–40
      analysis of simple portal frames, 33–34
      analysis of six-story frame, 35–40
  semi-rigid frame design, 16–26
    connection data base, 19–22
      Goverdhan data base, 19
      Kishi and Chen data base, 20–22
      Nethercot data base, 19–20
    design provisions and connection classifications, 16–19
    direct second-order analysis methods, 25–26
    simplified analysis/design method, 22–25
AISC, see American Institute of Steel Construction
AISC-LRFD, 71
  B_1/B_2 approach, 6, 22
  for beam-column design, 33

beam-column equations, 6, 9
bilinear interaction equations, 150
design specifications, 336
manual, 32
specifications, 12, 17, 22, 98
Alignment chart K factor, 10
Alignment charts, 8, 11
Allowable Stress Design (ASD), 2, 3, 91, 321
American Institute of Steel Construction (AISC), 1
Amplification factor, 22
Ang-Morris model, 99
Arc-length controls, 60
Arc-length method, 69, 71, 279
ASD, see Allowable Stress Design
Associated flow rule, 141
Australian calibration frames, 304–312
Australian specifications, 5
Australian Standard, 314
Automatic load-increment procedure, 167
Axial compressive strength, 7
Axial force-strain relationship, 157

Bauschinger effect, 323
B_1/B_2 amplification factors, 91
B_1/B_2 method, 25
BCIN, sample problems of, 244–246
Beam-column(s), 205, 294, 327
  BCIN for, 195
  elements, 165, 343
  equations, 24, 328
  I-section of, 217
  load deformation curves of, 199
  member, 4, 63, 64
  stability function, 141, 143, 159
  subassemblies, 27
  theory, 26, 261, 262, 316
  interaction equations, 6
  plastic-zone analysis of, 259
  ultimate strength interaction curves for, 295
Beam-to-column
  connections, 17–20, 47, 91–93, 103
  data base, 104
Benchmark problems and solutions, 321–374
  European calibration frames, 338–350

375

analytical assumptions, 343–347
  geometrical imperfections, 345–347
  residual stresses, 345
  stress-strain relation, 345
gable frame, 348–349
portal frame, 347–348
six-story two-bay frame, 349
Japanese calibration frames, 351–372
  full-size test of portal frame, 352–359
    loading procedures, 353
    test results, 355
    test specimens, 352–353
    theoretical analysis, 355–359
  one-quarter scaled test of portal frames, 359–372
    loading procedures, 365
    test results of one-story frames, 365
    test results of two-story frames, 365–370
    test specimens, 359–365
    theoretical analysis, 370
North American calibration frames, 328–337
  beam-columns, 329–332
  interaction curves for portal frames, 336–337
  portal frames, 333–335
  requirements for benchmark problems, 322–328
    comments on analytical assumptions, 325–328
      constitutive relations of material, 325
      geometrical imperfections, 327
      joint conditions, 328
      residual stresses, 325–327
    criteria for selecting calibration frames, 323–324
    physical attributes and behavioral phenomena of frames, 322–323
    required information, 324
Bernoulli-Euler hypothesis, 50
Bi-linear models, 68
Bifurcation, 60, 69
Bimoment effects, 141
Bounding surface model, 63
Bowing effect, 68, 76
  element, 143
  influence of on flexibly jointed frames, 73
  nonlinear terms resulting from, 53
  in second-order analysis, 71
  in two-story one-bay frame, 74
Brace structures, 13
Braced frames, 195, 215, 227, 327
  classification of, 14
  members, 7
Braced portal frame, 217

Bracing system, 13, 14
Broyden method, 235

Calibration frame
  analyzed as plane frame, 345
  criteria for selecting, 323
  for nonlinear elastic analysis, 35
  portal frame, 341
  requirements for, 329
  for second-order inelastic analysis, 322, 343
  six-story two-bay frame, 342
Canadian Limit State design specifications, 12
Cantilever column, 69, 70
Co-rotational formulations, 287
Co-rotational framework, 53
Column buckling, 225, 229, 230, 291
Column curve, 29, 219
Column end restraint coefficient, 9
Column strength formulas, 7
Column tangent modulus, 32
Compact cross-section, 143, 150
Compact sections, 324
Compatibility analysis, 259, 261
Computer-based analysis techniques, 43
Computer program, operation procedure of, 168
Concentrated plastic hinge theory, 260
Concentrated plasticity, 259, 260
Concentrated plasticity analysis, 333
Concentrated yielding, 323
Connection(s)
  behavior, 64, 75, 76
  classification, 16, 18
  curve, 23
  data, 19
  data bank, 48
  flexibility, 71, 72, 164, 167
  loading/unloading characteristics of, 73
  models, 60, 61, 77, 99
  moment-rotation behavior of, 164
  parameters, 171
  performance, 42
  rotational deformation of, 92
  stiffness, 24, 101
  tangent stiffness, 164, 165
  type, 99
  ultimate moment, 22
Conservative loading, 277
Constitutive model, 263
Convergence, 207, 223, 236, 288
Coordinate systems, 81
Corotational coordinate system, 142
CRC column strength equations, 157

INDEX **377**

CRC-E_t, 158
CRC tangent modulus, 36, 158, 162, 170
Critical behaviors, 60
Cross-section plastic strength, 32
Cubic B-spline, 98, 99
Cubic polynomials, 274
Curvature shortening, 53, 323
Cyclic hardening, 323
Cyclic loading
  connection behavior under, 63
  horizontal load for, 353
  load-rotation curves for, 355
  portal frame, 360, 361
  test data, 109
Cyclic plasticity effect, 323

Data bank, 21
Data files, 174
Dead load, 5
Deformation theory of plasticity, 26
Design procedure, 24
Direct stiffness matrix method, 48
Displacement
  control algorithm, 279
  incremental, 146
Distributed plasticity, 26, 32, 167, 283
  analysis, 259, 262
  effects, 32, 154, 160
Distributed yielding, 143
Double curvature bending, 16
Double web-angle
  connections, 92, 93, 100, 105–107
    collected test data for, 112–113, 115
    size parameters for, 114
    size parameters for top- and seat-angle connections with, 116
    size parameters for top- and seat-angle connections without, 119
Dynamic storage allocation, 169

Earthquake load, 5
ECCS, see European Convention for Constructional Steelworks
Effective length factor, 11, 12, 14
Effective tangent-modulus approach, 154
Elastic analysis, 6
Elastic buckling load factor, 307
Elastic model, 77
Elastic-plastic hinge, 26, 154, 167
  analysis, 28, 32, 140, 141, 153
  theory, 163, 333
Elastic shakedown, 323

Elastic unloading, 144, 262, 266, 323
Elasto-plastic analysis, 260, 261
Element
  basic tangent stiffness matrix, 146
  bowing effects, 143
  end, effect of plastification at, 159
  end forces, incremental, 146
  stiffness degradation function, 160
  tangent stiffness, 145
    matrix, 167
    relationship, 145
Elliptic integrals, 53
Elliptic integral solutions, 69, 71
End connections, frame element with, 165
End-plate connections, 93, 100
  extended on tension side, 123
  size parameters for, 122
End-restraint effects, 10
Engineering strain, 49
Enlarged geometric imperfection approach, 29
Equivalent initial out-of-straightness imperfections, 30
Equivalent uniform moment, 7
Euclidean norm, modified, 288
Euler buckling load, 12, 290
Euler integration, 278
Eurocode, 5
  design manual, 13
  procedures, notional load plastic hinge method with, 29
European calibration frames, 296, 325
European Convention for Constructional Steelworks (ECCS), 316, 322, 338, 345
Exponential mode, 101
Extended end-plate connections, 93, 98, 105, 106, 119–121

Factored loads, 1
Failure mechanism, 370
Finite difference method, 261
Finite displacements with finite strains, theory of, 69
Finite element
  analysis, 286, 287, 308
  approach, 316
  geometric stiffness, 145
  method, 98, 259
  nonlinear analysis, 262, 266
  programs, 169
Finite joint, 333
Finite shell elements, 26
Finite three-dimensional shell elements, 26
First-order analysis, 9, 24, 68, 72

First-order elastic analysis, 3, 315
First-order elastic-plastic hinge analysis, 141
First-order inelastic analysis, 217
First-order plastic collapse, 307
First-order plastic hinge analysis, 3
Fixed end force, 65
Flexible connections, 18, 20, 60, 63, 64, 78
Flexural strength, 7
Flexural-torsional buckling, 314
FLFRM, 48, 75, 77
flfrm.f, 77
Flow theory, 266
Flush-end connections, 123
Flush end-plate, 93
Flush end-plate connections, 98, 105, 106, 124
Force-displacement relationship
  incremental, 159
  modified element, 153
Force-point movement, equilibrium correction for, 154
Force-space plasticity, 141
Force-state parameter, 159
FORTRAN language, 77, 129, 168, 195
FR, see Fully restrained construction
Frame(s)
  element
    degrees of freedom numbering for, 173
    incremental force-displacement relationships of, 148
    kinematic relationships of, 147
    tangent stiffness matrix of, 148
  imperfections, 29
  laterally supported, 14
  laterally unsupported, 14
  stability, 91
FRAMH, sample problems of, 252–257
FRAMP, sample problems of, 246–251
Frye-Morris model, 99, 100, 102, 128
Fully restrained connections, 322
Fully restrained (FR) construction, 17, 47, 91, 93

Gable frame, 341, 350
Geometric imperfections, 158, 346
Geometric nonlinearity, 323
G factors, 24
Global displacement fields, 271
Global translational and rotational degrees of freedom, 146
Gradual plastification, 161
Gradual stiffness, 32, 155, 159, 165
Graphical interface, 260
Gravity loads, inclined members subjected to, 143

Green-Lagrange strain, 265
Green-Lagrange strain tensor, 263
Green strain, 50
Green strain tensor, 49

Hardening parameter, 266
Header plate connections, 98, 100, 105, 106, 125
  collected test data for, 126
  size parameters for, 127
Heaviside's step function, 101
Hermitian cubic polynomial functions, 272
High-rise building frames, 37
Hill's bifurcation theory, 69

In-plane strength curves, 33, 162
In-span loading, 143, 144
Incremental reversibility, 287
Incremental stiffness equations, 161
Independent hardening model, 60, 62, 63, 77
Inelastic analysis, 316
Inelastic bifurcation problems, 69
Inelastic effect, 321
Inelasticity
  combined effects of stability and, 1
  spread of, 259
Inelastic sectional stiffness, 333
Inelastic stiffness reductions, 156
Inelastic strength predictions, 163
Inextensional elastica, 71
Initial connection stiffness, 98
Initial deflection, 327
Initial imperfections, 154
Initial stiffness, 22, 62
Initial yield surface, 159, 167
inpex.d, 83
input.d, 77, 83
Irregular frames, 12
I-section column, 290, 291
Isotropic hardening rule, 266

Joint
  flexibility, 328
  panel, 359, 370
    local buckling in, 359–365
    zone, 333

K factor, 8, 9, 14, 24, 39
  equation, 11
  formula, 38

Kinematic hardening rule, 345
Kishi-Chen model, 99

Lagrangian formulations, 263
  total, 269, 287
  updated, 260, 263, 287
Large deformation theory, 143
Lateral buckling, 355, 365
Lateral-torsional buckling, 142, 143, 323
Lateral translation (LT), 7
Leaned column frames, 337
Leaner columns, 12
LeMessurier equation, 10
Limit load, 220, 225–227
  analysis, 174
  instability, 60, 69
Limit point, 68, 168, 279
Limit state
  design, 1
    method, 91
    philosophy, 312
    specifications, 1
Linear elastic analysis, 259
Linearized beam-column theory, 53, 54, 56, 57, 69, 75
Live load, 5, 15
Load
  application, sequence of, 298
  -control secant stiffness method, 67, 68, 77
  -displacement characteristics, comparison of, 35, 36
  combinations, 5, 15
  factor, 3–5, 307, 342
Loading cycle, 63
Loading path, 63
Load and Resistance Factor Design (LRFD), 2, 4, 47, 91, 151, 328
  column strength equations, 158, 172
  column strength formula, 32, 156
  elastic analysis, 35, 39
  E_t, 158
  format, 5
  method, 25
  second-order elastic/design approach, 39
  specifications, 5, 7
  tangent modulus, 170
Local buckling, 142, 323, 355, 365
Local effects, 38
LRFD, see Load and Resistance Factor Design
LT, see Lateral translation
Lui-Chen exponential model, 99, 100

makefile, 77
Material inelasticity, 288
Material nonlinearity, 323
Member
  buckling load, 7, 8
  capacity checks, 5, 28, 40, 142, 151
  curvature effects, 7
  design, effective length factors for, 14
  imperfections, 29
  maximum moment in, 15
  stiffness
    degradation, 163
    index, 12
Minimum residual displacement method, 279
Minimum total potential, principle of, 269
Modified element
  force-displacement relationship, 153
  tangent stiffness matrix, 153
Modified exponential model, 61, 75, 83, 99–101, 103, 128
Modified plastic hinge approach, 329
Modified secant stiffness equation, 63
Moment
  amplification methods, 13
  curvature
    curves, 362
    relations, 330
  magnifier, 7
  -rotation
    behavior, 60
    characteristics, 92, 103
    curves, 19, 67, 68, 77, 324
    relationship, 60, 73, 78, 84, 164–165
  -thrust curvature, 260, 261
Monotonic loading, 357, 358

Newmark's integration method, 195, 200, 204, 218, 259
Newmark's numerical method, 329
Newton-Raphson method, 60, 67, 167, 276, 279
Nodal residual force vector, 279
Nominal load, 5
Nominal plastic strength surface, 32
Nonconservative loading, 277, 278
Nonlinear analysis, 169, 261, 321
Nonlinear beam-column theory, 53–55, 57, 69, 77
Nonlinear equilibrium equations, 277
Nonlinear strain, 48
Nonlinear structural analysis, 164
Nonlinear theory, 65

Nonpositive definiteness, 168
Nonproportional load, 144
North American calibration frame, 303, 304, 306
Notional load, 30, 31, 163
Notional load plastic hinge, 29, 31, 35
No translation (NT), 7
Numerical algorithm, 167
Numerical instability, 56

outex.d, 83
Out-of-plane buckling, 41, 365
Out-of-plane deformations, 142
Out-of-plumbness, 35
Out-of-plumbness rule, 30
output.d, 77, 83
OUTPUT file, 174
Out-of-straightness imperfections, 30

Parabolic stiffness reduction function, 170
Partially restrained connections, 322
Partially restrained (PR) construction, 17, 25, 72, 91, 93–94
Partial plastification, 161
Path-dependent plasticity, 323
P-Δ
  amplification, 14
  effect, 12, 13, 323, 344, 347, 349
  moment, 8, 31
P-δ
  effect, 9, 10, 12, 13, 141, 323, 344
  moment amplification, 7
PD, see Plastic Design
PHINGE, 168, 174
Piecewise linear model, 68, 98
Pinned connection, 91
Pinned-joint model, 47
Piola-Kirchhoff stress tensor, second, 263
Pitched-roof portal frame, 299–301
Plastic curvature, 358
Plastic Design (PD), 2, 3
Plastic failure mechanism, 225, 237
Plastic hinge, 152, 207, 218, 241, 242, 365
  analysis, 37
  -based program, operating procedure of, 169
  concept, 260, 314
  formation, beam-column element per member to capture, 143
  method, 355, 370
  solutions, 297
  stiffness degradation, 32
  theory, 349
Plastic moment, 152, 358, 365

Plastic rotation, 141
Plastic section capacity, 151
Plastic strength surface, 141, 159, 161, 167
  equilibrium correction for force-point movement on, 154
  force-point movement on, 162
Plastic zone(s), 26
  analysis, 27, 33, 261, 262, 290, 294, 298, 299, 323
  approach, 328
  curves, 9
  finite element analysis, 259, 294, 297
  formulation, 282
  method, 34, 141, 296, 370
  results, 153
  solutions, 31, 40, 160, 297
  spread of, 196, 324, 343, 351
  strength curves, 155, 162
  theory, 195, 321, 333, 334, 343, 347
Plastic-zone analysis, of beam-columns and portal frames, 195–257
  analysis of in-plane beam-columns, 196–215
    analytical steps, 200–207
    flow chart of BCIN, 207
    principle of analysis, 196–200
      analytical conditions, 196–198
      analytical method, 198–200
    sample problems of BCIN, 213–215
    user's manual of BCIN, 208–213
      determining input data, 212–213
      input data, 208–210
      output data, 210–212
  analysis of portal braced frames, 215–231
    analytical steps, 218–220
    flow chart of FRAMP, 220
    principle of analysis, 215–218
    sample problems of FRAMP, 227–231
    user's manual of FRAMP, 220–227
      case with BAT = 0, 227
      frame failures, 225–227
      input data, 221–222
      output data, 222–225
  analysis of portal unbraced frame, 232–243
    analytical steps, 236–237
    flow chart of FRAMH, 237
    principle of analysis, 232–236
      second-order elastic analysis of unbraced frames, 232–235
      second-order inelastic analysis, 235–236
      structure and load of unbraced frame, 232
    sample problems for FRAMH, 241–243
    user's manual of FRAMH, 238–241
      frame failures, 241

INDEX **381**

input data, 238–240
output data, 240–241
Plastic-zone analysis, of frames, 259–319
analysis of beam-columns, 294–295
application to engineering practice, 312–315
current practice, 312–315
future research, 315
aspects of computer implementation, 286–290
computer requirements, 289–290
general, 286
material inelasticity, 288–289
node numbering and numerical integration, 286–287
nonlinear formulations and incremental-iterative strategies, 287–288
Australian calibration frames, 304–312
rigid-jointed truss, 304–307
stressed-arch frame, 307–312
cross-sectional analysis, 282–286
general, 282–283
numerical integration procedure, 283–285
numerical integration of stress resultants, 285–286
numerical integration of tangent modulus matrix, 286
European calibration frames, 296–303
general, 296–297
pitched-roof portal frame, 299–300
rectangular portal frame with fixed bases, 296–298
comments on sensitivity of analysis results to modeling assumptions, 298
six-story two-bay frame, 300–303
finite element formulation, 262–282
discretization of virtual work equation, 270–276
element geometric description, 270–274
equilibrium equations, 275–276
strain-displacement matrices, 274–275
general, 262–264
principle of virtual displacements, 269–270
solving nonlinear equilibrium equations, 276–279
strain-displacement relations, 264–266
stress-strain relations, 266–269
stress resultants, 267–269
transformation, condensation, and recovery of nodal variables, 279–282
condensation of nodal variables, 281
internal rotational releases, 282
recovery of nodal variables, 281–282
transformation of nodal variables, 279–281

historical sketch, 260–262
investigation of analysis parameters, 290–293
beam with fully constrained ends, 291–293
I-section column, 290–291
North American calibration frame, 303–304
Polynomial equation, 21
Polynomial model, 61, 75, 98, 100
Portal frame(s), 299, 341, 352, 365
actual section properties of, 355
full-size, 353
horizontal force-displacement behaviors of, 360
horizontal force-rotation behaviors of, 361
interaction curves for, 339, 340
lateral load-displacement behavior of, 335–338
material properties of, 356
plastic-zone analysis of, 259
spread of plastic-zones, 348
spread of yielded areas for, 347
test programs of full size, 354
test results of full-size, 359
unbraced, 306, 333
Post-buckling behavior, 71, 77, 262
Post-collapse behavior, 143
Post-ultimate response, 291
Power models, 99, 104
PR, see Partially restrained construction
Preliminary analysis/design, 13
Program INPUT, 174

Ramberg-Osgood curves, 288
Ramberg-Osgood equation, 288
Ramberg-Osgood stress-strain curves, 289
Redistribution, inelastic force, 4
Reference configuration, 263
Refined plastic hinge, 26, 142, 167
analysis, 33, 36, 162, 261, 289, 316
approach, 37, 38, 142, 158
concept, 312
method, 33
model, 32, 142, 170
Residual stress, 326, 332, 344, 365
distribution, 334
effects of, 158
pattern, 327, 343
Resistance factors, 5, 32, 151, 315
Reverse loading path, 63
Rigid connection, 17, 20, 91
Rigid frame, 47
Rigid-jointed frame, plastic zone analysis of, 307
Rigid-jointed roof truss, 305

Rigid-jointed truss, 304–307
  imperfect, 309, 310
  load-deflection response of, 308
Rigid joints, 47, 342
Rigid-plastic analysis, 3, 141
Roof load, 5
Rotation capacity, 18

Safety factor, 3
Scaling factor, 101
SCDB, see Steel Connection Data Bank
scdb.d, 129
Seat-angle connections, 93, 100, 105
  collected test data for, 117–118
  with double web angle, 106
  without double web angle, 106
Secant connection stiffness, 101
Secant stiffness, 60, 63, 65
  matrix, 65
  method, 60, 63, 67, 68, 77
Second-order analysis, 42, 48, 72
Second-order effect, 47, 321
Second-order elastic analysis, 4, 12, 15, 25, 199, 200, 202, 336
Second-order elastic analysis, of frames, 47–90
  computer program, 75–77
  modeling of semi-rigid connections, 60–63
  numerical examples, 69–75
  second-order theory for in-plane frames, 48–54
  solution procedure for nonlinear stiffness equations, 67–69
  stiffness equations for beam-column member, 54–60
  user's manual, 77–83
Second-order elastic moment, 7, 143
Second-order elastic-plastic frame analysis, 261
Second-order elastic-plastic hinge analysis, 28, 30, 141, 151
Second-order elastic plastic-hinge method, 174
Second-order elastic programs, 27
Second-order finite element analysis, 51
Second-order frame analysis, 75
Second-order inelastic analysis
  calibration frames for, 322
  consideration of system in-plane strength by, 42
  for in-plane beam-columns, 195
  linear limit of, 347
  partial plastification of members in, 260
  of portal unbraced frame, 232
  SP2D based on, 333
  tracing of moment-rotation curve using, 217
  unusual application of, 308
  verification of, 43
  $YQ1$ obtained by, 202
Second-order inelastic computer programs, 28
Second-order nonlinear structural analysis, 104
Second-order P-Δ, P-δ effects, 91
Second-order plastic hinge
  method, 347
  theory, 349
Second-order plastic hinge analysis, of frames, 139–194
  analysis of semi-rigid frames, 163–167
    modeling of connections, 163–164
    modification of element stiffness to account for end connections, 164–167
  assumptions and scope, 142–144
  modeling of elastic frame elements, 144–148
  modeling of elastic truss elements, 148–149
  numerical implementation, 167–168
  second-order elastic-plastic hinge analysis, 149–154
    cross-section plastic strength, 150–151
    illustrative example, 153–154
    modification of element stiffness for presence of plastic hinges, 151–153
  second-order plastic hinge based analysis program, 168–192
    examples, 174–192
    input instructions, 169–174
    program overall view, 168–169
  second-order refined plastic hinge analysis, 154–163
    illustrative example, 162–163
    tangent modulus approach, 155–158
    two-surface stiffness degradation model, 158–162
      effect of plastification at both ends, 161–162
      effect of plastification at end A only, 159–161
      effect of plastification at end B only, 161
Second-order plastic-zone analysis, 26, 141, 199
Second-order refined plastic hinge
  analysis, 32, 162
  method, 170, 174
  strength curves, 162
Second-order static plastic hinge analysis, 168
Second-order structural analysis, 101
Semi-rigid braced frame, 175
  factored load analysis of, 174
  limit load analysis of, 174
  structural modeling of, 175
Semi-rigid connections, 17, 20, 91–137, 171, 260, 328

connection data base, 104–105
modeling of connections, 98–104
  Frye-Morris polynomial model, 99–100
  general remarks, 98–99
  modified exponential model, 100–101
  three-parameter power model, 101–104
moment-rotation behavioral response of, 163
parameter definition for connection type, 105–127
  double web-angle connections, 107
  extended end-plate connections, 119–123
  flush end-plate connections, 123–125
  header-plate connections, 125–127
  single web-angle connections/single plate connections, 105–107
  top- and seat-angle connections, 114–119
  top- and seat-angle connections with double web angle, 107–114
steel connection data bank program, 127–134
  examples, 130–134
  outline of SCDB, 127–129
  user's manual for program SCDB, 129
types of semi-rigid connections, 92–98
  double web-angle connections, 93
  extended end-plate connections/flush end-plate connections, 93–98
  header-plate connections, 98
  single web-angle connections/single plate connections, 92
  top- and seat-angle connections with double web angle, 93
  top- and seat-angle connections, 93
Semi-rigid frame, 14, 24, 47, 69, 83
Semi-rigid supports, 359
Serviceability limit state conditions, 42
Shallow arc theory, 262
Shape functions, 271, 275
Shape parameter, 103, 104, 128, 163
Shear buckling, 353
Shear connection, 93
Side-sway instability, 6
Sign conventions, 81
Simple connection, 93
Simple framing, 17
Simple incremental solution, 167
Simplified analytical model, 21
Simplified second-order inelastic analysis, 329
Simpson's rule, 285, 286
Single curvature bending, 16
Single plate connections, 92, 106, 111
Single web-angle bolted beam-to-column connections, 108
Single web-angle connections, 92, 105, 106
Six-story frame, 300–303

comparison of bending diagrams and plastic hinge locations in, 38
comparison of load-displacement curves for, 37
Six-story two-bay frame, 300–303, 342, 350, 351
Small displacement theory, 77
Small strains, 48
Snow load, 5
Spatial beam-column behavior, 41
Spreaded yielding, 323
Spread of plasticity, 321, 324
Spread-of-plasticity analysis, 26, 162, 259–260
SSRC, see Structural Stability Research Council
Stability, 6
  functions, 58, 144, 145, 165
  problems, 259
Stability Design of Frames, 1
Stable equilibrium state, 199
Standardization constant, size parameter in, 100
Standardization parameter, 99
Steel beam-columns, 197
Steel Connection Data Bank (SCDB), 92, 127, 129, 130
Stiffness
  correction factor, 9
  degradation, 36, 355, 359
  equations, 54, 57
  reduction factor, 159
Story assemblage, buckling of, 9
Story buckling analysis, 11, 24
Story-buckling K factor, 10
Strain hardening, 307, 370
  effect of, 325
  factor, 359
  in modeling beam-column elements, 142–143
  in plastic-zone analysis, 302
  in zone analyses, 300
Strength curves, comparisons of, 34, 155, 162
Strength interaction curves, 150, 151
Strength limit state, 297
Strength surface, 153
Stressed-arch calibration frame, 310, 311
Stressed-arch frame, 308–312, 313
  load-deflection response of, 312
  midspan deflection for, 313
  system, 314
Strong-axis
  strengths, 151
  wide-flange sections bending about, 150
Structural Stability Research Council (SSRC), 1, 316, 317, 322
Structural stiffness matrix, 174

Structural systems, critical behaviors of, 68
Sway analysis, 24
System buckling analysis, 11, 12

Tangent connection stiffness, 104
Tangent modulus, 166, 267
  approach, 32
  matrix, 268, 286
  stiffness reduction, 157
Tangent stiffness
  of connections A and B, 165
  iterative procedures using, 60
  matrix, 148, 207, 276–278
  method, 200, 218
  programming for, 68
Theory of Beam-Columns, 1
Theory of finite displacements with finite strains, 69
Three-dimensional plastic-zone techniques, 27
Three-parameter model, 61, 75, 163
Three-parameter power model, 104
  moment-rotation relationship represented by, 103
  prediction equation, 128
  as simplified analytical model, 21
Three-point rule, 287
toframe, 77, 83
tofrex, 83
Top angle, 71
Top-angle connections, 93, 100, 105, 106
  collected test data for, 117–118
  with double web angle, 106
  without double web angle, 106
Total Lagrangian formulation, 263
Tri-linear stress-strain relation, 344
Truss element, 172, 174
  degrees of freedom numbering for, 174
  kinematic relationships of, 149
T-stub connection, 100
Tubular beam-columns, 262

Two-surface stiffness degradation model, 33, 160
Two-surface yield model, 142
Type 2 construction, 93
Type 3 framing, 98
Type FR construction, 47

Unbraced frames, 12, 14, 195, 241
Unbraced portal frame, 233, 305, 333
Unbraced semi-rigid frames, 22
Unbraced structures, 13
UNIX system, 77
Unloading
  connections, 75
  path, 62, 63
  of plastic hinges, 144
  process, 62
Unstable state, member in, 199
Updated Lagrangian formulation, 263

Variational calculus, 69
Virtual displacements, principle of, 263
Virtual work
  equation, 48, 269, 275
  principle of, 48, 51
  theorem of, 53

Wagner effect, 323
Weak-axis
  strength, 151
  wide-flange sections bending about, 151
Wind load, 5
Wu-Chen model, 99

Yield, partial, 161
Yielding, spread of, 309

Zone analysis, 300